Lecture Notes in Physics

Bisher erschienen/Already published

Vol. 1: J. C. Erdmann. Wärmeleitung in Kristallen, theoretische Grundlagen und fortgeschrittene experimentelle Methoden. II, 283 Seiten. 1969.

Vol. 2: K. Hepp, Théorie de la renormalisation. III, 215 pages. 1969.

Vol. 3: A. Martin, Scattering Theory: Unitarity, Analyticity and Crossing. IV, 125 pages. 1969.

Vol. 4: G. Ludwig, Deutung des Begriffs „physikalische Theorie" und axiomatische Grundlegung der Hilbertraumstruktur der Quantenmechanik durch Hauptsätze des Messens. 1970. Vergriffen.

Vol. 5: Schaaf, The Reduction of the Product of Two Irreducible Unitary Representations of the Proper Orthochronous Quantummechanical Poincare Group. IV, 120 pages. 1970.

Vol. 6: Group Representations in Mathematics and Physics. Edited by V. Bargmann. V, 340 pages. 1970.

Vol. 7: R. Balescu, J. L. Lebowitz, I. Prigogine, P. Résibois, Z. W. Salsburg, Lectures in Statistical Physics. V, 181 pages. 1971.

Vol. 8: Proceedings of the Second International Conference on Numerical Methods in Fluid Dynamics. Edited by M. Holt. 1971. Out of print.

Vol. 9: D. W. Robinson, The Thermodynamic Pressure in Quantum Statistical Mechanics. V, 115 pages. 1971.

Vol. 10: J. M. Stewart, Non-Equilibrium-Relativistic Kinetic Theory. III, 113 pages. 1971.

Vol. 11: O. Steinmann, Pertubation Expansions in Axiomatic Field Theory. III, 126 pages. 1976.

Vol. 12: Statistical Models and Turbulence. Edited by C. Van Atta and M. Rosenblatt. Reprint of the First Edition. VIII, 492 pages. 1975.

Vol. 13: M. Ryan, Hamiltonian Cosmology. VII, 169 pages. 1972.

Vol. 14: Methods of Local and Global Differential Geometry in General Relativity. Edited by D. Farnsworth, J. Fink, J. Porter, and A. Thompson. V, 188 pages.

Vol. 15: M. Fierz, Vorlesungen zur Entwicklungsgeschichte der Mechanik. V, 97 Seiten. 1972.

Vol. 16: H.-O. Georgii, Phasenübergang 1. Art bei Gittergasmodellen. IX, 167 Seiten. 1972.

Vol. 17: Strong Interaction Physics. Edited by W. Rühl and A. Vancura. V, 405 pages. 1973.

Vol. 18: Proceedings of the Third International Conference on Numerical Methods in Fluid Mechanics, Vol. I. Edited by H. Cabannes and R. Temam. VII, 186 pages. 1973.

Vol. 19: Proceedings of the Third International Conference on Numerical Methods in Fluid Mechanics, Vol. II. Edited by H. Cabannes and R. Temam. VII, 275 pages. 1973.

Vol. 20: Statistical Mechanics and Mathematical Problems. Edited by A. Lenard. VIII, 247 pages. 1973.

Vol. 21: Optimization and Stability Problems in Continuum Mechanics. Edited by P. K. C. Wang. V, 94 pages. 1973.

Vol. 22: Proceedings of the Europhysics Study Conference on Intermediate Processes in Nuclear Reactions. Edited by N. Cindro, P. Kulišic and Th. Mayer-Kuckuk. XIV, 329 pages. 1973.

Vol. 23: Nuclear Structure Physics. Proceedings 1973. Edited by U. Smilansky, I. Talmi, and H. A. Weidenmüller. XII, 296 pages. 1973.

Vol. 24: R. F. Snipes, Statistical Mechanical Theory of the Electrolytic Transport of Nonelectrolytes. V, 210 pages. 1973.

Vol. 25: Constructive Quantum Field Theory. The 1973 "Ettore Majorana" International School of Mathematical Physics. Edited by G. Velo and A. Wightman. III, 331 pages. 1973.

Vol. 26: A. Hubert, Theorie der Domänenwände in geordneten Medien. XII, 377 Seiten. 1974.

Vol. 27: R. K. Zeytounian, Notes sur les Ecoulements Rotationnels de Fluides Parfaits. XIII, 407 pages. 1974.

Vol. 28: Lectures in Statistical Physics. Edited by W. C. Schieve and J. S. Turner. V, 342 pages. 1974.

Vol. 29: Foundations of Quantum Mechanics and Ordered Linear Spaces. Advanced Study Institute, Marburg 1973. Edited by A. Hartkämper and H. Neumann. VI, 355 pages. 1974.

Vol. 30: Polarization Nuclear Physics. Proceedings 1973. Edited by D. Fick. IX, 292 pages. 1974.

Vol. 31: Transport Phenomena. Sitges International Schools of Statistical Mechanics, June 1974. Edited by G. Kirczenow and J. Marro. XIV, 517 pages. 1974.

Vol. 32: Particles, Quantum Fields and Statistical Mechanics. Proceedings 1973. Edited by M. Alexanian and A. Zepeda. V, 132 pages. 1975.

Vol. 33: Classical and Quantum Mechanical Aspects of Heavy Ion Collisions. Proceedings 1974. Edited by H. L. Harney, P. Braun-Munzinger, and C. K. Gelbke. VII, 311 pages. 1975.

Vol. 34: One-Dimensional Conductors GPS Summer School Proceedings, 1974. Edited by H. G. Schuster. VII, 371 pages. 1975.

Vol. 35: Proceedings of the Fourth International Conference on Numerical Methods in Fluid Dynamics, 1974. Edited by R. D. Richtmyer. V, 457 pages. 1975.

Vol. 36: R. Gatignol, Théorie Cinétique des Gaz à Répartition Discrète de Vitesses. II, 219 pages. 1975.

Vol. 37: Trends in Elementary Particle Theory. Proceedings 1974. Edited by H. Rollnik and K. Dietz. V, 472 pages. 1975.

Vol. 38: Dynamical Systems, Theory and Applications. Proceedings 1974. Edited by J. Moser. VI, 624 pages. 1975.

Vol. 39: International Symposium on Mathematical Problems in Theoretical Physics. Proceedings 1975. Edited by H. Araki. XII, 562 pages. 1975.

Vol. 40: Effective Interactions and Operators in Nuclei. Proceedings 1975. Edited by B. R. Barrett. XII, 339 pages. 1975.

Vol. 41: Progress in Numerical Fluid Dynamics. Proceedings 1974. Edited by H. J. Wirz. V, 471 pages. 1975.

Vol. 42: H II Regions and Related Topics. Proceedings 1975. Edited by D. Downes and T. L. Wilson. XII, 488 pages. 1975.

Vol. 43: Laser Spectroscopy. Proceedings 1975. Edited by S. Haroche, J. C. Pebay-Peyroula, T. W. Hänsch, and S. E. Harris. X, 466 pages. 1975.

Lecture Notes in Physics

Edited by J. Ehlers, München, K. Hepp, Zürich,
R. Kippenhahn, München, H. A. Weidenmüller, Heidelberg,
and J. Zittartz, Köln
Managing Editor: W. Beiglböck, Heidelberg

73

Invariant Wave Equations

Proceedings of the "Ettore Majorana"
International School of Mathematical Physics
Held in Erice, June 27 to July 9, 1977

Edited by
Giorgio Velo and Arthur S. Wightman

Springer-Verlag
Berlin Heidelberg GmbH 1978

Editors
Giorgio Velo
Istituto di Fisica "A. Righi"
Bologna/Italy

Arthur S. Wightman
Department of Physics
Princeton University
Princeton, NJ 08540/USA

ISBN 978-3-540-08655-0 ISBN 978-3-540-35929-6 (eBook)
DOI 10.1007/978-3-540-35929-6

Originally published by Springer-Verlag Berlin Heidelberg New York in 1978

2153/3140-543210

TABLE OF CONTENTS

INTRODUCTION

The present volume collects lecture notes from the session of the International School of Mathematical Physics "Ettore Majorana" on Invariant Wave Equations that took place in Erice (Sicily) from June 27 to July 9, 1977.

The School was sponsored by the Italian Ministry of Public Education, the Italian Ministry of Scientific and Technological Research, and the Regional Sicilian Government.

Invariant wave equations are involved in several quite different aspects of particle physics and quantum field theory :

1) Linear Lorentz invariant wave equations as descriptions of a single particle or of a single particle interacting with external fields.
2) Linear Euclidean invariant equations appearing as intermediates in the solution of Euclidean field theories.
3) Non-linear Lorentz invariant and non-linear Euclidean invariant equations appearing in semi-classical approximations to the solutions of quantum field theories.

Mathematically, these applications are related through the theory of hyperbolic and elliptic partial differential equations.

Linear Lorentz invariant equations have had a long and tangled history. The present volume should enable the beginner to understand the essential difficulties of the extended literature. For the expert it offers a survey of some recent advances which have brought the subject to a new stage.

The applications of invariant wave equations, linear and non-linear, in quantum field theory have seen a rapid development in the last several years and have much promise for the future. The lecture notes in the present volume offer a rich selection of results for which these applications have reached the stage of mathematical precision, as well as a survey of the mathematical results which provide general background.

INVARIANT WAVE EQUATIONS; GENERAL THEORY AND APPLICATIONS TO THE EXTERNAL FIELD PROBLEM

A. S. Wightman
Joseph Henry Laboratories of Physics
Princeton University
Princeton, New Jersey

TABLE OF CONTENTS

Chapter 1 : Introduction to the Physical Applications of Invariant Wave Equations.

The main purpose of these lectures is to give an introduction to the theory of invariant linear wave equations and their associated external field problems. Certain special equations (spin 0, 1/2, 1, 3/2, 2) will be discussed in more detail in other lectures by R. Seiler and D. Zwanziger. I will also describe in general terms how non-linear invariant equations arise in the theory of coupled fields. This latter discussion is intended to provide a general introduction to the lectures of Strauss, Strocchi, Parenti, Velo, Gervais, Fröhlich and Stora. Before entering into details, in this introduction I will survey the general ideas and strategy of what is to come.

1. Linear invariant wave equations and the external field problem

The first use of linear invariant wave equations in quantum mechanics was to describe particles ; the Klein-Gordon equation, the Dirac equation, and the Proca equation provide examples. The description came in two stages : first a one-body theory, and then a many-particle theory in terms of fields. In the 1930's this theory was generalized until it ultimately covered many general systems of equations describing families of particles with mass and spin spectra. From the beginning, theories were constructed to deal not only with free particles, but with particles moving in a given classical electromagnetic field - such a field will be referred to, for brevity, as external . Later on, other couplings to external scalar, tensor, etc. fields were considered. Of course, it is natural to regard such a theory as a limiting case of a theory of coupled quantized fields. In my opinion, most of what will be said here acquires physical significance mainly as a step toward understanding the theory of coupled quantized fields. The external field problem offers only a slight taste of the problems of the theory of coupled quantized fields.

For free particles, one can construct a consistent field theory for an essentially arbitrary invariant wave equation. However, from the earliest times there were difficulties with equations of higher spin in external fields. (By convention "higher spin" means $s > 1$.) These first appeared in Dirac's famous paper of 1936 in which consistent wave equations for single particles of arbitrary mass > 0 and spin were constructed [1]. Dirac proposed to extend his theory to external electromagnetic fields by the minimal substitution $\partial \to \partial - eiA$. It was pointed out by Fierz that this procedure gives rise to problems for $s > 1$ [2].

For spin 3/2, the difficulty is, in capsule, as follows. Consider the system of equations

$$(-\not{\pi} + m)\psi^{\mu}(x) = 0 \tag{1.1}$$

$$\pi_{\mu}\psi^{\mu}(x) = 0 \tag{1.2}$$

$$\gamma_{\mu}\psi^{\mu}(x) = 0 \tag{1.3}$$

Here $\not{\pi} = \gamma^{\mu}\pi_{\mu}$, $\pi_{\mu} = i\partial_{\mu} - qA_{\mu}$, and the γ_{μ} are the usual Dirac matrices satisfying

$$\gamma_{\mu}\gamma_{\nu} + \gamma_{\nu}\gamma_{\mu} = 2g_{\mu\nu} \text{ , } g_{\mu\nu} = \begin{cases} 0 & \mu \neq \nu \\ 1 & \mu = \nu = 0 \\ -1 & \mu = \nu = 1,2,3 \end{cases} \tag{1.4}$$

A_{μ} is the vector potential of the external electromagnetic field ; it is assumed infinitely differentiable for simplicity. For each value 0, 1, 2, 3 of the index ν, $\psi^{\nu}(x)$ is a four-component Dirac spinor. (With some changes in notation, these sixteen complex-valued functions, ψ^{ν}, provide Dirac's proposal for the description of a spin 3/2 particle). Now apply $-\not{\pi} + m$ to the second equation and use the identity

$$(-\not{\pi} + m)\pi_{\mu} = \pi_{\mu}(-\not{\pi} + m) - iq F_{\mu\nu}\gamma^{\nu} \tag{1.5}$$

to conclude that the three equations imply

$$F_{\mu\nu}(x)\gamma^{\nu}\psi^{\mu}(x) = 0 \tag{1.6}$$

Thus, at a space-time point x where $F_{\mu\nu}(x) \neq 0$, the wave function $\psi^{\mu}(x)$ satisfies two constraints, (1.3) and (1.6), while at a point where $F_{\mu\nu}(x) = 0$, only (1.3) holds. This should make it plausible that the coupling to the external electromagnetic field has reduced the number of spin states so that the equation does not provide a suitable description of a spin 3/2 particle. (A careful analysis has to show how the discontinuity in the number of constraints at zero field strength results in a discontinuity in the number of polarization states for a packet of well-defined momentum).

Fierz and Pauli proposed alternative equations for particles of higher spin in external electromagnetic fields, and worked out the details for spin 3/2 and 2 [3]. They a) introduced redundant components, which vanish by virtue of the equations of motion in the absence of an electromagnetic field, b) derived the equations from an action principle . For the Fierz-Pauli equations the number of constraints is continuous at zero field and therefore the difficulty that afflicted Dirac's equations is avoided. The trouble with the Fierz-Pauli equations is more subtle. Before describing it, I want to mention another difficulty already noted by Fierz and Pauli in an appendix to their paper, the phenomenon of <u>loss of constraints</u>. (In an analogous terminology, the difficulty with Dirac's equations would be gain of constraints). They showed that the degree of a constraint as a differential equation may be discontinuous at zero field strength. They gave an example in which the degree drops by one at zero field strength. As a result, there are solutions of the wave equation which do not have a well-defined limit as the field is turned off. This means that the equations in the presence of the external field are unsuitable to describe a particle of the desired spin because they have too many solutions. It turned out later on that a slight modification of the spin 2 equation of Fierz and Pauli, introduced by Wentzel [4], is an equation which shows this phenomenon of loss of constraint [5].

The paper of Fierz and Pauli proposed equations for each mass $m > 0$ and spin s which avoid the disasters of loss or gain of constraints in the presence of minimal electromagnetic coupling. Nevertheless, when S. Kusaka and J. W. Weinberg attempted to quantize the higher spin Fierz-Pauli equations, they found inconsistencies between the dynamics and the commutation relations [4]. Perhaps because this work was never published, the theory acquired a reputation for internal inconsistency without the reasons being well known.

Kusaka and Weinberg's discussion was based on an analysis of subsidiary conditions or constraints for relativistic wave equations. Their method is valid for all the equations of Fierz and Pauli but they made detailed calculations only for spin 3/2 and 2. They wrote the wave equations in first order form

$$(-i\beta^{\mu}\pi_{\mu} + m)\psi(x) = 0 \tag{1.7}$$

and argued that in order to yield a description of particles of spin > 1, the matrix β^0 has to have zero as a non-semi-simple eigenvalue i.e. when β^0 is brought to Jordan canonical form, the eigenvalue 0 has to be accompanied by some 1's above the diagonal

$$\begin{Bmatrix} 0 & 1 & 0 \ldots \\ 0 & 0 & 1 \\ \vdots & & \end{Bmatrix}$$

(That equations of spin ≤ 1 are distinguished by having diagonalizable β^0's appears to have been known at the time. It was spelled out in detail later on by E. Wild [7]). Because β^0 has an eigenvalue zero, one cannot make the coefficient of $\frac{\partial}{\partial t}$ in the wave equation equal to 1 by multiplying by $(\beta^0)^{-1}$. Furthermore there are subsidiary conditions on the wave function ; if χ is a non-vanishing vector such that $\chi\beta^0 = 0$ then

$$-\sum_{j=1}^{3} \chi \beta^{j} \pi_{j} \psi + m \chi \psi + \chi B \psi = 0 \qquad (1.8)$$

which is a non-trivial relation connecting ψ and its spatial derivatives at each time. Equation (1.8) is sometimes called a primary constraint. If by taking the time derivative of a primary constraint and using the wave equation (1.7), one arrives at a further relation between ψ and its spatial derivatives, it is called a secondary constraint. One of Kusaka and Weinberg's results is that a non-semi-simple eigenvalue zero for β^{o} always implies the existence of secondary constraints.

Viewed from the present, the pioneering work of Kusaka and Weinberg remains somewhat of an enigma. Their analysis of the subsidiary conditions led to the conclusion that the equal-time anti-commutator of the spin -3/2 field is non-local when the external field is time dependent. This result is in strong contrast to that of Johnson and Sudarshan [5] who computed the same anti-commutator starting from Schwinger's Action Principle and found that it is local for weak external fields but becomes non-positive whenever the magnetic field exceeds a certain critical value. Later analysis of the problem using methods independent of the canonical quantization formalism reproduce this result of Sudarshan and Johnson. It appears therefore that Kusaka and Weinberg's discussion of the subsidiary conditions led them to a method of quantization different from the one that is now standard. In any case, in the 1940's, wave equations of higher spin were regarded with suspicion. Many people seemed to be looking for reasons why the only elementary particles in Nature should be of spin 0, $\frac{1}{2}$, and 1 (At the time, nuclei with spin > 1 were not regarded as elementary particles) [9]. Only after the discovery of the spin 3/2 resonance in pion-nucleon scattering and the realization that it might make sense to regard it as an elementary particle, did attitudes toward higher spin equations change.

The basic result of Johnson and Sudarshan left open one question : could it be that the failure in positivity of the anti-commutator that they discovered reflects a break down in the canonical formalism rather than a fundamental flaw in the theory ? After all, in quantum field theories of coupled fields, the canonical formalism is often mathematically inconsistent because of infinite renormalizations.

The matter was taken up again by Velo and Zwanziger, who, in my opinion, greatly clarified the situation [10] [11]. They showed that the propagation character of the relativistic wave equation for a particle of spin 3/2 is altered by an external electromagnetic field : in the presence of the external field, the particle described by the equation moves faster than light. This statement has a quantitative expression in terms of the behavior of the retarded fundamental solution of the wave equation shown in Figure 1. When $A_\mu = 0$, $S_R(x,y;0) = S_R(x-y)$ has its support for the variable $x-y$ lying in or on the future light cone, is shown on the left hand side of the figure. As the field is turned on the support of $S_R(x,y;A_\mu)$ grows, looking for some suitably chosen A_μ like the right hand side of the figure. For sufficiently large fields, the support no longer lies in a half-space and the equation no longer describes wave propagation (≡ is no longer hyperbolic).

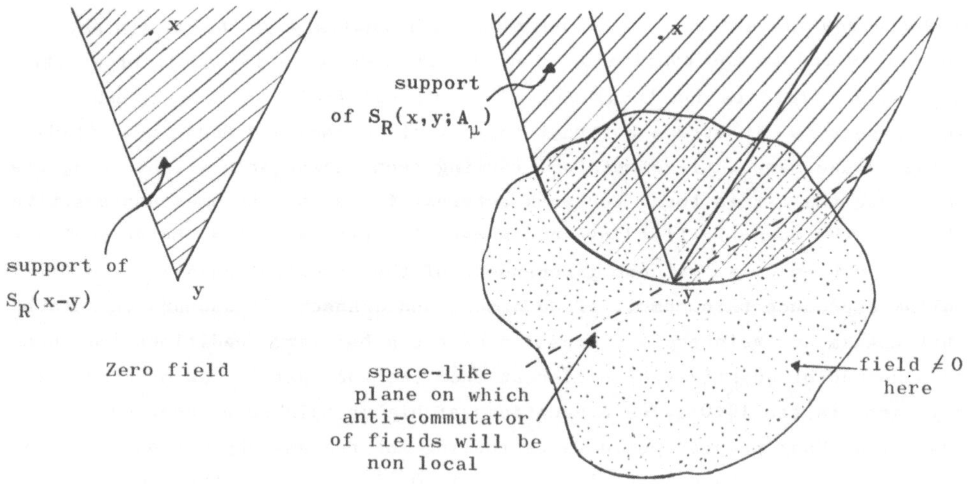

Figure 1

Acausality in the retarded fundamental solution $S_R(x,y;A_\mu)$. In the absence of an external field the retarded fundamental solution vanishes for x-y outside the future light cone. When the field is non-vanishing, the support of $S_R(x,y;A_\mu)$ in the variable x-y is larger than the light cone.

From these diagrams some simple facts about the anti-commutator of fields can be read off and thereby some light shed on the status of the anti-commutation relations. Since the anti-commutator is the difference of the retarded and advanced fundamental solutions,

$$S(x,y;A_\mu) = S_R(x,y;A_\mu) - S_A(x,y;A_\mu)$$

it will have a support in the union of the supports of the advanced and retarded solutions. For sufficiently weak fields

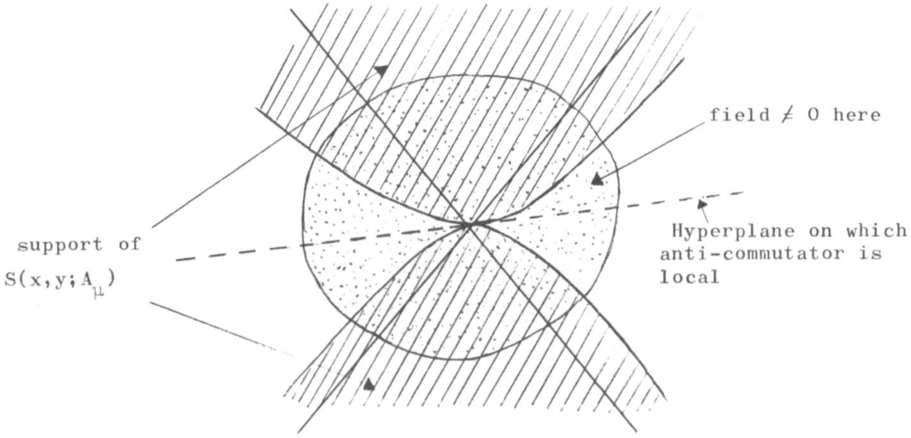

Figure 2
Support of the anti-commutator in the presence of external field.

there exist hyperplanes which cut the support only at the origin. For such hyperplanes, the anti-commutator is local and canonical commutator relations may make sense. However, for a space-like hyperplane cutting through the support, the anti-commutator is non-local, and the canonical formalism must fail.

Velo and Zwanziger found that the critical magnetic field at which the support of the retarded function no longer lies in a half space so there is no space-like hyperplane on which the anti-commutator of the fields is local is precisely that found by Johnson and Sudarshan for the failure of positivity. Thus the results of Johnson and Sudarshan can be understood as resulting from an instability in the propagation character of the spin 3/2 equation. For fields below the critical value, the theory makes sense in the given Lorentz frame, even though the canonical formalism is invalid in some other Lorentz frame.

To understand the full significance of the results of Velo and Zwanziger, it is necessary to relate the retarded and advanced fundamental solutions of the inhomogeneous wave equation to the quantized field solutions of the homogeneous wave equation. The external field problem in the general form discussed here involves the solution of the differential equation

$$[\beta^{\mu}\partial_{\mu} + \rho + B(x)]\,\psi(x) = 0 \tag{1.9}$$

for the unknown quantized field $\psi(x)$. Here β^{μ} and ρ are numerical matrices and $B(x)$ is a matrix whose entries are functions describing the given external field. The problem is therefore linear in the unknown quantized field. This suggests that it should be possible to convert it to a search for unknown functions. Such a procedure was implicit in the quantum field theory literature from the 1930's on, but typically it was entangled with the problem of defining a conserved current - a non-linear problem. The first explicitly linear treatment appears to have been that of A. Capri [12] [13]. Its main ideas are as follows. First, one replaces the differential equation (1.9) by a pair of integral equations

$$\psi(x) = \psi^{in}(x) - \int S_R(x-y)\,B(y)\,\psi(y)d^4y \tag{1.10}$$

$$\psi(x) = \psi^{out}(x) - \int S_A(x-y)\,B(y)\,\psi(y)d^4y \tag{1.11}$$

These equations are very convenient because one can impose the free field commutation relations on the in-field, ψ^{in}, and expect that the commutation relations of ψ and ψ^{out} should follow. Second, one introduces the smeared field

$$\psi(f) = \sum_{j=1}^{N} \int f_j(x)\,\psi_j(x)\,d^4x \tag{1.12}$$

The equations (1.10) and (1.11) then become

$$\psi(T_R f) = \psi^{in}(f) \qquad \psi(T_A f) = \psi^{out}(f) \tag{1.13}$$

where

$$(T_{\substack{R\\A}} f)(x) = f(x) + \int d^4y\,f(y)\,S_{\substack{R\\A}}(y-x)B(x) \tag{1.14}$$

Thus, if T_R and T_A map the space, $\mathcal{S}$, of test functions one to one onto itself and have continuous inverses, the fields ψ and ψ^{out} can be expressed in terms of ψ^{in} :

$$\psi(f) = \psi^{in}(T_R^{-1} f) \qquad \psi^{out}(f) = \psi^{in}(T_R^{-1} T_A f) \tag{1.15}$$

These formulae reduce the question of the existence of ψ and ψ^{out} to one of the existence and properties of the mappings T_R^{-1} and $T_R^{-1} T_A$. As will be shown in Chapter III

$$(T_{\substack{R\\A}}^{-1} f)(x) = f(x) - \int f(y) d^4y \, S_{\substack{R\\A}}(y,x;B)B(x) \tag{1.16}$$

where $S_{\substack{R\\A}}(x,y;B)$ is the weakly retarded (advanced) fundamental solution in the presence of the external field B. Weakly retarded (advanced) means vanishing rapidly outside the light cone. A further result of [14] and Chapter III is : For a large class of wave equations a quasi-local quantized field satisfying the wave equation (1.9) exists, if and only if weakly retarded and advanced fundamental solutions $S_{\substack{R\\A}}(x,y;B)$ exist having a certain continuity property. A quasi-local quantized field is one for which the (anti)commutator of fields vanishes rapidly for space-like separations of its points, but does not necessarily vanish identically. Combining this result with those of Velo and Zwanziger one sees that there exist wave equations and external fields for which there are quasi-local but not strictly local quantized field solutions, if there are any solutions at all. One can ask whether under these circumstances there is still a reasonable particle interpretation of the theory. The answer is yes under an additional condition on the fundamental solutions. It is shown in Chapter III that one can then establish the existence of an out-vacuum ψ_0^{out} and the associated family of out-particle states, as well as the unitarity of the S matrix.

If the external fields are weak, there are cases in which the existence of the required weakly retarded and advanced fundamental solutions can be proved (see [15] [16] and Seiler's lectures). However, for strong fields, those for which the hyperbolic character of the wave equation is destroyed, the problem of the existence of weakly retarded and advanced fundamental solutions has remained open for almost a decade. Recently, Gårding has shown in an example, that no such weakly but not strictly retarded and advanced solutions exist. The idea of the proof is sufficiently general that it seems likely that it should be possible to use it in other cases. What is already clear is that, in general for strong fields, no weakly but not strictly retarded and advanced fundamental solutions exist and the strong field Velo-Zwanziger phenomenon is a fatal illness for a quantized field coupled to the external field.

It is perhaps worth noting how the example given in Gårding's lectures arises for the PDK spin-zero wave equation in two-dimensional space-time if a symmetric tensor coupling is introduced. The Klein-Gordon equation

$$(\Box + m^2)\phi(x) = 0$$

is written as a first order system for a three component wave function $\psi = \{\phi, \phi^\mu\}, \mu = 0,1,2,3$ by requiring

$$\phi^\mu = \frac{1}{m}\partial^\mu\phi \qquad \partial_\mu\phi^\mu + m\phi = 0 \tag{1.17}$$

This system takes the form

$$(\beta^\mu\partial_\mu + m)\psi(x) = 0$$

if one introduces the matrices

$$\beta^o = \begin{Bmatrix} 0 & 1 & 0 \\ -1 & 0 & 0 \\ 0 & 0 & 0 \end{Bmatrix} \qquad \beta^1 = \begin{Bmatrix} 0 & 0 & 1 \\ 0 & 0 & 0 \\ 1 & 0 & 0 \end{Bmatrix}$$

These matrices satisfy

$$-(\beta^\mu)^* = \eta\,\beta^\mu\,\eta^{-1}$$

where

$$\eta = \begin{Bmatrix} 1 & 0 & 0 \\ 0 & 1 & 0 \\ 0 & 0 & -1 \end{Bmatrix} .$$

There are nine independent external field couplings corresponding to the real coefficients in the expansion

$$B(x) = \rho(x)\,1 + \sigma(x)\,\beta^\rho\beta_\rho + i\,A_\mu(x)\beta^\mu$$

$$+ \beta_\mu[\beta^\rho\beta_\rho, \beta^\mu] + F_{\mu\nu}\frac{i}{2}(\beta^\mu\beta^\nu - \beta^\nu\beta^\mu) + G_{\mu\nu}\frac{1}{2}[(\beta^\mu\beta^\nu + \beta^\nu\beta^\mu) - g^{\mu\nu}\beta^\rho\beta_\rho]$$

with the constraints $F_{\mu\nu} = -F_{\nu\mu}$, $G_{\mu\nu} = G_{\nu\mu}$, $G^\mu_\mu = 0$. $B(x)$ satisfies

$$B(x)^* = \eta\,B(x)\,\eta^{-1}$$

which guarantees the conservation law of the current

$$\partial_\mu\psi^+(x)i\beta^\mu\psi(x) = 0$$

with $\psi^{+}(x) = \overline{\psi(x)\eta}$. Consider now the effects of symmetric tensor coupling. It gives

$$B(x) = G_{oo}((\beta^{o})^{2} + (\beta^{1})^{2}) + G_{11}((\beta^{o})^{2} + (\beta^{(1)})^{2})$$

$$+ G_{o1}(\beta^{o}\beta^{1} + \beta^{1}\beta^{o})$$

$$= \begin{Bmatrix} 0 & 0 & 0 \\ 0 & -G_{oo} & -G_{o1} \\ 0 & G_{o1} & G_{oo} \end{Bmatrix}$$

For Gårding's examples one needs arbitrary elements in the second and third spots on the diagonal, and zero off diagonal terms. To obtain this one needs only add a suitable amount of the scalar coupling $\tau(x)[2 + \beta^{\rho}\beta_{\rho}]$ and set $G_{o1} = 0$. Thus, the non existence of weakly retarded and advanced fundamental solutions occurs in about the simplest example one could consider constructing.

In the years since acausality in the external field problem was discovered, several cures have been proposed, some involving new wave equations [17], some reinterpretations of known equations [18] [19] [20] [21] [22]. All these proposed theories successfully avoid acausality but at the cost of introducing either an indefinite metric or basic alterations in commutation relations. In [23], it was shown that the latter theories have no consistent particle interpretation. It is a plausible conjecture that, in fact, there are no consistent higher spin theories of massive particles in external fields. More in detail, the conjecture goes in three stages :

a) In multimass theories the scalar product arising from the conserved current regarded as a form on the positive energy solutions always is indefinite.

b) When theories for which the scalar product arising from the conserved current is indefinite regarded as a form on positive energy solutions are modified by redefinition of the scalar product, the commutation relations of the fields are altered so that they become non-local under perturbation by external fields.

c) Single mass theories with spins ≥ 3/2 and positive scalar products arising from the conserved current give rise to acausality in an external field. A proof of a) b) c) will not be attempted here but it seems to be within the reach of the techniques outlined here.

The above brief account of the present state of the external field problem would not be complete without some mention of the supersymmetric theories of spin 3/2 and 2 and their relatives [24]. In a sense these theories offer a way out of the difficulties of acausality. On the other hand, they suggest that to achieve this result, one must transcend the framework of the external field problem as presented here. More specifically,

a) The initial Lagrangians of the supersymmetric theories describe a massless spin 3/2 field coupled to a massless spin 2 (gravitational field), and possibly other fields. The spin 3/2 particle acquires a mass only by virtue of a comological term in Einstein's gravitational equations which explicitly breaks super symmetry.

b) The propagation cone of the spin 3/2 matter waves is by definition that fixed by the gravitational field, rather than the light cone of some fixed Minkowski space . From the point of view of any fixed Minkowski space one will in general have acausality. On the other hand, the equations remain hyperbolic no matter how strong the field ; that is progress compared to all previous models of spin 3/2 where sufficiently string fields destroyed hyperbolicity.

c) The new theory is totally different in spirit from those discussed earlier. In the new theories, consistency is a special miracle resulting from a combination of Einstein's theory of gravitation and the basic mechanisms of quantized gauge field theories.

2. Euclidean quantum field theory and the external field problem

The usefulness of the solution of the external field problem in the theory of coupled quantized fields arises as follows. According to the general theory of quantized fields, the physical content of the theory can be expressed in terms of vacuum expectation values of products of the fields. For example, for a single hermitean scalar field the required quantities are

$$(\psi_o, \phi(x_1)\ldots\phi(x_n)\psi_o)$$

By virtue of the temperedness of the fields regarded as operator-valued distributions and the spectral properties of the energy momentum operator of the theory and the local commutativity of the field ϕ, it follows that this tempered distribution is, in fact, an analytic function when the points $x_1 \ldots x_n$ are all separated by space-like intervals. Furthermore, the analytic function can be continued analytically to the so-called Schwinger points

$$(ix_1, \vec{x}_1)(ix_2^o, \vec{x}_2)\ldots(ix_n^o, \vec{x}_n)$$

where $(\vec{x}_j - \vec{x}_k)^2 + (x_j^o - x_k^o)^2 \neq 0$, $j \neq k$. The resulting functions, commonly called Schwinger functions, can be regarded as analytic functions of n real four-dimensional-vector variables $y_1 \dots y_n$; $y_j^o = ix_j^o$, $\vec{y}_j = \vec{x}_j$ defined and analytic except at points of coincidence where for some $j \neq k$, $y_j = y_k$. The Lorentz invariance of the original theory implies the O_4 invariance of the Euclidean theory.

It was a beautiful idea of Nakano and Schwinger that these analytic functions could perhaps be defined in the points of coincidence in such a manner that there would be a theory of Euclidean fields whose vacuum expectation values would yield the given Schwinger functions [25] [26].

There is a series of important general results of Euclidean field theory that give sufficient conditions that a) a set of Schwinger functions defined at non-coincident points, yields a unique Minkowski space quantum field theory satisfying the usual axioms, b) that a Euclidean field theory yields such an acceptable Minkowski space theory [27] [28].

Now in Euclidean field theory there is a solution of Lagrangian field theories by quadratures in terms of functional integrals, the Euclidean Gell-Mann-Low formula. If cutoffs are introduced, this formula can be given a mathematical meaning and gives non-perturbative expressions for the cutoff Schwinger functions [29]. One of the main themes of current constructive field theory is the study of the limits of these expressions as the cutoffs are removed. To see how the external field problem enters these deductions, consider a Yukawa interaction between a spinor field ψ and a pseudo-scalar field, ϕ. The corresponding Schwinger functions, $S(x_1 \dots x_r; y_1 \dots y_s; z_1 \dots z_t)$, are the analytic continuation of the Green's functions

$$(\psi_o, (\prod_{j=1}^{r} \psi(x_j) \prod_{k=1}^{s} \psi^+(y_k) \prod_{\ell=1}^{t} \phi(Z_\ell))_+ \psi_o)$$

where ψ_o is the physical vacuum and $(\quad)_+$ the time-ordering operation. The Euclidean Gell-Mann-Low formula for S is, formally,

$$S(x;y;z) = Z^{-1} \iiint \mathcal{D}\psi \mathcal{D}\psi^+ \mathcal{D}\phi \prod_j \psi(x_j) \prod_k \psi^+(y_k)$$

$$\prod_\ell \phi(Z_\ell) \exp \int \mathcal{L}(\psi, \psi^+, \phi) d^4x$$

$$Z = \iiint \mathcal{D}\psi \mathcal{D}\psi^+ \mathcal{D}\phi \exp [\int \mathcal{L}(\psi, \psi^+ \phi) d^4x]$$

where $\mathcal{L}$ is the Lagrangian. I will not explain the meaning to be attributed to the integration over the Fermi fields ψ and ψ^+; that is done in the references. The main point is that they appear only quadratically in the exponent because the Yukawa interaction is $\int \psi^+(x)\gamma^5\psi(x)\phi(x)d^4x$, and the resulting Gaussian integrals can be done to yield

$$S(\underline{x},\underline{y},\underline{z}) = Z^{-1}\int \mathcal{D}\phi \det\{S(x_j,y_k;g\gamma^5\phi)\}$$
$$\exp[\int \mathcal{L}_o(\phi)d^4x] \det(1+S^{(o)*}g\gamma^5\phi)$$

$$Z = \int \mathcal{D}\phi \det(1+S^{(o)}*g\gamma^5\phi) \exp[\int \mathcal{L}_o(\phi)d^4x]$$

Here $S(x,y;g\gamma^5\phi)$ is the Euclidean fundamental solution for the Dirac equation in the external field ϕ

$$(\gamma^{E\mu}\partial_\mu + m + g\gamma^5\phi(x))S(x,y;g\gamma^5\phi) = \delta(x-y)1$$

$$S(x,y;g\gamma^5\phi)(-\gamma^{E\mu}\overleftarrow{\partial}_\mu + m + g\gamma^5\phi(y)) = \delta(x-y)1$$

and $\det(1+S^{(o)}*g\gamma^5\phi)$ is the Fredholm determinant of the integral operator, K, whose kernel is

$$K(x,y) = S^{(o)}(x,y)g\gamma^5\phi(y)$$

$S^{(o)}$ is the Euclidean fundamental solution of the free Dirac equation $S^{(o)}(x,y) = S(x,y;0)$. Here we see the second important application of solutions of the external field problem : as an intermediate in the theory of Yukawa coupling, expressing the effect on the Bose field ϕ of the presence of the Fermi field ψ,ψ^+.

One feature of this application is very different from that of the preceding application to the description of particles. Here the field, ϕ, has its smoothness properties determined by the measure on function space associated with the free Bose field, and, as a consequence, with probability one ϕ is very rough, being nowhere a function [32]. In the rigorous discussion of Y_2, the Yukawa model in two-dimensional space time, much ink is spent on controlling the resulting functional analytic difficulties.

3. Euclidean quantum field theory and non-linear invariant wave equations

Up to this point, I have talked about invariant systems of linear differential equations. How do non-linear invariant systems creep into quantum field theory ? You will hear a number of answers to this question from Gervais and Fröhlich. Let me give yet another, probably the oldest [33]. Consider the generating functional for the Schwinger functions in $P(\phi)_2$ theory

$$W(f) = \int \exp \phi(f)d\mu(\phi)$$

where

$$d\mu(\phi) = Z^{-1} \exp \int \mathcal{P}(\phi(x))\, d^2x \, d\mu_o(\phi)$$

$\mathcal{P}$ being a polynomial of even degree bounded below, $d\mu_o$ the free field measure of mass m_o. Then

$$S(x_1 \ldots x_n) = \frac{\delta^n}{\delta F(x_1) \ldots \delta f(x_n)} W(f) \Big|_{f=0}$$

There is an important approximation to W obtained by replacing $S(x_1 \ldots x_n)$ by the lowest non-trivial contribution to its perturbation series in $\mathcal{P}$ and resumming ; it is called the tree approximation, $W^{Tree}(f)$. Then $\Phi(x) = \frac{\delta W^{Tree}}{\delta f(x)}$ satisfies the differential equation

$$(-\Delta + m_o^2)\Phi(x) = -\mathcal{P}'(\Phi(x)) + f(x)$$

where $\mathcal{P}'$ is the derivative of $\mathcal{P}$. Thus, the tree approximation to the generating functional of the Schwinger functions of the coupled field theory problem yields a classical field satisfying the classical non-linear differential equation of the theory.

The tree approximation is the lowest term in a loop expansion, and one of the standard approaches to the study of symmetry breaking is via the symmetry properties of the classical non-linear equation.

One method for realizing such an approach is the following : look for solitary wave or soliton solutions of the classical non-linear equation ; decorate them with appropriate quantum corrections to obtain solutions of the quantum field theory problem. The lectures of Gervais provide a systematic way of deriving such procedures in various model quantum field theories.

A second method for passing from solitary wave solutions of the classical non-linear wave equation to solutions of quantum field theory is to recognize the relation between the solitary wave solutions and topological quantum numbers and to base a construction of new superselection sectors on the existence of topological quantum numbers. For a detailed exposition of this point of view see [34] [35], and Fröhlich's lectures. This method seems to be the most satisfactory in principle.

4. Euclidean quantum field theory and instantons

The last of the applications of non-linear invariant wave equations to be mentioned is the theory of instantons [36]. These arise in Euclidean field theory when one attempts to approximate the functional integral solution of a Lagrangian field theory by methods related to steepest descent. To arrive at the notion of instantons in a typical case consider the Euclidean version of the $(\lambda\phi^4)_1$ theory (the anharmonic oscillator !) It has a Euclidean action

$$I(t_2,t_1;\phi) = \int_{t_1}^{t_2} d\tau\left[\frac{1}{2}\left(\frac{d\phi}{d\tau}\right)^2 + V(\phi(\tau))\right]$$

where

$$V(\phi) = \frac{m^2}{2}\phi^2 + \lambda\phi^4 + \text{const.}$$

The measure on the Euclidean field ϕ formally indicated by

$$Z^{-1}\mathcal{D}\phi \exp[-I(t_2,t_1,\phi)] \tag{4.1}$$

then has stationary phase for those ϕ for which

$$I(t_2,t_1;\phi) < \infty$$

$$\delta I = 0$$

A function ϕ satisfying these conditions for $t_2 = +\infty$, $t_1 = -\infty$, and not vanishing identically is called an <u>instanton</u>. (With $\pm\infty$ interchanged it is called an anti-instanton). Since $\delta I = 0$ implies the Euler-Lagrange equations

$$\frac{d^2\phi}{dx^2} = -m^2\phi - 4\lambda\phi^3 ,$$

the instanton can be visualized in terms of Newtonian motion in the "upside down" potential $-V(\phi)$. When both m^2 and λ are positive, only $\phi = 0$ satisfies the equation and has finite action. When $m^2 < 0$ and $\lambda > 0$, $V(\phi)$ has a double minimum and there is an instanton as shown in the Figure 3 (the constant in V has been adjusted to make V = 0 the minimum value). Since the two minima ϕ_-, ϕ_+ describe two alternative classical ground states, the instanton can be thought of as a solution tunneling from one vacuum to the other. The derivative of the instanton solution with respect to τ is positive and goes to zero rapidly for large $|\tau|$. Thus, the instanton can be regarded as fairly well localized. Since translating an instanton gives another the center of instanton can be located anywhere.

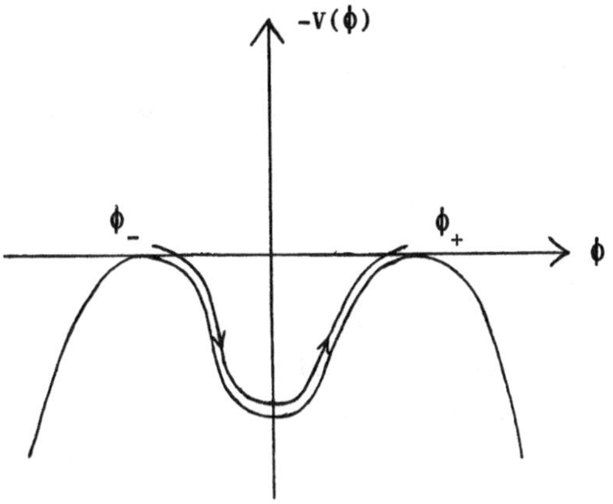

Figure 3

The instanton

The physical significance of the instanton arises from the fact that it can be used to evaluate functional integrals approximately. For example, the expression $<\phi_-|\exp[-HT]|\phi_->$ whose behavior for large T is determined by the lowest eigenvalues of H can be written as a functional integral using the measure (4.1) according to the Feynman-Kac formula. The functional integral can be written approximately as a sum of contributions from n instantons and n anti-instantons as shown in the figure

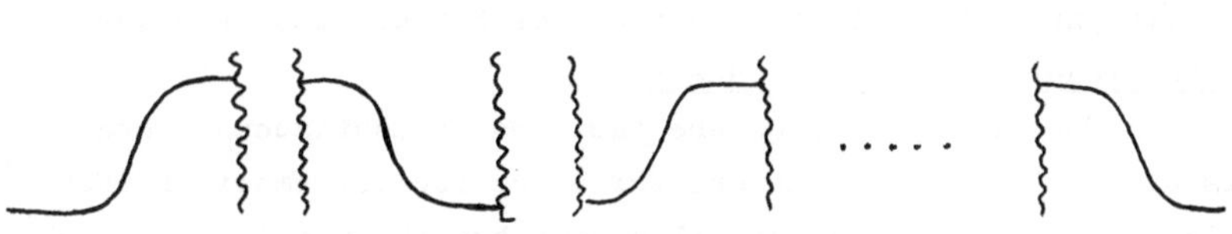

Figure 4

A configuration of n instantons and n anti-instantons.

These configurations do not make the action stationary except in the limit where the instantons are all infinitely far from one another. For a systematic method which in principle permits one to calculate corrections to the approximate formula see the lectures of Gervais. The contributions from the individual instantons are calculated by making a Gaussian approximation to the integrand around the instanton . Corrections to that approximation can also be computed.

When one generalizes the discussion of $(\phi^4)_1$ to higher dimensions two striking features emerge. There are no instantons for theories whose Euclidean action takes the form $\frac{1}{2}\int(\partial_\mu\phi)(\partial_\mu\phi)d^dx + V(\phi)$, $d \geq 2$ with $V \geq 0$. There are instantons in non-abelian gauge theories and they are connected with the existence of non trivial homotopy classes of mappings associated with gauge transformations. The beautiful theory of instantons in gauge theories is the subject of the lectures of Stora.

There is still no mathematically rigorous theory which treats the systematic expansion of functional integrals and yields the expressions given by the theory of instantons as a first approximation. In view of this, it is perhaps worth while to call attention to one remarkable feature of the situation. Instantons are in general given by very smooth functions. One may expect, in fact, that in general they will be defined by analytic functions. On the other hand, the behavior of the typical field in the functional integrals appearing in field theory is very rough ; with probability one it will be a tempered distribution which is nowhere a measure. Thus, the theory has to deal with a situation in which the field which yields stationary action is analytic but the typical field in the "neighborhood" is very rough. That is not paradoxical, but it is remarkable.

5. Open problems

The following is a miscellaneous collection of problems raised during the school. The reader is referred to other lectures for further lists. (In particular, to the lectures of Strauss and Zwanziger).

Physical significance of Gevray classes

The Gevray classes are families of continuous linear functionals defined on test functions which are locally smoother than C^∞. The test functions have derivatives of all orders but their values at each point are restricted in their growth with order. Thus, a generalized function in a Gevray class may be locally rougher than an ordinary distribution lying in $\mathcal{D}'$. When one admits coefficients in Gevray test function spaces and solutions in Gevray classes some of the instabilities of hyperbolicity disappear. For example, as Gårding will describe, the occurrence of lower order terms in an operator may not destroy hyperbolicity. Can it actually happen for invariant wave equations that hyperbolicity would be destroyed for external fields in $\mathcal{D}$ but not for external field in a Gevray class test function space ? The first step toward a solution of this problem has been taken by

Bellissard and Seiler who remarked that the Leray-Ohya theory is applicable to the Fierz-Pauli equation for particles of spin 3/2 [37].

Current operators

Proofs of the existence of quantized fields solving the external field problem which are based on the existence of weakly advanced and retarded fundamental solutions do not require a knowledge of current operators. These bilinear quantities require singular limiting operations e.g.

$$j^{\mu}(x) = \lim_{x \to y} \left[\psi^{+}(x)\, i\beta^{\mu}\psi(y) - \text{Counter Term}^{\mu}(x,y)\right] .$$

The legitimacy of this definition and analogous definitions for other bilinear quantities depends on the singularities of the fundamental solutions in the external field. Although the proofs has been sketched [38], it seems that the details have never been written out. A clean discussion would be valuable for another reason : ambiguities in the definition of such bilinear quantities have been a source of confusion in the context of the Hawking evaporation process of black hole physics. A clear non-perturbative treatment for the present problem would be a good starting point for the black hole problem.

Gauge invariance and the Cauchy problem

Theories for which the unknown wave function is a gauge field or a gauge dependent quantity will be mapped by gauge transformation into physically equivalent theories. Can initial data in the Cauchy problem be grouped into gauge equivalence classes and the initial value problem be solved for equivalence classes ? Is the occurrence of the Velo-Zwanziger phenomenon gauge invariant ?

The significance of acausality for coupled fields

The external field problem for linear invariant wave equations discussed here introduces the external field in such a way that linearity in the unknown function is preserved. It is plausible that appropriate non-linear terms in the unknown wave function might stabilize the situation and prevent the loss of hyperbolicity. Examples of this phenomemon are known [39]. More systematic investigation would be worthwhile as a preliminary to the study of coupled field problems. Coupled field problems for higher spin in four-dimensional space-time are rather inaccessible at the moment because they lead to non-renormalizable theories which cannot be dealt with effectively with present techniques. Can one find a model of coupled quantized fields in two-dimensional space-time which is treatable and is at least interestingly analogous to coupled higher spins in four-dimensions?

REFERENCES

[1] P.A.M. Dirac, Relativistic Wave Equations, Proc.Roy.Soc.Lond. 155A (1936) 447-459.

[2] M. Fierz, Über die relativistische Theorie kraftfreier Teilchen mit beliebigem Spin, Helv. Phys. Acta 12 (1939) 3-37.

[3] M. Fierz and W. Pauli, On Relativistic Waves Equations for Particles of Arbitrary Spin in an Electromagnetic Field, Proc. Roy. Soc. Lond. 173A (1939) 211-232.

[4] G. Wentzel, Quantum Field Theory, p.205, Wiley-Interscience, New York, (1969).

[5] P. Federbush, Minimal Electromagnetic Coupling for Spin Two Particles, Nuovo Cimento 19 (1961) 512-513.

[6] J. W. Weinberg, Univ. Cal. Berkeley Thesis, Studies in the Quantum Field Theory of Elementary Particles, 1943 unpublished.
S. Kusaka and J. W. Weinberg, Charged Particles of Higher Spin unpublished.

[7] E. Wild, On first order waves equations for elementary Particles without subsidiary conditions, Proc.Roy.Soc.Lond. 191A (1947) 253-268.

[8] K. Johnson and E. Sudarshan, The Impossibility of a Consistent Theory of Charged Higher Spin Fermi Fields, Ann. of Phys. 13 (1961) 126-145.

[9] For a review of developments up to the end of the 1940's see E. M. Corson, Introduction to Tensors, Spinors, and Relativistic Wave Equations, Blackie, London 1953.

[10] G. Velo and D. Zwanziger, Propagation and Quantization of Rarita-Schwinger Waves in an External Electromagnetic Potential, Phys. Rev. 186 (1969) 1337-1341.

[11] G. Velo and D. Zwanziger, Non causality and other defects of Interaction Lagrangians for Particles with Spin 1 and Higher, Phys. Rev. 188 (1969) 2218-2222.

[12] A. Capri, Electron in a Given Time Dependent Electromagnetic Field, J. Math. Phys. 10 (1969) 575-580.

[13] A. S. Wightman, Partial Differential Equations and Relativistic Quantum Field Theory, Lectures in Differential Equations, Vol.II, A. K. Aziz Editor, van Nostrand, Princeton New Jersey 1969.

[14] A. S. Wightman, Relativistic Wave Equations as Singular Hyperbolic Systems, pp.441-477 in Partial Differential Equations, Proceedings of Symposia in Pure Math., Vol.XXIII, Amer. Math. Soc., Providence, Rhode

Island 1973.

[15] G. Velo, Anomalous Behavior of a Massive Spin Two Charged Particle in an External Electromagnetic Field, Nuclear Physics B43 (1972) 389-401.

[16] G. Velo, An Existence Theorem for a Massive Spin One Particle in an External Tensor Field, Annales de l'Institut Henri Poincaré 22, (1975) 249-255.

[17] G. Iversen, Some Remarks on the Supermultiplet Theory, pp.44-64 in Troubles in the External Field Problem, Tracts in Mathematical and Natural Sciences, Vol. 4, Gordon and Breach 1971.

[18] W. J. Hurley, Relativistic Wave Equations for Particles with Arbitrary Spin, Phys. Rev. D4 (1971) 3605-3616.

[19] W. J. Hurley, Consistent Description of Higher Spin Fields, Phys. Rev. Letts. 29 (1972) 1475-1477.

[20] W. J. Hurley, Invariant Bilinear Forms and the Discrete Symmetries for Relativistic Arbitrary Spin Fields, Phys. Rev. D10 (1974) 1185-1200.

[21] R. A. Krajcik and M. M. Nieto, Bhaba First Order Wave Equations I...VII, Part VII is Phys. Rev. D15 (1977) 445-452; it contains references to the six earlier papers.

[22] R. A. Krajcik and M. M. Nieto, Foldy Wouthuysen Transformations in an Indefinite Metric Space I...IV, part IV is Phys. Rev. D15 (1977) 426-432

[23] A. Wightman, Instability Phenomena in the External Field Problem for Two Classes of Relativistic Wave Equations, pp.423-460 in Studies in Mathematical Physics, Essays in Honor of Valentine Bargmann, eds. E. Lieb, B. Simon, A. Wightman, Princeton Press 1976.

[24] S. Deser and B. Zumino, Broken Super Symmetry and Super Gravity, Physics Letts. 38 (1977) 1433-1436.

[25] T. Nakano, Quantum Field Theory in Terms of Euclidean Parameters, Prog. Theor . Phys. 21 (1959) 241-259.

[26] J. Schwinger, On the Euclidean Structure of Relativistic Quantum Field Theory, Proc. Nat. Acad. Sci. U. S. A. 44 (1958) 956-965.

[27] E. Nelson, Construction of Quantum Fields from Markoff Fields, Jour. Fcnal. Anal. 12 (1973) 97-112.

[28] K. Osterwalder and R. Schrader, Axioms for Euclidean Green's Functions Commun. Math. Phys. 31 (1973) 83-112, II ibid 42 (1975) 281-305; K. Osterwalder, Euclidean Green's Functions and Wightman Distributions, pp.71-93 in Constructive Quantum Field Theory, Lecture Notes in Physics # 25 Springer Verlag Berlin 1973, eds. G. Velo and A. Wightman.

[29] The following account of the Euclidean Gell-Mann-Low formula is deliberately abbreviated since the preceding school of mathematical physics treated the matter in some detail. See, in particular [30], [31].

[30] E. Seiler, Non Perturbative Renormalization in the Yukawa Model in Two Dimensions pp 415-433 in Renormalizarion Theory eds. G. Velo and A. Wightman, D.Reidel Dordrecht Holland 1976, serves as an introduction to the papers : E. Seiler Schwinger Functions for the Yukawa Model in Two Dimensions with Space-Time Cutoff Commun. Math. Phys. 42 (1975) 163-182;,
B. Simon On Finite Mass Renormalizations in the Two Dimensional Yukawa Model Jour. Math. Phys. 16 (1975) 2289-2293.
E. Seiler and B. Simon Bounds in the Yukawa Quantum Field Theory : Upper Bound on the Pressure, Hamiltonian Bound and Linear Lower Bound Commun. Math. Phys. 45 (1975) 99-114.
E. Seiler and B. Simon Nelson's Symmetry and All That in the Y_2 and $(\phi^4)_3$ Quantum Field Theories, Annals of Physics 97 (1976) 420-518.
O. McBryan, Finite Mass Renormalizations in the Euclidean Yukawa_2 Field Theories, Commun. Math. Phys. 44 (1975) 237-243.
O. McBryan, Volume Dependence of Schwinger Functions in the Yukawa_2 Quantum Field Theory, Commun. Math. Phys. 45 (1975) 279-294.
O. McBryan, Convergence of the Vacuum Energy Density, -Bounds and Existence of Wightman Functions for the Yukawa_2 Model pp.237-252 in Les Méthodes Mathématiques de la Théorie Quantique des Champs, 1975.

[31] A. S. Wightman, Orientation pp.1-24 in Renormalization Theory, Proceedings of the NATO Advanced Study Institute held at Erice 1975, eds. G. Velo and A. S. Wightman, D. Reidel Dordrecht Holland 1976.

[32] M. C. Reed, Functional Analysis and Probability Theory, pp.2-43 and P. Colella and O. Lanford, Appendix : Sample Field Behavior for the Free Markov Random Field pp.44-70 in Constructive Quantum Field Theory.

[33] I learned it from K. Symanzik in the early 1960's. See Green's Functions and the Quantum Theory of Fields pp 490-531 in Lectures in Theoretical Physics III, Boulder 1960, ed. W. Britten, et al. Interscience Publishers N. Y. 1961. There it is derived for the generating functional of the Green's functions.

[34] J. Fröhlich, New Super-Selection Sectors ("Soliton States") in Two Dimensional Bose Quantum Field Theory Models, Comm. Math. Phys. 47 (1976) 269-310.

[35] J. Fröhlich, Phase Transitions, Goldstone Bosons and Topological Super Selection Rules, Acta Physica Austriaca Suppl. XV (1976) 133-269.

[36] In addition to the lectures of Gervais, I would recommend S. Coleman Classical Lumps and their Quantum Descendents, Erice Lectures 1975, for a general account of this subject.

[37] J. Bellissard and R. Seiler, On the Fierz-Pauli Equation for Particles with Spin $\frac{3}{2}$, Lett. al Nuovo Cimento 5 (1972) 221-225.

[38] A. S. Wightman, The Dirac Equation, pp 95-115 in Aspects of Quantum theory, eds. A. Salam and E. P. Wigner, Cambridge University Press London, 1972 p.105.

[39] G. Vélo, Restrictions on the Interactions between Vector Mesons, Nuclear Physics B65 (1973) 427-444.

Chapter II : The General Theory of Invariant Wave Equations

1. General requirements

I will consider systems of equations for a finite number of unknown complex valued functions and treat only the case of invariance under the connected part of the Poincaré group or, more generally, its covering group, ISL(2,ℂ), the inhomogeneous special linear group in two dimensional complex space. Since equations of higher order than one can be reduced to those of first order by introducing derivatives as new unknowns, it suffices to consider first order systems. Explicitly, they will be written in the form

$$(\beta^{\mu}\partial_{\mu} + \rho)\psi(x) = 0 \tag{1.1}$$

Here β^{μ}, $\mu = 0,1,2,3$, and ρ are numerical $K \times N$ matrices and ψ is a column vector whose rows are complex functions or possibly distributions.

What should one mean by saying that (1.1) is invariant ? A possible answer consistent with many examples appearing in mathematical physics is this : (1.1) is invariant if there exists a representation $A \mapsto S(A)$ of $SL(2,\mathbb{C})$ such that

$$(\mathcal{V}(a,A)\psi)(x) = S(A)\psi(\Lambda(A^{-1})(x-a)) \tag{1.2}$$

is a solution of (1.1) whenever ψ is. If (1.2) is inserted in (1.1), it becomes

$$(\beta^{\mu}\partial_{\mu} + \rho)(\mathcal{V}(a,A)\psi)(x) = (\beta^{\mu}S(A)\partial_{\mu} + \rho\, S(A))\psi(\Lambda(A^{-1})(x-a))$$

or if we introduce $y = \Lambda(A^{-1})(x-a)$ and note that $\partial_{x\mu} = \Lambda(A^{-1})^{\nu}{}_{\mu}\partial y_{\nu}$

$$[\Lambda(A^{-1})^{\nu}{}_{\mu}\, \beta^{\mu}S(A)\, \partial_{\nu} + \rho\, S(A)]\psi(y) = 0 \tag{1.3}$$

A sufficient condition that (1.1) imply (1.3) is that there exist another representation $A \mapsto S_1(A)$ of $SL(2,\mathbb{C})$ such that

$$\begin{aligned} S_1(A)^{-1}\beta^{\mu}\, S(A) &= \Lambda(A)^{\mu}{}_{\nu}\, \beta^{\nu} \\ S_1(A)^{-1}\rho\, S(A) &= \rho \end{aligned} \tag{1.4}$$

Thus, it is natural to make the definition

Definition : The wave equation

$$(\beta^{\mu}\partial_{\mu}+\rho)\,\psi(x) = 0$$

is said to be invariant under ISL(2,$\mathbb{C}$) or, for brevity, relativistically invariant if there exist representations $A \mapsto S(A)$ and $A \mapsto S_1(A)$ of SL(2,$\mathbb{C}$) such that (1.4) holds for all $A \in SL(2,\mathbb{C})$.

Up to this point, the class of functions ψ in which we seek solutions of the basic wave equation has not been specified. When, later on, a scalar product is introduced, it will be natural to consider only those ψ for which $(\psi,\psi) < \infty$. On the other hand, there are occasions when it is useful to consider ψ a distribution for which (ψ,ψ) is meaningless. For the moment, it suffices to consider solutions for which the Fourier transform, $\hat{\psi}$, exists, in some sense. Then $\hat{\psi}$ satisfies

$$(-i\beta^{\mu}p_{\mu} + \rho)\hat{\psi}(p) = 0 \tag{1.5}$$

and, as usual, passage to momentum space has reduced the problem to solving a system of algebraic equations. For all the classes of $\hat{\psi}$ we will have to do with, multiplication of $\hat{\psi}$ by an infinitely differentiable function, φ, of compact support yields $\varphi\hat{\psi}$ again in the class. Therefore, multiplying (1.5) by φ yields that $\varphi\hat{\psi}$ is an admissible solution if $\hat{\psi}$ is. Thus, in studying solutions of (1.5) it suffices to treat the case of $\hat{\psi}$ of compact support.

With this preliminary out of the way, I now turn to the notion of the mass spectrum of the equation. There appears here a basic distinction between the cases in which the number, K, of equations is less than the number, N, of components of the unknown, $\hat{\psi}$, and the cases in which it is greater or equal to N. If K = N, then $(-i\beta^{\mu}p_{\mu} + \rho)$ has an inverse unless the algebraic equation

$$\det(-i\beta^{\mu}p_{\mu} + \rho) = 0 \tag{1.6}$$

for p is satisfied. The inverse is an infinitely differentiable function of p when p does not satisfy (1.6). Thus, the inverse may be applied to (1.5) and one may draw the conclusion that $\hat{\psi}$ vanishes off the set of p that satisfy (1.6). If K > N, the same argument can be repeated with any K element subset of the equations (1.5) and the same conclusion follows : $\hat{\psi}$ must have its support on an algebraic set. This time the algebraic set is all p for which

$$\operatorname{rank}(-i\beta^{\mu}p_{\mu} + \rho) < N \tag{1.7}$$

a set which is the intersection of all those obtained by setting equal to zero a determinant of an $N \times N$ matrix formed from some rows of $(-i\beta^{\mu}p_{\mu} + p)$. On the other hand, if $K < N$ the wave equation (1.5) will have non-trivial smooth solutions $\hat{\psi}$ with support in any compact set of momentum space.

Lorentz invariance restricts the solutions of (1.6) and (1.7). When $K = N$, (1.4) implies

$$\begin{aligned} \det(-i\beta^{\mu}p_{\mu} + \rho) &= \det(S_1(A)^{-1}(-i\beta^{\mu}p_{\mu} + \rho)S(A)) \\ &= \det(-i\Lambda(A)^{\mu}{}_{\nu}\beta^{\nu}p_{\mu} + \rho) \\ &= \det(-i\,\beta^{\mu}[\Lambda(A)^{\nu}{}_{\mu}\,p_{\nu}] + \rho) \end{aligned} \tag{1.8}$$

so $\det(-i\beta^{\mu}p_{\mu} + \rho)$ is a polynomial in p, invariant under the action of the restricted Lorentz group. The fundamental theorem on vector invariants says that it is a polynomial in p^2 and hence of the form

$$Q(p^2) = Q_o \prod_{k=1}^{\ell} (-p^2 + m^2)$$

where Q_o and m^2 are some complex numbers. The set of numbers $\{m_1^2, \ldots, m_{\ell}^2\}$ is called the <u>mass spectrum</u> of the theory. If $K > N$, an analogous argument yields the Lorentz invariance of the algebraic set where rank $[-i\beta^{\mu}p_{\mu} + \rho] < N$ and an analogous definition of the mass spectrum can be made.

The existence of the mass spectrum for equations with $K \geq N$ has a physical interpretation : wave functions satisfying such a wave equation can provide a quantum theory of particles capable of existing only in superpositions of states of definite mass belonging to the spectrum. For that reason, in the following, attention will be restricted to the case $K \geq N$. For the further elaboration of the physical interpretation, one has to distinguish the cases $m_k^2 \geq 0$, $m_k^2 < 0$ and $\mathrm{Im}\, m_k^2 \neq 0$. For $m_k^2 \geq 0$ the real momenta satisfying $p^2 = m_k^2$ are time-like with positive energy $p^o \geq 0$ or negative energy $p^o \leq 0$. The former are directly interpreted as the momenta of physical particles. The latter appear naturally in the expansion of second quantized fields. For $m_k^2 < 0$, the corresponding real momenta are space-like. Attempts have been made to interprete these as the momenta of tachyons (particles moving faster than light) but here they will simply be regarded as non-physical. For $\mathrm{Im}\, m_k^2 \neq 0$, $p^2 = m_k^2$ can only be satisfied by complex p, which will also be regarded as non-physical. (In section two, we will give an independent argument for the non-physical character of complex momenta).

There are two straightforward methods of eliminating non-physical momenta from a theory. One can insist that the wave equation have a mass spectrum such that $m_k^2 \geq 0$, $k = 1,\dots,\ell$. Alternatively, one can permit complex or negative m_k^2 but require that the corresponding solutions make zero contribution to the scalar product. As has been remarked in Chapter I, the latter procedure has always led to difficulties in external field problems although as will be seen in the following it works well for free fields. For the former, we will introduce a precise definition.

Definition : A relativistically invariant equation

$$(\beta^\mu \partial_\mu + \rho)\psi(x) = 0 \tag{1.1}$$

is called proper, if

1) The number, K, of equations is greater than or equal to the number N, of components of ψ, and the mass spectrum $\{m_1^2,\dots,m_\ell^2\}$ of the equation satisfies

$$m_k^2 \geq 0 \qquad k = 1,\dots,\ell$$

2) There exists a positive sesquilinearform $(.,.)$ defined on the subset, V, of smooth positive energy solutions, ψ, of (1.1). $(.,.)$ is positive

$$(\psi,\psi) \geq 0$$

for all $\psi \in V$.

3) If $A \mapsto S_1(A)$ and $A \mapsto S(A)$ are the representations of $SL(2,\mathbb{C})$ associated with β^μ and ρ according to (1.4), then the representation of $ISL(2,\mathbb{C})$ defined by

$$(\mathcal{V}(a,A)\psi)(x) = S(A)\psi(\Lambda\ (A^{-1})(x-a))$$

leaves the scalar product invariant :

$$(\mathcal{V}(a,A)\phi, \mathcal{V}(a,A)\psi) = (\phi,\psi)$$

for all $\phi,\psi \in V$.

The above definitions naturally suggest the problem : classify all proper relativistically invariant wave equations. As a first step in its solution consider one such theory specified by

$$\beta^\mu, \rho, A \mapsto S_1(A), A \mapsto S(A),\ (.,.),\ (a,A) \mapsto \mathcal{V}(a,A)$$

and suppose that V_1 and V are non-singular K × K and N×N matrices respectively. Then another such theory is defined by

$$\begin{aligned}
&\beta^{(1)\mu} = V_1 \beta^{\mu} V^{-1} \qquad \rho^{(1)} = V_1 \rho V^{-1} \\
&S_1^{(1)}(A) = V_1 S_1(A) V_1^{-1} \qquad S^{(1)}(A) = VS(A)V^{-1} \\
&(\phi,\psi) = (V\phi, V\psi)^{(1)} \\
&(\mathcal{V}^{(1)}(a,A)\psi(x) = S^{(1)}(A)\psi(\Lambda(A)^{-1}(x-a))
\end{aligned} \tag{1.10}$$

and we call the two theories <u>equivalent</u>.

In the equivalence class of any relativistically invariant equation there is another for which ρ has an especially simple form, arrived at using the following Lemma .

<u>Lemma</u> : Let ρ be a linear transformation from $\mathbb{C}^K$ to $\mathbb{C}^N$. There exists a non singular K×K matrix, E, and a non-singular N×N matrix, F, such that

$$\rho = E\rho^{(1)} F \tag{1.11}$$

where $\rho^{(1)}$ is the K×N matrix

$$\rho^{(1)} = \begin{Bmatrix} 1 & 0 & 0 & 0 & \ldots 0 \\ 0 & 1 & 0 & 0 & \\ 0 & & & & \\ \vdots & & & & \\ 0 & & & & \end{Bmatrix} \tag{1.12}$$

The number of 1's along the diagonal being r, the rank of ρ, i.e. the linear dimension of the range of ρ .

Since E and F are non-singular we can introduce

$$\begin{aligned}
&\beta^{\mu} = E\beta^{(1)\mu}F, \; S_1^{(1)}(A) = E^{-1}S_1(A)E, \; S^{(1)}(A) = FS(A)F^{-1} \\
&\psi^{(1)}(x) = F\psi(x) \qquad (\phi^{(1)},\psi^{(1)})^{(1)} = (\phi,\psi)
\end{aligned} \tag{1.13}$$

and obtain from (1.1)

$$[\beta^{(1)\mu}\partial_{\mu} + m\rho^{(1)}]\,\psi^{(1)}(x) = 0 \tag{1.14}$$

Here the notation has been changed slightly to introduce a factor, m, so that both $\beta^{(1)\mu}$ and $\rho^{(1)}$ are dimensionless, and all the components of ψ have the same dimension. Sometimes, this procedure seems inadvisable. For example, the wave equation for a massless scalar field, ϕ :

$$\Box \phi = 0$$

may be rewritten in terms of the five component wave function, $\psi = \{\phi, \partial_\mu \phi\}$ as

$$(\beta^\mu \partial_\mu + \rho)\psi(x) = 0$$

with

$$\rho = \left\{\begin{array}{c|cccc} 0 & 0 & 0 & 0 & 0 \\ \hline 0 & 1 & 0 & 0 & 0 \\ 0 & 0 & 1 & 0 & 0 \\ 0 & 0 & 0 & 1 & 0 \\ 0 & 0 & 0 & 0 & 1 \end{array}\right\} \quad \beta^o = \left\{\begin{array}{ccccc} 0 & 1 & 0 & 0 & 0 \\ -1 & 0 & 0 & 0 & 0 \\ 0 & 0 & 0 & 0 & 0 \\ 0 & 0 & 0 & 0 & 0 \\ 0 & 0 & 0 & 0 & 0 \end{array}\right\} \quad \beta^1 = \left\{\begin{array}{ccccc} 0 & 0 & 1 & 0 & 0 \\ 0 & 0 & 0 & 0 & 0 \\ 1 & 0 & 0 & 0 & 0 \\ 0 & 0 & 0 & 0 & 0 \\ 0 & 0 & 0 & 0 & 0 \end{array}\right\}$$

$$\beta^2 = \left\{\begin{array}{ccccc} 0 & 0 & 0 & 1 & 0 \\ 0 & 0 & 0 & 0 & 0 \\ 0 & 0 & 0 & 0 & 0 \\ 1 & 0 & 0 & 0 & 0 \\ 0 & 0 & 0 & 0 & 0 \end{array}\right\} \quad \beta^3 = \left\{\begin{array}{ccccc} 0 & 0 & 0 & 0 & 1 \\ 0 & 0 & 0 & 0 & 0 \\ 0 & 0 & 0 & 0 & 0 \\ 0 & 0 & 0 & 0 & 0 \\ 1 & 0 & 0 & 0 & 0 \end{array}\right\}$$

These results may be summarized .

<u>Theorem</u> : In each equivalence class of proper relativistically invariant wave equations there is one in which the $K \times N$ matrix ρ takes the form

$$\rho = \left\{\begin{array}{cccc} 1 & 0 & 0 \cdots & 0 \\ 0 & 1 & 0 & \\ \vdots & & & \\ 0 & 0 & & \end{array}\right\} \tag{1.15}$$

with r ones on the diagonal and zeros everywhere else. Here r is the rank of ρ and $r \leq N$.

For a wave equation for which ρ has this standard form, $S_1(y)$ and $S(y)$ are partly triangular because

$$\rho\, S(g) = S_1(g)\, \rho$$

with ρ of the form (1.15) implies

$$S(y) = \left\{\begin{array}{c|c} A(y) & 0 \\ \hline B(y) & C(y) \end{array}\right\} \qquad S_1(y) = \left\{\begin{array}{c|c} A(y) & D(y) \\ \hline 0 & E(y) \end{array}\right\} \tag{1.16}$$

where the $r \times r$ matrices in the upper left hand corners are equal, as the notation indicates. A further simplification can be achieved if we use the remaining freedom in the choice of E and F ; the submatrices B(g) and D(g) can be taken zero. The point is that the pairs of non-singular E and F satisfying

$$E \rho F = \rho$$

may be read off from (1.16) : E is partially triangularized in the same way as $S_1(g)$ and F as S(g) and the $r \times r$ submatrices in the upper left hand corner of E and F are reciprocal. Among the admissible E is any non-singular $K \times K$ matrix leaving invariant the subspace of $\mathbb{C}^k$ spanned by vectors of the form $\binom{x}{0}$ where the last K-r components vanish. Among the admissible F is any non-singular matrix leaving invariant the subspace of $\mathbb{C}^N$ spanned by vectors of the form $\binom{0}{y}$ where the first r components vanish. Now the finite-dimensional representations of SL(2,$\mathbb{C}$) are completely reducible so there must exist non singular linear transformations E and F of the form just described that bring S(g) and $S_1(g)$ into the partially diagonalized form for which B(g) = D(g) = 0, via

$$S(g) \mapsto E^{-1} S(g) E \qquad S_1(g) \mapsto F S_1(g) F^{-1}$$

the fact that the form of the $r \times r$ submatrices is linked in no way impedes this operation since E and F can be chosen arbitrarily on their respective invariant subspaces. In summary, we have

Corollary : In each equivalence class of relativistic wave equations there is one for which ρ is in the standard form (1.15) and the $N \times N$ matrix S(g) and the $K \times K$ matrix $S_1(g)$ take the form

$$S(g) = \left\{\begin{array}{c|c} A(y) & 0 \\ \hline 0 & C(g) \end{array}\right\} \qquad S_1(g) = \left\{\begin{array}{c|c} A(g) & 0 \\ \hline 0 & E(g) \end{array}\right\} \tag{1.17}$$

Here A(g) is a $r \times r$ matrix, if ρ has r ones on the diagonal.

The theory of relativistic wave equations with $K > N$ or with $K = N$ and ρ singular but not zero has not been worked out as completely as that for $K = N$ and ρ non singular. In sections two and three of this chapter, I will continue treating the general case. However, in the remaining sections at various points only the better known case will be discussed. It would be interesting to have the full theory for $K = N$ and ρ singular, since it is appropriate for a general phenomenological theory of leptons. [1].

2. Finding all β^μ, ρ, given the representations of $SL(2,\mathbb{C})$

The problem of finding all β^μ and ρ satisfying (1.4) given the representations $A \mapsto S(A)$ and $A \mapsto S_1(A)$ has had a long history beginning with E. Majorana's pioneering study of 1932 [2]. The version given here follows Gårding's thesis of 1944 [3]. Similar ideas occur in the work of I. M. Gelfand and Y. Yaglom [4] and F. Bruhat [5].

Gårding put the problem this way. Given is a group G with elements g and three representations of it, $g \mapsto S^{(i)}(g)$, $i = 1,2,3$ in finite dimensional vector spaces V_i, $i = 1,2,3$, respectively. Suppose $v \mapsto X(v)$ is a linear mapping of V_3 into $L(V_2,V_1)$ the set of linear transformations from V_2 to V_1. The $X(v)$ are said to form a tensor set relative to $\{S^{(1)},S^{(2)},S^{(3)}\}$ if

$$S^{(1)}(g)X(v)S^{(2)}(g^{-1}) = X(S^{(3)}(g)v) \qquad (2.1)$$

The problem is: given $\{S^{(1)},S^{(2)},S^{(3)}\}$, find all tensor sets. Clearly, the problems of finding β^μ and ρ satisfying (1.4) are obtained by taking $G = SL(2,\mathbb{C})$ and choosing as the representation $S^{(3)}$ either $A \mapsto \Lambda(A)$ or the trivial one-dimensional representation $A \mapsto 1$.

To solve the problem Gårding uses the notion of homomorphic mapping defined as follows. Given two vector spaces V and W and in them two representations of a group G : $g \mapsto R(g)$ in V and $g \mapsto S(g)$ in W, one calls a linear mapping, T, of V into W homomorphic if the correspondence it defines is invariant under the simultaneous action of R and S, i.e.

$$TR(g)x = S(g)Tx$$

for all $x \in V$ and $g \in G$ or what is the same thing

$$TR(g) = S(g)T$$

Such a T is also called sometimes, an intertwining operator for the representations R and S.

Now a tensor set $X(v)$ relative to $\{S^{(1)}, S^{(2)}, S^{(3)}\}$ defines a linear transformation from $V_2 \otimes V_3$ into V_1 obtained by extending the formula

$$T x_2 \otimes x_3 = X(x_3)x_2$$

for $x_2 \in V_2$, $x_3 \in V_3$ by linearity and, conversely, every T defines a family $X(x_3)$. Further if T is homomorphic from $V_2 \otimes V_3$ to V_1, what that means for X is precisely (2.1) :

$$T S^{(2)}(g) \otimes S^{(3)}(g) = S^{(1)}(g)T$$

when applied on $x_2 \otimes x_3$ yields

$$X(S^{(3)}(g)x_3)S^{(2)}(g)x_2 = S^{(1)}(g)X(x_3)x_2$$

as required.

Thus we have the lemma

Lemma : Let T be a homomorphic mapping of $V_2 \otimes V_3$ into V_1, these vector spaces being equipped with the representations $g \mapsto S^{(2)}(g) \otimes S^{(3)}(g)$ and $g \mapsto S^{(1)}(g)$ of the group G respectively. Then T defines a tensor set relative to $\{S^{(1)}, S^{(2)}, S^{(3)}\}$.

Conversely, every such tensor set is obtained in this way.

Now the problem of finding homomorphic mappings of a vector space V into a vector space W, given representations $g \mapsto R(g)$ and $g \mapsto S(g)$ in V and W respectively is one with a simple solution provided the representations R and S are completely reducible i.e. are sums of irreducible representations [6]. That is the case for $SL(2,\mathbb{C})$.

The first step is to decompose the representation spaces into subspaces invariant under the group representations. There exist two families of subspaces $\{V_k, k \in K\}$, $\{W_\ell, \ell \in L\}$ of V and W respectively, such that

1) $V_k \cap V_{k'} = \{0\}$ for $k \neq k' \in K$

$W_\ell \cap W_{\ell'} = \{0\}$ for $\ell \neq \ell' \in L$

2) $V = \sum_{k \in K} V_k$, $W = \sum_{\ell \in L} W_\ell$

3) $R(g)V_k \subset V_k$ for $k \in K$, $g \in G$

$S(g)W_\ell \subset W_\ell$ for $\ell \in L$, $g \in G$

4) $R(g)$ restricted to V_k is irreducible for $k \in K$

$S(g)$ restricted to W_ℓ is irreducible for $k \in L$.

These decompositions of V and W are in part unique. If we consider the subspace $\hat{V}_\alpha$ spanned by all those V_k, $k \in K$ such that the restriction of R to V_k is equivalent to a given irreducible representation of G labelled by α, then $\hat{V}_\alpha$ is uniquely determined by R. A similar argument holds for the analogously defined $\hat{W}_\alpha$. On the other hand, the decomposition of $\hat{V}_\alpha$ into a sum of V_i is not necessarily unique since if A is any non-singular mapping of $\hat{V}_\alpha$ onto itself that commutes with the restriction of R to $\hat{V}_\alpha$, then $\{AV_i; V_i \subset \hat{V}_\alpha\}$ gives another admissible decomposition.

The basic theorem on homomorphic mappings says that a homomorphic mapping T maps the subspace $\hat{V}_\alpha$ belonging to the irreducible representation of G labelled by α into the subspace $\hat{W}_\alpha$ of W belonging to the same irreducible representation. T maps a given $V_i \subset \hat{V}_\alpha$ either into zero or bijectively onto an invariant subspace of $\hat{W}_\alpha$. There exists a non-singular B_α mapping $\hat{W}_\alpha$ onto itself and commuting with the restriction of S to $\hat{W}_\alpha$, such that each TV_i with $V_i \subset \hat{V}_\alpha$ that is not identically zero is some $B_\alpha W_j$ with $W_j \subset \hat{W}_\alpha$.

Thus the construction of all homomorphic mappings proceeds as follows. Decompose the representations R and S into irreducible representations of G labeled by the index α with multiplicities $n_\alpha(R)$ and $n_\alpha(S)$ respectively. For each α decide which of the subspaces of V that carry the irreducible representation α are to be mapped into zero. If the number of such subspaces is n'_α it must satisfy

$$n'_\alpha \le n_\alpha(R) \qquad n_\alpha(R) - n'_\alpha \le n_\alpha(S)$$

For each such choice take a mapping, B_α, of $\hat{W}_\alpha$ into itself which commutes with the restriction of S to $\hat{W}_\alpha$.

The matrix of T that results can be written

$$T = BP$$

where the restriction of T to $\hat{V}_\alpha$ is

$$T \restriction \hat{V}_\alpha = B_\alpha P_\alpha$$

with

$$
P_\alpha = \begin{array}{c|c|c|c|c|c}
 & V_1 & V_2 & \cdots & V_{n_\alpha(R)-n'_\alpha} \;\cdots & V_{n_\alpha(R)} \\
\hline
W_1 & \mathbb{1} & 0 & \cdots & 0 & \\
\hline
W_2 & 0 & \mathbb{1} & & & 0 \\
\hline
\vdots & \vdots & & \mathbb{1} & & \\
\hline
W_{n_\alpha(S)} & & 0 & & & 0
\end{array}
\tag{2.2}
$$

where there are n'_α zeros and $n_\alpha(R) - n'_\alpha$ ones on the diagonal that begins in the upper left hand corner. The matrix B_α is of the form

$$
B_\alpha = \begin{array}{c|c|c|c|c}
 & W_1 & W_2 & \cdots & W_{n_\alpha(S)} \\
\hline
W_1 & b_{11}\mathbb{1} & b_{12}\mathbb{1} & \cdots & b_{1n_\alpha(S)}\mathbb{1} \\
\hline
W_2 & b_{21}\mathbb{1} & b_{22}\mathbb{1} & \cdots & \\
\hline
\vdots & \vdots & \vdots & & \\
\hline
W_{n_\alpha(S)} & b_{n_\alpha(S)1}\mathbb{1} & & & b_{n_\alpha(S)n_\alpha(S)}\mathbb{1}
\end{array}
\tag{2.3}
$$

For the matrix ρ, the theorem in homomorphic mappings is applicable as stated because ρ is an intertwining operator for S and S_1

$$\rho S(g) = S_1(g)\rho$$

One has only to diagonalize $S(g)$ and $S_1(g)$ and label the subspaces $\hat{V}_\alpha$ and $\hat{W}_\alpha$ carrying a given irreducible representation with the same indices α. Then

$$\rho = BP$$

where

$$\rho \restriction \hat{V}_\alpha = B_\alpha P_\alpha$$

with B_α and P_α given by (2.2) and (2.3).

For β^μ the argument is more complicated because one has first to reduce the tensor product $S \otimes [\frac{1}{2},\frac{1}{2}]$. Suppose the matrix $\mathcal{U}$ brings it to diagonal form. Then we have

$$BP[\mathcal{U}\, S \otimes [\tfrac{1}{2},\tfrac{1}{2}]\mathcal{U}^{-1}] = S_1 BP \;.$$

If the rows of $\mathcal{U}$ are adapted to the labels in the diagonalized form of $S \otimes [\frac{1}{2},\frac{1}{2}]$ and the column labels corresponding to the basis in the space acted on by $[\frac{1}{2},\frac{1}{2}]$ are displayed explicitly, one has

$$\beta^{\mu} = B\,P\,\mathcal{U}^{\mu}$$

where $\mathcal{U}^{\mu}$ can be expressed in terms of vector coupling coefficients and Pauli matrices as follows

$$\mathcal{U}^{\mu}([k,\ell]\alpha_k\dot{\alpha}_\ell \;;[j_1,j_2]\alpha_{j_1}\dot{\alpha}_{j_2}) =$$

$$\sum_{\alpha_{\frac{1}{2}}\,\dot{\alpha}_{\frac{1}{2}}} (k\,\alpha_k | j_1\,\alpha_{j_1}\,\tfrac{1}{2}\,\alpha_{\frac{1}{2}})\overline{(\ell\,\dot{\alpha}_\ell | j_2\,\dot{\alpha}_{j_2}\,\tfrac{1}{2}\,\dot{\alpha}_{\frac{1}{2}})}\,\sigma^{\mu}_{\alpha_{\frac{1}{2}}\,\dot{\alpha}_{\frac{1}{2}}}$$

The method used to find ρ works equally well if one seeks an $N\times K$ matrix η satisfying

$$S(A)^*\eta = \eta\, S_1(A^{-1})$$

Such an η defines a homomorphic mapping of the representation space of $A \mapsto S_1(A)$ into the representation space of $A \mapsto S(A^{-1})^*$. η is represented by a matrix

$$\eta = BP$$

where P and B are of the form given in (2.2) and (2.3) respectively. In some of the applications of η, one has to satisfy in addition

$$\eta\beta^{\mu} = -(\beta^{\mu})^*\eta^*$$

$$\eta\rho = \rho^*\eta^*$$

In general, these conditions impose further restrictions on η.

<u>Example</u> : Dirac's spin $\frac{1}{2}$ theory .

Here $S_1 = S \cong [\frac{1}{2},0] \oplus [0,\frac{1}{2}]$ and so the matrix $\mathcal{U}$ satisfies

$$\mathcal{U}\{[\tfrac{1}{2},0] \otimes [\tfrac{1}{2},\tfrac{1}{2}] \oplus [0,\tfrac{1}{2}] \otimes [\tfrac{1}{2},\tfrac{1}{2}]\}\mathcal{U}^{-1}$$
$$= [1,\tfrac{1}{2}] \oplus [\tfrac{1}{2},0] \oplus [0,\tfrac{1}{2}] \oplus [\tfrac{1}{2},1]$$

The homomorphic mappings of this representation into $[\frac{1}{2},0] \oplus [0,\frac{1}{2}]$ map the representation spaces of $[1,\frac{1}{2}]$ and $[\frac{1}{2},1]$ into zero. As far as the representation spaces of $[\frac{1}{2},0]$ and $[0,\frac{1}{2}]$ are concerned, they are either mapped into zero or isomorphically onto themselves. Since the irreducible representations $[\frac{1}{2},0]$ and $[0,\frac{1}{2}]$ occur with multiplicity 1 the matrices B_α are multiples of the identity. In general β^μ takes the form

$$\beta^\mu = \begin{pmatrix} B_{[\frac{1}{2},0]} & 0 \\ 0 & B_{[0,\frac{1}{2}]} \end{pmatrix} \begin{pmatrix} 0 & \sigma_\mu \\ \sigma^\mu & 0 \end{pmatrix}$$

Similarly

$$\rho = \begin{pmatrix} B^r_{[\frac{1}{2},0]} & 0 \\ 0 & B^r_{[0,\frac{1}{2}]} \end{pmatrix}$$

Apart from an over all normalization factor, there is one parameter free in each of these matrices. It can be fixed if one assumes invariance under space inversion. That gives $B_{[\frac{1}{2},0]} = B_{[0,\frac{1}{2}]}$ and $B^r_{[\frac{1}{2},0]} = B^r_{[0,\frac{1}{2}]}$.

For an alternative approach to the material of this section which has the advantage of being well adapted to wave equations with unknown wave functions having an infinite number of components, see [7].

3. Finding all invariant sesquilinear forms on positive energy solutions

To make a quantum mechanical theory one needs a Hilbert space of state vectors. Therefore, given a family of solutions of an invariant wave equation one has to introduce a scalar product defined on pairs of members of the family. The traditional method to obtain such a scalar product is to integrate a conserved current over a space-like hyperplane. However, there is no reason, a priori, why there should not exist other admissible scalar products. A general theory has to discuss all such.

To pose the problem is to list the required properties of the scalar product. To some extent these properties have to be implicitly defined, since the class of solutions on which the form is defined is not fixed until the scalar product itself is known.

Let V_1 be a vector space of solutions of the invariant wave equation in question. The elements of V_1 will be assumed to be elements of $\mathcal{D}'(\mathbb{R}^1)$ i.e. distributions but not necessarily tempered, because we want to investigate the role of complex masses in the mass spectrum and a typical solution with complex mass, say a plane wave, u(p)exp ip.x, grows exponentially in absolute value as $|x| = \sqrt{\Sigma_{\mu=0}^{3} X_\mu^2}$ approaches infinity. The elements of V are assumed to be equipped with a transformation law under ISL(2,$\mathbb{C}$)

$$(\mathcal{V}(a,A)F)(x) = S(A)F(\Lambda(A^{-1})(x-a))$$

so we assume that $F \in V_1$ implies $\mathcal{V}(a,A)F \in V_1$. Furthermore, we will assume $F \in V_1$ implies $\sigma * F \in V_1$ where σ is any infinitely differentiable function of compact support and $*$ denotes convolution.

The scalar product is denoted $(\cdot,\cdot)$. It is assumed to have the standard properties :

$$(\phi,\psi) = \overline{(\psi,\phi)} \qquad \text{(\underline{hermitian})}$$

$$(\phi,\psi_1+\psi_2) = (\phi,\psi_1)+(\phi,\psi_2) \qquad (\phi,\alpha\psi) = \alpha(\phi,\psi)$$

$$(\phi_1+\psi_2,\psi) = (\phi_1,\psi)+(\phi_2,\psi) \qquad (\alpha\phi,\psi) = \bar{\alpha}(\phi,\psi) \quad \text{(\underline{sesquilinear})}$$

$$(\phi,\phi) \geq 0 \qquad \text{(\underline{non-negative})}$$

for all ϕ, $\phi_1,\phi_2,\psi,\psi_1,\psi_2 \in V_1$. Given such a scalar product on V_1, one defines the null space V_o as the set of all $\phi \in V_1$ such that $(\phi,\phi) = 0$. Then V_1/V_o is the vector space of equivalence classes of vectors of V_1, two vectors being equivalent if their difference lies in V_o. The scalar product is constant on equivalence classes and so one can regard it as defined on vectors of V_1/V_o. If $[\phi]$ is the equivalence class of ϕ, $\phi \in V$, then

$$([\phi],[\phi]) = 0$$

implies

$$[\phi] = 0 \quad \text{i.e.} \quad \phi \in V_o$$

so V_1/V_o is a prehilbert space. Its completion defines the Hilbert space of states, $\mathcal{H}$, we want to consider

$$\mathcal{H} = \overline{V_1/V_o}$$

Most of our calculations will be on V_1 rather than $\mathcal{H}$.

It will be assumed that $(.,.)$ has a weak form of continuity : $(\sigma * \phi , \tau * \psi)$ is separately continuous in σ and τ for each ϕ and ψ in V_1. The nuclear theorem then asserts that there is a distribution in two variables such that

$$(\sigma * \phi, \tau * \psi) = \iint d^4 x\, d^4 y\, \overline{\sigma(x)} K(\phi,\psi;x,y)\tau(y)$$

The scalar product is required to be invariant under ISL(2,$\mathbb{C}$) in the sense that

$$(\mathcal{U}(a,A)\phi, \mathcal{U}(a,A)\psi) = (\phi,\psi)$$

Since the action of $\mathcal{U}(a,1)$ on $\sigma * \phi$ is

$$(\mathcal{U}(a,1)\sigma * \phi)(x) = (\sigma * \phi)(x-a) = (\sigma_a * \phi)(x)$$

where

$$\sigma_a(x) = \sigma(x-a)$$

the invariance of the scalar product under translation implies

$$K(\phi,\psi;x-a,y-a) = K(\phi,\psi,x,y).$$

This means $K(\phi,\psi;x,y)$ depends only on $x-y$, so we make a change in notation, replacing $K(\phi,\psi;x,y)$ by $K(\phi,\psi;x-y)$.

The positivity requirement says that

$$\iint d^4 x\, d^4 y\, \overline{\sigma(x)} K(\phi,\phi;\, x-y)\sigma(y) \geq 0$$

i.e. $K(\phi,\phi;x)$ is a distribution of positive type. The Bochner-Schwartz theorem then says that it can be extended by continuity from test function in $\mathcal{D}$ to those in $\mathcal{S}$ so as to become a tempered distribution, and, furthermore, that tempered distribution is the Fourier transform of a positive measure of slow increase. Explicitly

$$K(\phi,\phi;x) = (2\pi)^{-2}\int d^4p\, e^{-ip\cdot x}\hat{K}(\phi,\phi;p)$$

where $\hat{K}(\phi,\phi;p)d^4p$ is a positive measure of slow increase.

Having obtained this form for $K(\phi,\phi;.)$ we obtain an analogous form for $K(\phi,\psi;.)$ by a two step polarization. First

$$\begin{aligned}(\sigma * \phi,\sigma * \psi) &= \tfrac{1}{4}((\sigma * (\phi+\psi),\sigma * (\phi+\psi)) \\ &\quad - (\sigma * (\phi-\psi),\sigma * (\phi-\psi)) \\ &\quad + i(\sigma * (\phi+i\psi),\sigma * (\phi+i\psi)) \\ &\quad - i(\sigma * (\phi-i\psi),\sigma * (\phi-i\psi))) \\ &= (2\pi)^2\int d^4p\, \overline{\hat{\sigma}(p)}\hat{K}(\phi,\psi;p)\hat{\sigma}(p)\end{aligned} \tag{3.1}$$

$\hat{K}(\phi,\psi;p)d^4p$ being a complex measure of slow increase. Second,

$$(\sigma * \phi,\tau * \psi) = (2a)^2\int d^4p\, \overline{\hat{\sigma}(p)}\hat{K}(\phi,\psi;p)\hat{\tau}(p)$$

is obtained from (3.1) by polarization in σ.

Up to this point, the differential equation has not been used. It implies, for each $\phi,\psi \in V_1$

$$(\phi,(\beta^\mu\partial_\mu+\rho)\psi) = ((\beta^\mu\partial_\mu+\rho)\phi,\psi) = 0$$

which imply immediately

$$\hat{K}(\phi,(\beta^\mu\partial_\mu+\rho)\psi;p) = \hat{K}((\beta^\mu\partial_\mu+\rho)\phi,\psi;p) = 0$$

Now, notice that $(\beta^\mu\partial_\mu+\rho)\psi = 0$ implies $\det(\beta^\mu\partial_\mu+\rho)\psi = 0$ where $\det(\beta^\mu\partial_\mu+\rho)$ is the single partial differential operator obtained by computing the determinant of the matrix of partial differential operator. (If β^μ and ρ are not square matrices being $K\times N$ with $K > N$, then the statement holds for any $N\times N$ submatrix formed of rows).

<u>Proof</u> : If $\mathrm{Cof}(\beta^\mu\partial_\mu+\rho)](\beta^\mu\partial_\mu+\rho)$ is the matrix of cofactors of the matrix $(\beta^\mu\partial_\mu+\rho)$ we have

$$[\mathrm{Cof}(\beta^\mu\partial_\mu+\rho)](\beta^\mu\partial_\mu+\rho) = \det(\beta^\mu\partial_\mu+\rho)\mathbb{1}$$

so

$$(\sigma * \phi, \tau * \det(\beta^{\mu}\partial_{\mu} + \rho)\psi) = 0 = (\sigma * \det(\beta^{\mu}\partial_{\mu} + \rho)\phi, \ \tau * \psi)$$

and therefore since

$$\tau * \det(\beta^{\mu}\partial_{\mu} + \rho)\psi = (\det(\beta^{\mu}\partial_{\mu} + \rho)\tau) * \psi$$

we get

$$\det(-i\beta^{\mu}p_{\mu} + \rho)K(\phi,\psi;p) = 0$$

Thus, $\hat{K}(\phi,\psi;p)$ vanishes unless $\det(-i\beta^{\mu}p_{\mu} + \rho) = 0$. The conclusion is, the necessary changes having been made to cover the non-square case, that the support of $\hat{K}$ is contained in the mass spectrum of the equation. Furthermore, since only real momenta, p, appear in the support of $\hat{K}$ any complex masses which may occur as zeros of $\det(-i\beta^{\mu}p_{\mu} + \rho)$ cannot contribute to the scalar product. This argument, which is based on positivity of the scalar product and a weak continuity property (continuity of $(\sigma * \phi, \tau * \psi)$ in σ and τ) does not exclude space like momenta, $p^2 < 0$, or negative time-like momenta $p^2 < 0$ and $p^o < 0$ from the support of $\hat{K}$. We have excluded these possibilities by explicit assumption in our definition of proper wave equation; V_1 is supposed to contain only positive energy solutions of the invariant wave equation.

Next, invariance of the scalar product under SL(2,$\mathbb{C}$) will be expressed as a property of $\hat{K}$. Note that

$$\begin{aligned} S(A)(\sigma * \phi)(\Lambda(A^{-1})x) &= \int \sigma(\Lambda(A^{-1})x - y)S(A)\phi(y)d^4y \\ &= \int \sigma(\Lambda(A^{-1})(x-y))S(A)\phi(\Lambda(A^{-1})y)d^4y \\ &= (\sigma_A * \mathcal{U}(a,A)\phi)(x) \end{aligned}$$

thus,

$$\begin{aligned} (\sigma * \phi, \tau * \psi) &= (\mathcal{U}(0,A)\sigma * \phi, \ \mathcal{U}(0,A)\tau * \psi) \\ &= (\sigma_A * \mathcal{U}(0,A)\phi, \tau_A * \mathcal{U}(0,A)\psi) \end{aligned}$$

which, expressed in terms of the kernel, $\hat{K}$, is

$$\hat{K}(\mathcal{U}(0,A)\phi, \mathcal{U}(0,A)\psi;\Lambda(A)p) = \hat{K}(\phi,\psi;p)$$

To reduce the problem further, we introduce an additional assumption. We admit that V_1 contains solutions whose Fourier transform is of the form $\hat{\sigma}(p)u(p)$ where $\hat{\sigma}$ is a C^∞ function of compact support and $u(p)$ is a standard solution of

$$(-i\beta^\mu p_\mu + \rho)u(p) = 0 \tag{3.2}$$

depending in a piecewise smooth way on p. It is convenient, in part, to work with a projection $\Pi(p)$ onto all solutions of (3.2). Then u(p) can be chosen as

$$u(p) = \Pi(p)u$$

where u is an arbitrary element of $\mathbb{C}^N$. The conventional choice of $\Pi(p)$ satisfies

$$S(A)\Pi(p)S(A)^{-1} = \Pi(\Lambda(A)p) \quad .$$

If we write

$$\phi(x) = \frac{1}{(2\pi)^2}\sum_j \int d\Omega_{m_j}(p)\hat{\sigma}(p)\Pi(p)u \; \exp(-ip.x)$$

$$\psi(x) = \frac{1}{(2\pi)^2}\sum_j \int d\Omega_{m_j}(p)\tilde{\tau}(p)\Pi(p)v \; \exp(-ip\cdot x)$$

then $\hat{K}(\phi,\psi;p)$ becomes a measure depending anti-linearly on the N-component vector u and linearly on v . Thus

$$\hat{K}(\widehat{\Pi(p)u}, \widehat{\Pi(p)v}; p) = \bar{u}\mathcal{D}(p)v$$

where $\mathcal{D}(p)$ is an $N\times N$ matrix whose entries are complex measures. The transformation law of $\mathcal{D}(p)$ under $SL(2,\mathbb{C})$ follows from that of $\hat{K}$:

$$S(A)^*\mathcal{D}(p)S(A) = \mathcal{D}(\Lambda(A^{-1})p)$$

We have a lemma

Lemma : Let $\mathcal{D}$ be an $N\times N$ matrix of complex measures supported by a finite number of positive energy hyperboloids and perhaps the light cone (≡ the mass spectrum). Suppose for some representation $A \mapsto S(A)$ of $SL(2,\mathbb{C})$

$$S(A)^*\mathcal{D}(p)S(A) = \mathcal{D}(\Lambda(A^{-1})p)$$

Then the entries $\mathscr{D}_{jk}$ are absolutely continuous with respect to the invariant measures on the mass spectrum i.e.

$$\mathscr{D}_{jk}(p) = \sum_{\ell} d_{jk}(p)\, 2\,\delta(p^2 - m_{\ell}^2)\ \theta\ (p^o) d^4 p \quad .$$

The matrix of functions $\{d_{jk}\}$ satisfies

$$S(A)^* d(p) S(A) = d(\Lambda(A^{-1})p \quad .$$

The analysis of the functions d_{jk} follows a pattern familiar from the theory of induced representations. Consider for a fixed p, the corresponding stability group consisting of all $A \in SL(2,\mathbb{C})$ such that $\Lambda(A)p = p$. When p is time-like this subgroup is isomorphic to SU(2) so for simplicity we may take $p = \{m,\vec{o}\}$. Then $S(A)^* = S(A)^{-1}$ and $d(\{m,\vec{o}\})$ commutes with all S(A), $A \in SU(2)$. d(p) for all other p on the same hyperboloid can be determined by boosting i.e. if A_p is the boost

$$A_p = [\frac{2}{m}(p^o + m)]^{-1/2}[\frac{m + \underset{\sim}{p}}{m}] = \sqrt{\frac{\underset{\sim}{p}}{m}}$$

defines d(p) by

$$d(\Lambda_p\{m,\vec{o}\}) = S(A_p^{-1})^* d(\{m,\vec{o}\}) S(A_p^{-1}) \qquad (3.3)$$

Since $S(A_p^{-1})$ has entries which are polynomials in the matrix elements of A^{-1}, the matrix elements of d(p) are polynomials in the components of p divided by a power of $\frac{p^o + m}{m}$. In fact, these powers in the denominator are cancelled by factors in the numerator. To see this one brings S(A) to redueed form by an equivalence $S(A) \mapsto TS(A)T^{-1}$ so that it has its irreducible constituents on the diagonal and zeros elsewhere. Now the irreducible constituent [j,k] evaluated at $\sqrt{\tilde{p}/m}$ is homogeneous of degree (j+k) in the components of p. Since d({m,o}) commutes with the representation of SU(2) it cannot have non-vanishing matrix elements between the representation spaces of [j,k] and [j',k'] where j+k+j'+k' is half odd integral. Thus, the non vanishing terms contributing to $d(\Lambda(A_p)\{m,\vec{o}\})$ according to (3.3) (are sums of terms homogeneous of integer degree in p i.e. polynomials.

d(p) is defined uniquely all over the hyperboloid $p^2 = m^2$ $p^o > 0$ in terms of $d(\{m,\vec{o}\})$ by (3.3). So defined it satisfies the relation (3.3) for <u>all</u> $A \in SL(2,\mathbb{C})$, provided only that d({m,o}) commutes with the representation, $A \mapsto S(A)$ of SU(2). The calculation is elementary and familiar so it will be omitted. The only further restriction on d(p) is that the replacement of u by $\Pi(p)u$ should not affect the scalar product. For p = {m,o} this means

$$d(\{m,\vec{o}\}) = \Gamma(\{m,\vec{o}\})^* d(\{m,\vec{o}\}) = d(\{m,\vec{o}\})\Gamma(\{m,\vec{o}\}) \tag{3.4}$$

From this there follows

$$\begin{aligned} d(p) &= S(A_p)^* d(\{m,\vec{o}\}) S(A_p) = S(A_p)^* \Gamma(\{m,o\})^* S(A_p^{-1})^* S(A_p)^* d(\{m,\vec{o}\}) S(A_p) \\ &= \Gamma(p)^* d(p) \end{aligned} \tag{3.5}$$

and

$$\begin{aligned} d(p) &= S(A_p)^* d(\{m,o\}) S(A_p) S(A_p)^{-1} \Gamma(\{m,o\}) S(A_p) \\ &= d(p)\Gamma(p) \end{aligned} \tag{3.6}$$

For $m = 0$ and, therefore, p light-like, there is an analogous discussion, with the vector $\{m,o\}$ replaced by an arbitrarily chosen vector, q ; with SU(2) replaced by the stability group, G_q, of q ; and with the boosts (3.3) replaced by some suitable representatives of the cosets of G_q in $SL(2,\mathbb{C})$. With these changes, the argument goes through.

The standard theory in which the β^μ are square and there exists a matrix η satisfying

$$-(\beta^\mu)^* = \eta\,\beta^\mu\,\eta^{-1} \tag{3.7}$$

has a d(p)

$$d(p) = \eta\Gamma(p) \tag{3.8}$$

It is hermitean by virtue of the relation

$$\Gamma(p)^* = \eta\Gamma(p)\eta^{-1} \tag{3.9}$$

but in general it is not positive, and consequently the sesquilinear form it defines is not in general physically acceptable. It would be convenient to have some simple sufficient conditions for the positivity. At the moment, what can be said is this : i) $\eta\Gamma(p)$ is not positive for any known multi-mass equation ii) for a single mass equation there are many examples that are positive, Fierz-Pauli equations for example, and some that are indefinite.

There is a method of obtaining a d(p) even when no η exists.

For simplicity, consider first the case in which $S(A)^* = S(A^*)$. Then the proposed form is defined by

$$d(p) = S(\tilde{p}/m)\,\Pi(p) \qquad (3.10)$$

where $\tilde{p} = p^0\mathbb{1} - \vec{p}\,\vec{\tau}$ and $\vec{\tau}$ are the Pauli matrices. It satisfies

$$d(p) = d(p)\Pi(p) = \Pi(p)^* d(p)$$

by virtue of

$$\Pi(p)^* S(\tilde{p}/m) = S(\tilde{p}/m)\,\Pi(p) \qquad (3.11)$$

This relation combined with

$$S(\tilde{p}/m)^* = S(\tilde{p}/m^*) = S(\tilde{p}/m) \qquad (3.12)$$

implies that $d(p)$ is hermitean. Furthermore

$$S(\tilde{p}/m) \geq 0 \qquad (3.13)$$

because

$$S(\tilde{p}/m) = S(\sqrt{\tilde{p}/m})^2 \qquad (3.14)$$

Actually $S(\tilde{p}/m)$ is strictly positive because it has an inverse $S(\underset{\sim}{p}/m)$. The positivity of $S(\tilde{p}/m)\Pi(p)$ as a form on the solutions of the generalized Dirac equation is equivalent to the positivity of $S(\tilde{p}/m)$. This completes the proof that (3.10) defines an acceptable $d(p)$.

The restriction of this discussion to representations satisfying $S(A)^* = S(A^*)$ is actually inessential. We know $S(A)^* = S(A^*)$ for the irreducible representations $[j, j_2]$ in standard form and that every finite dimensional representation is equivalent to a direct sum of such. Thus $S(A) = V\ S_1(A)V^{-1}$ where $S_1(A)^* = S_1(A^*)$. Therefore, $S(A)^* = (V^{-1})^* S_1(A)^* V^* = (V^{-1})^* V^{-1} S(A^*) VV^*$ and so if we write $d(p) = (VV^*)^{-1} S(\tilde{p}/m)$ we can go through the same calculations as before to conclude the validity of the transformation law under $SL(2,\mathbb{C})$. That $d(p)$ is hermitean follows from the identity

$$S(\tilde{p}/m)^* = (VV^*)^{-1} S(\tilde{p}/m) VV^*$$

To see that it is positive note that

$$(VV^*)^{-1}S(\tilde{p}/m) = (V^{-1})^* S_1(\tilde{p}/m)V^{-1}$$

the right hand side is positive since $S_1(\tilde{p}/m)$ is positive.

A complete classification of the admissible d(p) will not be undertaken. Progress can be made by exploiting the fact that d(p) depends polynomially on p

$$d(p) = \sum_n d_{\mu_1 \ldots \mu_n} p^{\mu_1} \ldots p^{\mu_n}$$

The transformation law of d(p) under $SL(2,\mathbb{C})$ then implies that

$$S(A)^* d_{\mu_1 \ldots \mu_n} S(A) = \Lambda(A)_{\mu_1}{}^{\nu_1} \ldots \Lambda(A)_{\mu_n}{}^{\nu_n} d_{\nu_1 \ldots \nu_n}$$

and this is an equation all of whose solutions can be determined by the method described in section two.

4. <u>Mass spectrum</u> ; <u>Minimal equation for $\not{p}$ in the standard case</u> ; <u>Projections onto definite mass and spin</u>.

In section two, the most general β^μ and ρ consistent with given representations $A \mapsto S_1(A)$ and $A \mapsto S(A)$ of $SL(2,\mathbb{C})$ have been found :

$$\beta^\mu = B P \mathcal{U}^\mu \qquad \rho = B'P'$$

Apart from the discrete choices available in P and P', one has only the free parameters of the matrices B and B'. The determinants whose zeros define the mass spectrum are polynomials in p^2 and the matrix elements of B and B'. For example, for square matrices (K = N), the polynomial defining the mass spectrum is

$$Q_o \prod_{j=1}^{\ell} (-p^2 + m_j^2) = \det(-i\beta^\mu p_\mu + \rho) = 0$$

Thus, the symmetric functions of the masses $m_1^2 \ldots m_\ell^2$ are rational functions in the matrix elements of B and B'. (They may have a polynomial in the denominator because Q_o may be a non trivial polynomial in matrix elements of B and B'). <u>Thus the set of theories with given representations of $SL(2,\mathbb{C})$ and given mass spectrum is a union of algebraic surfaces in the space whose coordinates are the matrix elements of B and B'</u>.

The natural next step in the theory would be to describe the further restrictions on the parameters B and B' arising from the assignment of a spin spectrum and the positivity of some chosen scalar product. However, the general theory is not in such a state that this program can be carried out without further assumptions. Therefore for most of the rest of this section we deal with the special case of standard theories, in which the matrices β^μ are square, ρ is non-singular and may therefore be taken as $m\mathbb{1}$ with $m > 0$. Furthermore, it is assumed that there exist matrices η, B, C such that

$$\begin{aligned} -(\beta^\mu)^* &= \eta\, \beta^\mu \eta^{-1} \\ (\beta^\mu)^T &= B\, \beta^\mu B^{-1} \\ \overline{\beta^\mu} &= C\, B^\mu\, C^{-1} \end{aligned} \tag{4.1}$$

We will develop the standard theories far enough so that for them, the above mentioned next step in the general theory can be carried out explicitly.

The simplicity of the equation

$$(-\not{p} + m\,\mathbb{1})\, u = 0 \tag{4.2}$$

arises from the fact that the mass spectrum is directly related to the eigenvalue spectrum of $i\beta^o$. The solution of (4.2) can be reduced by a boost to the solution of

$$(-i\beta^o E + m)u(\{E,\vec{0}\}) = 0 \tag{4.3}$$

Thus, if λ is an eigenvalue of $i\beta^o$ different from zero :

$$i\,\beta^o u = \lambda\, u \tag{4.4}$$

there is a corresponding value of E, namely, $\frac{m}{\lambda}$ and conversely. It is physically important to note that the eigenvalue zero of $i\beta^o$ does not yield a solution of (4.3).

There is no a priori reason for $i\beta^o$ to be diagonalizable, and in fact it was early recognized that to describe particles of mass $m > 0$ and spin $\geq 3/2$, it is natural to use a non-diagonalizable $i\beta^o$ [8]. However, it was also recognized that the eigenvalues of $i\beta^o$ different from zero have to be semi-simple i.e. when $i\beta^o$ is brought to Jordan canonical form, its eigenvalues different from zero which have multiplicity greater than zero are not accompanied by 1's above the diagonal. The first explicit proof was given by Harish Chandra [9]. He based it on three assumptions : i) the irreducibility of the β^μ ii) their transformation law under $SL(2,\mathbb{C})$ iii) the hypothesis that the wave equation should imply that every component of the wave function satisfy $(\Box + m^2)\psi_j(x) = 0$. Later on there was a proof by Speer that replaced i) and iii) by the assumption that the current form $u(p)^+ i\beta^o u(p)$, with $u(p)^+ = \overline{u(p)}\eta$, is positive on positive energy solutions of (4.3) [10]. For a general mass spectrum it becomes :

Lemma : Let β^μ, $\mu = 0,1,2,3$ be $N \times N$ matrices satisfying

$$S(A)^{-1}\beta^\mu S(A) = \Lambda(A)^\mu{}_\nu\, \beta^\nu \tag{4.5}$$

where $A \mapsto S(A)$ is an $N \times N$ matrix representation of $SL(2,\mathbb{C})$. Suppose the generalized Dirac equation

$$(-\not{p} + m\,\mathbb{1})u = 0 \tag{4.6}$$

has a mass spectrum $m_1^2, \ldots, m_\ell^2$ and that the current form $u(p)^+ i\beta^o u(p)$ is strictly positive or strictly negative for all positive energy solutions of (4.3) of each mass m_j $j = 1, \ldots, \ell$, but may have different signs for different masses. Then all non-zero eigenvalues of $i\beta^o$ are semi-simple and its minimal polynomial has the form

$$\prod_{j=1}^{\ell}\left[(i\beta^o)^2 - \frac{m^2}{m_j^2}\right](i\beta^o)^q \tag{4.7}$$

More generally, for an arbitrary complex vector p

$$\prod_{j=1}^{\ell}\left[(\not{p})^2 - \frac{m^2}{m_j^2}p^2\right](\not{p})^q = 0 \tag{4.8}$$

Here $q = 0$, if $i\beta^o$ is non singular. Otherwise it is one more than the longest run of ones above the diagonal in the Jordan canonical form of $i\beta^o$.

The statement is slightly stronger but the proof is the same as in [11] p.460. Instead of repeating it, I give the proof of a closely related formula, that for the projection operator $E_{m_k,+}(p)$ onto the positive energy solutions of (4.6) that have mass m_k.

Again for simplicity one may make the calculation in the rest system. Then for $\lambda_k > 0$

$$\left(\frac{i\beta^o + \lambda_k}{2\lambda_k}\right) \prod_{\lambda_j \neq \pm\lambda_k} \frac{i\beta^o - \lambda_j}{\lambda_k - \lambda_j} \left(\frac{i\beta^o}{\lambda_k}\right) q$$

clearly acts as the identity on a solution of $i\beta^o u = \lambda_k u$ and annihilates all solutions belonging to distinct eigenvalues. Boosting it one gets an operator

$$\left[\frac{\frac{\not{p}}{\sqrt{p^2}} + \lambda_k}{2\lambda_k}\right] \prod_{j\neq k} \left[\frac{\frac{\not{p}}{\sqrt{p^2}} - \lambda_j}{\lambda_k - \lambda_j}\right] \left(\frac{\not{p}}{\lambda_k\sqrt{p^2}}\right)^q$$

Combining terms in λ_j and $-\lambda_j$, it becomes

$$\left[\frac{\frac{\not{p}}{\sqrt{p^2}} + \lambda_k}{2\lambda_k}\right] \prod_{\substack{j=1\\ j\neq k}}^{\ell} \frac{\left(\frac{\not{p}^2}{p^2} - \frac{m^2}{m_j^2}\right)}{\left(\frac{m^2}{m_k^2} - \frac{m^2}{m_j^2}\right)} \left(\frac{\not{p}}{\lambda_k\sqrt{p^2}}\right)^q$$

Summing over $\lambda_k > 0$, one has

$$\Pi(p) = \sum_{k=1}^{\ell} \frac{\frac{\not{p}}{\sqrt{p^2}} + \lambda_k}{2\lambda_k} \prod_{\substack{j=1\\ j\neq k}}^{\ell} \frac{\left(\frac{\not{p}^2}{p^2} - \frac{m^2}{m_j^2}\right)}{\left(\frac{m^2}{m_k^2} - \frac{m^2}{m_j^2}\right)} \left(\frac{\not{p}}{\lambda_k\sqrt{p^2}}\right)^q$$

There is an analogous formula for negative energies. The important feature of this formula is that it shows that $\Pi(p)$ is a polynomial in $\not{p}$ with real coefficients. Thus one has immediately

$$\Pi(p)^* = \eta\, \Pi(p)\eta^{-1}$$

$$\overline{\Pi(p)} = C\, \Pi(-p)C^{-1}$$

$$\Pi(p)^T = B\, \Pi(p)B^{-1}$$

The minimal equation (4.8) can be used to obtain a number of other useful expressions. For example, the so-called Klein-Gordon divisor is a matrix $d_{KG}(p)$ depending on a complex vector p such that

$$d_{KG}(p)(-\not{p} + m) = \prod_{j=1}^{\ell} (-p^2 + m_j^2)\mathbb{1}$$

The Klein-Gordon divisor appears in the commutation relations for fields in the standard case

<u>Lemma</u> [12] :

$$d_{KG}(p) = \frac{1}{m}\prod_{j=1}^{\ell}(-p^2+m_j^2)\sum_{r=o}^{q-1}\left(\frac{\not p}{m}\right)^r$$

$$+\ \frac{1}{m^2}(\not p+m)\left(\frac{\not p}{m}\right)^q\Big\{\sum_{r=o}^{\ell-1}\left(\frac{\not p^2}{m^2}\right)^r\Big[(p^2)^\ell-(p^2)^{\ell-1}\sum_{j=1}^{\ell}m_j^2$$

$$+(p^2)^{\ell-2}\sum_{j_1<j_2}m_{j_1}^2,m_{j_2}^2-\ldots(-1)^r(p^2)^{\ell-r}\sum_{j_1<j_2<\ldots<j_r}m_{j_1}^2\ldots m_{j_r}^2$$

$$-\prod_{j=1}^{\ell}(-p^2+m_j^2)\Big]\Big\}$$

The proof involves only a straightforward calculation of $d_{KG}(p)(-\not p+m)-\prod_{j=1}^{\ell}(-p^2+m_j^2)\mathbb{1}$ followed by a cancellation which uses the minimal equation for $\not p$.

It is of some interest to relate the Klein-Gordon divisor to the projection operators, $E_{m_{k,+}}(p)$. A direct comparison yields

$$d_{KG}(p)\big|_{p^2=m_j^2} = \frac{2m_j^2}{m}\prod_{\substack{k=1\\k\neq j}}^{\ell}(m_k^2-m_j^2)E_{m_{j,+}}(p)\big|_{p^2=m_j^2}$$

This shows that if one expands the inverse (for p^2 not in the spectrum)

$$(-\not p+m)^{-1} = \Big[\prod_{j=1}^{\ell}(-p^2+m_j^2)\Big]^{-1}d_{KG}(p)$$

$$= \sum_{j=1}^{\ell}\big[-p^2+m_j^2\big]^{-1}\Big[\prod_{k\neq j}(-m_j^2+m_k^2)\Big]^{-1}d_{KG}(p)/_{p^2=m_j^2}$$

that one gets

$$(-\not p+m)^{-1} = \sum_{j=1}^{\ell}\frac{2m_j^2}{(-p^2+m_j^2)}\frac{1}{m}E_{m_{j,+}}(p)$$

As an application, one sees that if one takes the difference between retarded and advanced boundary conditions one gets

$$(-\not p+m)_k^{-1}-(-\not p+m)_A^{-1} = \sum_{j=1}^{\ell}\delta(-p^2+m_j^2)E_{m_{j,+}}(p)$$

Here the left hand side is a distribution obtained from the boundary values :

$$(-\not{p} + m)^{-1}_{\substack{R\\A}} = \lim_{\varepsilon \to 0+} \left[-(p \pm i\varepsilon(1,\vec{0})) + m\right]^{-1}$$

and the identities

$$[-p^2 + m^2]^{-1}_{\substack{R\\A}} = P[-p^2 + m^2]^{-1} \pm \pi i \, \delta(-p^2 + m^2) \operatorname{sgn} p^0$$

The general theory of invariant wave equations is rather incomplete because no analogues are known of the above simple formulae for $\Gamma(p)$, d_{KG}, etc. We will see in section five that the construction of a quantized field satisfying a general wave equation is incomplete for lack of such information.

5. Quantized fields for general wave equations

The construction of a quantized field satisfying an invariant wave equation goes in two steps : one first constructs a Fock space based on the single particle theory of the invariant wave equation and then defines the field in terms of the annihilation and creation operators of the Fock space. For the Fock space itself one has the choice of symmetry (Bose-Einstein statistics) or anti-symmetry (Fermi-Dirac statistics) and the choice of a theory with or without anti-particles. The theory without anti-particles yields a so-called Majorana field, the theory with anti-particles a so-called charged field suitable for coupling to an external electromagnetic field. The formulae are sufficiently familiar that they need only a brief review. Here the emphasis will be on those features needed for the treatment of a general wave equation.

The symmetric ($\varepsilon = s$) and anti-symmetric ($\varepsilon = a$) Fock spaces based on a single particle space $\mathcal{H}^{(1)}$ are the direct sums

$$\mathfrak{F}_\varepsilon(\mathcal{H}^{(1)}) = \bigoplus_{n=0}^{\infty} \mathcal{H}_\varepsilon^{(n)} \qquad (5.1)$$

where $\mathcal{H}_\varepsilon^{(0)}$ is the one-dimensional complex Hilbert space and $\mathcal{H}_\varepsilon^{(1)} = \mathcal{H}^{(1)}$ and $\mathcal{H}_\varepsilon^{(n)}$ is the n-fold tensor product

$$\mathcal{H}_\varepsilon^{(n)} = (\mathcal{H}^{(1)})_\varepsilon^{\otimes n}$$

symmetrized or anti-symmetrized according to the subscript ε.

For the theory with anti-particles the state space is

$$\mathcal{H} = \mathfrak{F}_\varepsilon(\mathcal{H}^{(1)}) \otimes \mathfrak{F}_\varepsilon(\mathcal{H}^{(\bar{1})}) \tag{5.2}$$

where $\mathcal{H}^{(\bar{1})}$ is the Hilbert space of states for one anti-particle. The theory without anti-particles has a state space

$$\mathcal{H} = \mathfrak{F}_\varepsilon(\mathcal{H}^{(1)}) \tag{5.3}$$

The annihilation and creation operators on $\mathfrak{F}_\varepsilon(\mathcal{H}^{(1)})$ are given by

$$a^+(\Phi)\Psi = \sqrt{N}\,(\Phi \otimes \Psi)_\varepsilon \tag{5.4}$$

$$a(\chi)\Psi = \sqrt{N+1}\,\langle\chi, \Psi\rangle_\varepsilon \tag{5.5}$$

Here $\Phi \in \mathcal{H}^{(1)}$, N is the operator defined by

$$(N\Phi)^{(n)} = n\Phi^{(n)} \tag{5.6}$$

and $\chi \in \mathcal{H}^{(1)'}$, the dual of $\mathcal{H}^{(1)}$, which consists of the complex linear functionals on $\mathcal{H}^{(1)}$. The bracket $\langle\chi,\Psi)$ means that χ should be evaluated on the first argument of Ψ. The operations indicated in the definition certainly make sense as they stand on vectors Ψ for which only a finite number of components $\Psi^{(n)}$ are non-zero. They can be extended by passing to the closure. But that will play little role in what follows. The annihilation and creation operators satisfy the (anti-) commutation relations

$$[a(\chi), a(\Sigma)]_{\pm} = 0 \;, \; [a(\chi), a^+(\Phi)]_{\pm} = \langle\chi,\Phi\rangle \mathbb{1} \tag{5.7}$$

where the commutator, $[\;,\;]_-$ holds for $\varepsilon = s$ and the anti-commutator, $[\;,\;]_+$, holds for $\varepsilon = a$.

The relation between the a^+'s and the adjoints of the a's is

$$a(\chi) = [a^+(J\chi)]^* \tag{5.8}$$

Here J is the anti-linear mapping of $\mathcal{H}^{(1)'}$ into $\mathcal{H}^{(1)}$ given by

$$\langle\chi,\Phi\rangle = (J\chi, \Phi) \tag{5.9}$$

(By the Riesz representation theorem every continuous linear functional on $\mathcal{H}^{(1)}$ is of this form).

If $\{a,A\} \mapsto \mathcal{V}^{(1)}(a,A)$ is the unitary representation of ISL(2,$\mathbb{C}$) in $\mathcal{H}^{(1)}$ given by the one particle theory, then

$$\mathcal{V}(a,A) = \Gamma(\mathcal{V}^{(1)}(a,A))$$

is the corresponding representation in $\mathfrak{F}_\varepsilon(\mathcal{H}^{(1)})$ defined by

$$(\mathcal{V}(a,A)\Psi)^{(n)} = [\mathcal{V}^{(1)}(a,A)^{\otimes n}]\Psi^{(n)} \tag{5.10}$$

Straightforward computation shows that

$$\mathcal{V}(a,A)a^{+}(\Phi)\mathcal{V}(a,A)^{-1} = a^{+}(\mathcal{V}^{(1)}(a,A)\Phi) \tag{5.11}$$

and consequently

$$\begin{aligned}\mathcal{V}(a,A)a(\chi)\mathcal{V}(a,A)^{-1} &= \mathcal{V}(a,A)[a^{+}(J\chi)]^{*}\mathcal{V}(a,A)^{-1} \\ &= a(J^{-1}\mathcal{V}^{(1)}(a,A)J\chi)\end{aligned} \tag{5.12}$$

The Majorana field operator is defined on $\mathfrak{F}_\varepsilon(\mathcal{H}^{(1)})$ by

$$\psi(f) = a(\Pi_{+} f) + a^{+}(\Pi_{-} f) \tag{5.13}$$

Here f is a test function for the field, Π_{+} is a mapping of the test function space into $\mathcal{H}^{(1)}$, and Π_{-} is a mapping of the test function space into $\mathcal{H}^{(1)}$. In the presence of anti-particles the analogous formula is

$$\psi(f) = a(\Pi_{+} f) + b^{+}(\Pi_{-} f) \tag{5.14}$$

where $a(\Pi_{+} f)$ is short hand for $a(\Pi_{+} f)\otimes \mathbb{1}$ and $b^{+}(\Pi_{-} f)$ for

$$\left\{\begin{matrix} \mathbb{1} & \varepsilon = s \\ (-\mathbb{1})^{N} & \varepsilon = a \end{matrix}\right\} \otimes a^{*}(\Pi_{-} f),$$

and Π_{+} maps the test function space into $\mathcal{H}^{(1)'}$ while Π_{-} maps it into $\mathcal{H}^{(\bar{1})}$. Here the a's (anti-) commute with b's and b^{*}'s while satisfying the usual (anti)-commutation relations among themselves.

The test functions are supposed to be C^{∞} functions of fast decrease and to have N components if the field has N components :

$$\psi(f) = \sum_{j=1}^{N} \int d^4x\, f_j(x)\,\psi_j(x)$$

the transformation law of the field can be expressed in terms of the transformation law of the test functions $f \mapsto \{a,A\}f$

$$\mathcal{V}(a,A)\psi(f)\,\mathcal{V}(a,A)^{-1} = \psi(\{a,A\}f) \tag{5.15}$$

where

$$(\{a,A\}f)(x) = S(A^{-1})^T f(\Lambda(A^{-1})(x-a))$$

From (5.11) (5.12) and (5.14), we see that

$$\Pi_- \{a,A\}f = \mathcal{V}^{(\bar{1})}(a,A)\,\Pi_- f \tag{5.16}$$

and

$$\Pi_+ \{a,A\}f = J^{-1}\,\mathcal{V}^{(1)}(a,A)\,J\,\Pi_+ f \tag{5.17}$$

are sufficient to insure the validity of the transformation law (5.15) of the field in the presence of anti-particles. For a Majorana field the $\mathcal{V}^{(\bar{1})}$ is replaced by $\mathcal{V}^{(1)}$.

As far as the (anti-)commutation relations of ψ are concerned, in the presence of anti-particles

$$[\psi(f),\psi(g)]_{\pm} = 0 \tag{5.18}$$

holds for all test functions f and g. On the other hand, the (anti-)commutation relation between $\psi(f)$ and $\psi(g)^*$ requires a calculation

$$\begin{aligned}
[\psi(f),\psi(g)^*]_{\pm} &= [a(\Pi_+ f), a(\Pi_+ g)^*]_{\pm} \\
&\quad + [b^+(\Pi_- f), (b^+(\Pi_- g))^*]_{\pm} \\
&= (\langle \Pi_+ f, J\Pi_+ g\rangle \pm \langle J^{-1}\Pi_- g, \Pi_- f\rangle)\mathbb{1}
\end{aligned} \tag{5.19}$$

For a Majorana field, the calculation of the commutator of the field with itself yields

$$
\begin{aligned}
[\psi(f),\psi(g)]_{\pm} &= [a(\Pi_+ f), a^+(\Pi_- g)]_{\pm} \\
&\quad + [a^+(\Pi_- f), a(\Pi_+ g)]_{\pm} \\
&= (\langle \Pi_+ f, \Pi_- g\rangle \pm \langle \Pi_+ g, \Pi_- f\rangle)\, \mathbb{1}
\end{aligned}
\tag{5.20}
$$

The problem is now to choose the mappings Π_+ and Π_- so that the transformation laws (5.16) and (5.17) under ISL(2,$\mathbb{C}$) are satisfied, and so that the right hand sides (5.19) and (5.20) of the commutation relations become local in f and g i.e. become such that when the supports of f and g are space-like separated the (anti-)commutator vanishes. In the standard case the general expressions for Π_+ and Π_- are well known and can be written in terms of the square matrices C and η used in the definitions of charge conjugation and the current [11]. In [12] some wave equations were treated for which no η matrix exists. The formulae proposed here are generalizations from these cases. We do not expect to be able to construct a Majorana field for general representations of SL(2,$\mathbb{C}$). The easiest way to see why is to note that a Majorana field is characterized by a linear relation between the field $\psi(x)$ and its adjoint $\psi(x)^*$, say

$$\psi(x) = C^{-1}\psi(x)^* \tag{5.21}$$

where C is some non-singular matrix. If this equation is multiplied before with $\mathcal{V}(a,A)$ and behind with $\mathcal{V}(a,A)^{-1}$ it yields

$$S(A)\,\psi(x) = C^{-1}\overline{S(A)}\;\psi(x)^*$$

Multiplying the preceding equation by S(A) and comparing one gets

$$S(A)C^{-1} = C^{-1}\,\overline{S(A)} \tag{5.22}$$

which is a non-trivial restriction on S(A). (It implies that an irreductible representation $[j_1,j_2]$ occurs in S(A) with the same multiplicity as the irreductible representation $[j_2,j_1]$). On the other hand, one expects to be able to construct charged fields without such restrictions because for the above mentioned examples it has already been shown to be possible.

The main point to recognize is that if the field transforms with the representation $A \mapsto S(A)$ of SL(2,$\mathbb{C}$), the single particle states,

$\psi(x)^* \psi_o$, will transform according to $\overline{S(A)}$. This suggests that $\mathcal{H}^{(\bar{1})}$ be chosen as the Hilbert space spanned by $\Phi(p)$ satisfying

$$(-p + \rho)\Phi(p) = 0 \tag{5.23}$$

equipped with the scalar product

$$(\Phi,\psi) = \sum_j \int d\Omega_{m_j}(p)\, \overline{\Phi(p)}\, d(p)\varphi(p) \tag{5.24}$$

and the representation of ISL(2,$\mathbb{C}$)

$$(\mathcal{V}^{(1)}(a,A)\Phi)(p) = \exp(ip.a)S(A)\Phi(\Lambda(A^{-1})p) \tag{5.25}$$

while, on the other hand, $\mathcal{H}^{(1)}$ is chosen as the Hilbert space spanned by $\Phi(p)$ satisfying

$$[-\bar{p} + \bar{\rho}]\Phi(p) = 0 \tag{5.26}$$

equipped with the scalar product

$$(\Phi,\psi) = \sum_j \int d\Omega_{m_j}(p)\, \overline{\Phi(p)}\,\overline{d(p)}\psi(p) \tag{5.27}$$

and the representation

$$(\mathcal{V}^{(1)}(a,A)\Phi)(p) = \exp(ip.a)\overline{S(A)}\,\Phi(\Lambda(A^{-1})p) \tag{5.28}$$

Notice that $\mathcal{V}^{(1)}$ and $\mathcal{V}^{(1)}$ are both unitary by virtue of the relation

$$S(A)^* d(p)\; S(A) = d(\Lambda(A^{-1})p) \tag{5.29}$$

The elements, χ, of the dual of $\mathcal{H}^{(1)}$ are given by

$$\langle\chi,\Phi\rangle = \sum_j \int d\Omega_{m_j}(p)\, \chi(p)\, \overline{d(p)}\, \Phi(p) \tag{5.30}$$

where

$$\Pi(p)\chi(p) = \chi(p)$$

This suggests the formula

$$(\Pi_+ f)(p) = \sqrt{\pi}\, \Pi(p)\, B^{-1}\, f(p) \tag{5.31}$$

where B is a non-singular matrix to be chosen so that the transformation law (5.17) holds, and the wave equation as well. If we compare

$$(\Pi_+\{a,A\}f)(p) = \sqrt{\pi}\, \Pi(p)B^{-1}\exp(-ip\cdot a)S(A^{-1})^T\hat{f}(\Lambda(A^{-1})p)$$

with

$$(J^{-1}\, \mathcal{V}^{(1)}(a,A)J\, \Pi_+ f)(p) = \overline{\exp(ip.a)\overline{S(A)}\sqrt{\pi}\, \Pi(\Lambda(A^{-1})p)B^{-1}\hat{f}(\Lambda(A^{-1})p)}$$

we see that to obtain (5.17), we need

$$B^{-1}S(A^{-1})^T = S(A)B^{-1} \tag{5.32}$$

On the other hand, if h is a K-component test function and the N-component test function f is chosen as $f = (-\beta^{\mu T}\partial_\mu + p^T)h$, then the equations of motion require

$$0 = (\Pi_+ f)(p) = \sqrt{\pi}\, \Pi(p)B^{-1}(-\, i\beta^{\mu T}p_\mu + \rho^T)\hat{h}(p) \tag{5.33}$$

which is guaranteed by

$$\Pi(p)B^{-1} = B^{-1}\Pi(p)^T \tag{5.34}$$

The analogous discussion for Π_- begins with the proposed formula

$$(\Pi_- f)(p) = \sqrt{\pi}\Pi(p)F^{-1}\hat{f}(-p) \tag{5.35}$$

where F is a non-singular matrix. Π_- yields the correct transformation law (5.16) provided

$$\begin{aligned}(\Pi_-\{a,A\}f)(p) &= \sqrt{\pi}\Pi(p)\ F^{-1}\exp ip.a\ S(A^{-1})^T\hat{f}(-\Lambda(A^{-1})p)\\ &= \exp(ip.a)S(A)\sqrt{\pi}\,\Pi(\Lambda(A^{-1})p)\,F^{-1}\hat{f}(-\Lambda(A^{-1})p)\\ &= (\mathcal{V}^{(1)}(a,A)\,\Pi_- f)(p)\end{aligned}$$

which is true provided

$$F^{-1}S(A^{-1})^T = S(A)F^{-1} \tag{5.36}$$

The equations of motion require

$$0 = (\Pi_- f)(p) = \sqrt{\pi}\,\Pi(p)F^{-1}(+\,i\,\beta^{\mu T}p_\mu + \rho^T)\hat{h}(-p)$$

which is true provided

$$\Pi(p)F^{-1} = F^{-1}\Pi(-p)^T \tag{5.37}$$

The two pieces of the commutator are computed as follows

$$\begin{aligned}
\langle \Pi_+ f, J\Pi_+ g\rangle &= (J\Pi_+ f, J\Pi_+ g)\\
&= \sum_j \int d\Omega_{m_j}(p)(\Pi_+ f)(p)\overline{d(p)}\,(\overline{\Pi_+ g})(p)\\
&= \pi\sum_j \int d\Omega_{m_j}(p)\hat{f}(p)(B^{-1})^T\Pi(p)^T\,\overline{d(p)}\;\overline{\Pi(p)}\;\overline{B^{-1}}\;\overline{\hat{g}(p)}\\
&= \pi\sum_j \int d\Omega_{m_j}(p)\hat{f}(p)[B^T)^{-1}\overline{d(p)}(\bar{B})^{-1}]\,\overline{\hat{g}(p)}
\end{aligned} \tag{5.38}$$

where we have used

$$\Pi(p)^* d(p) = d(p)\Pi(p) = d(p) \tag{5.39}$$

$$\langle J^{-1}\Pi_- g, \Pi_- f\rangle = (\Pi_- g, \Pi_- f)$$

$$\begin{aligned}
&= \pi\sum_j \int d\Omega_{m_j}(p)\,(\overline{\Pi(p)F^{-1}\hat{g}(-p))}\,d(p)\Pi(p)F^{-1}\hat{f}(-p)\\
&= \pi\sum_j \int d\Omega_{m_j}(p)\hat{f}(-p)[(F^*)^{-1}\,d(p)\,F^{-1}]^T\,\overline{\hat{g}(-p)}
\end{aligned} \tag{5.40}$$

where (5.39) has again been used. To obtain the conditions for locality we have to write there two expressions in coordinate space. The first is

$$\langle \Pi_+ f, J\Pi_+ g\rangle = \iint d^4x\, d^4y\, f(x) \sum_j (B^T)^{-1}\,\overline{d(-i\partial_x)}(\bar{B})^{-1}\,\frac{1}{i}\Delta^{(+)}(m_j, x-y)\overline{g(y)} \tag{5.41}$$

where

$$\Delta^{(+)}(m,x) = \frac{i}{2(2\pi)^3}\int d\Omega_m(p)\exp(-ip.x)$$

The second is

$$<J^{-1}\sqcap\_g, \sqcap\_f> =$$

$$- \iint d^4 x\, d^4 y\, f(x) \sum_j [(F^*)^{-1} d(i\partial x)\, F^{-1}]^T \frac{1}{i}\, \Delta^{(-)}(m_j, x-y)\overline{g(y)} \qquad (5.42)$$

where

$$\Delta^{(-)}(m,x) = \frac{-i}{2(2\pi)^3} \int d\Omega_m(p) \exp(ip.x) \ .$$

If the identity

$$(B^T)^{-1}\, \overline{d(-i\partial_x)}\, (\bar{B})^{-1} = (-1)^\sigma [(F^*)^{-1} d(i\partial_x) F^{-1}]^T \qquad (5.43)$$

holds where σ is zero if $A \mapsto S(A)$ is single valued as a representation of the homogeneous Lorentz group and is one if $A \mapsto S(A)$ is double valued, then

$$[\psi(f), \psi(g)^*]_{\pm} = \iint d^4x\, d^4y\, f(x) \sum_j (B^T)^{-1} \overline{d(-i\partial_x)}\, (\bar{B})^{-1} \frac{1}{i} \Delta(m_j, x-y) g(y)$$

and the field is local.

In the standard case $d(p) = \eta \sqcap(p)$ B is the matrix which carries β^μ into its transpose

$$(\beta^\mu)^T = B\, \beta^\mu\, B^{-1}$$

and $(B^T)^{-1}\overline{\eta \sqcap(p)}(\bar{B})^{-1} = (B^T)^{-1}\bar{\eta}\, \bar{B}^{-1} \sqcap(p)^* \eta\, \eta^{-1} = (B^T)^{-1}\bar{\eta}\, \bar{B}^{-1}\, \eta \sqcap(p)\, \eta^{-1} =$ $\sqcap(p)\eta^{-1}$ because $(B^T)^{-1}\bar{\eta}(\bar{B})^{-1}\eta = 1$. On the other hand, in the standard case, $F = C^T \eta$ so $(F^*)^{-1} = (\eta^* \bar{C})^{-1} = C\eta^{-1}$ since $\bar{C} = C^{-1}$ and $\eta = \eta^*$. Consequently

$$[(F^*)^{-1} d(p)\, F^{-1}]^T = [(\eta^T)^{-1} (\eta^T C) \sqcap(p) (C^T \eta)^{-1}]^T$$

$$= (-1)^\sigma \sqcap(-p)\, \eta^{-1}$$

because $(C^T \eta)^T = (-1)^\sigma (C^T \eta)$ and $(C^T\eta)\sqcap(p)(C^T \eta)^{-1} = \sqcap(-p)^T$. Thus, in the standard case,

$$[\psi(f), \psi(g)^*]_{\pm} = \iint d^4 x\, d^4 y\, f(x) \sum_j \frac{1}{i} S(m_j, x-j)\eta^{-1}\overline{g(y)}$$

where

$$S(m_j, x) = \sqcap(i\partial)\, \Delta(m_j, x) \quad .$$

The determination of B and F satisfying the identities (5.32) (5.34) (5.36) (5.37) and (5.43) for a completely general wave equations appears to require an extension of the work of section four to the general case. For any wave equation for which such B and F can be found the formulae of the present section provide a local relativistic field satisfying the wave equation.

REFERENCES

[1] A. S. Wightman, The Dirac Equation pp.109-113 in Aspects of Quantum Theory, Eds. A. Salam and E. Wigner, Cambridge University Press 1972.

[2] E. Majorana, Teoria Relativistica di Particelle con Momento Intrinseco Arbitrario, Nuovo Cimento 9 (1932) 335-344.

[3] L. Gårding, On a Class of Linear Transformations Connected With Group Representations, Medd. Lunds Mat. Sem. 6 (1944) n°6 1-125.

[4] I. M. Gelfand and Y. Yaglom, General Relativistically Invariant Equations and Infinite Dimensional Representations of the Lorentz group, J. Exp. and Theo. Phys. USSR 18 (1948) 703-733.
Pauli's Theorem for General Relativistically Invariant Equations ibid. 1096-1104.

[5] F. Bruhat, Sur les Représentations Induites des Groupes de Lie, Bull. Soc. Math. de France, 84 (1956) 97-205.

[6] B. L. van der Waerden, Modern Algebra, Vol.II, F. Ungar Publishing Co. New York 1950 exp. § 119.

[7] I. M. Gelfand, P. A. Minlos and Z. Ya. Shapiro, Representations of the notation and Lorentz groups and their applications, Pergamon Press Oxford 1963 exp. Part II, Chap. II.

[8] E. Wild, On First Order Wave Equations for Elementary Particles without Subsidiary Conditions, Proc. Roy. Soc. Lond. 191A (1947) 253-268.

[9] Harish-Chandra, On Relativistic Wave Equations, Phys. Rev. (2) 71 (1947) 793-805.

[10] E. Speer, Generalized Feynman Amplitudes, Annals of Math. Studies # 62 Appendix C Princeton Press Princeton, New Jersey 1969.

[11] A. S. Wightman, Relativistic Wave Equations as Singular Hyperbolic Systems, pp.441-477 in Partial Differential Equations, Proc. of Symposia in Pure Math. Vol. XXIII Amer. Math. Soc. 1973.

[12] A. S. Wightman, Instability Phenomena in the External Field Problem for Two Classes of Relativistic Wave Equations pp.423-460 in Studies in Mathematical Physics Essays in Honor of Valentine Bargmann, Princeton Series in Physics, Princeton University Press, Princeton, New Jersey 1976.

Chapter III : The External Field Problem

1. Introduction

The present chapter develops the external field problem in a form which covers the possibility that the field is quasi-local, a possibility which we know must be reckoned with in general as a result of the fundamental discovery of Velo and Zwanziger.[1] The formalism introduced by A. Capri [2] and used in [3] to discuss the stability of representations of the Poincaré group turns out to be well adapted to this purpose. The only necessary modification of Capri's formalism is to consider weakly retarded and weakly advanced fundamental solutions of the c-number problem. That is, one looks for tempered distributions $S_{\substack{R\\A}}(x,y;B)$ satisfying

$$[\beta^{\mu} \frac{\partial}{\partial x^{\mu}} + m + B(x)]S_{\substack{R\\A}}(x,y;B) = \delta(x-y)\ 1 \tag{1.1}$$

$$-\frac{\partial}{\partial y^{\mu}} S_{\substack{R\\A}}(x,y;B)\beta^{\mu} + S_{\substack{R\\A}}(x,y;B)[m + B(y)] = \delta(x-y)\ 1 \tag{1.2}$$

which are such that for the weakly retarded fundamental solutions S_R and for each positive integer n, for each pair of test functions $f,g \in \mathcal{S}$ and for each vector, $\ell \notin \overline{V}_+$ (the future light cone), there is a constant $c(f,g,n,\ell)$ such that

$$\left|\int f(x)S_R(x + \tau\ell,y;B)g(y)dxdy\right| \leq \frac{c(f,g,n,\ell)}{1 + |\tau|^n} \tag{1.3}$$

for all real τ. For the weakly advanced fundamental solutions $S_A(x,y;B)$ an analogous relation holds with $\ell \notin \overline{V}_-$, the past light cone. (1.1) and (1.2) define the c-number problem associated with the Yang-Feldman equations

$$\psi(x) = \psi^{in}(x) - \int S_R(x-y)B(y)\psi(y)dy \tag{1.4}$$

$$\psi(x) = \psi^{out}(x) - \int S_A(x-y)B(y)\psi(y)dy \tag{1.5}$$

We are going to consider the case in which

$$(\beta^{\mu}\partial_{\mu} + m)u(x) = 0 \tag{1.6}$$

is a proper wave equation in the sense of [3] and the external field $B(x)$ is an $N \times N$ matrix infinitely differentiable in x and of fast decrease in all directions in space-time. The only further restriction on $B(x)$ which we will make is

$$B(x)^* = \eta B(x)\eta^{-1} \tag{1.7}$$

where η is the hermitizing matrix

$$-\beta^{\mu *} = \eta \beta^{\mu} \eta^{-1} \tag{1.8}$$

Of course, if $S_R(x,y;B)$ is retarded i.e. if it vanishes except when $x - y \in \overline{V}_+$, it is also weakly retarded, and an analogous statement holds for the weakly advanced fundamental solution $S_A(x,y;B)$.

With one technical assumption to be discussed later, it is possible to show that the quantized field defined by Capri's method exists and is quasi-local, if and only if unique weakly retarded and weakly advanced solutions of the c-number problem (1.1) and (1.2) exist. The argument is described in Section 2 of the present chapter.It is a very important advantage of this construction that it does not depend on the validity of the canonical formalism for the theory in the presence of the external field. In fact, in general, the canonical commutation relations will not be valid for the solutions constructed.

As you can learn from Gårding's lectures it can happen that no weakly retarded and advanced fundamental solutions exist. When this happens the methods of this chapter are totally ineffective.

The existence of ψ having been reduced to a c-number problem, it is natural to ask about the existence and uniqueness of the out vacuum and the existence and unitarity of the S operator. An explicit discussion of these points is worthwhile even in the case of local fields since existing results are confined to special cases [4], [5], [6]. (See also the results in the contributions to this volume of Seiler and Ruijsenaars.) In Section 3, conditions on the fundamental solutions are located which are sufficient to guarantee that the theory goes through. If these rather reasonable conditions on the weakly advanced and weakly retarded solutions are admitted the quantum field theory of the external field problem has a certain coherence (a reasonable scattering theory exists) whether the field is local (that is so if and only if the weakly retarded solution of the c-number problem is, in fact retarded) or only quasi-local, provided it exists at all.

2. Connection Between the Existence of a Quasi-Local Field Satisfying the Yang-Feldman Equation and the Weakly Retarded and Advanced Solutions of the c-number Equations

Recall that in Capri's method [2] one smears the Yang-Feldman equations (1.4) and (1.5) with a test function f and rewrites them

$$\psi(T_R f) = \psi^{in}(f) \tag{2.1}$$

$$\psi(T_A f) = \psi^{out}(f) \tag{2.2}$$

with

$$T_R f(x) = f(x) + \int f(y)dy\ S_R(y-x)B(x) \tag{2.3}$$

$$T_A f(x) = f(x) + \int f(y)dy\ S_A(y-x)B(x) \tag{2.4}$$

The test function is assumed to be an infinitely differentiable function of fast decrease; the set of such equipped with the standard topology is denoted $\mathcal{S}$. [7] Then assuming that T_R is a one-to-one mapping of $\mathcal{S}$ onto itself, one defines ψ and ψ^{out} in terms of ψ^{in} by

$$\psi(f) = \psi^{in}(T_R^{-1} f) \tag{2.5}$$

and

$$\psi^{out}(f) = \psi^{in}(T_R^{-1} T_A f) \tag{2.6}$$

Elementary computations then yield for $\psi^+(f) = \psi(\bar{\eta}\bar{f})^*$ and $\psi^{out\,+}(f) = \psi^{out}(\bar{\eta}\bar{f})^*$

$$\psi^+(f) = \psi^{in+}(T_R'^{-1} f) \tag{2.7}$$

$$\psi^{out+}(f) = \psi^{in+}(T_R'^{-1} T_A' f) \tag{2.8}$$

where

$$(T_R' f)(x) = f(x) + B(x) \int S_A(x-y)dy\ f(y) \tag{2.9}$$

$$(T_A' f)(x) = f(x) + B(x) \int S_R(x=y)dy\ f(y) \tag{2.10}$$

The first Lemma is elementary and is contained in [2].

Lemma 1

$T_R f$, $T_A f$, $T_R' f$, $T_A' f$ regarded as functions of B and f define continuous maps of $\mathcal{S} \times \mathcal{S}$ into $\mathcal{S}$. If for B fixed they map $\mathcal{S}$ one-to-one onto $\mathcal{S}$, the inverses T_R^{-1}, T_A^{-1}, $T_R'^{-1}$ are continuous.

It follows from this Lemma that for $\psi(f)$ and $\psi^{out}(f)$ to be well defined as fields it suffices that the maps T_R and T_A be one-to-one maps (≡ bijections) of $\mathcal{S}$ onto itself.

To verify that T_R is a bijection one has to show two things. First that $\ker T_R$, the set of all $f \in \mathcal{S}$ such that $T_R f = 0$, contains only $f(x) = 0$ and secondly that $\operatorname{Im} T_R$, the set of all $h \in \mathcal{S}$ such that there is a g satisfying $h = T_R g$ is all of $\mathcal{S}$. The next two Lemmas relate these two properties to the solution of the Cauchy problem at $\pm\infty$. To state them we need a definition: $\mathcal{O}_M$ is the set of all infinitely differentiable functions on space-time such that they and all their derivatives are bounded in absolute value by polynomials.

Lemma 2

$\ker T_{\substack{R\\A}} = 0$ if and only if the function $f = 0$ is the only $f \in \mathcal{O}_M$ that satisfies

$$(\beta^{\mu}\partial_{\mu} + m + B)f(x) = 0 \tag{2.11}$$

and vanishes at $\begin{Bmatrix}+\infty\\-\infty\end{Bmatrix}$ in the following sense: $f(x+\tau\ell)$ approaches zero faster than any negative power of τ as $\tau \to \begin{Bmatrix}+\infty\\-\infty\end{Bmatrix}$ for each plus time-like or plus light-like vector ℓ and any x.

Lemma 3

Im $T_{\substack{R\\A}} = \mathscr{S}$ if and only if the inhomogeneous Cauchy problem with null data at $\begin{Bmatrix}+\infty\\-\infty\end{Bmatrix}$ has a solution in $\mathcal{O}_M$ i.e. for each $k \in \mathscr{S}$ there exists an $h \in \mathcal{O}_M$ such that

$$[\beta^{\mu}\partial_{\mu} + m + B]h = k \tag{2.12}$$

and $h(x+\tau\ell)$ approaches zero faster than any negative power of τ as $\tau \to \begin{Bmatrix}+\infty\\-\infty\end{Bmatrix}$ for each plus time-like or plus light-like vector, ℓ, and any x.

The proof of these Lemmas is given in Appendix A. The conditions given in Lemmas 2 and 3 are also necessary and sufficient for ker $T_{\substack{R\\A}}' = 0$ and Im $T_{\substack{R\\A}}' = \mathscr{S}$ because there is a bijection of J of $\mathscr{S}$ onto $\mathscr{S}$

$$(J^{-1}f)(x) = \bar{\eta}\,\bar{f}(x) \qquad (Jf)(x) = \eta^{-1}\,\bar{f}(x) \tag{2.13}$$

such that

$$T_R' = J\,T_R\,J^{-1} \qquad T_A' = J\,T_A\,J^{-1}\ . \tag{2.14}$$

Let us suppose that T_R and T_A are bijections of $\mathscr{S}$ onto $\mathscr{S}$. Then we can define distributions $S_{\substack{R\\A}}(x,y;B)$ by the equations

$$\iint f(x)S_R(x,y;B)g(y)dxdy = \iint (T_R^{-1}f)(x)S_R(x-y)g(y)dxdy \tag{2.15}$$

$$\iint f(x)S_A(x,y;B)g(y)dxdy = \iint f(x)S_A(x-y)(T_R'^{-1}g)(y)dxdy \tag{2.16}$$

Elementary manipulations given in [8] show that the tempered distributions so defined are fundamental solutions i.e. they satisfy the differential equations (1.1) and (1.2). Furthermore, we have

Lemma 4

$S_R(x,y;B)$ as defined by (2.15) is weakly retarded and $S_A(x,y;B)$ as defined by (2.16) is weakly advanced.

The proof of this Lemma is given in Appendix A.

The defining equations of $S_{\substack{R\\A}}(x,y;B)$ do not permit one to draw the conclusion that $S_R(x,y;B)$ is retarded and $S_A(x,y;B)$ advanced under the hypotheses we have made. Whether that is, in fact, the case will depend on the β^{μ} and on B. As was already remarked by Capri [2], for spin one-half and the Dirac equation with an arbitrary $B(x)$, the stronger statement holds as a consequence of known results in the theory of partial differential equations. On the other hand, from the work of Velo and Zwanziger one knows examples for which a weakly retarded fundamental solution cannot be retarded, if it exists at all. In any case, whenever T_R and T_A are bijections of $\mathcal{S}$ onto $\mathcal{S}$ the field defined by (2.5) is quasi-local, since it satisfies

$$[\psi(x),\psi^+(y)]_\pm = \frac{1}{i} S(x,y;B) \mathbb{1} \tag{2.17}$$

where

$$S(x,y;B) = S_R(x,y;B) - S_A(x,y;B) \quad .$$

Although $S(x,y;B)$ need not vanish for $x-y$ space-like, it approaches zero rapidly as $x-y$ goes to infinity in a space-like direction. Furthermore, the field defined by (2.5) satisfies the field equation

$$\psi(h) = 0 \tag{2.18}$$

for all h of the form $h=(-\beta_\mu{}^T\partial^\mu + m + B^T)f$ with $f \in \mathcal{S}$. The proof of these last results is contained in [3] or [8].

The $S_{\substack{R\\A}}$ defined in (2.15) and (2.16) are not arbitrary tempered distributions in the variables x and y. The linear mappings

$$f \to \int f(x)S_{\substack{R\\A}}(x,y;B)dx \; , \qquad\qquad f \to \int S_{\substack{R\\A}}(x,y;B)f(y)dy \tag{2.19}$$

are continuous from $\mathcal{S}$ to $\mathcal{O}_M$. This can be seen as follows. Since by the definition

$$\int f(x)S_R(x,y;B)dx = \int (T_R{}^{-1}f)(x)S_R(x-y)dx \tag{2.20}$$

and the right-hand side is the convolution of a function in $\mathcal{S}$ by a tempered distribution, the statement for $S_R(x,y;B)$ and the first map follows from a standard theorem of distribution theory (see, for example, [9] p 42). Similarly, the definition yields

$$\int S_A(x,y;B)g(y)dy = \int S_A(x-y)(T_R{}'^{-1}g)(y)dy \tag{2.21}$$

which shows that $S_A(x,y;B)$ after smearing in its second argument with f, lies in $\mathcal{O}_M$ and its value is continuous in f. The corresponding statements for $S_R(x,y;B)$ in its first argument and $S_A(x,y;B)$ in its second argument, follow by similar reasoning once one has recognized the following elementary identities

$$\begin{aligned}&\iint (T_R{}^{-1}f)(x)S_R(x-y)g(y)dxdy\\ &\quad = \iint f(x)S_R(x-y)(T_A{}'^{-1}g)(y)dxdy\end{aligned} \tag{2.22}$$

and

$$\begin{aligned}&\iint f(x)S_A(x-y)(T_R{}'^{-1}g)(y)dxdy\\ &\quad = \iint (T_A{}^{-1}f)(x)S_A(x-y)g(y)dy\end{aligned} \tag{2.23}$$

proved in [8]. These regularity properties of the advanced and retarded functions will be of considerable importance in the next section.

Using $S_{\substack{R\\A}}(x,y;B)$ it is easy to give explicit formulae for T_R^{-1} and T_A^{-1}

$$(T_{\substack{R\\A}}^{-1}f)(x) = f(x) - \int f(y)dy\ S_{\substack{R\\A}}(y,x;B)B(x) \tag{2.24}$$

The proof goes as follows for T_R^{-1}. Define

$$g_R(x) = \int f(y)S_R(y-x)dy \quad . \tag{2.25}$$

Then

$$(-\beta_\mu{}^T\partial^\mu + m)g_R(x) = f(x) \tag{2.26}$$

and, from the definition of T_R

$$T_Rf = [-\partial^\mu g_R(x)\beta_\mu + g_R(x)(m + B(x))] \quad . \tag{2.27}$$

Furthermore, $g_R(x) \to 0$ as $x \to +\infty$ in the sense of Lemma 2. (The proof is part of the proof of Lemma 2.) Such a g_R is given by

$$g_R(x) = \int (T_Rf)(y)S_R(y,x;B)dy \tag{2.28}$$

and by Lemma 2 it is unique. Thus,

$$f(x) = (-\beta_\mu{}^T\partial^\mu + m)g_R(x) = [-\partial^\mu g_R(x)\beta_\mu + g_R(x)(m + B(x))] - g_R(x)B(x) \tag{2.29}$$

i.e.

$$f(x) = T_Rf(x) - \int (T_Rf(y)dy\ S_R(y,x;B)B(x) \tag{2.30}$$

which is (2.24) if f is replaced by $T_R^{-1}f$.

The whole of the preceding may be summarized in the following theorem.

<u>Theorem</u>

The formulae (2.5) and (2.6)

$$\psi(f) = \psi^{in}(T_R^{-1}f) \quad , \quad \psi^{out}(f) = \psi^{in}(T_R^{-1}T_Af)$$

with ψ^{in} a free field satisfying the equation (1.6)

$$(\beta^\mu\partial_\mu + m)\psi^{in}(x) = 0$$

and the commutation relations

$$[\psi^{in}(x),\psi^{in}(y)]_\pm = 0 \quad , \quad [\psi^{in}(x),\psi^{in+}(y)]_\pm = \frac{1}{i}\,S(x-y)1 \tag{2.31}$$

define solutions of the Yang-Feldman equations (1.4) and (1.5) provided that T_R^{-1} and T_A^{-1} are continuous mappings of $\mathcal{S}$ onto $\mathcal{S}$. The fields ψ and ψ^{out} so defined are tempered operator-valued distributions. ψ is quasi-local, satisfying the commutation relations

$$[\psi(x),\psi(y)]_{\pm} = 0 \quad , \quad [\psi(x),\psi^{+}(y)]_{\pm} = \frac{1}{i} S(x,y;B) \quad . \tag{2.32}$$

T_R^{-1} and T_A^{-1} are continuous mappings of $\mathcal{S}$ into $\mathcal{S}$ if and only if T_R and T_A are bijections of $\mathcal{S}$ onto $\mathcal{S}$. This condition is in turn equivalent to the requirement of the existence of unique weakly retarded and advanced tempered fundamental solutions $S_R(x,y;B)$ and $S_A(x,y;B)$, solutions of the differential equations (1.1) and (1.2), restricted by the further requirement that the mappings (2.19) are continuous from $\mathcal{S}$ to $\mathcal{O}_M$.

Appendix A: Proofs of Lemmas 2,3,4

Proof of Lemma 2

It is convenient to discuss $T_{\substack{R\\A}}'$ instead of $T_{\substack{R\\A}}$. Since they are related by the bijection J of $\mathcal{S}$ onto $\mathcal{S}$, $\ker T_{\substack{R\\A}} = 0$ is equivalent to $\ker T_{\substack{R\\A}}' = 0$ and $\operatorname{Im} T_{\substack{R\\A}} = \mathcal{S}$ is equivalent to $\operatorname{Im} T_{\substack{R\\A}}' = \mathcal{S}$. For brevity, only T_R' will be discussed explicitly. The treatment of T_A' runs parallel.

If $f \in \ker T_R'$ then

$$f(x) + B(x) \int S_A(x-y)dy\, f(y) = 0 \tag{A.1}$$

Now note that if we write

$$g(x) = \int S_A(x-y)dy\, f(y) \tag{A.2}$$

then $g \in \mathcal{O}_M$,

$$(\beta^\mu \partial_\mu + m)g(x) = f \tag{A.3}$$

and (A.1) can be written

$$(\beta^\mu \partial_\mu + m + B(x))g(x) = 0 \quad . \tag{A.4}$$

Thus, g is an element of $\mathcal{O}_M$ satisfying the (homogeneous) generalized Dirac equation with external field B. It is not identically zero if f is not identically zero. (The easiest way to see that is to Fourier transform (A.2) and get $\hat{g}(p) = [-\not{p} + m]^{-1}\hat{f}(p)$ which implies immediately that if $\hat{g} = 0$, then $\hat{f} = 0$ except on the mass shell. By continuity, $\hat{f} = 0$ identically.)

Furthermore, g vanishes at $+\infty$ in the sense that for each positive integer k, and each vector $\ell \in \overline{V}_+$, there is a constant C such that

$$|g(x + \tau\ell)| \leq \frac{C}{1+\tau^{\kappa}} \tag{A.5}$$

To see this, note that g does not change if $f(y)$ is replaced by $\psi(x-y)f(y)$ where ψ is infinitely differentiable, $\psi(x-y) = 1$ for $x-y \in \overline{V}_-$ and $\psi(x-y) = 0$

for $x-y$ outside the larger cone $\overline{V}_{-,\varepsilon}$ obtained by translating $\overline{V}_-$ forward in time by ε . Then,

$$|g(x)| = |S_A(\psi\tilde{f}_x)| < C\,\|\psi\tilde{f}_x\|_{r,s} \tag{A.6}$$

where the semi-norm $\|\ \|_{r,s}$ is defined by

$$\|h\|_{r,s} = \sum_{\substack{\alpha,\beta \\ |\alpha|\leqslant r|\beta|\leqslant s}} \sup_{\gamma=1\ldots N} \sup_{x\in\mathbb{R}^4} |x^\alpha D^\beta h_\gamma(x)| \tag{A.7}$$

with α and β defined by the sequences $\alpha_0\alpha_1\alpha_2\alpha_3$ and $\beta_0\beta_1\beta_2\beta_3$ of non-negative integers respectively and

$$x^\alpha = x_0^{\alpha_0}\, x_1^{\alpha_1}\, x_2^{\alpha_2}\, x_3^{\alpha_3} \quad , \quad D^\beta = \left(\frac{1}{i}\frac{\partial}{\partial x^0}\right)^{\alpha_0} \cdots \left(\frac{1}{i}\frac{\partial}{\partial x^3}\right)^{\alpha_3} \tag{A.8}$$

and $\tilde{f}(\xi) = f(-\xi)$, $\tilde{f}_x(\xi) = \tilde{f}(\xi-x) = f(x-\xi)$. The validity of (A.6) is a consequence of the fact that S_A is a tempered distribution; any tempered distribution satisfies an inequality of the form (A.6) for some suitably chosen r,s by definition (see, for example [9] p 25).

The norm on the right-hand side of (A.6) is bounded by a sum of terms of the form

$$\sup_{\xi\in x-V_{-,\varepsilon}} |(x-\xi)^\alpha[D^\beta\psi(x-\xi)][D^\gamma f(\xi)]| \tag{A.9}$$

If x approaches infinity along any straight line whose direction lies outside $\overline{V}_+$, then the region over which the sup is taken runs away to plus infinity. The sup of $|D^\gamma f(\xi)|$ then falls off faster than $(1+|x|)^{-k}$ for every integer k . (With suitable choice of ψ , all its relevant derivatives will be uniformly bounded in the region in question.) Thus, $g(x+\tau\ell)\to 0$ faster than any inverse power of τ for $\ell\notin\overline{V}_+$. Notice that what has been proved is stronger than the statement of the Lemma since the fall-off holds also for space-like directions. This completes the first half of the proof.

Conversely, suppose one is given a $g\in\mathcal{O}_M$ satisfying (A.4) and vanishing at plus infinity. Then we define an f by (A.3). f is necessarily in $\mathcal{S}$, because the differential equation (A.4) displays it as a product of $B\in\mathcal{S}$ and $g\in\mathcal{O}_M$. Thus we can invert the differential equation (A.3) and express g as the advanced potential of f as in (A.2). Then f provides a solution of (A.1) which is not identically zero. Notice that in asserting that g is necessarily of the form (A.2), we are using the uniqueness theorem for the Cauchy problem with null data at plus infinity for the equation without external forces.

Proof of Lemma 3

The proof will again be carried out only for T_R' . It runs parallel to the proof of Lemma 2.

If g is a solution of (2.12) which is an element of $\mathcal{O}_M$, the infinitely differentiable function f defined by $(\beta^\mu \partial_\mu + m)g = f$ belongs to $\mathcal{S}$ since it is of the form $k - Bg$. Thus, using the fact that g vanishes at $+\infty$, we may invert the differential equation to obtain

$$g(x) = \int S_A(x-y)f(y)dy \quad . \tag{A.10}$$

Then the differential equation (2.12) may be rewritten

$$f(x) + B(x) \int S_A(x-y)f(y)dy = k(x) \tag{A.11}$$

Thus, given $k \in \mathcal{S}$, we have found an $f \in \mathcal{S}$ mapped into k by the transformation T_R'.

Conversely, given an $f \in \mathcal{S}$ satisfying (A.11), one can define a $g \in \mathcal{O}_M$ by (A.10) and recover a solution of (2.12). That g will vanish at $+\infty$, follows from the argument used in Lemma 2.

Proof of Lemma 4

The first step is to show that $F_R(x,y) = S_R(x-y)$ the retarded function in the absence of external fields is weakly retarded. That means: for any $f,g \in \mathcal{S}$ any $x \in \mathbb{R}^4$, any $\ell \notin \overline{V}_+$, and any positive integer n, there exists a constant C such that

$$\int f(x + \tau\ell)S_R(x-y)g(y)dxdy < \frac{C}{1+\tau^n} \tag{A.12}$$

for all $\tau > 0$. The left-hand side is

$$\sum_{r,s} S_R(\{-\tau\ell,1\}f_r * \tilde{g}_s)_{rs} \tag{A.13}$$

where $(\{a,1\}f)(x) = f(x-a), \tilde{g}(x) = g(-x)$ and $(f*g)(\xi) = \int f(\xi-x)g(x)dx$. (A.13) is precisely of the form estimated in (A.6) and (A.9) so the required argument for (A.12) goes just as in Lemma 2.

The weak retardedness of $S_R(x-y)$ having been established the corresponding statement for $S_R(x,y;B)$ follows immediately from the definition (2.15).

3. Existence and Uniqueness of Ψ_0^{out} and the Unitarity of the S-Matrix

According to Capri, the ψ^{out} defined by (2.6) satisfies the free field equation

$$\psi^{out}(h) = 0$$

for all h of the form $h = (-\beta_\mu^T \partial^\mu + m)f$ with $f \in \mathcal{S}$, and the free field commutation relations

$$[\psi^{out}(f), \psi^{out+}(g)]_\pm = \iint f(x)i^{-1}S(x-y)g(y)dxdy \tag{3.1}$$

$$[\psi^{out}(f), \psi^{out}(g)]_\pm = 0 \tag{3.2}$$

For proofs see [3] and [8]. He also argued that there is a unique vector Ψ_0^{out} such that

$$\psi^{out}(f)\Psi_0^{out} = 0 \quad \text{and} \quad \psi^{out\,c}(f)\Psi_0^{out} = 0 \tag{3.3}$$

for all f such that the Fourier transform of f

$$\hat{f}(p) = \frac{1}{(2\pi)^2} \int e^{-ip\cdot x} f(x)dx \tag{3.4}$$

vanishes for $p^0 < 0$. Then there is a unitary operator S unique up to a phase factor such that

$$\psi^{out}(f) = S^{-1} \psi^{in}(f) S \tag{3.5}$$

for all $f \in \mathcal{S}$ and

$$S\Psi_0^{out} = \Psi_0^{in} \quad . \tag{3.6}$$

Given that Ψ_0^{out} exists and is unique up to a factor, the proof that S exists is unitary and unique up to a phase is standard (see, for example, [10]). Thus, the crucial part of the argument is that which shows that Ψ_0^{out} exists and is unique. The version of it appearing in [2] does not appear to stand close scrutiny. Let me reconsider the question in detail.

Consider first the map $T_R^{-1} T_A$ which appears in the definition of ψ^{out}. It satisfies

$$(T_R^{-1}T_A f)(x) = f(x) - T_R^{-1}(\int f(y)dy\ S(y-x)B(x))$$

as one sees by subtracting (2.3) from (2.4) to obtain

$$(T_A f)(x) = (T_R f)(x) - \int f(y)dy\ S(y-x)B(x)$$

and then applying T_R^{-1}. If one now introduces the formula (2.24) for T_R^{-1} one gets

$$\begin{aligned}(T_R^{-1}T_A f)(x) &= f(x) - \int f(y)dy\ S(y-x)B(x) \\ &+ \int f(z)S(z-y)B(y)S_R(y,x;B)B(x)dzdy\end{aligned} \tag{3.7}$$

This formula displays explicitly the fact that $T_R^{-1}T_A f$ is f plus terms which depend only on the value of $\hat{f}(p)$ restricted to the mass shell and projected onto the subspace satisfying the transposed Dirac equation of the appropriate momentum. Such a formula is precisely what is needed to obtain explicit expressions for the annihilation and creation operators of the out field in terms of those of the in field. To see this, note that expressed in terms of annihilation and creation operators ψ^{in} is

$$\psi^{in}(f) = a^{in}(\Pi_+ f) + b^{in*}(\Pi_- f) \tag{3.8}$$

where a^{in} and b^{in} are annihilation operators of the particles and anti-particle wave functions and Π_+ is a map carrying test functions into the dual space of the space of particle wave functions. More precisely, $\Pi_+ f$ is obtained from f by first passing to its Fourier transform, then restricting the Fourier transform to the positive energy mass shell, and finally projecting it onto solutions of the transposed generalized Dirac equation of the appropriate (positive) energy momentum p

$$(\Pi_+ f)(p) = \sqrt{\pi}\, \hat{f}(p)\Pi(p) \quad . \tag{3.9}$$

The map Π_- is defined by

$$(\Pi_- f)(p) = \sqrt{\pi}\, \Pi(p)(C^T\eta)^{-1}\hat{f}(-p) \tag{3.10}$$

and therefore $\Pi_- f$ is obtained from f by first passing to its Fourier transform, restricting it to the negative energy mass shell, transforming it with $(C^T\eta)^{-1}$ and then projecting it onto solutions of the generalized Dirac equation of energy momentum p. Note that $\Pi_- f$ lies in the Hilbert space of positive energy solutions of the generalized Dirac equation, while $\Pi_+ f$ lies in its dual. The linear functional $\chi = \Pi_+ f$ is related to a vector by an anti-linear bijection J :

$$\langle\chi,\Phi\rangle = (J\chi,\Phi) \tag{3.11}$$

where J is just the Fourier transform of the J appearing in (2.13)

$$(J\chi)(p) = \eta^{-1}\,\overline{\chi(p)} \quad . \tag{3.12}$$

With this notation we have

$$a(\chi) = a^+(J\chi)^* \tag{3.13}$$

and the same for b. The operation J is closely related to charge conjugation. We define the charge conjugate ${}^c f$ of a test function f by

$$({}^c f)(x) = C^T\,\overline{f(x)} \quad , \tag{3.14}$$

this definition being arranged so that

$$\psi^c(f) = \psi({}^c f)^* \quad . \tag{3.15}$$

Then

$$b^+(\Pi_- f)^* = b(J^{-1}\Pi_- f) = b(\Pi_+ {}^c f) \quad . \tag{3.16}$$

Clearly, one can recover $a^{in}(\Pi_+ f)$ from $\psi^{in}(f)$ by writing f as the sum of f_+ and f_- where f_+ vanishes on the negative mass shell and f_- on the positive. Then

$$a^{in}(\Pi_+ f) = \psi^{in}(f_+) \tag{3.17}$$

and

$$b^{in+}(\Pi_- f) = b^{in}(\Pi_+ {}^c f)^* = \psi^{in}(f_-) \tag{3.18}$$

or

$$b^{in}(\Pi_+ f) = \psi^{in}((^c f)_-)^* \tag{3.19}$$

Applying these formulae to ψ^{out}, we obtain

$$a^{out}(\Pi_+ f) = a^{in}(\Pi_+ T_R^{-1} T_A f_+) + b^{in+}(\Pi_- T_R^{-1} T_A f_+) \tag{3.20}$$

$$b^{out}(\Pi_+ f) = b^{in}(\Pi_+ (T_R^{c})^{-1} T_A^{c} f_+) + a^{in+}(\Pi_- (T_R^{c})^{-1} T_A^{c} f_+) \tag{3.21}$$

In deriving (3.21) the identity $b^{out}(\Pi_+ f) = \psi^{out}((^c f)_-)^* = \psi^{in}(T_R^{-1} T_A (^c f)_-)^*$ has been used as well as the fact that $(^c f)_- = {}^c(f_+)$ so $\Pi_+{}^c(T_R^{-1} T_A (^c f)_-) = \Pi_+{}^c[(T_R^{-1} T_A)^c (f_+)] = \Pi_+ (T_R^{c})^{-1} (T_A^{c}) f_+$ where T_R^{c} and T_A^{c} are the maps T_R and T_A with the external field $B(x)$ replaced by the charge conjugate external field $C^{-1}\overline{B(x)}C$. Equation (3.7) makes the legitimacy of the formulae (3.20) and (3.21) explicit, because it shows that the right-hand sides depend only on $\Pi_+ f$.

By virtue of the commutation relations (3.1) and (3.2) the out annihilation and creation operators satisfy

$$\begin{aligned}
&[a^{out}(\Pi_+ f), a^{out+}(\Pi_- g)]_\pm = \langle \Pi_+ f, \Pi_- g\rangle \mathbb{1} \\
&[b^{out}(\Pi_+ f), b^{out+}(\Pi_- g)]_\pm = \langle \Pi_+ f, \Pi_- g\rangle \mathbb{1} \\
&[a^{out}(\Pi_+ f), a^{out}(\Pi_+ g)]_\pm = [b^{out}(\Pi_+ f), b^{out}(\Pi_+ g)]_\pm = \\
&= [a^{out}(\Pi_+ f), b^{out}(\Pi_+ g)]_\pm = [a^{out}(\Pi_+ f), b^{out+}(\Pi_- g)]_\pm = 0
\end{aligned} \tag{3.22}$$

and these relations in turn imply

$$\langle \Pi_+ T_R^{-1} T_A f_+, \Pi_- (T_R^{c})^{-1} T_A^{c} g_-\rangle \pm \langle \Pi_+ (T_R^{c})^{-1} T_A^{c} g_-, \Pi_- T_R^{-1} T_A f_+\rangle = \langle \Pi_+ f, \Pi_- g\rangle \tag{3.23}$$

$$\langle \Pi_+ (T_R^{c})^{-1} (T_A^{c}) f_+, \Pi_- T_R^{-1} T_A g_-\rangle \pm \langle \Pi_+ T^{-1} T_A g_-, \Pi_- (T_R^{c})^{-1} T_A^{c} f_+\rangle = \langle \Pi_+ f, \Pi_- g\rangle \tag{3.24}$$

$$0 = \langle \Pi_+ T_R^{-1} T_A f_+, \Pi_- (T_R^{c})^{-1} (T_A^{c}) g_+\rangle \pm \langle \Pi_+ (T_R^{c})^{-1} (T_A^{c}) g_+, \Pi_- T_R^{-1} T_A f_+\rangle \tag{3.25}$$

These conditions on the mapping $T_R^{-1} T_A$ arising from the commutation relations of ψ^{out} will play an important role in what follows.

The correspondence leading from a^{in}, b^{in} to a^{out}, b^{out} is a Bogoliubov transformation and there are well known necessary and sufficient conditions for the existence of Ψ_0^{out} when a^{in} and b^{in} are in the Фок representation, as they are here by construction [11] [12]. Roughly speaking, these conditions say that the "off-diagonal" parts of the Bogoliubov transformation (i.e. the transformation $f_+ \to \Pi_- T_R^{-1} T_A f_+$ associated with the argument of b^{in+} in (3.20) and the transformation $(f_+) \to \Pi_-(T_R{}^c)^{-1}(T_A{}^c) f_+$ associated with the argument of a^{in*} in (3.21) have to define Hilbert-Schmidt operators. A direct proof is very instructive since it brings out explicitly the relationship between the structure of the vacuum state and the Hilbert-Schmidt property. Therefore, even though it amounts to an elaboration in our context of a known result, we will devote the rest of this section to a detailed discussion.

Expressed as a vector in the ФОК space of the in field, the out vacuum, $\Psi_0{}^{out}$, is given by a set of amplitudes.

$$\Psi_0{}^{out} = \{\Psi_0{}^{out(n_1,n_2)}(p_1 \ldots p_{n_1}; q_1 \ldots q_{n_2}); n_1, n_2 = 0,1,2,\ldots\} \tag{3.26}$$

where the p's and q's are a shorthand for both momentum and spin variables. The condition that $a^{out}(\Pi_+ f)$ annihilate the out vacuum is

$$[a^{in}(\Pi_+ T_R{}^{-1} T_A f_+) + b^{in*}(\Pi_- T_R{}^{-1} T_A f_+)]\Psi_0^{out} = 0 \tag{3.27}$$

that is

$$\sqrt{n_1+1} \int d\mu_+(p)(\Pi_+ T_R{}^{-1} T_A f_+)(p)\Psi_0{}^{out(n_1+1,n_2)}(p, p_1 \ldots p_{n_1}; q \ldots q_{n_2})$$
$$+ \frac{(-1)^{n_1}}{\sqrt{n_2}} \sum_{j=1}^{n_2} (-1)^{j+1} (\Pi_- T_R{}^{-1} T_A f_+)(q_j)\Psi_0{}^{out(n_1,n_2-1)}(p_1 \ldots p_{n_1}; q_1 \ldots \hat{q}_j \ldots q_n) = 0 \tag{3.28}$$

(The equation has been written for Fermi-Dirac statistics. For Bose-Einstein statistics, the factors $(-1)^{n_1}$ and $(-1)^{j+1}$ are to be replaced by 1. $d\mu_+(p)$ is the invariant measure on the positive energy mass shell.) Similarly, the condition that $b^{out}(\Pi_+ f)$ annihilate the out vacuum is

$$[b^{in}(\Pi_+(T_R{}^c)^{-1}(T_A{}^c)f_+) + a^{in*}(\Pi_-(T_R{}^c)^{-1}(T_A{}^c)f_+)]\Psi_0{}^{out} = 0 \tag{3.29}$$

which says

$$\sqrt{n_2+1}\,(-1)^{n_1} \int d\mu_+(q)(\Pi_+(T_R{}^c)^{-1}(T_A{}^c)f_+)(q)\Psi_0{}^{out(n_1,n_2+1)}(p_1 \ldots p_{n_1}; q, q_1 \ldots q_{n_2})$$
$$+ \frac{1}{\sqrt{n_1}} \sum_{j=1}^{n_1} (-1)^{j+1} (\Pi_-(T_R{}^c)^{-1}(T_A{}^c)f_+)(p_j)\,\Psi_0^{out(n_1-1,n_2)}(p_1 \ldots \hat{p}_j \ldots p_{n_1}; q_1 \ldots q_{n_2}) = 0 \tag{3.30}$$

The nature of the solutions of (3.28) and (3.30) depends very much on the properties of the mappings

$$f_+ \to \Pi_+ T_R{}^{-1} T_A f_+ \quad , \quad f \in \mathcal{S} \tag{3.31}$$

and

$$f_+ \to \Pi_+(T_R{}^c)^{-1}(T_A{}^c)f_+ \quad , \quad f \in \mathcal{S} . \tag{3.32}$$

If the range of these transformations is dense in the dual of the one-particle subspace of ΦοΚ space, the following argument shows that Ψ_0^{out} is uniquely determined up to an overall phase factor and the functions $\Psi_0^{out(n_1,n_2)} = 0$ if $n_1 \neq n_2$. Let $n_2 = 0$ in (3.28) and $n_1 = 0$ in (3.30) and conclude that $\Psi_0^{out(n_1,0)}$, $n_1 > 0$ and $\Psi_0^{out(0,n_2)}$, $n_2 > 0$ are annihilated by all functionals in the range of (3.30) and (3.32) respectively. From the assumed denseness they are therefore zero. Applying the same argument step by step, we get $\Psi_0^{out(n_1+k,k)} = 0 = \Psi_0^{out(k,n_2+k)}$ for $n_1,n_2 > 0$. For the amplitudes with $n_1 = n_2$, this argument yields no conclusion. However, it can be used to obtain a uniqueness statement. Consider first the equations for the one pair and no particle amplitudes

$$
\begin{aligned}
\int d\mu_+(p)\,[\eta^{-1}\overline{(\Pi_+T_R^{-1}T_Af_+)(p)}]^+\,\Psi_0^{out(1,1)}(p;q)& \\
+ (\Pi_-T_R^{-1}T_Af_+)(q)\Psi_0^{out(0,0)} &= 0
\end{aligned}
\tag{3.33}
$$

$$
\begin{aligned}
\mp \int d\mu_+(q)\,[\eta^{-1}\overline{(\Pi_+(T_R^c)^{-1}(T_A^c)f_+)(q)}]^+\,\Psi_0^{out(1,1)}(p;q)& \\
+ (\Pi_-(T_R^c)^{-1}(T_A^c)f_+)(p)\Psi_0^{out(0,0)} &= 0
\end{aligned}
\tag{3.34}
$$

where the upper sign holds for fermions and the lower for bosons and the first term of each equation has been rewritten as a scalar product. If $\Psi_0^{out(0,0)} = 0$, then $\Psi_0^{out(1,1)} = 0$ also. If $\Psi_0^{out(0,0)} \neq 0$ it can be normalized to be $\mathbb{1}$. Then $\Psi_0^{out(1,1)}$ is uniquely determined because the difference between two solutions $\Psi_0^{out(1,1)}$ of (3.33) and (3.34) would have to be orthogonal to a dense set. This argument works also to show the uniqueness of $\Psi_0^{out(n+1,n+1)}$ given $\Psi_0^{out(n,n)}$. Thus, Ψ_0^{out} is uniquely determined up to an overall factor. In fact, it is a striking and well-known fact that the n pair amplitude can be written in terms of the one pair amplitude:

$$
\Psi_0^{out(n,n)}(p_1\ldots p_n;q_1\ldots q_n) =
\begin{cases}
\dfrac{(-1)^{\frac{n(n-1)}{2}}}{n!}\det\left\{\Psi_0^{out(1,1)}(p_j;q_k)\right\} \\[2ex]
\dfrac{1}{n!}\,\mathrm{perm}\left\{\Psi_0^{out(1,1)}(p_j,q_k)\right\}
\end{cases}
\quad j,k = 1\ldots n
\tag{3.35}
$$

where the determinant is to be chosen for Fermi-Dirac and the permanent for Bose-Einstein statistics.

Up to this point, the argument has shown that the amplitudes $\Psi_0^{out(n_1,n_2)}$ are uniquely determined up to an overall factor, but it has not been proved that they define a vector in ΦOK space; for this the norm $||\Psi_0^{out}||$ must be finite. The condition on $\Psi_0^{out(1,1)}$ that this should hold is derived by a straightforward calculation reminiscent of Fredholm theory

$$||\Psi_0^{out}||^2 = \sum_{n=0}^{\infty} ||\Psi_0^{out(n,n)}||^2_{\mathcal{H}(n,n)}$$

$$||\Psi_0^{out(n,n)}||^2_{\mathcal{H}(n,n)} = (n!)^{-2} \int\ldots\int \left(\prod_{k=1}^{n}\prod_{\ell=1}^{n} d\mu_+(p_k)d\mu_+(q_\ell)\right)$$

$$\sum_{\substack{i_1\ldots i_n\\ j_1\ldots j_n}} \sigma(i_1\ldots i_n)\sigma(j_1\ldots j_n) \prod_{r=1}^{n} \overline{\Psi_0^{out(1,1)}(p_r,q_{i_r})}$$

$$\prod_{r=1}^{n}(\eta)_{\alpha_r\beta_r} \prod_{r=1}^{n}(\eta)_{\alpha_{i_r}\beta_{i_r}} \prod_{s=1}^{n} \Psi_0^{out(1,1)}(p_s,q_{j_s})$$

$$= [n!]^{-2} \int\ldots\int \prod_{\ell=1}^{n} d\mu_+(q_\ell) \sum_{\substack{i_1\ldots i_n\\ j_1\ldots j_n}} \sigma(i_1\ldots i_n)\sigma(j_1\ldots j_n)$$

$$\prod_{k=1}^{n} \int d\mu_+(p_k)[\eta^T\Psi_0^{out(1,1)}(p_k,q_{i_k})^*\eta\Psi_0^{out(1,1)}(p_k,q_{j_k})]_{\alpha_{i_k}\beta_{j_k}}$$

$$= (n!)^{-1} \int\ldots\int \prod_{\ell=1}^{n} d\mu_+(q_\ell) \sum_{j_1\ldots j_n} \sigma(j_1\ldots j_n)$$

$$\prod_{k=1}^{n} \int d\mu_+(p_k)[\eta^T\Psi_0^{out(1,1)}(p_k,q_k)^*\eta\Psi_0^{out(1,1)}(p_k,q_{j_k})]_{\alpha_k\beta_{j_k}}$$

$$= (n!)^{-1} \int\ldots\int \prod_{\ell=1}^{n} d\mu_+(q_\ell)\det[K(q_j,q_k)]$$

where $K(q_j,q_k)$ is the $N \times N$ matrix

$$K(q_j,q_k) = \int d\mu_+(p)[\eta^T\Psi_0^{out(1,1)}(p,q_j)^*\eta\Psi_0^{out(1,1)}(p,q_k)].$$

The calculation has been made for Fermi-Dirac statistics. For Bose-Einstein statistics the signature factors σ are absent and therefore the result contains the permanent instead of the determinant. Summation on n yields formally

$$||\Psi_0^{out}||^2 = \begin{cases} \det(1+K) \\ [\det(1-K)]^{-1} \end{cases} \qquad (3.36)$$

where the upper alternative holds for Fermi-Dirac statistics, the lower for Bose-Einstein statistics.

Under the assumption $\mathrm{tr}\, K < \infty$, the power series in λ for $\det(1 + \lambda K)$ defines an entire function of λ . Since $\mathrm{tr}\, K$ is just $||\Psi_0^{\mathrm{out}(1,1)}||^2_{\mathcal{H}(1,1)}$ we have the statement that for Fermi-Dirac statistics $||\Psi_0^{\mathrm{out}(1,1)}||_{\mathcal{H}(1,1)} < \infty$ implies $||\Psi_0^{\mathrm{out}}|| < \infty$. On the other hand, for Bose-Einstein statistics $\mathrm{tr}\, K < \infty$ does not assure the legitimacy of the expansion of $[\det(1 + \lambda K)]^{-1}$ in a series in λ for $\lambda = -1$; K might have the eigenvalue $+1$. However, (3.36) is correct with the additional proviso that $\det(1 + \lambda K)$ is non-vanishing in some disc in the complex λ plane with center at the origin and radius greater than one. We will see later that this proviso is always satisfied as a consequence of the canonical commutation relations for the out annihilation and creation operators.

It should be emphasized that up to this point no argument has been offered to show there is any one-pair amplitude, $\Psi_0^{\mathrm{out}(1,1)}$, satisfying (3.33) and (3.34) . Since the discussion remaining is rather long, it is convenient to summarize what has been obtained so far in a proposition.

Proposition 3.1

If the sets

$$\mathcal{R}_+ = \{\eta^{-1}\overline{(\Pi_+ T_R^{-1} T_A f_+)}\ ;\ f \in \mathcal{S}\} \tag{3.37}$$

and

$$\mathcal{R}_+^c = \{\eta^{-1}\overline{(\Pi_+ (T_R^c)^{-1} T_A^c f_+)}\ ;\ f \in \mathcal{S}\} \tag{3.38}$$

are dense in the set of single particle states, the out vacuum, Ψ_0^{out} , is uniquely determined, up to a normalization factor, in terms of the one-pair amplitude, $\Psi_0^{\mathrm{out}(1,1)}$; the n-pair amplitude, $\Psi_0^{\mathrm{out}(n,n)}$, is the determinant (Fermi-Dirac statistics) or permanent (Bose-Einstein statistics) built of the one-pair amplitudes according to (3.35). If a one-pair amplitude, $\Psi_0^{\mathrm{out}(1,1)}$, satisfying (3.33) and (3.34) exists, it is unique, given the normalization $\Psi_0^{\mathrm{out}(0,0)} = 1$.

The normalizability of Ψ_0^{out}: $||\Psi_0^{\mathrm{out}}||^2 < \infty$ is guaranteed for Fermi-Dirac statistics by $||\Psi_0^{\mathrm{out}(1,1)}||^2_{\mathcal{H}(1,1)} < \infty$ since, in that case

$$||\Psi_0^{\mathrm{out}}||^2 = \det(1 + K)\ . \tag{3.39}$$

For Bose-Einstein statistics

$$||\Psi_0^{\mathrm{out}}||^2 = [\det(1 - K)]^{-1} \tag{3.40}$$

which is finite provided the entire function, $\det(1 + \lambda K)$, has no zeros for λ in a disc of radius, $1 + \varepsilon$, centered at the origin, for some $\varepsilon > 0$.

Remarks

1) If one of $\mathcal{R}_+$ and $\mathcal{R}_+{}^c$ is dense in the single particle states the other need not be, because the conditions on $\mathcal{R}_+$ and $\mathcal{R}_+{}^c$ are on the solution in two different potentials, one the charge conjugate of the other.

2) We will prove that (3.37) and (3.38) hold for Bose-Einstein statistics and that a square integrable $\Psi_0{}^{out(1,1)}$ exists, satisfying (3.33) and (3.34) directly from the canonical commutation relations. Thus, for Bose-Einstein statistics, Proposition 3.1, so supplemented, will provide a complete description of the out vacuum, $\Psi_0{}^{out}$, and, hence, of the scattering matrix. For Fermi-Dirac statistics, there is an alternative possibility: (3.37) and (3.38) may fail. Then the structure of the out vacuum will be different from that given by (3.35). Nevertheless, a unique out vacuum will exist and the S matrix will exist and be unitary. The altered structure of $\Psi_0{}^{out}$ is implicit in the Shale-Stinespring characterization of the unitary implementability of Bogoliubov transformations for the canonical anti-commutation relations, but it was first made explicit in the context of the external field problem by Labonté as will be recounted below. [11] [14]

3) It is worth noting that the formula (3.36) is invariant under an equivalence $K \to AKA^{-1}$ where A is any bounded operator with bounded inverse. That is true because $\det(1 + \lambda K)$ depends on K only through the quantities $\mathrm{tr}\, K^n$.

In order to complete the analysis of $\Psi_0{}^{out}$, it is necessary to study the mappings $f_+ \to \Pi_\pm T_R{}^{-1} T_A f_+$ and $f_+ \to \Pi_\pm (T_R{}^c)^{-1}(T_A{}^c) f_+$ more closely. Up to this point, they have been defined only for $f \in \mathcal{S}$, but we know that, in fact, two f's which yield the same $\Pi_+ f$ give the same image under the maps. This suggests that one consider the maps as defined on the subset $\{\Pi_+ f,\ f \in \mathcal{S}\}$ of the (dual of the) one-particle space and attempt to extend them by continuity to the whole space.

To carry this out in detail it is convenient to abandon field theory temporarily and to study what we will call <u>one-body theory</u> in which the unknown $\psi(x)$ is taken as an N component smooth function on space time. (More precisely, we will assume $\psi \in \mathcal{O}_M$ the set of smooth functions all of whose derivatives are polynomially bounded.) With this alternative interpretation the equations (2.5) for the Heisenberg picture field and (2.6) for the out field in terms of the in field still make sense. We consider in particular in fields of the form

$$\psi^{in}(x) = (2\pi)^{-3/2} \int d\mu(p) \exp[-ip\cdot x] \Phi^{in}(p) \tag{3.41}$$

where $d\mu(p)$ is the invariant measure on the mass shell including both positive and negative energies

$$d\mu(p) = d\mu_+(p) + d\mu_-(p) \tag{3.42}$$

(3.42)

$\Phi^{in}(p)$ is a solution of the generalized Dirac equation in momentum space

$$(-\not p + m)\Phi^{in}(p) = 0 \tag{3.43}$$

which is smooth and rapidly decreasing $(\in \mathcal{S})$ on the mass shell. The one-body scattering operator $\mathcal{S}$ is defined by

$$\psi^{out}(f) = (\mathcal{S}\,\psi^{in})(f) = \psi^{in}(T_R^{-1}T_A f) \quad . \tag{3.44}$$

We are going to show that if ψ^{in} is of the form (3.41) so is ψ^{out} .

It should be emphasized that this one-body scattering problem will involve both positive and negative energies, and does not possess a consistent physical interpretation in which the wave function is interpreted directly as a probability amplitude in the sense of quantum mechanics (Klein-Paradox). Nevertheless, it constitutes a perfectly respectable scattering theory.

We have

$$\begin{aligned} \psi^{out}(f) &= \sum_{\alpha=1}^{N} \int \psi_\alpha^{in}(x)(T_R^{-1}T_A f)_\alpha(x)d^4x \\ &= \sqrt{2\pi}\int d\mu(p)\sum_{\alpha=1}^{N}\Phi_\alpha^{in}(p)(\widehat{T_R^{-1}T_A f})_\alpha(p) \end{aligned} \tag{3.45}$$

where by virtue of (3.7)

$$\begin{aligned} (\widehat{T_R^{-1}T_A f})(p) &= \frac{1}{(2\pi)^2}\int e^{-ip\cdot x}\, d^4x(T_R^{-1}T_A f)(x) \\ &= \hat{f}(p) - \frac{i}{4\pi}\int f(q)d\mu(q)\varepsilon(q^0)\Pi(q)\hat{B}(q-p) \\ &+ \frac{i}{2(2\pi)^3}\int \hat{f}(q)d\mu(q)\varepsilon(q^0)\Pi(q)\hat{B}(q-r)d^4r\ \hat{S}_R(r,s;B)d^4s\ \hat{B}(s-p) \end{aligned} \tag{3.46}$$

where

$$\begin{aligned} B(x) &= (2\pi)^{-2}\int \hat{B}(p)d^4p\ \exp[ip\cdot x] \\ S(x) &= \frac{i}{2(2\pi)^3}\int d\mu(q)\varepsilon(q^0)\Pi(q)\exp[-iq\cdot x] \end{aligned} \tag{3.47}$$

and

$$S_R(x,y;B) = (2\pi)^{-4}\iint \hat{S}_R(r,s;B)d^4r\ d^4s\ \exp[-r\cdot x + s\cdot y] \quad .$$

Thus the scattering operator $\mathcal{S}$ regarded as acting on $\Phi^{in}(p)$ is of the form

$$\mathcal{S} = 1 + i\,\mathcal{T} \tag{3.48}$$

where $\mathcal{T}$ is an integral operator

$$(\mathcal{T}\Phi^{in})(p) = \int \mathcal{T}(p,q)d\mu(q)\Phi^{in}(q)$$

with the kernel

$$\mathcal{T}(p,q) = -\frac{1}{4\pi}\,\varepsilon(p)\Pi(p)\hat{B}(p-q)\Pi(q)$$
$$+\frac{1}{16\pi^3}\iint \varepsilon(p)\Pi(p)\hat{B}(p-r)d^4r\;\hat{S}_R(r,s;B)d^4s\;\hat{B}(s-q)\Pi(q) \tag{3.49}$$

Consider first the smoothness properties of $\mathcal{T}(p,q)$. $\mathcal{T}(p,q)$ is infinitely differentiable in p and q . That is evident for the first term in (3.49) because $\Pi(p)$ is a polynomial in p and $B(x)$ has been assumed to be in $\mathcal{S}$, so $\hat{B}(p-q)$ is also. For the second term, it is a consequence of the fact that $\hat{S}_R(p,q;B)$ is tempered in p and q so that a convolution with B in p from the left and with B in q from the right yields an infinitely differentiable function.

Next consider the rate of decrease of $\mathcal{T}(p,q)$ in one argument, the other being held fixed. For q fixed the first term is clearly in $\mathcal{S}$ as a function of p and vice versa. For the second term the argument is more involved. When tested in one variable by a test function in $\mathcal{S}$, by assumption $S_R(x,y;B)$ becomes of class $\mathcal{O}_M$ in the other. Consequently, $\hat{S}_R(p,q;B)$ is in $\mathcal{O}_c'$ in p when smeared in q with a test function in $\mathcal{S}$, and the same is true in q when it is smeared in p with a test function in $\mathcal{S}$. (Recall that $\mathcal{O}_c'$ is by definition the set of Fourier transforms of $\mathcal{O}_M$ regarded as a space of tempered distributions.) Since the convolution of a function in $\mathcal{S}$ by a distribution in $\mathcal{O}_c'$ is again in $\mathcal{S}$ we have that

$$\int \hat{B}(p-r)dr\;\hat{S}_R(r,s;B)ds\;\hat{B}(s-q) \tag{3.50}$$

is a function of p and q which is in $\mathcal{S}$ in each variable separately, the other being held fixed.

To complete the argument that

$$\int \mathcal{T}(p,q)d\mu(q)\Phi^{in}(q) \tag{3.51}$$

is infinitely smooth and of rapid decrease on the mass shell if Φ^{in} is infinitely smooth and of rapid decrease there, we use first the estimate

$$|(\hat{B}(p-q))_{jk}| \leq \frac{C_n}{[1+\sum_{j=1}^{3}(\vec{p}-\vec{q})^2]^n} \tag{3.52}$$

which holds for all integers n with a suitably chosen C_n . Each derivative of $\hat{B}$ satisfies a similar estimate. The integrals

$$\int \Pi(p)\hat{B}(p-q)\Pi(q)d\mu(q)\Phi^{in}(q) \tag{3.53}$$

and

$$\int \hat{B}(p-q)d\mu(q)\Phi^{in}(q) \tag{3.54}$$

appearing in the last two terms of (3.51), then converge for each p, because the estimate (3.52) shows that $(\Pi(p)\hat{B}(p-q)\Pi(q))_{jk}$ and $(\hat{B}(p-q))_{jk}$ are integrable in q with respect to the measure μ. The same holds for the integrals differentiated with respect to p^μ under the integral sign. Since the derivatives with respect to p of $(\Pi(p)\hat{B}(p-q)\Pi(q))_{jk}$ and $(\hat{B}(p-q))_{jk}$ are continuous in p uniformly in q, differentiation under the integral sign is permitted and we have completed the proof that the integral (3.51) is infinitely differentiable in p.

The rapid decrease in p for all directions in $\mathbb{R}^4$ is established by an argument very similar to the proof of Lemma 3.2 of [13]. One has the slight refinement of (3.52)

$$|\hat{B}(p-q)_{jk}| \leq \frac{C_L}{[1+||p-q||]^L} \leq \frac{C_L}{[1+\big|\,||p||-||q||\,\big|]^L} \tag{3.55}$$

where $||p|| = [(p^0)^2 + \vec{p}^{\,2}]^{\frac{1}{2}}$ is the Euclidean norm. It suffices to prove the result for the contribution of a single hyperboloid, say $q^2 = m^2$, to (3.51). Thus $||q|| = [m^2 + 2\vec{q}^{\,2}]^{\frac{1}{2}}$ in the integrand and so

$$\begin{aligned} \left|\int \hat{B}(p-q)d\Omega_m(q)\Phi(q)\right| &\leq \sum_{k=1}^{N} \int |\hat{B}_{jk}(p-q)|\frac{d^3q}{q^0}|\Phi_k(q)| \\ &\leq 4\pi\, C_L\, D_M \int_m^\infty \frac{dt}{[1+\big|\,||p||-t\,\big|]^L[1+t]^L} \end{aligned} \tag{3.56}$$

where we have used

$$|\Phi^{in}(q)| \leq \frac{D_L{}^2}{[1+||q||]^L} \; \frac{1}{[m^2 + 2\vec{q}^{\,2}]} \tag{3.57}$$

and $q/q^0 \leq 1$. (3.57) holds for any integer L with a suitable choice of D_L. The integral in the last term of (3.56) was proved in [13] to be less than a constant times $[C + ||p||]^{-(L-1)}$, so the proof is complete.

The rapid decrease is used in two slightly different ways for the two terms of (3.49). For (3.53) one has immediately the required rapid decrease on the mass shell. For the second term of (3.49) what is needed is rapid decrease of (3.54) in all of $\mathbb{R}^4$. That shows that the function (3.54) is in $\mathcal{S}$, so

$$\int \hat{B}(p-r)dr\; \hat{S}(r,s;B)ds \int \hat{B}(s-q)d\mu(q)\Phi^{in}(q) \tag{3.58}$$

is $\hat{S}$ with its second argument tested with a function in $\mathcal{S}$, namely (3.54). By the above argument given in connection with the discussion of (3.50), this expression (3.58) is of rapid decrease in p. This completes the proof that $\widetilde{\mathcal{T}}$ maps rapidly decreasing C^∞ functions on the mass shell into rapidly decreasing C^∞ functions on $\mathbb{R}^4$ (which are therefore also C^∞ and rapidly decreasing when restricted to the mass shell).

Next we turn to a study of $\mathcal{S}$ as an operator in a Hilbert space constructed from the Φ^{in}. We introduce a scalar product into the space of Φ^{in} as follows. Split Φ^{in} into positive and negative energy parts

$$\Phi^{in} = \Phi_+^{in} + \Phi_-^{in} \tag{3.59}$$

The scalar product is defined

$$(\Phi^{in},\Psi^{in}) = (\Phi_+^{in},\Psi_+^{in})_+ + (-1)^\sigma(\Phi_-^{in},\Psi_-^{in})_- \tag{3.60}$$

where

$$(\Phi_\pm^{in},\Psi_\pm^{in})_\pm = \int d\mu_\pm(p)\Phi_\pm^{in}(p)^+\Psi_\pm^{in}(p) \tag{3.61}$$

and $(-1)^\sigma$ is -1 if the representation of $SL(2,\mathbb{C})$ is univalent as a representation of the Lorentz group (this is the case usually associated with Bose statistics), while it is $+1$ if the representation of $SL(2,\mathbb{C})$ is double-valued regarded as a representation of the Lorentz group (Fermi Statistics). The scalar product (3.60) is positive definite because $\Phi^{in}(p)^+\Phi^{in}(p)$ is positive in the latter case and negative in the former. (That is a consequence of the fact that $\Phi_+^{in}(p)^+\Phi_+^{in}(p)$ is positive for all positive energy solutions, that all negative energy solutions are obtainable as charge conjugates $\Phi_-(p) = C^{-1}\bar{\Phi}_+(p)$ of suitable positive energy solutions, and finally that

$$C^T\eta = (-1)^\sigma\eta^T C \quad . \tag{3.62}$$

The scalar product (3.55) has been defined on smooth rapidly decreasing solutions. Now we complete in the norm associated with (3.55) to get the space of all square integrable solutions

$$\mathcal{H} = \mathcal{H}_+ \oplus \mathcal{H}_- \tag{3.63}$$

which splits in the indicated way into a direct sum of the positive and negative energy solutions. Correspondingly the operator S splits into a matrix of operators

$$S = \begin{Bmatrix} S_{++} & S_{+-} \\ S_{-+} & S_{--} \end{Bmatrix} \tag{3.64}$$

where S_{++} maps $\mathcal{H}_+$ into itself, S_{+-} maps $\mathcal{H}_-$ into $\mathcal{H}_+$, S_{--} maps $\mathcal{H}_-$ into $\mathcal{H}_-$ and S_{-+} maps $\mathcal{H}_+$ into $\mathcal{H}_-$. These mappings have only been defined on the dense subsets of smooth rapidly decreasing solutions. They can be extended by continuity to all vectors of the appropriate $\mathcal{H}_\pm$, under appropriate circumstances as we will now relate.

We begin with a Lemma about some of the eight constituent operators of $S_{\pm\pm}$. Let the mappings $\Phi_\pm \to A_{1+\pm}\Phi_\pm$ and $\Phi_\pm \to A_{2+\pm}\Phi_\pm$ of (the above-mentioned dense subsets of) $\mathcal{H}_\pm$ into $\mathcal{H}_+$, and the mappings $\Phi_\pm \to A_{1-\pm}\Phi_\pm$ and $\Phi_\pm \to A_{2-\pm}\Phi_\pm$

of (the above-mentioned dense subsets of) $\mathcal{H}_{\pm}$ into $\mathcal{H}_{-}$ be defined by

$$(A_{1+\pm}\Phi_{\pm})(p) = \frac{-1}{4\pi}\int \Pi(p)\hat{B}(p-q)\Pi(q)\Phi_{\pm}(q)d\mu_{\pm}(q)$$

$$(A_{2+\pm}\Phi_{\pm})(p) = \frac{1}{2(2\pi)^3}\iiint drds\ \Pi(p)\hat{B}(p-r)\hat{S}_R(r,s;B)\hat{B}(s-q)\Pi(q)\Phi_{\pm}(q)d\mu_{\pm}(q) \qquad (3.65)$$

$$(A_{1-\pm}\Phi_{\pm})(p) = \frac{1}{4\pi}\int \Pi(p)\hat{B}(p-q)\Pi(q)\Phi_{\pm}(q)d\mu_{\pm}(q)$$

$$(A_{2-\pm}\Phi_{\pm})(p) = \frac{-1}{2(2\pi)^3}\iiint drds\ \Pi(p)\hat{B}(p-r)\hat{S}_R(r,s;B)\hat{B}(s-q)\Pi(q)\Phi_{\pm}(q)d\mu_{\pm}(q)$$

The following Lemma is established by direct elementary estimates in Appendix B.

Lemma 3.1

A_{1++} and A_{1--} are bounded and A_{1+-} and A_{1-+} are Hilbert-Schmidt i.e. satisfy $\operatorname{tr} A^*A < \infty$.

The Lemma describes, A_1 , the first Born approximation to S ; it makes no statement about the behavior of the operators A_2 . As will be seen shortly, a necessary condition for the existence of the out vacuum is that S_{+-} and S_{-+} be Hilbert-Schmidt, or what is the same thing, that A_{2+-} and A_{2-+} be Hilbert-Schmidt. Since a direct calculation using

$$\hat{S}_A(s,r;B) = \eta^{-1}\ \hat{S}_R(r,s;B)^*\eta$$

and

$$\Pi(p)^* = \eta\Pi(p)\eta^{-1}$$

shows that $(A_{2+-})^*$ is a mapping from $\mathcal{H}_{+}$ to $\mathcal{H}_{-}$ given by

$$((A_{2+-})^*\Phi_{+})(p) = \frac{-i}{2(2\pi)^3}\iint drds\ \Pi(p)\hat{B}(p-r)\hat{S}_A(r,s;B)\hat{B}(s-q)\Pi(q)\Phi_{+}(q)d\mu_{+}(q) \qquad (3.66)$$

and similarly $(A_{2-+})^*$ is a mapping from $\mathcal{H}_{-}$ to $\mathcal{H}_{+}$ given by

$$((A_{2-+})^*\Phi_{-})(p) = \frac{i}{2(2\pi)^3}\iint drds\ \Pi(p)\hat{B}(p-r)\hat{S}_A(r,s;B)\hat{B}(s-q)\Pi(q)\Phi_{-}(q)d\mu_{-}(q), \qquad (3.67)$$

we have

$$\operatorname{tr}[(A_{2+-})^*(A_{2+-})] = [2(2\pi)^3]^{-2}\int d\mu_{-}(q)d\mu_{+}(p)\iint dr_1ds_1dr_2ds_2\operatorname{tr}[\Pi(q)\hat{B}(q-r_1)$$
$$\hat{S}_A(r_1,s_1;B)\hat{B}(s_1-p)\Pi(p)\hat{B}(p-r_2)\hat{S}_R(r_2,s_2;B)\hat{B}(s_2-q)] \qquad (3.68)$$

and

$$\mathrm{tr}[(A_{2-+})^*(A_{2-+})] = [2(2\pi)^3]^{-2} \int d\mu_+(q)d\mu_-(p)$$

$$\iint dr_1 ds_1 dr_2 ds_2 \ \mathrm{tr}[\Pi(q)\hat{B}(q-r_1)\hat{S}_A(r_1,s_1;B)\hat{B}(s_1-p)\Pi(p)$$
$$\hat{B}(p-r_2)\hat{S}_R(r_2,s_2;B)\hat{B}(s_2-q)] \tag{3.69}$$

What guidance does perturbation theory provide on the finiteness of these traces? In Appendix B, it is shown that every individual term of the perturbation series for A_{2+-} and A_{2-+} is Hilbert-Schmidt i.e. yields a finite trace in (3.68) and (3.69). However, as one knows from the work of Velo and Zwanziger, the perturbation series may not be a good guide to the properties of the exact solution. (Every term of the perturbation series for $S_R(x,y;B)$ is strictly retarded even when the exact solution itself is not.) For special cases, the indication provided by the perturbation series can be supported by exact results. For spin one-half, the original treatment of Schwinger shows that Fredholm methods yield the Hilbert-Schmidt property for A_{2+-} and A_{2-+}. (See [15]; and the lectures of R. Seiler).

For spin zero and special interactions there are results of Schroer, Seiler, and Swieca [4].

In the general case considered here the problem can be expressed in terms of the properties of the kernels

$$\iint d^4r \ d^4s \ \Pi(p)\hat{B}(p-r)\hat{S}_{\substack{R\\A}}(r,s;B)B(s-q)\Pi(q) \tag{3.70}$$

We have shown that they are infinitely smooth functions of p and q, and furthermore, that they are in $\mathcal{S}$ in the variable p for fixed q and in $\mathcal{S}$ in the variable q for fixed p. The question is their behavior in p and q together for $p^0 = \pm\sqrt{m_j^2 + \vec{p}^2}$ and $q^0 = \mp\sqrt{\vec{q}^2 + m_k^2}$, m_j and m_k being any two points of the mass spectrum. In the following, it will be assumed that the fundamental solutions are such that each matrix element of these kernels belongs to in p and q together. (This is our <u>second regularity assumption on</u> $S_{\substack{R\\A}}(x,y;B)$ the first being (2.19).) This assumption is consistent with the evidence from the perturbation series and the special cases of spin zero, one-half. It is much stronger than what is necessary to guarantee that A_{2+-} and A_{2-+} are Hilbert-Schmidt; that would require only square-integrability i.e. (3.68) and (3.69) finite.

Given the Hilbert-Schmidt property of A_{2+-} and A_{2-+} the Hilbert-Schmidt property of S_{+-} and S_{-+} follows immediately. The next step in the argument is to prove that the boundedness of A_{2++} and A_{2--} follow from the relations (3.23)...(3.25) that express the content of the commutation relations for the out-field.

The formal idea behind the argument is that the one-body scattering operator,

S , ought to satisfy a type of unitarity condition that reflects the conservation of the scalar product:

$$\int d\Sigma_\mu(x)\psi^{out\,+}(x)i\beta^\mu\,\psi^{out}(x) = \int d\Sigma_\mu(x)\psi^{in\,+}(x)i\beta^\mu\,\psi^{in}(x) \tag{3.71}$$

i.e. since both ψ^{in} and ψ^{out} are solutions of the free equation (1.6)

$$\int d\mu(p)\Phi^{out}(p)^+\,\Phi^{out}(p) = \int d\mu(p)\Phi^{in\,+}(p)\Phi^{in}(p) \quad . \tag{3.72}$$

This relation says that S is isometric in the Hilbert space equipped with the (indefinite when σ is 1) scalar product defined by omitting the $(-1)^\sigma$ in (3.59) i.e.

$$\begin{Bmatrix} S_{++} & S_{+-} \\ S_{-+} & S_{--} \end{Bmatrix}^* \begin{Bmatrix} 1 & 0 \\ 0 & (-1)^\sigma \end{Bmatrix} \begin{Bmatrix} S_{++} & S_{+-} \\ S_{-+} & S_{--} \end{Bmatrix} = \begin{Bmatrix} 1 & 0 \\ 0 & (-1)^\sigma \end{Bmatrix} \tag{3.73}$$

Not only is S isometric in this sense but it has an inverse defined by

$$\psi^{out}(f) \longrightarrow \psi^{out}(T_A^{-1}T_R f)$$

Thus (3.73) also holds with S interchanged with S^* . Written explicitly (3.73) is

$$\begin{aligned} [S_{++}]^* S_{++} + (-1)^\sigma [S_{-+}]^* S_{-+} &= 1 \\ [S_{+-}]^* S_{+-} + (-1)^\sigma [S_{--}]^* S_{--} &= (-1)^\sigma \\ [S_{++}]^* S_{+-} + (-1)^\sigma [S_{-+}]^* S_{--} &= 0 \\ [S_{+-}]^* S_{++} + (-1)^\sigma [S_{--}]^* S_{-+} &= 0 \end{aligned} \tag{3.74}$$

The first two of these relations imply that if S_{++} is given on the smooth solutions of the generalized Dirac equation which are rapidly decreasing at inifinity on each space-like hyperplane, then it can be extended to all of $\mathcal{H}_+$, and the same for S_{--} on $\mathcal{H}_-$. For Fermi-Dirac statistics this conclusion follows immediately from (3.74) without any appeal to the second regularity assumption, because (3.74) implies

$$\begin{aligned} ||S_{++}|| &\leq 1 \qquad & ||S_{-+}|| &\leq 1 \\ ||S_{+-}|| &\leq 1 \qquad & ||S_{--}|| &\leq 1 \end{aligned} \tag{3.75}$$

For Bose-Einstein statistics (3.74) implies

$$\begin{aligned} ||S_{++}|| &\leq 1 + ||S_{-+}|| \\ ||S_{--}|| &\leq 1 + ||S_{+-}|| \end{aligned} \tag{3.76}$$

so that under the second regularity assumption, which implies that $\mathcal{S}_{-+}$ and $\mathcal{S}_{+-}$ are Hilbert-Schmidt and therefore, in particular, bounded, $\mathcal{S}_{++}$ and $\mathcal{S}_{--}$ are bounded.

It remains to justify (3.72). That could be done by a direct appeal to the conservation law of the current plus an argument that

$$\int_{\Sigma} d\Sigma_{\mu}(x)\psi^{+}(x)i\beta^{\mu}\psi(x)$$

converges to the left-hand side of (3.71) as the space-like hyperplane $\Sigma \to +\infty$ and the right-hand side as $\Sigma \to -\infty$. (When the external field, $B(x)$, has compact support the required convergence is evident since $\psi = \psi^{out}$ for hyperplanes later than the support of B and $\psi = \psi^{in}$ for hyperplanes earlier.) The hermiticity requirement (1.7) on $B(x)$ is necessary and sufficient to insure the conservation of current. Alternatively one can recognize the content of the relations (3.23)... (3.25) that follows from the CCR (or CAR) for the out field. Written out (3.23) reads

$$\int d\mu_{+}(p)\hat{f}(p)\Pi(p)(C^{T}\eta)^{-1}\hat{g}(-p) = \iiint d\mu_{+}(p)d\mu_{+}(q)d\mu_{+}(r)\{\hat{f}(p)\mathcal{S}_{++}(p,q)\Pi(q)$$

$$(C^{T}\eta)^{-1}[\mathcal{S}^{c}_{--}(-r,-q)]^{T}\hat{g}(-r) \pm \hat{f}(p)\mathcal{S}_{+-}(p,-q)[(C^{T}\eta)^{T}]^{-1}\Pi(q)^{T}\mathcal{S}^{c}_{-+}(-r,q)^{T}\hat{g}(-r)\} \tag{3.77}$$

Now an elementary calculation using

$$\hat{S}_{R}(r,s;B^{c}) = C^{-1}\,\overline{\hat{S}_{R}(-r,-s;B)}\,C \tag{3.78}$$

shows that

$$\begin{aligned} C^{-1}\,\overline{\mathcal{S}(p,q;B)}\,C &= \mathcal{S}(-p,-q;B^{c}) \\ &\equiv \mathcal{S}^{c}(-p,-q;B) \end{aligned} \tag{3.79}$$

and therefore

$$\begin{aligned} &(C^{T}\eta)^{-1}\,\mathcal{S}^{c}_{--}(-r,-q)^{T}(C^{T}\eta) \\ &= \eta^{-1}[C\mathcal{S}^{c}_{--}(-r,-q)C^{-1}]^{T}\eta \\ &= \eta^{-1}[\mathcal{S}_{++}(r,q)]^{*}\eta = (\mathcal{S}^{*})_{++}(q,r) \end{aligned} \tag{3.80}$$

Similarly, using

$$(C^{T}\eta)^{T} = (-1)^{\sigma}C^{T}\eta \tag{3.81}$$

we get

$$[(C^{T}\eta)^{T}]^{-1}\Pi(q)^{T} = (-1)^{\sigma}\Pi(-q)(C^{T}\eta)^{-1} \tag{3.82}$$

and

$$(C^T\eta)^{-1}[S^c_{-+}(-r,q)]^T C^T\eta$$

$$= (-1)^\sigma \eta^{-1} S_{+-}(r,-q)^*\eta \tag{3.83}$$

$$= (S^*)_{-+}(-q,r)$$

Thus, regarded as a relation between quadratic forms defined on a dense set of $\mathcal{H}_+$, (3.77) may be written

$$S_{++} S_{++}^* + (-1)^\sigma S_{+-} S_{+-}^* = 1 \tag{3.84}$$

which is the first of the equations (3.74) with S and S^* interchanged. Similarly, (3.24) becomes

$$S^c_{++} S^c_{++}{}^* + (-1)^\sigma S^c_{+-} S^c_{+-}{}^* = 1 \tag{3.85}$$

which when translated in terms of the matrix elements of S is

$$S_{-+} S_{-+}^* + (-1)^\sigma S_{--} S_{--}^* = (-1)^\sigma \tag{3.86}$$

i.e. the second equation of (3.74) with S and S^* interchanged.

Finally, (3.25) may be written

$$0 = \iiint d\mu_+(p)d\mu_+(q)d\mu_+(r)$$

$$\{f(p)S_{++}(p,q)\Pi(q)(C^T\eta)^{-1}[S^c_{+-}(r,-q)]^T g(r) \tag{3.87}$$

$$+ (-1)^\sigma f(p)S_{+-}(p,-q)[(C^T\eta)^T]^{-1}\Pi(q)^T[S^c_{++}(r,q)]^T g(r)\}$$

which leads to

$$0 = S_{++} S_{-+}^* + (-1)^\sigma S_{+-} S_{--}^* \tag{3.88}$$

The adjoint of this relation is

$$0 = S_{-+} S_{++}^* + (-1)^\sigma S_{--} S_{+-}^* \tag{3.89}$$

This completes the derivation of half the relations for S, namely, those obtained from (3.74) by interchanging S with S^*. To derive the relations (3.74) from the CCR (CAR) one has only to note that the inverse of the Bogoliubov transformation (3.20) and (3.21) leading from the in- to the out-creation and annihilation operators is the Bogoliubov transformation

$$a^{in}(\Pi_+ f) = a^{out}(\Pi_+ T_A{}^{-1} T_R f_+) + b^{out+}(\Pi_- T_A{}^{-1} T_R f_+) \tag{3.90}$$

$$b^{in}(\Pi_+ f) = b^{out}(\Pi_+ (T_A{}^c)^{-1}(T_R{}^c) f_+) + a^{out+}(\Pi_- (T_A{}^c)^{-1}(T_R{}^c) f_+) \tag{3.91}$$

and that

$$\eta^{-1}[\mathcal{S}(p,q)]^*\eta = \mathcal{S}^{-1}(q,p) \tag{3.92}$$

as follows from

$$\eta^{-1}S_R(x,y;B)^*\eta = S_A(y,x;B) \tag{3.93}$$

and

$$\begin{aligned}(T_A^{-1}T_Rf)(x) &= f(x) + \int f(y)dy\ S(y-x)B(x) \\ &- \int f(z)dz\ S(z-y)B(y)dy\ S_A(y,x;B)B(x)\end{aligned} \tag{3.94}$$

The result of this argument is that $\mathcal{S}$ is unitary on the Hilbert space $\mathcal{H}$ for Fermi-Dirac statistics, and pseudo-unitary for Bose-Einstein statistics. By pseudo-unitary we mean that (3.74) and the analogous relations with $\mathcal{S}$ and $\mathcal{S}^*$ interchanged hold with $(-1)^\sigma = -1$.

With this additional information available we can now return to the determination of Ψ_0^{out}. Notice that the definition of the Bogoliubov transformation (3.20) and (3.21) can be extended to any $\Phi \in \mathcal{H}_+$

$$a^{out}(\Phi) = a^{in}(\Phi\,\mathcal{S}_{++}) + b^{in+}(\mathcal{S}^{c\,*}_{-+}J^{-1}\,{}^c\Phi) \tag{3.95}$$

$$b^{out}(\Phi) = b^{in}(\Phi\,\mathcal{S}^c_{++}) + a^{in+}(\mathcal{S}^*_{-+}J^{-1}\,{}^c\Phi) \tag{3.96}$$

and that the sets of vectors $\mathcal{R}_+$, $\mathcal{R}_+^c$ occurring in (3.37) and (3.38) can be replaced by the ranges of $\mathcal{S}^*_{++}$ and $\mathcal{S}^{c*}_{++}$ respectively. (We have here written $\Phi\,\mathcal{S}_{++}$ for the element of $\mathcal{H}_+'$ given by

$$\begin{aligned}(\Phi\mathcal{S}_{++})(p) &= \int d\mu_+(q)\,\Phi(q)\,\mathcal{S}_{++}(p,q)_- \\ &= \int d\mu_+(q)[\eta^{-1}\mathcal{S}_{++}(p,q)^*\overline{\Phi(q)}]^+\end{aligned} \tag{3.97}$$

i.e. the linear functional whose representative function is $(J\,\mathcal{S}^*_{++}\,\Phi)(p)$. ${}^c\Phi$ is defined by

$$({}^c\Phi)(p) = C^T\,\overline{\Phi(p)} \tag{3.98}$$

and J by (3.12).) We now prove that for Bose-Einstein statistics these ranges are all of $\mathcal{H}_+$ and $\mathcal{S}^{-1}_{++}$ and $\mathcal{S}^{c\,-1}_{++}$ exist and are bounded operators. The arguments are parallel in the two cases so we concentrate on $\mathcal{S}^*_{++}$ and $\mathcal{S}^{-1}_{++}$.

To prove that S^{*}_{++} has a dense range is equivalent to proving that a vector annihilated by its adjoint is necessarily zero : $S_{++}\Phi = 0$ implies $\Phi = 0$. Now by virtue of the CCR, the first equation (3.74) applied to Φ yields

$$-S^{*}_{+-}\, S_{+-}\,\Phi = \Phi$$

which implies $\Phi = 0$. (Take the scalar product with Φ. The left-hand side is then negative, the right-hand side positive ; therefore both are zero). The parallel argument with S_{++} and S^{*}_{++} interchanged shows S_{++} has a range which is dense. It therefore has a densely defined inverse. The inverse is bounded, because, again according to (3.74),

$$\| S_{++}\Phi\|^2 = \|\Phi\|^2 + \| S_{-+}\Phi\|^2$$

so setting $\psi = S_{++}\Phi$

$$\| S^{-1}_{++}\psi \| \le \|\psi\| \tag{3.100}$$

Now, since S_{++} is bounded and everywhere defined on $\mathcal{K}_+$, its inverse is a closed operator, so boundedness of S^{-1}_{++} on the range of S_{++} implies that, in fact, S^{-1}_{++} is defined on all of $\mathcal{K}_+$. Equation (3.100) says its bound $\|S^{-1}_{++}\|$ is less than or equal to 1.

This information is enough to yield the existence and uniqueness of the pair amplitude

$$\Psi_o^{out(1,1)}(p,q) = -(S^{-1}_{++}\; S_{+-}(p,-q)(\eta^T C)^{-1}\Psi_o^{out(0,0)} \tag{3.101}$$

because (3.33) can be written

$$\begin{aligned}&\int d\mu_+(p)\,\Phi(p)\,S_{++}(p,q)\Psi_o^{out\,(1,1)}(p;q)\\ &+ \int d\mu_+(p)\Phi(p)S_{+-}(p,-q)(\eta^T C)^{-1}\Psi_o^{out(o,o)} = 0\end{aligned} \tag{3.102}$$

The pair amplitude can also be determined from the charge conjugate equation (3.34) which can be written

$$\begin{aligned}&-(-1)^{\sigma}\int d\mu_+(p)\int d\mu_+(q)\Phi(p)\,S^{c}_{++}(p,q)\Psi_o^{out(1,1)}(p_1,q)\\ &+ \int d\mu_+(q)\,\Phi(q)\,S^{c}_{+-}(q,-p_1)(\eta^T C)^{-1}\Psi_o^{out(o,o)} = 0\end{aligned} \tag{3.103}$$

which implies

$$\Psi_0^{out(1,1)}(p,q) = (-1)^{\sigma}(C^T\eta)^{-1}[S^{c\,-1}_{++}S^{c}_{+-}(q,-p_1)]^T\Psi_0^{out(0,0)} \tag{3.104}$$

The two expressions (3.101) and (3.104) for the pair amplitude are equal by virtue of an elementary calculation using (3.79). In fact, their equality is equivalent to the identity (3.88) which expresses part of the content of the CCR (or CAR) for the out field.

The operator K occurring in Proposition (3.1) is given by

$$K(p,q) = \int d\mu_+(r)$$
$$\eta^T[S^{-1}_{++}S_{+-}(r,p)(\eta^T C)^{-1}]^*\eta[S^{-1}_{++}S_{+-}(r,q)(\eta^T C)^{-1}]$$
$$= (-1)^{\sigma}(\eta^T C)[S^{-1}_{++}S_{+-}]^*[S^{-1}_{++}S_{+-}](p,q)(\eta^T C)^{-1}$$

i.e. up to a sign, $(-1)^{\sigma}$, and equivalence by the matrix $\eta^T C$, K is $[S^{-1}_{++}S_{+-}]^*[S^{-1}_{++}S_{+-}]$. Since as the product of a bounded operator, S^{-1}_{++}, and a Hilbert-Schmidt operator S_{+-}, the operator $S^{-1}_{++}S_{+-}$ is Hilbert-Schmidt, K is of the trace class. That shows $\det(1+\lambda K)$ exists and is entire. It remains to show that it has no zeros in λ in a circle of radius strictly greater than one. To prove this it suffices to demonstrate that $S^{-1}_{++}S_{+-}$ has norm strictly less than one. (A slightly tricky point arises here. There is a distinction between traces taken as in the power series for (3.38): $\operatorname{tr} K^n = \sum_{\alpha=1}^{n} \int d\mu_+(p)(K^n)(p,p)_{\alpha\alpha}$ and those taken in $\mathcal{H}$:

$\operatorname{tr} K^n = \int d\mu_+(p)\sum_i \Phi^{(i)}(p)^+(K^n)(p,p)\Phi^{(i)}(p)$ where $\Phi^{(i)}(p)$ are an orthonormal set of solutions of the generalized Dirac equation of momentum p. However, these two are equal and as a result in the following calculation we may work in the Hilbert space $\mathcal{H}$, use the scalar product containing the η, and exploit properties like $S^{-1}_{++}(S^{-1}_{++})^* \geq 0$ where the adjoint is computed in the scalar product of $\mathcal{H}$, which contains η).

To obtain a statement on the norm of $S^{-1}_{++}S_{+-}$ multiply (3.84) from the right by $(S^{-1}_{++})^*$ and the left by S^{-1}_{++} to obtain

$$[S^{-1}_{++}S_{+-}][S^{-1}_{++}S_{+-}]^* = 1 - S^{-1}_{++}[S^{-1}_{++}]^* \tag{3.105}$$

Since the operator on the left-hand side is of trace class, its spectrum is purely discrete and has zero as its only limit point. To show that $S^{-1}_{++}S_{+-}$ has norm strictly less than 1 it suffices to show that the left-hand side does not have the eigenvalue 1. If Φ were such an eigenvector, we would have

$$S_{++}^{-1}[S_{++}^{-1}]^{*}\Phi = 0 \tag{3.106}$$

which violates what has been proved already, that S_{++}^{*} is a bijection of $\mathcal{H}_+$ onto itself. Thus in Proposition 3.1 $\|K\| < 1$ so $\|\Psi_0^{out}\| < \infty$ and we have completed the proof that a unique out vacuum and a unitary S operator exists for Bose-Einstein statistics. This argument for the existence of the vacuum was first given by R. Seiler [6] in the context of a theory of spin zero mesons but, as we have seen, it extends without essential modification to our present circumstances. We summarize the result in the following proposition.

Proposition 3.2 : Let there exist unique weakly retarded and advanced fundamental solutions of (1.1) satisfying the regularity assumptions (2.19) and (3.69)ff. For Bose-Einstein statistics, the sets $\mathcal{R}_+$ and $\mathcal{R}_+^c$ defined in (3.37) and (3.38) are always dense. The out vacuum, Ψ_0^{out}, exists and is uniquely determined up to an overall phase. The pair amplitude is given in terms of S_{++} and S_{+-} by (3.101). It always leads to a kernel in Proposition 3.1 satisfying $\|K\|_{\mathcal{H}_+} < 1$. The S operator exists, is unitary and is uniquely defined up to an overall phase.

For Fermi-Dirac statistics, we have already seen that the CAR imply $\|S_{++}\| \leq 1$ and $\|S_{+-}\| \leq 1$, but the preceding argument for the existence of S_{++}^{-1} fails because of a crucial change in sign in (3.99). What happens here is that by virtue of

$$S_{++}S_{++}^{*} + S_{+-}S_{+-}^{*} = 1 \tag{3.107}$$

if $S_{++}^{*}\Phi = 0$, then Φ is an eigenvector of $S_{+-}S_{+-}^{*}$ of eigenvalue 1. Since by our second regularity assumption $S_{+-}S_{+-}^{*}$ is of trace class we know that the eigenvalue 1 is isolated and of finite multiplicity. If 1 has multiplicity zero then S_{++}^{-1} exists and is bounded because

$$\begin{aligned}\|S_{++}^{-1}\psi\| &= \|[1 - S_{-+}^{*}S_{-+}]^{-1/2}\psi\| \\ &\leq [1-\lambda_{max}]^{-1/2}\|\psi\|\end{aligned} \tag{3.108}$$

Then the formula (3.101) for the pair amplitude is again valid and Ψ_0^{out} exists without further ado.

It does not appear to be possible in general to exclude the occurrence of the eigenvalue 1 for $S_{+-}S_{+-}^{*}$. Such an occurrence in no way affects the existence and uniqueness of the out vacuum, Ψ_0^{out}, and the existence of a unitary S-operator, S, because the theorem of Shale and Stinespring assures us that the Hilbert-Schmidt property for S_{+-} is sufficient for the existence and uniqueness of Ψ_0^{out} and S. However, when $S_{+-}S_{+-}^{*}$ has 1 as eigenvalue the structure of the vacuum is strongly affected. In particular, as we shall see

$$(\Psi_0^{out}, \Psi_0^{in}) = 0 \tag{3.109}$$

Suppose the eigenvectors of $S_{+-}S_{+-}^{*}$ of eigenvalue 1 form a subspace, V. Then V is of finite dimension, k. The subspace V^c consisting of the eigen vectors of the operator of $S_{+-}^{c}(S_{+-}^{c})^{*}$ belonging to the eigenvalue 1 is also of finite dimension, say k^c. In general one might have $k \neq k^c$. Furthermore, the equation (3.28) written out in terms of S_{++} shows

$$\sqrt{n_1+1}\int d\mu_+(p)\int d\mu_+(r)\,\Phi(r)\,S_{++}(r,p)\,\Psi_0^{out(n_1+1,n_2)}(pp_1\ldots p_{n_1},q_1\ldots q_{n_2})$$
$$+\frac{(-1)^{n_1}}{\sqrt{n_2}}\sum_{j=1}^{n_2}(-1)^{j+1}\int d\mu_+(r)\Phi(r)\,S_{+-}(r,-q_j)(\eta^T C)^{-1} \tag{3.110}$$
$$\Psi_0^{out(n_1,n_2-1)}(p_1\ldots p_{n_1},q_1\ldots,\hat{q}_j,\ldots,q_{n_2}) = 0$$

For all Φ of the form $\eta^T\bar{\psi}$ with $\psi \in V$ the first term vanishes, and consequently we are faced with the algebraic problem of finding a skew symmetric tensor whose exterior product with any vector $S_{+-}^{c}\Psi^c$, such that $\Psi \in V$ is zero. The solution of this problem is $\Psi_0^{out(n_1,n_2)} = 0$ for $n_2 < k$, while for $n_2 \geq k$ $\Psi_0^{out(n_1,n_2)}$ is the exterior product of the skew symmetric tensor belonging to the subspace with any skew symmetric tensor associated with complementary subspaces [16].

The analogue of (3.110) arising from (3.30) is

$$\sqrt{n_2+1}(-1)^{n_1}\int d\mu_+(q)\int d\mu_+(r)\,\Phi(r)\,S_{++}^{c}(r,q)\Psi_0^{out(n_1,n_2+1)}(p_1\ldots p_{n_1};qq_1\ldots q_{n_2})$$
$$+\frac{1}{n_1}\sum_{j=1}^{n_1}(-1)^{j+1}\int d\mu_+(r)\,\Phi(r)\,S_{+-}^{c}(r,-p_j)(\eta^T C)^{-1}\Psi_0^{out(n_1-1,n_2)} \tag{3.111}$$
$$(p_1\ldots\hat{p}_j\ldots p_{n_1};\,q_1\ldots q_{n_2})$$

Here we set $\Phi = \eta^{T}\bar{\psi}$ with $\psi \in V^{c}$ and conclude that in its particle arguments $p_1 \dots p_n$, $\psi_o^{out(n_1,n_2)}$ must vanish for $n_1 < k^{c}$ and for $n_1 \geq k^{c}$ be an antisymmetric tensor which is the exterior product of the subspace spanned by $S_{+-}\psi^{c}$ $\psi \in V^{c}$ with an antisymmetric tensor associated with complementary subspaces.

The condition (3.110) with $\Phi = \eta^{T}\bar{\psi}$ and $\psi \in V^{\perp}$, the orthogonal complement of V, determines the dependence of $\Psi_o^{out(k^c+n,k+n)}$ on $(q_1 \dots q_{k+n})$ uniquely up to an overall factor independent of n. Similarly, its dependence on $p_1 \dots p_{k^c+n}$ is uniquely determined from (3.111) with $\Phi = \eta^{T}\bar{\psi}$ and $\psi \in V^{c\perp}$. The result is the formula

$$\Psi_o^{out(n+k^c,n+k)}(p_1 \dots p_{n+k^c}; q_1 \dots q_{n+k}) = \frac{1}{\sqrt{(n+k)!}}\,\frac{1}{\sqrt{(n+k^c)!}}\,\mathcal{A}(p_1 \dots p_{n+k^c})\,\mathcal{A}(q_1 \dots q_{n+k^c}) \prod_{r=1}^{k^c}[S_{+-}\Phi_r^{c}](p_r) \prod_{s=1}^{k}[S^{c}_{+-}\Psi_s^{c}](q_s) \prod_{t=1}^{n}\psi(p_{k^c+t}; q_{k+t}) \tag{3.112}$$

where $\mathcal{A}(p_1 \dots p_{n+k^c})$ is the antisymmetrizer, $\Phi_1 \dots \Phi_{k^c}$ is an orthonormal set in V^c; $\Psi_1, \dots, \Psi_k$ is an orthonormal set in V and ψ is uniquely determined up to an over all constant.

Thus, for Fermi-Dirac statistics the situation can be summarized as follows.

<u>Proposition 3.3</u> : Let there exist unique retarded and advanced fundamental solutions of (1.1) satisfying the regularity assumptions (2.19) and (3.70)ff. For Fermi-Dirac statistics the out vacuum, Ψ_0^{out}, exists and is uniquely determined up to an overall phase. The S operator exists, is unitary and is uniquely determined up to an overall phase.

If the sets $\mathcal{R}_+$ and $\mathcal{R}_+^{c}$ defined by (3.37) and (3.38) are dense, the pair amplitude is expressed in terms of S_{++} and S_{+-} by (3.101) just as for Bose-Einstein statistics.

If the sets $\mathcal{R}_+$ and $\mathcal{R}_+^{c}$ are not dense, their orthogonal complements are finite dimensional linear manifolds V and V^c of dimension k and k^c respectively.

$$\psi_0^{out(n_1,n_2)} = 0$$

for $n_1 < k^c$ or $n_2 < k$ or $n_1 - k^c \neq n_2 - k$. The remainder of the amplitudes are given up to a common factor by (3.112).

Remarks : 1) In the course of the arguments leading to Proposition 3.1, 3.2, and 3.3, we have recovered the theorems of Shale [12] and Shale and Stinespring [11] in our context.

2) The attentive reader of [15] [6] [14] will recognize that their arguments have required only technical modifications to cover the general wave equation.

3) The theory of this section determines the scattering operator and the out vacuum only up to phase factors which can depend on the external field. While such phase factors have no direct physical significance within the scattering theory of the external field problem itself, they affect the Green's functions and current operator and are fundamental in applications to problems of coupled quantized fields. It is an advantage of the technique which uses the retarded and advanced solutions that one can obtain the unitarity of the S operator and the existence and uniqueness of the out vacuum without having to settle the proper definition of these phase factors. However, it only postpones the problem.

4) For a proof that in the Dirac theory of spin $\frac{1}{2}$, the retarded and advanced fundamental solutions satisfy the conditions of this section and consequently that the S matrix is unitary, see the lectures of Seiler.

Appendix B.

Proof of Lemma 5 : The boundedness of the operators A is most easily seen if they are carried over to x-space by Fourier transformation. For this purpose it is convenient to introduce a Hilbert space

$$\mathcal{K} = \mathcal{K}_{+} \oplus \mathcal{K}_{-} \tag{B.1}$$

where

$$\mathcal{K}_{\pm} = L^2(\mathbb{R}^3, d\mu_{\pm}) + \ldots + L^2(\mathbb{R}^3, d\mu_{\pm}) \tag{B.2}$$

with N summands. $\mathcal{K}_{\pm}$ has the scalar product

$$(\Phi,\psi)_{\pm} = \int \overline{\Phi(p)}\, \psi(p)\, d\mu_{\pm}(p) \tag{B.3}$$

as well as the sesquilinear form

$$\langle\Phi,\psi\rangle_{+} = \int \Phi(p)^{+}\psi(p)\, d\mu_{+}(p) \tag{B.4}$$

and

$$\langle\Phi,\psi\rangle_{-} = (-1)^{\sigma} \int \Phi(p)^{+}\psi(p)\, d\mu_{-}(p)$$

respectively. The form

$$\langle\cdot,\cdot\rangle = \langle\cdot,\cdot\rangle_{+} + \langle\cdot,\cdot\rangle_{-} \tag{B.5}$$

is indefinite on $\mathcal{K}$ but on the closed subspace $\mathcal{H}$ consisting of all $\Phi \in \mathcal{K}$ that satisfy

$$\Pi(p)\Phi(p) = \Phi(p) \tag{B.6}$$

for almost all p on the mass shell, it is strictly positive. $\mathcal{H}$ contains Φ that are smooth and of rapid decrease on the mass shell and satisfy (B.6) ; the scalar product $\langle\cdot\,,\cdot\rangle$ coincides there with (3.60). On $\mathcal{H}$, the norms corresponding to the scalar products $\langle\cdot,\cdot\rangle_{\pm}$ and $(\cdot,\cdot)_{\pm}$ are equivalent i.e. there exist strictly positive constants c and d such that

$$c[\Phi,\Phi)]^{1/2} \leq [\langle\Phi,\Phi\rangle]^{1/2} \leq d[(\Phi,\Phi)]^{1/2} \qquad \text{(B.7)}$$

for all $\Phi \in \mathcal{K}$. (The hermitizing matrix, η, is hermitean and strictly positive on the positive solutions of (B.6)

$$\overline{\Phi(p)}\,\eta\,\Phi(p) \geq c\overline{\Phi}(p)\Phi(p)$$

That such a $c > 0$ exists is one the defining properties of a proper wave equation. On the other hand,

$$|\overline{\Phi(p)}\eta\Phi(p)| \leq \|\eta\|\overline{\Phi(p)}\Phi(p).$$

The argument is similar for the negative energy solutions, if (3.62) is brought to bear). Thus the $\mathcal{K}$ we have just defined coincides with the space defined in (3.63).

The passage to functions on space-time is made via the Fourier transform

$$\Phi_{\pm}(x) = (2\pi)^{-3/2}\int \Phi(p)\,d\mu_{\pm}(p)\exp - ip.x \qquad \text{(B.8)}$$

The integrals converge in the mean and define distribution solutions of the differential equation. Now look at $\Phi_{\pm}(x)$ restricted to a space-like hyperplane, described in an appropriate coordinate system by $x^0 = 0$. The integrals define functions with the property

$$\sqrt[4]{1-\Delta}\,\Phi_{\pm}(x) \in L^2(\mathbb{R}^3, d^3x) \qquad \text{(B.9)}$$

(Proof : $[m^2+\vec{p}^2]^{-1/4}\Phi(p) \in L^2(\mathbb{R}^3, d^3p)$ for each m in the mass spectrum but, in fact| it doesn't matter which $m > 0$ is used. Thus, passing to the three-dimensional Fourier transform, we get that

$$\sqrt[4]{m^2-\Delta}\,\Phi_{\pm}(x) = (2\pi)^{-3/2}\int [m^2+\vec{p}^2]^{-1/4}\Phi(p)\exp -ip.x\, d^3p \qquad \text{(B.10)}$$

is square integrable in $\vec{x}$, i.e.

$$\Phi_{\pm}(0,\vec{x}) \in H_{1/2}(\mathbb{R}^3, d^3x)\oplus\ldots\oplus H_{1/2}(\mathbb{R}^3, d^3x) \qquad \text{(B.11)}$$

with N summands. Here $H_{\alpha}(\mathbb{R}^3, d^3x)$ is the Sobolev space consisting of all complex valued functions, $f(\vec{x})$, such that $[1-\Delta]^{\alpha/2}f(\vec{x})$ is square integrable with respect to d^3x.

The condition (B.11) restricts the admissible data in the initial value problem for the wave equation but there are further restrictions implied by (B.6). To display them explicitly it is convenient to write

$$\Phi(x) = \sum_{j=1}^{n} \Phi^{(j)}(x)$$
$$\Phi^{(j)}(x) = \Phi_{+}^{(j)} + \Phi_{-}^{(j)} \qquad (B.12)$$

where $\Phi_{\pm}^{(j)}(x)$ is the contribution to Φ of mass m_j and positive (negative) energy. The differential equation (1.6) provides an initial condition on the $\Phi_{\pm}^{(j)}$

$$(\mp i\beta^{o}[m_j^2 - \Delta]^{1/2} + \vec{\beta}.\nabla + m)\,\Phi_{\pm}^{(j)}(0,\vec{x}) = 0 \qquad (B.13)$$

Furthermore, the scalar product can be written in terms of the $\Phi_{\pm}^{(j)}$.

$$(\Phi,\chi) = \int d^3x\, \Phi_{+}(0,\vec{x})^{+} i\beta^{o}\chi_{+}(0,\vec{x})$$
$$+ (-1)^{\sigma} \int d^3x\, \Phi_{-}(0,\vec{x})^{+} i\beta^{o}\chi_{-}(0,\vec{x}) \qquad (B.14)$$
$$= \sum_{j=1}^{n} \int d^3x[\Phi_{+}^{(j)}(0,\vec{x})^{+} i\beta^{o}\chi_{+}^{(j)}(0,\vec{x})$$
$$+ \Phi_{+}^{(j)}(0,x)^{+} i\beta^{o}\chi_{-}^{(j)}(0,x)]$$
$$= \sum_{j=1}^{n} \frac{1}{m_j} \int d\Omega_{m_j}(p)[\Phi_{+}(p)^{+}\chi_{+}(p)$$
$$+ (-1)^{\sigma}\, \Phi_{-}(p)^{+}\chi_{-}(p)]$$
$$= (\Phi_{+},\chi_{+})_{+} + (-1)^{\sigma}(\Phi_{-},\chi_{-})_{-}$$

It is easy to see that the transformation law

$$(U(a,A)\Phi)(x) = S(A)\Phi(\Lambda(A^{-1})(x-a)) \qquad (B.15)$$

defines a continuous unitary representation of ISL(2,$\mathbb{C}$) in the space of initial data, but this fact will not be needed for the arguments under discussion here, so the proof will be omitted.

The preceding discussion is summarized in the following proposition.

Proposition : The Hilbert space of admissible initial data dor the wave equation (1.6) consists of all Φ of the form

$$\Phi = \sum_{j=1}^{n} \Phi^{(j)} \quad , \quad \Phi^{(j)} = \Phi_{+}^{(j)} + \Phi_{-}^{(j)} \tag{B.12}$$

with $\Phi_{\pm}^{(j)}$ satisfying

$$\text{a)} \quad \Phi_{\pm}^{(j)} \in H_{1/2}(\mathbb{R}^3, d^3x) + \ldots H_{1/2}(\mathbb{R}^3, d^3x) \tag{B.11}$$

$$\text{b)} \quad (\mp i\beta^o \sqrt{m_j^2 - \Delta}\, \Phi_{\pm}^{(j)} + \vec{\beta}.\nabla + m)\Phi_{\pm}^{(j)} = 0 \tag{B.13}$$

The scalar product is

$$\begin{aligned}(\Phi,\chi) &= \sum_{j=1}^{n} [(\Phi_{+}^{(j)}, \chi_{+}^{(j)})_{+} + (-1)^{\sigma}(\Phi_{-}^{(j)}, \chi_{-}^{(j)})-] \\ &= (\Phi_{+}, X_{+})_{+} + (-1)^{\sigma}(\Phi_{-}, X_{-})_{-}\end{aligned} \tag{B.14}$$

with

$$(\Phi_{\pm}^{(j)}, \chi_{\pm}^{(j)}) = \int d^3x\, \Phi_{\pm}^{(j)}(0,\vec{x}) i\beta^o \chi_{\pm}^{(j)}(0,\vec{x})$$

and

$$\Phi(x) = (2\pi)^{-3/2} \int d\mu(p)\Phi(p) \exp - ip.x \tag{B.8}$$

The transformation law (B.15) under ISL(2,$\mathbb{C}$), which includes a description of the temporal evolution of the system is a continuous unitary representation in this Hilbert space.

Now we turn to a description of the operators A_1 in $\mathcal{K}$. For simplicity, only the case A_{1++} will be considered ; the argument runs parallel for A_{1--}. A_{1+-} and A_{1-+} will be treated separately.

Consider the form

$$(\Phi, A_{1++}\psi)_{+} = -\frac{i}{4\pi} \int d\mu_{+}(p)\, \Phi(p)^{+}\Pi(p)B(p-q)\Pi(q)\psi(q)d\mu_{+}(q) \tag{B.16}$$

It is well defined for all infinitely smooth Φ and ψ of rapid decrease on the positive energy mass shell by virtue of the discussion around equations (3.51) to (3.57). If we use (B.6) and introduce the Fourier transform (B.8) we can evaluate it as

$$\frac{-i}{4\pi} \sum_{j,k=1}^{n} \frac{1}{m_j m_k} \iint d\Omega_{m_j}(p)\, \Phi(p)^+ \hat{B}(p-q) \psi(q)\, d\Omega_{m_k}(q)$$

$$= -\frac{i}{2} \int d^4x \sum_{j,k=1}^{n} \Phi_+^{(j)}(x)^+ B(x)\, \psi_+^{(k)}(x) \tag{B.17}$$

This last expression can be estimated as follows

$$|(\Phi, A_{1++}\psi)_+| \leq \frac{1}{2} \int dx^o [\sup_{r,s=1,2,\ldots N} \sup_{\vec{x}} |B(x)_{rs}|]$$

$$[\sum_{j,k=1}^{N} \int d^3x |\Phi_+^{(j)}(x^o,\vec{x})_r^+ \psi_+^{(k)}(x^o,\vec{x})_s|] \tag{B.18}$$

The integral over $\vec{x}$ can be estimated by two applications of Schwarz's inequality as

$$\leq \|\eta\| \int d^3x [\sum_r |\Phi_+^{(j)}(x^o,\vec{x})_r|^2]^{1/2} [\sum_s |\psi_+^{(k)}(x^o,\vec{x})_s|^2]^{1/2}$$

$$\leq \|\eta\| [\int d^3x \sum_r |\Phi_+^{(j)}(x^o,\vec{x})|^2]^{1/2} [\int d^3x \sum_r |\psi_+^{(k)}(x^o,\vec{x})|^2]^{1/2} \tag{B.19}$$

By Parseval's theorem

$$\int d^3x \sum_r |\Phi_+^{(j)}(x^o,\vec{x})_r|^2 = m_j^{-2}(2\pi)^{-3} \int d^3p \sum_r |\frac{\Phi(p)_r}{[m_j^2 + \vec{p}^2]^{1/2}}|^2$$

$$\leq \frac{c_j}{m_j} \int d\Omega_{m_j}(p)\, \Phi(p)^+ \Phi(p) \tag{B.20}$$

for some constant c_j independent of Φ and x^o. Thus

$$|(\Phi, A_{1++}\psi)_+| \leq \|\eta\| \sum_{j,k=1}^{n} \sqrt{c_j c_k}$$

$$[\int dx^o \sup_{r,s=1,\ldots,N} \sup_{x} |B(x^o,\vec{x})_{rs}|]$$

$$[\frac{1}{m_j} \int d\Omega_{m_j}(p)\, \Phi^{(j)}(p)^+ \Phi^{(j)}(p)]^{1/2}$$

$$[\frac{1}{m_k} \int d\Omega_{m_k}(p)\, \psi^{(k)}(p)^+ \psi^{(k)}(p)]^{1/2}$$

$$\leq [\sum_{j=1}^{n} |c_j|^2] \|\eta\| [\int dx^o \sup_{r,s=1,\ldots,N} \sup_{x} |B(x^o,\vec{x})_{rs}|] (\Phi,\Phi)_+^{1/2} (\psi,\psi)_+^{1/2}$$

This inequality shows A_{1++} is uniformly bounded on a dense subset of $\mathcal{H}$ and therefore bounded on $\mathcal{H}$.

There remain A_{1+-} and A_{1-+}. We are now going to prove them Hilbert-Schmidt, which will imply immediately that they are bounded. For A_{1+-} the required estimate is

$$\iint d\mu_+(p)\,d\mu_-(q)\,\mathrm{tr}[\Pi(p)\hat{B}(p-q)\Pi(q)\hat{B}(q-p)] < \infty \tag{B.22}$$

since the adjoint A^*_{1+-} of A_{1+-} is A_{1-+}. (Do not forget the dagger in the definition of the scalar product, nor that $B(-p)^* = \eta B(p)\eta^{-1}$). Since $\hat{B}\in\mathcal{S}$ for every integer $n > 0$ there exist constants C_{ab} such that

$$|\hat{B}(r)_{ab}| \le \frac{C_{ab}}{[1+\sum_{\mu=o}^{3} r_\mu^2]^n} \tag{B.23}$$

Because, in the computation of (B.22), p runs over the positive energy mass shell, while q runs over the negative energy, we have, when $p^2 = m_j^2$ and $q^2 = m_k^2$

$$\sum_{\mu=o}^{3} (p_\mu - q_\mu)^2 \ge ([m_j^2 + \sum_{\ell=1}^{3} p_j^2] + [m_k^2 + \sum_{\ell=1}^{3} q_\ell^2]^{1/2})^2 \tag{B.24}$$

Thus, the estimate (B.23) implies (B.22) and A_{1+-} and hence A_{1-+} is Hilbert-Schmidt, and the proof of Lemma 5 is complete.

It is worth emphasizing that this last estimate, (B.24), does not work if p and q have the same sign of energy. Consequently, A_{1++} and A_{1--} need not be Hilbert-Schmidt.

The Hilbert-Schmidt property of A_{1+-} and A_{1-+} holds for all terms of the perturbation series for A_{2+-} and A_{2-+}. To prove this, it suffices to show that estimates of the form

$$|\hat{B}_1(r_j,(q)| \le \frac{C_n}{[1+\sum_{\mu=o}^{3} (r_j^\mu - q^\mu)^2]^n}\,, \tag{B.25}$$

valid for all positive integer n with some choice of C_n, imply an estimate of the same form for

$$\int d^4_{r_j}\; \hat{B}(r_{j+1} - r_j)[-\not{r}_j + m]_R^{-1}\,\hat{B}_1(r_j,q) \tag{B.26}$$

as a function of r_{j+1} and q. Then the result follows by induction.

To obtain the required estimate one may study the integral

$$I = \int \left[1 + \sum_{\mu=0}^{3} (p^\mu - r^\mu)^2\right]^{-k} \rho(r)\left[-r^2 + m^2\right]_R^{-1}\left[1 + \sum_{\nu=0}^{3} (r^\nu - q^\nu)^2\right]^{-\ell}$$

Now

$$\frac{1}{A^k} = \frac{1}{(k-1)^2} \int_0^\infty e^{-sA} s^{k-1} ds$$

for $A > 0$, so I may be written

$$I = \frac{1}{(k-1)!} \frac{i}{(\ell-1)!} \int_0^\infty \int_0^\infty \int_0^\infty ds_1 ds_2 ds_3 \; s_1^{k-1} s_3^{\ell-1} \int d^4r$$

$$\exp\left[-s_1\left(1 + \sum_{\mu=0}^{3} (p^\mu - r^\mu)^2\right) - i s_2 (-r^2 + m^2)_R - s_3\left(1 + \sum_{\nu=0}^{3} (r^\nu - q^\nu)^2\right)\right]\rho(r)$$

$$= \frac{\pi^2 i \, \rho\left(\frac{1}{i}\frac{\partial}{\partial k}\right)}{(k-1)!(\ell-1)!} \int_0^\infty \int_0^\infty \int_0^\infty ds_1\, ds_2\, ds_3 \frac{s_1^{k-1} \; s_3^{\ell-1}}{\left[s_1 + s_3 - is_2\right]^{1/2}\left[s_1 + s_3 + is_2\right]^{3/2}}$$

$$\exp\Big[- s_1 - s_3 - i s_2 m^2 - s_1 \sum_{\mu=0}^{3} (p^\mu)^2 - s_3 \sum_{\nu=0}^{3} (q^\nu)^2$$

$$+ \frac{(s_1 p^o + s_3 q^o + ik^o/2)^2}{s_1 + s_3 - is_2} + \frac{(s_1 \vec{p} + s_3 \vec{q} + i\frac{\vec{k}}{2})^2}{s_1 + s_3 + is_2}\Big] \Big|_{k=o}$$

If one carries out the differentiations under the integral sign and then introduces the scaling variables $s_1 = tx_1$, $s_2 = tx_2$, $s_3 = tx_3$,

$$ds_1 ds_2 ds_3 = t^2 dt \, dx_1 dx_2 dx_3 \, \delta(1 - x_1 - x_2 - x_3)$$

one can carry out the t integration. The remaining integral is over a compact domain in x_1 x_2 x_3 space and it has a p and q dependence from which an estimate of the form (B.25) follows. A similar argument works for A_{2+-}.

REFERENCES

[1] G. Velo and D. Zwanziger, Propagation and quantization of Rarita-Schwinger waves in an external potential, Phys. Rev. 186 (1969) 1337-1341, Noncausality and other defects of interaction Lagrangians for particles with spin one and higher, Phys. Rev. 188 (1972), 2218-2222.

[2] A. Capri, External Field Problem for Higher Spin Particles, Princeton thesis 1967, unpublished, Electron in a given time-dependent electromagnetic field, Jour. Math. Phys. 10 (1969) 575-580.

[3] A. Wightman, The stability of representations of the Poincaré group pp.291-314 in "Symmetry Principles at High Energy, Fifth Coral Gables Conference ", A. Perlmutter, C. A. Hurst, and B. Kursunoglu, Eds. W. A. Benjamin, New York, 1968.

[4] B. Schroer, R. Seiler, and J. A. Swieca, Problems of stability for quantum fields in external time-dependent potentials, Phys. Rev. D2 (1970) 2927-2937.

[5] P. Minkowski and R. Seiler, Massive vector mesons in external fields, Phys. Rev. D4 (1971) 359-366.

[6] R. Seiler, Quantum theory of particles with spin zero and one-half in external fields, Comm. in Math. Phys. 25 (1972) 127-151.

[7] L. Schwartz, Theorie des Distributions II, Hermann, Paris, 1952.

[8] A. S. Wightman, Partial Differential equations and relativistic quantum field theory, pp.1-52 in "Lecture Series in Differential Equations, Volume II, Part V, Differential Equations of Mathematical Physics" A. K. Aziz, Ed. van Nostrand, 1969.

[9] L. Gårding and J. L. Lions, Functional Analysis, Nuovo Cim. Supp 14 (1959) 9-66.

[10] R.F. Streater and A.S. Wightman "PCT, Spin and Statistics and All that" p.125, W. A. Benjamin, New York, 1964, or
A. S. Wightman and S. Schweber, Configuration space methods in relativistic quantum field theory I, Phys. Rev. 98 (1955) 815-837 especially p.825.

[11] D. Shale and W. F. Stinespring, Spinor representations of infinite orthogonal groups, Jour. Math. and Mech. 14 (1965) 315-322.

[12] D. Shale, Linear symmetries of free boson fields, Trans. Amer. Math. Soc. 103 (1965) 149-167.

[13] A.S. Wightman and L.Gårding, Fields as operator valued distributions in relativistic quantum theory, Arkiv för Fysik 28 (1964) 129-184.

[14] G. Labonté, On the nature of "strong" Bogoliubov transformations for fermions, Comm. in Math. Phys. 36 (1974) 59-72.

[15] J. Schwinger, Theory of quantized fields V, Phys. Rev. 93 (1954) 615-28.

[16] H. Flanders, Differential forms with applications to the physical sciences, Academic Press, New York, 1963.

MATHEMATICS OF INVARIANT WAVE EQUATIONS

Lectures delivered at the 1977 "Ettore Majorana" school of Invariant Wave Equations

Lars Gårding

Lund University
Lund, Sweden

INTRODUCTION

Linear invariant wave equations have the form

$$Pu = A_1\partial_1 u+\dots+A_n\partial_n u+Bu=0,\ \partial_k=\partial/\partial x_k ,$$

where $A_1,\dots,A_n,B$ are square matrices, functions of n real variables $x_1,\dots,x_n$. The invariance is the requirement that when L is a Lorentz transformation, the map $u(x) \to u(Lx)$ preserves the manifold of solutions. This leads to algebraic properties of the coefficients which shall not concern us here. A second requirement is that, as for the wave equation $\partial_1^2u-\partial_2^2u-\dots-\partial_n^2u=0$, disturbances propagate with a finite velocity in all directions, the variable x_1 standing for time. One way of expressing this requirement is that P should have a fundamental solution $E(x,y)$,

$$P(x,\partial_x)E(x,y)=\delta(x-y)$$

such that the distributions

$$x\to E(x,y)$$

vanish outside a proper cone with its vertex at y, e.g

$$x_1-y_1 \geq c((x_2-y_2)^2+\dots+(x_n-y_n)^2)^{1/2}$$

for some $c>0$, at least when x is sufficiently close to y. This is the requirement of hyperbolicity. For equations with constant coefficients, there is a very precise theory of this concept. The essential parts of it will be given in Chapter 1. Chapter 2 treats the Cauchy problem for hyperbolic equations with constant coefficients. The wave equation gets special treatment and there are a few words on a quasi-linear variant often studied in physics. In Chapter 3, the theory is extended to the wave operator with variable coefficients. The chapter ends with a section on the dangers for scattering theory of abandoning hyperbolicity. Chapter 4, added in the interest of general education, is a brief outline of the microlocal theory of the propagation of singularities.

CHAPTER 1 HYPERBOLICITY

To get some perspective on hyperbolic differential operators we start by studying hyperbolic convolution operators. Specializing to constant coefficient differential operators we then arrive at the hyperbolic ones and their fundamental solutions. Here the main properties are given (with some proofs left out) and the chapter ends with a section on the dangers of adding lower order terms. They may destroy the hyperbolicity.

1.1 The convolution algebra associated with a cone. Let $K \subset R^n$ be a convex closed cone with its vertex at the origin and proper in the sense that $K \cap -K$ is the origin. Let $\underline{A}(K)$ be the linear space of distributions f such that supp $f \subset C+K$ where C is a compact set varying with f and let $\underline{M}(K)$ be the linear space of distributions g such that supp $g \cap (C-K)$ is compact for every compact set C. (See the figure below.)

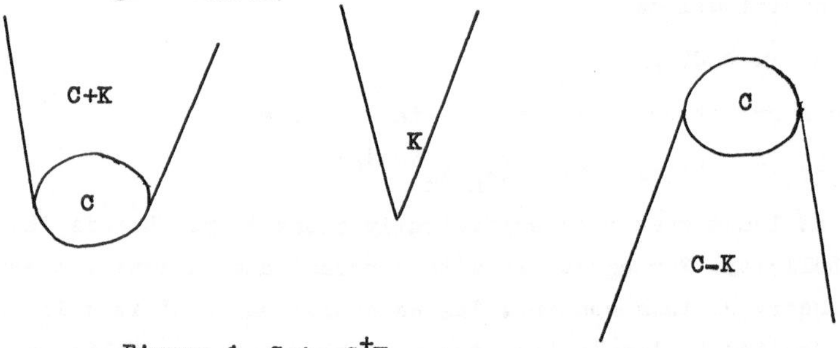

Figure 1. Sets $C \pm K$.

Using that K and -K have only the origin in common, it is a simple exercise to show that $(C+K) \cap (C-K)$ is compact for every compact C and hence that $\underline{A}(K) \subset \underline{M}(K)$. Further, if $f \in \underline{A}(K)$ and $g \in \underline{M}(K)$ and x belongs to some compact set C, the distributions $y \to f(x-y)g(y)$ (suitably interpreted) vanish outside (C - supp f) $\cap$ supp g which is a compact set. Hence the convolution $(f*g)(x) = \int f(x-y)g(y)dy$ is a well-defined distribution. Since its support is contained in suppf + supp g, $f*g$ belongs to $\underline{M}(K)$. A routine verification shows that $\underline{A}(K)$ is an associative and commutative algebra under convolution with unit element $\delta(x)$ and that $\underline{M}(K)$ is an $\underline{A}(K)$-module.

Let $\underline{A}_b(K)$ be the set of f in $\underline{A}(K)$ with supp $f \subset b+K$. Then $A(K) = \underline{A}_0(K)$ is a subalgebra. Let H(f) be the closed convex hull of the support of f and put $H_K(f) = H(f)+K$. According to a theorem by Titchmarsh and Lions (see Gårding |5|) $H_K(f*g) = H_K(f)+H_K(g)$ for all f and g in $\underline{A}(K)$.

1.2 Hyperbolic elements. An f in $\underline{A}(K)$ is said to be hyperbolic if it is invertible, i.e. if $f*g=\delta$ for some g in $\underline{A}(K)$, necessarily unique. Every hyperbolic f has a unique vertex $b=b(f)\in$ supp f with the property that $H_K(f)\subset b+K$. In fact, since $H_K(f)+H_K(g)=K$, there is a $b\in H_K(f)$ such that $-b\in H_K(g)$ and hence $H_K(f)$ is contained in b+K. If supp f does not contain b, i.e. if supp $f\subset b+C+K$ where C is a compact convex subset of K not containing the origin, then $H_K(f)+H_K(g)\subset$ C+K does not contain the origin, a contradiction. Hence b is in supp f. If b and b' are both vertices of f, then $b\in b'+K$ and conversely so that b=b'. Translating f so that b(f)=0, we restrict ourselves in the sequel to the algebra A(K).

Let B(K) be the ideal of A(K) consisting of elements vanishing in neighborhoods of the origin. If $g\in B(K)$ vanishes in such a neighborhood G, then $g*\ldots*g$ (k factors) vanishes in the set $k(G\cap K)$ and hence also in kG which tends to R^n as k tends to infinity. Hence $\delta+g$ is hyperbolic with inverse $\delta-g+g*g-g*g*g+\ldots$. Moreover, since $(f+g)^{-1}=f^{-1}*(\delta+f^{-1}*g)^{-1}$ when f is invertible, $f\in A(K)$ and f+g are invertible at the same time. Writing f=g+h where $g\in A(K)$ has compact support and h is in B(K), this reduces the analysis of hyperbolic elements to those of compact support. If the support is the origin, then $f(x)=P(D)\delta(x)$ where P(D) is a differential operator, i.e. a polynomial in the gradient $D=\partial/i\partial x$. The inverse E of f is then a distrbution E such that $P(D)E(x)=\delta(x)$, i.e.a fundamental solution of P. Such differential operators are said to be hyperbolic with respect to K.

Examples. When n=1, K is the positive or negative half-axis and when $f\in A(K)$, its Fourier-Laplace transform

$$\zeta \to F(\zeta)=\int_K e^{-ix\zeta}f(x)dx,\quad \zeta=\xi+i\eta,$$

is entire analytic. When K is the positive half-axis and, e.g., $1/F(\zeta)$ is bounded when $\eta\leq -t_0$ is large enough negative and γ is any line $\eta=\text{const}<-t_0$, then, by the Paley-Wiener-Schwartz theorem, the distribution

$$g(x) = (2\pi)^{-1}\int_\gamma F(\zeta)^{-1}e^{ix\zeta}d\zeta$$

is independent of γ and vanishes outside K. It is clearly a convolution inverse of f and hence f is hyperbolic. Theorem 1.1 below extends this observation to a criterion for hyperbolicity in many variables which is both necessary and sufficient. - Any differential operator $P(D)\neq 0$ with constant coefficients in one variable is hyperbolic with respect to both half-axes for $F(\zeta)=P(\zeta)$ when $f=P(D)\delta$ and $P(\zeta)^{-1}$ is bounded off the zeros of the polynomial P. The corresponding fundamental solutions $E(x)=g(x)$ can then be computed explicitly as a sum of residues. The standard example of a hyperbolic differential operator in many variables is the wave operator $c^{-2}D_1^2-D_2^2-\ldots-D_n^2$. It has one fundamental solution with support in the cone $x_1\geq 0, c^2x_1^2-x_2^2-\ldots-x_n^2\geq 0$ and another one with support in the opposite cone. We shall see later that there are many examples of hyperbolic operators of all orders.

Matrix-valued convolution operators. We can extend the notion of hyperbolicity by replacing A(K) by the set $A_m(K)$ of square matrices of order m with elements in A(K). Then an f in $A_m(K)$ and its determinant det f in A(K), taken in the sense of convolution, are hyperbolic at the same time. In fact, if $f \in A_m(K)$ and $f*g=\delta I$ where I is the unit matrix of order m, then det f $*$ det g = δ so that det $f \in A(K)$ is invertible. Conversely, let F be the matrix of convolution minors of f. Then $f*F$ = I det f , and if det f has a convolution inverse h, then $f*g=I\delta$ where $g=F*h$ so that f is invertible.

1.3 Hyperbolic elements with compact supports. Let $\Gamma \subset R^n$ be the dual cone of K, i.e. the set of η in R^n such that $x\eta = x_1\eta_1+\ldots+x_n\eta_n > 0$ for all $x\neq 0$ in K. It is open and convex. By the direct part of the Paley-Wiener theorem for distributions, the Fourier-Laplace transform

$$F(\zeta) = \int e^{-ix\zeta} f(x)dx, \; \zeta=\xi+i\eta, \; \xi \text{ and } \eta \text{ real},$$

of an f in A(K) with compact support is an entire function such that, for some numbers C and N,

$$(1) \qquad |F(\zeta)| \leq C(1+|\zeta|)^N e^{h(\eta)}$$

where $h(\eta)$ = max $x\eta$ on supp $f\subset K$ is continuous, homogeneous and ≤ 0 on $-\Gamma$. If supp f does not contain the origin, then, in addition, h is negative on $-\Gamma$. In the theorem below, h_o denotes a function with precisely these properties (see |3|).

Theorem 1.1 An element f of A(K) with compact support is hyperbolic if and only if there is a function h_o and numbers C, N such that

$$(2) \qquad |F(\zeta)^{-1}| \leq C(1+|\zeta|)^N$$

when $\eta = \text{Im}\,\zeta$ is in $-\Gamma$ and ζ is further restricted so that

$$(3) \qquad C(1+|\zeta|)^N e^{h_o(\eta)} < 1 .$$

Proof. Assume that $f*g=\delta$ for some g in A(K) and choose φ in C_o^∞ equal to 1 close to the origin and put $g_1=g\varphi$, $g_2=f*g(1-\varphi)$. Then $f*g_1+g_2=\delta$ where the first term on the left and hence also g_2 has compact support. Letting capital letters denote Fourier-Laplace transforms, this gives $F(\zeta)G_1(\zeta) = 1-G_2(\zeta)$. Estimating G_1 and G_2 according to (1), requiring that $|G_2(\zeta)|<1/2$, noting that the origin is not in the support of g_2 and adjusting the constants gives the desired estimate (2),(3) of F. Conversely, when (2) and (3) hold, we are going to construct the desired inverse g of f by the formula

$$g(x) = (2\pi)^{-n} \int_\gamma F(\zeta)^{-1} e^{ix\zeta} d\zeta_1\ldots d\zeta_n ,$$

i.e.

$$(4) \qquad g(\varphi) = (2\pi)^{-n} \int_\gamma F(\zeta)^{-1} \Phi(-\zeta) d\zeta_1\ldots d\zeta_n$$

where Φ is the Fourier-Laplace transform of φ. The $\gamma = \gamma(\vartheta)$ is an n-chain in the region (3) defined by $\xi \to \zeta = \xi + i\vartheta\log(2+|\xi|)$ where ϑ is in $-\Gamma$. Replacing an arbitrary ϑ in $-\Gamma$ by $t\vartheta$ where $t>0$ and substituting in the left side of (3) gives

$$C(1 + |\xi+it\vartheta\log(2+|\xi|)|)^N(2+|\xi|)^{th_0(\vartheta)}.$$

This expression is majorized by

$$\underline{O}(1)(1+t)^N(2+|\xi|)^{N+th_0(\vartheta)}$$

uniformly in t and locally uniformly in ϑ and hence $\gamma(t\vartheta)$ lies in the region defined by (3) for any ϑ in a compact part of $-\Gamma$ when t is large enough. Moreover, by (2), the integrand of (4) is majorized by

(5) $$\underline{O}(1)\ (1+t)^{N+n}(2+|\xi|)^{N-M-th_1(\vartheta)}|d\xi_1\ldots d\xi_n|$$

on $\gamma(t\vartheta)$. Here $M>N$ is arbitrary and $h_1(\eta)$ equals min $x\eta$ on the support of φ. In fact,

$$\Phi(-\xi-i\eta) = \underline{O}(1)((1+|\xi+i\eta|)^{-M}e^{h_1(\eta)}$$

for any M and on $\gamma(t\vartheta)$ we have $|d\zeta_1\ldots d\zeta_n| = \underline{O}((1+t)^n)|d\xi_1\ldots d\xi_n|$ locally uniformly in ϑ. Hence the right side of (4) with $\gamma = \gamma(t\vartheta)$ converges uniformly for ϑ in compact parts of $-\Gamma$ and t large enough but bounded and under these circumstances it defines a distribution g. Moreover, within the given restrictions, g is independent of the choice of ϑ and t. In fact, the integrand is a closed differential form and on the (n+1)-chain

$$s,\xi \to \xi+i((1-s)t_1\vartheta_1 + st_t\vartheta_2)\log(2+|\xi|)$$

where $0\leq s\leq 1$ and $\vartheta_1, \vartheta_2 \in -\Gamma$, it is majorized by

$$\underline{O}(1)\ (1+|\xi|)^{-m}(|d\xi_1\ldots d\xi_n| + \sum_{j=1}^{n} |dsd\xi_1\ldots d\xi_{j-1}d\xi_{j+1}\ldots d\xi_n|)$$

for any $m>0$ provided t_1 and t_2 are large enough. Hence the desired independence follows from Stokes's formula. Finally, if x_0 is outside K, there is a ϑ in $-\Gamma$ with $x_0\vartheta>0$. Choosing the support of φ close enough to x_0, we then have $h_1(\vartheta)>0$. Letting $t\to\infty$, it follows from (5) that $g(x)=0$ close to x_0. Hence g is in A(K) and this finishes the proof since

$$(g*f)(\varphi) = (2\pi)^{-n}\int_\gamma \Phi(-\zeta)d\zeta_1\ldots d\zeta_n$$

where the integral does not change if γ is replaced by R^n and then it equals $\varphi(0)$.

Propagation cones and hyperbolicity cones. Let us say that a distribution f is hyperbolic if f and one of its convolution inverses f^{-1} are in A(K) for some closed convex proper cone K with its vertex at the origin. A minimal K with this property, denoted by $K(f,f^{-1})$, is called a propagation cone of f. Its open dual, denoted by $\Gamma = \Gamma(f,f^{-1})$ is called the hyperbolicity cone of f. We shall think of K as lying in physical space R^n with Latin coordinates x,y,... and Γ as lying in <u>momentum space</u> R^n with Greek coordinates $\xi,\eta,\ldots$. The scalar product $x\xi=x_1\xi_1+\ldots+x_n\xi_n$ is a pairing of the two spaces. The cone Γ is the set of all ξ in momentum space such that $x\xi>0$ for all $x\neq 0$ in K. By the duality of cones, K then consists of all x in physical space such that $x\xi\geq 0$ for all ξ in Γ.

If a propagation cone $K(f,f^{-1})$ minus the origin is contained in some half-space $x\vartheta>0$, it is uniquely determined by f and ϑ and will be denoted by $K(f,\vartheta)$. In fact, if f has two convolution inverses g and h with supports contained in some convex closed proper cone K' on which $x\vartheta>0$ for $x\neq 0$, we can choose K' so that it also contains the support of f and then $g=g*\delta=g*f*h=\varepsilon*h=h$. By Theorem 1.1, if f has compact support, the hyperbolicity cone $\Gamma(f,\vartheta)$ dual to $K(f,\vartheta)$ is the maximal open convex cone containing ϑ on which the implication (3)=>(2) holds with suitably chosen C,N and h_o.

1.4 <u>Hyperbolic differential operators with constant coefficients</u>. Let P(D), $D=\partial/i\partial x$, be a differential operator with constant coefficients allowed to be square matrices. We say that P is hyperbolic if the corresponding distribution with support at the origin, $P(D)\delta(x)$, has a convolution inverse E with support in some proper convex closed cone K. In other words, P is hyperbolic if it has a fundamental solution E(x),

$$P(D)E(x) = I\delta(x) \text{ ,(I a unit matrix),}$$

vanishing outside such a cone K. Since the Fourier-Laplace transform of $P(D)\delta$ is the characteristic polynomial $P(\zeta)$ of P(D), Theorem 1.1 and the note at the end of section 1.2 permit us to characterize a hyperbolic differential operator with propagation cone K in terms of the behavior of the characteristic determinant $\det P(\zeta)$ far away in $R^n-i\Gamma$, where Γ is the corresponding hyperbolicity cone. But the criterion of Theorem 1.1 can now be simplified as follows:

A differential operator P(D) has a fundamental solution with support in a cone K if and only if there is a locally bounded positive function $t_o(\eta)$ defined in the dual cone Γ such that

$$\text{(6)} \qquad \xi \text{ real, } \eta\in\Gamma,\ t>t_o(\eta)\log(e+|\xi|) \Rightarrow \det P(\xi-it\eta)\neq 0.$$

In order to pass to the proof, note that, putting $\operatorname{Im}\zeta=-t\eta$, (2) says that $-th_o(-\eta)\geq c\log(e+|\xi-it\eta|)$ for some c. Hence (3) and the hypothesis of (6) are actually equivalent. Since the conclusion of (6) is weaker than (2), (6) is certainly a necessary condition. To prove the sufficiency, let $a(\zeta)$ = pr det $P(\zeta)$ be the principal part of $\det P(\zeta)$. We shall see that $a\neq 0$ in Γ. In fact, let $a(\eta)=0$, $\eta\in\Gamma$, and choose a real ξ with $a(\xi)\neq 0$. Then the polynomial $s,t \to \det P(s\xi-it\eta)$ has degree m=deg a in s and degree <m in t. Hence there is a Puiseux series $s_o(t)=ct^b(1+\underline{o}(1))$ with $c\neq 0$, b rational <1 such that

$$\det P(s_o\xi-it\eta) = \det P(\operatorname{Re} s_o\xi - it(\eta - t^{-1}\operatorname{Im} s_o\xi)) = 0$$

for all large real t. But this contradicts (6) since $t/\log(e+|\mathrm{Re}\, s_0\xi|)$ tends to ∞ as $t \to \infty$. Now, since $a(\eta)\neq 0$ everywhere in Γ, we can write $P(\xi-it\eta)$ as

$$P(\xi-it\eta) = a(\eta)\prod_1^m(-it+t_k),\quad t_k = t_k(\eta,\xi),$$

where, according to (6), $\mathrm{Im}\, t_k \leq t_0(\eta)\log(e+|\xi|)$. Hence, if, e.g.,

$$t > (t_0(\eta)+|a(\eta)|^{-1/m})\log(e+|\xi|),$$

all $|-it+t_k|$ are $\geq |a(\eta)|^{-1/m}$ so that $|P(\xi-it\eta)|\geq 1$. Hence the requirements of Theorem 1.1 hold and P(D) has a fundamental solution with support in K.

Using the Seidenberg-Tarsky lemma, one can prove that the logarithmic factor of (6) can be replaced by 1 (see Hörmander |6|). But the ultimate simplification of the criterion is the fact that it can be replaced by the simple requirement that (pr means "principal part of")

(7) $$\mathrm{pr}\det P(\vartheta)\neq 0$$

(8) $$\xi \text{ real},\ t>t_0 \Rightarrow \det P(\xi-it\vartheta)\neq 0$$

for just one ϑ. In order to make this statement more precise, we shall now state a few definitions.

Let $\mathrm{hyp}(\vartheta)$ be the set of polynomials $P(\zeta)$ with the properties (7) and (8) and let $\mathrm{Hyp}(\vartheta)$ be the set of scalar homogeneous polynomials in $\mathrm{hyp}(\vartheta)$ and let $\mathrm{Hyp}(\vartheta,m)$ be the set of polynomials in $\mathrm{Hyp}(\vartheta)$ of degree m.

Note that if $P=a\in\mathrm{Hyp}(\vartheta)$, then $a(\vartheta)\neq 0$ and $a(\xi-it\vartheta)\neq 0$ for all real ξ and all sufficiently large t. Since a is homogeneous, this means simply that

(9) $$a(\xi+i\vartheta)\neq 0$$

for all real ξ. In particular, if the degree of a is m, all the m zeros of the polynomials

$$t \to a(\xi+t\vartheta) = a(\vartheta)\prod_1^m(t+t_k(\xi))$$

are real for all real ξ. Hence the polynomial $a(\xi)/a(\vartheta)$ has real coefficients.

When $P\in\mathrm{hyp}(\vartheta)$, let $\Gamma=\Gamma(P,\vartheta)$, the common hyperbolicity cone of P and $a = \mathrm{pr}\det P$, be the component of the complement of the real conical hypersurface $A: a(\xi)=0$ that contains ϑ. Finally, let $K = K(P,\vartheta)$, the common propagation cone of P and a, be the cone in physical space dual to Γ, i.e. the set of $x\in R^n$ such that $x\xi\geq 0$ for all ξ in Γ. (These cones have been defined differently earlier, but just wait for an explanation).

We can now list the basic properties of hyperbolic differential operators with constant coefficients.

__Theorem__ 1.2 When $P\in\mathrm{hyp}(\vartheta)$, its hyperbolicity cone $\Gamma=\Gamma(P,\vartheta)$ is open and convex and P and $a = \mathrm{pr}\det P$ are in $\mathrm{hyp}(\eta)$ for all η in $\pm\Gamma$ and $\Gamma(P,-\vartheta) = -\Gamma(P,\vartheta)$. There is a positive function $h(\eta)$ from Γ of homogeneity 1 such that the functions

$$\xi \to P(\xi\pm i\eta)^{-1}$$

are uniformly bounded when $h(\eta)>1$. The integral

(10) $$E(x) = (2\pi)^{-n} \int e^{ix\zeta} P(\zeta)^{-1} d\zeta$$

where $\zeta = \xi - i\eta$, $h(\eta)>1$, is independent of the choice of η, it is a fundamental solution of $P(D)$ and the convex hull of its support is the propagation cone $K=K(P,\vartheta)$.

Note. The hyperbolicity cone Γ and the propagation cone K are the algebraically defined ones above. Since any fundamental solution with support in K must be $E(x)$ and K is the convex hull of the support of E, $K=K(P,\vartheta)$ is actually the propagation cone as defined at the end of section 1. Hence, by the duality of cones, our two definitions of the hyperbolicity cone are equivalent.

Proof. The first two sentences have algebraic proofs (see Hörmander |6| or Gårding |4|), too long to be given here. That $E(x)$ is a fundamental solution whose support is contained in K is proved as the corresponding part of Theorem 1.1 . If the convex hull of the support of E were less than K, then $\det P(\zeta)$ would satisfy the hyperbolicity criterion (6) in a cone bigger than $\Gamma = \Gamma(P,\vartheta)$ which is impossible since $a=0$ on the boundary of Γ. This proves the last part of the theorem.

The reader is now advised to forget everything given so far except Theorem 1.2 and the definitions preceding it. Next, we shall give some badly needed examples of hyperbolic polynomials and the corresponding cones and fundamental solutions. To begin with, we restrict ourselves to $Hyp(\vartheta)$, the scalar homogeneous case.

1) $a=const\neq 0$, $\Gamma = R^n$, $K=0$, $E=1/a$.
2) $a=(b\xi)^m$, a positive power of a real linear form $b\xi$ such that $b\vartheta\neq 0$, Γ the half-space $(b\xi)b\vartheta>0$, K the closed half-ray $x=rb$, $r\geq 0$. When $b\xi=\xi_1$ and $b\vartheta>0$ and $a=i^m\xi_1{}^m$, then $E(x)=\Theta(x_1)x_1^{m-1}\delta(x_2)\ldots\delta(x_n)/m!$ (Θ is the Heaviside function).
3) $a = a(\xi) = \sum a^{jk}\xi_j\xi_k$, a quadratic form with Lorentz signature $+-\ldots-$, Γ the open cone $a(\xi)>0$ containing ϑ, K the dual cone $a^*(x) = \sum a_{jk}x_jx_k \geq 0$ containing an y with $y\vartheta>0$, the matrices (a_{jk}) and (a^{jk}) being inverses of each other. When $n=4$, then $E(x) = (4\pi)^{-1}\delta(a^*(x))\Theta(x\vartheta)(\det(a_{jk}))^{1/2}$. When $a(\xi) = \xi_1{}^2-\xi_2{}^2-\xi_3{}^2-\xi_4{}^2$, E is the classical fundamental solution of the wave operator.
4) When $n=2$, every $a\in Hyp(\vartheta)$ is a product of real linear factors $b(\xi)=b_1\xi_1+b_2\xi_2$ such that $b(\vartheta)\neq 0$ and conversely.

Further examples of arbitrarily high order can be constructed at will since, by (9), factors and products of polynomials in $Hyp(\vartheta)$ are in $Hyp(\vartheta)$. Note that $\Gamma(ab,\vartheta) = \Gamma(a,\vartheta)\cap\Gamma(b,\vartheta)$ so that $K(ab,\vartheta)$ is the convex hull of $K(a,\vartheta)\cup K(b,\vartheta)$. Since the poynomials $t\to da(t\vartheta+\xi)/dt$ have only real zeros when ξ is real and $a\in Hyp(\vartheta)$, differentiation in the ϑ direction maps $Hyp(\vartheta,m)$ into $Hyp(\vartheta,m-1)$.

<u>Localization and the wave front surface.</u> When $a(\xi)$ is a homogeneous polynomial and $\eta \in R^n$, let $a_\eta(\xi)$, the localization of a at η, be the lowest homogeneous term $\neq 0$ in the Taylor expansion of a at η, $a(\xi+\eta)=a_\eta(\xi)+$ higher terms. Examples: when $a(\eta)\neq 0$, then $a_\eta(\xi)=a(\eta)$ is constant, when $a(\eta)=0$ but grad $a(\eta)\neq 0$, then $a_\eta(\xi)=\xi \operatorname{grad} a(\eta)$ is linear, when the hypersurface $A: a(\xi)=0$ is regular at η, then a_η is a constant times a power of such a linear form. At singular points of A, the localization $\xi \to a_\eta(\xi)$ may be a complicated polynomial. In any case, it can be shown that if $a \in \operatorname{Hyp}(\vartheta)$, then $a_\eta \in \operatorname{Hyp}(\vartheta)$ for all η and that $\Gamma(a_\eta,\vartheta)\supset\Gamma(a,\vartheta)$ so that $K(a_\eta,\vartheta)\subset K(a,\vartheta)$. The union $W=W(a,\vartheta)$ of all the local propagation cones $K_\eta=K(a_\eta,\vartheta)$ with $\eta\neq 0$ turns out to be a semi-analytic set of codimension >0 of the propagation cone $K=K(a,\vartheta)$. It is called the wave front surface because the fundamental solution $E(P,\vartheta,x)$ of any $P\in \operatorname{hyp}(\vartheta)$ with a = pr det P is real analytic outside W. In all but exceptional cases its singular support is actually equal to W. Projective figures of wave front surfaces are given in Figure 2 below. Physically, W represents the wave fronts issuing from a δ shock at the origin in an elastic medium where wave propagation is governed by the operator P. When P=a is the wave operator, W is just the boundary of the propagation cone K. In the general case, there will be also slow waves inside K. -Proofs of all these statements can be found in [1].

<u>Strong hyperbolicity.</u> When $a \in \operatorname{Hyp}(\vartheta)$ has the property that

$$\xi \text{ real} \neq 0 \Rightarrow \operatorname{grad} a(\xi)\neq 0, \tag{11}$$

we say that a is strongly hyperbolic and write $a \in \operatorname{Hyp}^o(\vartheta)$. In this case, the real conical hypersurface $A: a(\xi)=0$ is non-singular outside the origin. Hence, if $\eta\neq 0$, every localization $a_\eta(\xi)$ is constant or linear and every local propagation cone is a half-ray. Figure 2 has projective pictures of Γ, S, K, W when n=3 for some a in $\operatorname{Hyp}(\vartheta)$.

If we factor

$$a(\xi+t\vartheta) = a(\vartheta)\prod_1^m (t+t_k(\xi)), \tag{12}$$

(11) means precisely that at most one of the t_k vanishes for any given $\xi\neq 0$. Hence, in view of the homogeneity (take $|\xi|=1$), (11) is equivalent to the inequality

$$a(\xi+it\vartheta)^{-1} = \underline{O}(\ |t|^{-1}(|t|+|\xi|)^{1-m}) \tag{13}$$

for real t and ξ when t is large.

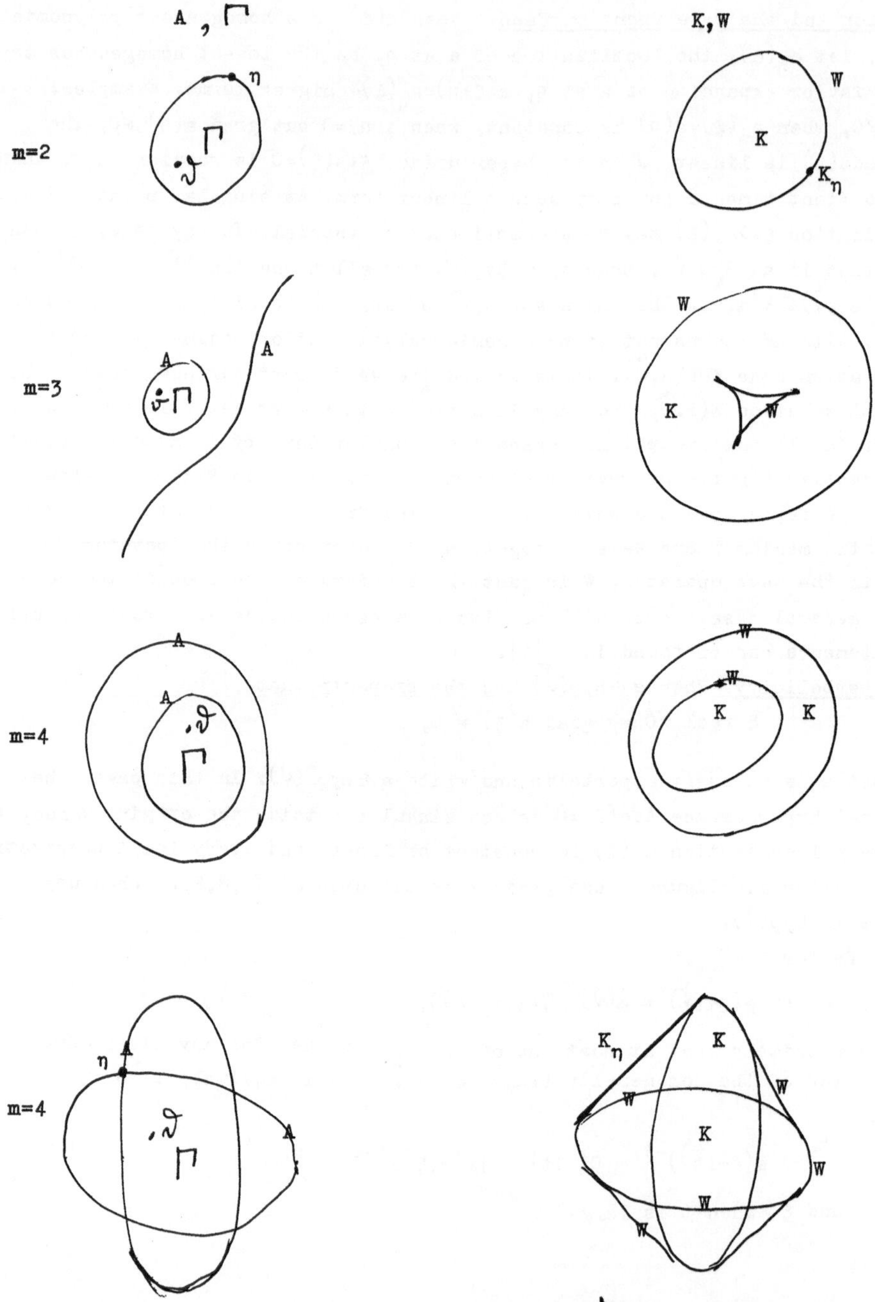

Figure 2. In the first three cases, $a \in \mathrm{Hyp}^{\circ}(\vartheta)$, in the last case, η is a singular point of A and K_η the corrsponding local propagation cone. The figures are projective.

<u>Systems</u>. The invariant wave equations of physics correspond to first order linear operators

$$P = A_1D_1+\dots+A_nD_n + B = A(D)+B$$

whose coefficients are N by N matrices. When $A_1,\dots,A_n$ are hermitian and $A(\vartheta)>0$ or $A(\vartheta)<0$ for some ϑ, then $P\in \mathrm{hyp}(\vartheta)$ for then pr det $P(\vartheta) = \det A(\vartheta) \neq 0$ and

$$A(\xi+it\vartheta)^{-1} = (A(\xi)+itA(\vartheta))^{-1} = \underline{O}(|t|^{-1})$$

and hence

$$P(\xi+it\vartheta) = A(\xi+it\vartheta)(I+A(\xi+it\vartheta)^{-1}B)$$

is invertible when ξ and t are real and t is large. The same conclusion holds when a = pr det P is strongly hyperbolic of degree N. In fact, then $a(\vartheta)\neq 0$ and a satisfies (13) so that $A(\xi+it\vartheta)^{-1} = \underline{O}(|t|^{-1})$. Unfortunately, these two cases cover very few invariant wave equations. For most of them $P\in \mathrm{hyp}(\vartheta)$ but $A(\vartheta)$ is not invertible.

1.5 <u>The dangers of lower order terms</u>. Let $a \in \mathrm{Hyp}(\vartheta,m)$ and suppose that deg b<m. When is P=a+b in $\mathrm{hyp}(\vartheta)$? Since $a(\vartheta)\neq 0$, we only have to verify that $P(\xi+it\vartheta)\neq 0$ for all real ξ when t is real and large enough. Writing $P=a(1+a^{-1}b)$, it suffices for this that

(14) $$a(\xi+it\vartheta)^{-1}b(\xi+it\vartheta) = \underline{O}(|t|^{-1}),$$

uniformly for all real ξ. When a is strongly hyperbolic, (13) shows that (14) holds for all b of degree <m, but if a is only in $\mathrm{Hyp}(\vartheta,m)$, (14) is a condition on b. According to Svensson |12|, **this** condition is also necessary for P to be in $\mathrm{hyp}(\vartheta)$. (Another condition, both necessary and sufficient, is simply that $a(\xi+i\vartheta)^{-1}b(\xi+i\vartheta)$ be a bounded function of ξ.)

According to (14), all b of degree <m such that a+b is in $\mathrm{hyp}(\vartheta)$ form a linear space, but this space gets smaller as a degenerates in the sense that the polynomials $t\to a(\xi+t\vartheta)$ acquire multiple zeros.

<u>Example</u>. Let $a\in \mathrm{Hyp}(\vartheta,m)$ with $\vartheta=(1,0)$ be a polynomial in two variables t,s. Then

$$a(t,s) = a(\vartheta)\prod_1^m(t+c_ks)$$

with real c_k and any P=a+b with principal part a has a factorization

$$P(t,s) = a(\vartheta)\prod_1^m (t+c_ks+h_k(s))$$

where, for large s, the $h_k(s)$ are Puiseux series of the form

$$\sum_{m-1}^{-\infty} b_ks^{k/m}.$$

It is easy to see that $P\in \mathrm{hyp}(\vartheta)$, i.e. $P(t,s)\neq 0$ for all real s and all sufficiently large Im t if, and only if, all $h_k(s)$ are bounded, i.e. there are now positive powers in the series above. It follows, e.g., that if a(t,s) has a has a factor L^j where $L=t+cs$, then $b=\underline{O}(s^{m-j})$ when $L=0$. More precisely,

(15) $$L^j \text{ divides } A \Rightarrow b= \underline{O}(1+|L(t,s)|^{j-1}(|t|+|s|)^{m-j}) .$$

On the other hand, in a conical neighborhood of the ray $L(\mathrm{Re}\, t,s)=0$ and for complex t,

$$a(t,s)^{-1} = \underline{O}(|L(t,s)|^{-j}(|t|+|s|)^{j-m})$$

and hence $b/a \to 0$ in this neighborhood as $\mathrm{Im}\, t \to \infty$. Since, outside of such neigborhoods, $a^{-1} = \underline{O}((|t|+|s|)^{-m})$, (15) is both necessary and sufficient for P to be in $\mathrm{hyp}(\vartheta)$.

Note. If

$$a = a(\vartheta)\, L_1^{m_1} \ldots L_p^{m_p}$$

with different linear factors $L_1,\ldots,L_p$, it is easy to see that (15) is equivalent to the following condition (Anneli Lax |10|)

$$\prod_1^p L_k^{(m_k-j)_+} \quad \text{divides } P_{m-j}$$

where $P=P_m+\ldots+P_o$ is a decomposition of P into homogeneous parts.

Example. Let $a(\zeta) = \zeta_1^2-\zeta_2^2-\ldots-\zeta_n^2 \in \mathrm{Hyp}(\vartheta,2)$, $(\vartheta=(1,0,\ldots,0))$ be the wave polynomial and let b have degree <4. Then $P=a^2+b \in \mathrm{hyp}(\vartheta)$ if and only if $b=La+c$ where L is linear and c has degree <3. In fact, consider the polynomial

$$t,s \to P(\zeta) = a(\zeta)^2 + b_3(\zeta) + c(\zeta)$$

where $\zeta = t\vartheta+s\xi$ and b_3 is homogeneous of degree 3. According to (15), $b_3(t\vartheta+s\xi)=0$ when ξ is real, t and s are arbitrary complex and $a(t\vartheta+s\eta)=0$. Since a is irreducible and the real hypersurface $a(\xi)=0$ has codimension 1, this proves that $a(\zeta)$ divides $b(\zeta)$.

Systems. Consider $P=A_1D_1+\ldots+A_nD_n+B=A(D)+B$ where the coefficients are N by N matrices and suppose that $P\in \mathrm{hyp}(\vartheta)$ so that pr det $P(\vartheta)\neq 0$ and $P(t\vartheta+\xi)$ is invertible for all real ξ when $\mathrm{Im}\, t$ is large enough. When $A(\vartheta)$ is non-singular, the situation is comparatively simple and entirely analogous to the scalar case. In fact, then pr det P = det A so that A is in $\mathrm{hyp}(\vartheta)$. Hence P and A have the same propagation cones, $K(P,\vartheta)=K(A,\vartheta)$. Further, according to Svensson |12| , $P=A+B$ with $A\in \mathrm{hyp}(\vartheta)$ and B constant is in $\mathrm{hyp}(\vartheta)$ if and only if the spectral radius of the matrix $A(t\vartheta+\xi)^{-1}B$ tends to zero as $\mathrm{Im}\, t \to \infty$, uniformly in ξ. (Note that this condition is also sufficient).

When $P\in \mathrm{hyp}(\vartheta)$ but $A(\vartheta)$ is singular, the situation is much more delicate and it is of regular occurrence that addition of a constant matrix changes the propagation cone or destroys the hyperbolicity. One example is the following one. The ordinary wave equation $(D_1^2-D_2^2-\ldots-D_n^2)u=0$ can be written as the system

$$D_k u_o = u_k \;,\; -D_1u_1+D_2u_2+\ldots+D_nu_n = 0 \;,\; (u_o=u,\; k=1,\ldots,n)$$

or, in matrix form,

$$(1.16)\qquad A(D)U = \begin{pmatrix} 0 & -D_1 & D_2 & \dots & D_n \\ D_1 & 1 & 0 & \dots & 0 \\ & & & & \\ D_n & 0 & & \dots & 1 \end{pmatrix} \begin{pmatrix} u_o \\ \\ \\ u_n \end{pmatrix} = 0 .$$

If $B=(b_{jk})$ with $b_{jk}=0$ when j or k=0, the operator

$$P(D)=A(D)+B$$

is in general not hyperbolic. In fact, if, e.g., n=2, then

$$\det P(D) = (1+b_{11})D_1{}^2 -(1+b_{22})D_2{}^2 +(b_{21}-b_{12})D_1D_2$$

is essentially any polynomial of degree 2. When B is real, it is hyperbolic with respect to some ϑ only when it is not elliptic. And when it is hyperbolic, its propagation cone varies with B.

Note. Our definition of a hyperbolic operator P (the existence of a distribution E with support in a cone such that $PE=\delta$) depends, of course, on what we mean by a distribution. We have taken the ordinary kind but if we allow generalized distributions, continuous functionals from certain Gevray spaces, the homogeneous hyperbolic operators remain the same but the space of admissible lower order terms depends on the choice of Gevrey space (see Larsson |9|) . In the case of hyperfuctions (see|11|), all lower order terms are permitted.

CHAPTER 2 THE CAUCHY PROBLEM FOR CONSTANT COEFFICIENTS

We shall make a few general remarks about the Cauchy problem for hyperbolic differential operators with constant coefficients and then make a closer study in case of the wave equation. The chapter finishes with some words on the quasi-linear equation $\Delta u+F'(u)=0$.

2.1 **Cauchy's problem for C^∞ functions.** Let $P\in hyp(\vartheta,m)$ be a scalar operator and let $E=E(P,\vartheta,x)$ be the corresponding fundamental solution given by (1.10). We shall see that if $v\in C^\infty(x\vartheta\geq 0)$, then

$$u(x) = E*v(x) = \int_{x\vartheta>0} E(x-y)v(y)dy$$

belongs to the same class. It is also the unique solution of Cauchy's problem with vanishing Cauchy data on $x\vartheta=0$,

$$(1)\qquad Pu=v \text{ when } x\vartheta>0\ ,\ u = \underline{O}((x\vartheta)^m).$$

To begin with, since $E(x-y)=0$ except when $y\in x-K$ where K is the propagation cone of P, the values of v far away do not influence $E*v$ on a given compact set. Hence it suffices to consider the case when v has compact support. We can also choose our variables so that $x\vartheta=x_1$. In terms of the Laplace transform

$$V(\zeta) = \int_{x_1>0} e^{-ix\zeta}v(x)dx\ ,$$

we can write $u(x)$ as

$$u(x) = (2\pi)^{-n}\int e^{ix\zeta}P(\zeta)^{-1}V(\zeta)d\zeta$$

where $\zeta=\xi-it\vartheta$ with ξ real and t large positive. When

$$(2)\qquad v(x) = x_1^{\,N} f(x)\ ,\ f\in C_0^\infty(x_1\geq 0),$$

then by integrations by parts $V = \underline{O}(|\zeta|^{-N-1})$ so that u is of class C^N when $x_1\geq 0$. When

$$(3)\qquad v(x) = x_1^{\,k}g(x')\ ,\ x' =(0,x_2,\dots,x_n),$$

then

$$(4)\qquad u(x) = (2\pi)^{-n}\int G(\zeta')H_k(x_1,\zeta')e^{ix'\zeta'}d\zeta'$$

where G is the Fourier transform of g and

$$H_k(x_1,\zeta') = \int_\gamma e^{-x_1\zeta_1}\zeta_1^{\,-k-1}\ P(\zeta)^{-1}d\zeta_1$$

with γ a big loop around all the zeros of $\zeta_1 \to P(\zeta)$. Since these zeros have bounded imaginary parts, $u\in C^\infty(x_1\geq 0)$. Now, by Taylor's formula, v is a sum of terms of the type (3) and one term of the type (2) with arbitrarily large N. Hence $u\in C^\infty(x_1\geq 0)$. To compute the derivatives of u, note that when v has the form (2), arbitrarily high derivatives of u vanish when $x_1=0$ and N is sufficiently large and that, when v has the form (3), a derivative $D_1^{\,j}u(x)$ has the form

(4) with k replaced by k-j and that $H_{k-j}(0,\zeta')=0$ when $j-k-m<0$, i.e. $j<k+m$. Hence, in all cases, $D_1{}^j u=0$ when $x_1=0$ and $j<m$. Hence (1) follows. To see that the solution is unique, note that it u solves the homogeneous equation, then (by the equation itself) $u=\underline{O}(x_1{}^N)$ for all N so that u, **extended by 0 for $x_1<0$**, is a C^∞ solution in all of space. If $E_- = E(P,-\vartheta,x)$ is the fundamental solution of P with support in -K, we then have $u=E_- * Pu=0$. This completes the proof.

It follows from (1) that the complete Cauchy problem ,

$$Pu=v \text{ when } x\vartheta>0,\ u-w = \underline{O}((x\vartheta)^m)$$

with v in $C^\infty(x\vartheta\geq 0)$ and w, representing the Cauchy data on the hyperplane $x\vartheta=0$, in the same space, has the unique solution

$$u = E*v + w - E*Pw \in C^\infty(x\vartheta\geq 0).$$

<u>Systems</u>. Let

$$P(D)=A(D)+B = A_1D_1+\ldots+A_nD_n + B$$

be an operator in hyp(ϑ) whose coefficients are N by N matrices . In particular,

$$a = \text{pr det } P \in \text{Hyp}(\vartheta,m)$$

for some m. Then $A(\vartheta)$ has rank m. In fact, this rank is the degree of the polynomials

$$t \to \det P(t\vartheta+\xi) = \det(tA(\vartheta)+P(\xi)).$$

Further, if Q is the matrix of minors of P, then

$$A(\vartheta)Q(\xi+it\vartheta) = \underline{O}(\ |t|^{m-1})$$

and hence also

(5) $$A(\vartheta)(P(\xi+it\vartheta)^{-1} = \underline{O}(|t|^{-1})$$

for fixed real ξ when t is real and large. In fact, this is certainly true when

$$A(\vartheta) = \begin{pmatrix} I & 0 \\ 0 & 0 \end{pmatrix}$$

and I is the unit matrix of order m. Since the hyperbolicity condition, the form of P and the inequality (5) are invariant under multiplication of P on the right and on the left by constant invertible matrices, (5) holds as stated. Precisely as in the preceding section one now proves

<u>Theorem</u> 2.1 Let $P(D)=A(D)+B\in \text{hyp}(\vartheta)$ and suppose that $V,W\in C^\infty(x\vartheta\geq 0)$. Then the Cauchy problem

$$P(D)U=V \text{ when } x\vartheta>0,\ A(\vartheta)U-A(\vartheta)W= \underline{O}(x\vartheta)$$

has the unique solution

$$U = E*V + W - E*PW$$

where $E=E(P,\vartheta,x)$ is the fundamental solution of P with respect to ϑ.

<u>Example</u>. When $P=A(D)$ is given by (1.16) and $\vartheta=(1,0,\ldots,0)$, then $A(\vartheta)U$ has only two components $\neq 0$, $u_0=u$ and $u_1=D_1u$, regular Cauchy data for the wave operator.

2.2 Cauchy's problem for distributions. Let $P \in \mathrm{hyp}(\vartheta)$ be an N by N matrix and let a superscript t denote transposition. Then, obviously,

$$P^t(D) = P(-D)^t$$

belongs to $\mathrm{hyp}(-\vartheta)$ and the distribution

$$E^t(x) = E(-x)^t$$

where $E=E(P,\vartheta,x)$, has its support in $-K=-K(P,\vartheta)$ and the property that

$$E^t * P^t F(x) = F(x)$$

for every C^∞ function F with N components vanishing for large $x\vartheta$, and

$$E^t * G(x) = \int E^t(z)G(x-z)dz$$

depends only on the values of G in the cone x+K (see Figure 3)

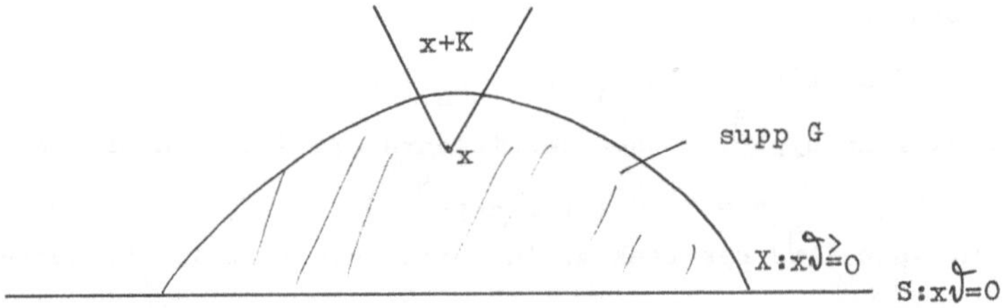

Figure 3. The half-space $X:x\vartheta \geq 0$ and the support of $G \in \mathcal{E}(X)$.

It follows that, e.g.,

$$P^t : \mathcal{E}(X) \to \mathcal{E}(X)$$

is a linear homeomorphism, where X is the closed half-space $x\vartheta \geq 0$ and $\mathcal{E}(X)= C_0^\infty(X)$ (with N components), equipped with the Schwartz topology. Hence, if we let $\mathcal{E}'(X)$ denote the linear space dual to $\mathcal{E}(X)$,

$$P : \mathcal{E}'(X) \to \mathcal{E}'(X)$$

is also a linear homeomorphism. Here P stands for the adjoint of P^t, operating on smooth functions F in $\mathcal{E}'(X)$ by putting $PF(x) = P(F(x)\Theta(x\vartheta))$ and applying the rules of differentiation. To explain this further, write the value $\langle F,G\rangle$ of a distribution $F\in\mathcal{E}'(X)$ at $G\in\mathcal{E}(X)$ as

$$\langle F,G\rangle = \int_X F(x).G(x)dx$$

where $F(x).G(x)=\sum F_k(x)G_k(x)$, $F_1,\dots,F_N$ and $G_1,\dots,G_N$ being the components of F and G. When F is a smooth function, then by integrations by parts,

$$\langle F,P^tG\rangle = \int_X PF(x).G(x)dx + \int_S B(F,G)dS$$

where $S=\partial X$ is the hyperplane $x\vartheta=0$ and $B(F,G)$ is a bilinear form in the derivatives of F and G. The right side is, by definition, the value of PF at $G\in\mathcal{E}(X)$ when $F\in\mathcal{E}'(X)$ is a smooth function and P is taken in the distributional sense in the closed half-space X. When F vanishes close to $S=\partial X$, the boundary term venishes. In any case, it follows that if $V\in\mathcal{E}'(X)$ vanishes close to S, then PU=V has a unique solution $U\in\mathcal{E}'(X)$ such that supp $U\subset$ supp V + K. The proper substitute for Cauchy's problem is the following one: given a distribution on X,

$$G \to \int_X V(x).G(x)dx + \int_S B(W,G)dS$$

where, close to S, V and W are functions in the time variable $t=x\vartheta$ whose values are distributions on S, V is continuous and W sufficiently differentiable, then the equation

$$PU=V \text{ when } t>0,\ B(U,.)=B(W,.)$$

has a unique solution U, sufficiently differentiable in t for B to be defined. In the scalar case, the condition that B(U,.)=B(W,.) is equivalent to $U-W=\underline{O}((x\vartheta)^m)$ where m=deg P.

2.3 The energy inequality and Cauchy's problem for the wave operator. Let

$$P = \Delta = \partial_1{}^2-\partial_2{}^2-\dots-\partial_n{}^2$$

be the wave operator. The Cauchy problem

$$Pu=v \text{ when } x_1>0,\ u-w = \underline{O}(x_1{}^2)$$

is naturally connected with certain classes of functions which we now describe. A block $V=V_T$ will be the cut-off light cone

$$0\leq x_1 \leq (T-(x_2{}^2+\dots+x_n{}^2)^{1/2}),\ (T\geq 1),$$

or a translate of it parallel to the plane $x_1=0$. Let S_t, V_t and M_t be, respectively, the intersection of V with the hyperplane $x_1=t$ and the intersections of V and ∂V with the band $0<x_1\leq t$ (see Figure 4)

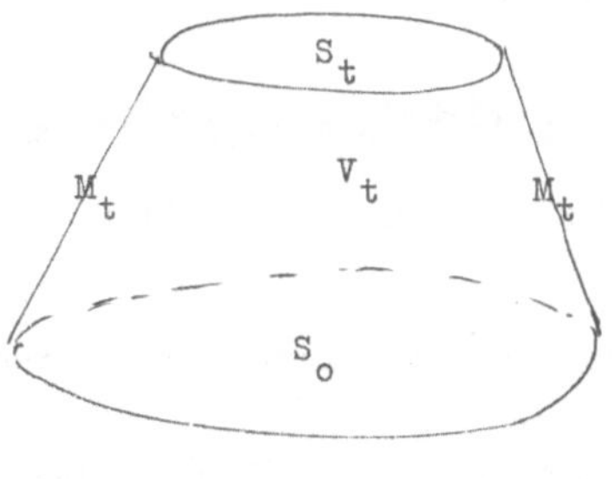

Figure 4. A block with solid part V_t, bottom S_o, top S_t, mantle M_t .

When w is a smooth function, put

$$|D^{j,k}w,S_t| = \left(\int\sum|\partial^{\alpha}w(x)|^2 dS_t\right)^{1/2}$$

where $\alpha_1 \leq j$ and $|\alpha| \leq j+k$. Also,

$$|D^{j,k}w,V_t|_p = \left(\int_0^t |D^{j,k}w,S_\tau|^p d\tau\right)^{1/p} \tag{6}$$

where $1 \leq p \leq \infty$.

<u>Definition</u>. Let $\mathscr{C}^{j,k}$ and $\mathscr{L}^{j,k}$ be the completions of $C^\infty(x_1 \geq 0)$ with respect to the seminorms

$$u \to |D^{j,k}u,V_t|_\infty$$

and the seminorms

$$u \to |D^{j,k}u,V_t|_1$$

respectively. Let $\mathscr{L}^{j,k}_\infty$, equipped with the seminorms (6), be the weak closure of $\mathscr{C}^{j,k}$ under the restriction that these seminorms be bounded. Finally, let $\mathscr{B}^{j,k}$ be the completion of $C^\infty(x_1 \geq 0)$ with respect to the seminorms

$$u \to |D^{j,k}u,S_0| .$$

<u>Note</u>. The space $\mathscr{B}^{j,k}$ is represented by j functions of n-1 variables ,

$$x_2,\dots,x_n \to \partial_1^r u(0,x_2,\dots,x_n) ,$$

each with locally square integrable derivatives of order $\leq j+k-r$. The spaces $\mathscr{B}^{j,k}$, $\mathscr{L}^{j,k}$, $\mathscr{L}^{j,k}_\infty$ can be thought of as functions $x_1 \to u(x)$ with values in $\mathscr{B}^{j,k}$ which are, respectively, continuous, locally integrable and locally essentially bounded. - Similar spaces can of course be defined for $x_1 \leq 0$ or in all of space.

The connection between the spaces above and the wave equation depends on the classical energy inequality.

Lemma 2.1 (Energy inequality). Let V be a block, let $j>0$ and $k \geq 0$ be integers and let $u \in C^\infty(x_1 \geq 0)$. Then

$$|D^{j,k}u,V_t|_\infty \leq ce^t(|D^{1,j+k-1}u,S_0|+|D^{j-2,k}u,S_0|+|D^{j-1,k}u,V_t|_1) \tag{7}$$

where $c=c_{j,k}$ only depends on j and k and the middle term on the right occurs only when $j>1$.

Proof. Since Δ is real, it suffices to prove (7) for real u. We start from the energy identity

$$\tfrac{1}{2}\partial_1(u_1^2+\dots+u_n^2+u^2) - \partial_2(u_1u_2)-\dots-\partial_n(u_1u_n)=u_1(\Delta u + u) \tag{8}$$

where $u_k = \partial_k u$. Integrating over V_t and noting that the boundary term on the mantle M_t, viz. (ν is the outer normal)

$$\tfrac{1}{2}\nu_1(u_1^2+\dots+u_n^2+u^2) - \nu_2u_1u_2-\dots-\nu_nu_1u_n$$

is ≥ 0 when $\nu_1>0$, $\nu_1^2-\dots-\nu_n^2=0$, we get

$$|D^1u,S_t|^2 \leq |D^1u,S_o|^2 + |D^1u,V_t|_2^2 + 2|D^1u,V_t|_\infty\,|\Delta u,V_t|_1$$

where $D^k=D^{k,o}$. Putting $f(t) = |D^1u,V_t|_\infty$ and $g(t) = |\Delta u,V_t|_1$ and noting that $2f(t)g(t) \leq 2^{-1}f(t)^2 + 2g(t)^2$, this gives

$$f(t)^2 \leq 2f(0)^2 + 2\int_0^t f(\tau)^2 d\tau + 4g(t)^2.$$

Hence, by Gronwall's lemma, i.e. the implication that

$$h(t) = A(t) + B\int_0^t h(s)ds \Rightarrow h(t) \leq A(t)(1+Be^{Bt})$$

where $t\geq 0$, $0\leq A(t)\uparrow$ and $B\geq 0$, we get

$$f(t)^2 \leq (2f(0)^2+4g(t)^2)(1+2e^{2t})$$

so that (7) follows when $j=1$, $k=0$. Applying it to all derivatives $\partial^\alpha u$ with $|\alpha|<j+k$, $\alpha_1<j$, adding and using the equivalence

$$|D^{j,k}u,V_t|_p \sim \sum |D^1\partial^\alpha u,V_t|_p$$

we get

$$|D^{j,k}u,V_t| \quad \leq ce^t\,(|D^{j,k}u,S_o| + |D^{j-1,k}\Delta u,V_t|_1)$$

and a simple argument shows that

$$|D^{j,k}u,S_o| \leq |D^{1,j+k-1}u,S_o| + c|D^{j-2,k}\Delta u,S_o|$$

when $j>1$.

We can now prove

Theorem 2.2 When $j>0$, the Cauchy problem

(9) $\qquad \Delta u = v\in\mathcal{L}^{j-1,k}$, $u=w\in\mathcal{B}^{1,j+k-1}$ when $x_1=0$,

has a unique solution $u\in\mathcal{C}^{j,k}$. It satisfies the inequality of the lemma.

Proof. Let (v_m) and (w_m) be sequences of C^∞ functions approximating v and w so that, as m tends to infinity,

(10) $\qquad |D^{j-1,k}(v-v_m),V_t|_1 \to 0$ and $|D^{1,j+k-1}(w-w_m),S_o| \to 0$

for all blocks and let (u_m) be the sequence of the corresponding solutions of Cauchy's problem with data v_m and w_m. The existence of these solutions is proved in 2.1 . Letting $m\to\infty$ and noting that (10) implies that

$$|D^{j-2,k}(v-v_m),V_t|_\infty \to 0,$$

by virtue of the lemma, we get a solution $u\in\mathcal{C}^{j,k}$ satisfying (9) at least in the weak sense, i.e.

$$\int_{x_1>0} u(x)\Delta f(x)dx + \int_{x_1=0} (\partial_1 w(x)f(x) - w(x)\partial_1 f(x))dx_2\ldots dx_n = \\ = \int_{x_1\geq 0} v(x)f(x)dx$$

for all f in C_o $(x_1\geq 0)$. To prove uniqueness note that if $u\in\mathcal{C}^1$ satisfies this equation with v=w=0, then

$$\int_{x_1>0} u(x)\Delta f(x)=0$$

for all f in $C_0^\infty(x_1 \geq 0)$. Since Δ is surjective from this space to itself, u=0. This uniqueness proof works also e.g. for locally integrable solutions of (9) when v,w are locally integrable.

Note. The energy inequality is easy to adopt to hyperbolic second degree operators with variable coefficients. It can also be used for an existence proof of Cauchy's problem. All this will be done in the next chapter.

2.4 Cauchy's problem for the equation $\Delta u+F'(u)=0$. Quasilinear wave equations of the type $\Delta u+P(u)=0$ with P a polynomial and u real have been guinea pigs of quantum field theory. We shall say a few words about them based upon the analogy with the ordinary differential equation

(11) $$u''+F'(u)=0\ ,\quad u=u(t),$$

with Cauchy data at t=0. When F' is a C^1 function, the usual method of successive approximations show that there are unique solutions. Since $u'^2+2F(u)$ is constant for every solution, a simple argument shows that they extend to all of R when $u^{-2}F(u)$ is bounded from below for large u. When F' is any continuous function with the same property, there are global solutions obtained by approximating F' by smoother functions and picking convergent subsequences among the corresponding solutions. But there might be many solutions. E.g., u=0 and $u=t^4$ are both solutions of (11) when $F'=12|u|^{1/2}$ and have the same Cauchy data at t=0.

The general case, Δu replacing u", has similar features but there the difficulties arise when F(u) is large positive for large u. In the first place, successive approximations starting with F(u)=0 and Cauchy data in $\mathcal{B}^1$ show that the equation $\Delta u+F'(u)=0$ has a unique solution u in $\mathcal{E}^1$ provided

$$|F'(u),S_t| \leq c_0(1+|D^1u,S_t|) \text{ and } |F'(u)-F'(v),S_t| \leq c_1|D^1(u-v),S_t|$$

for some c_0, c_1, functions u and v in $\mathcal{E}^1$ and all blocks of fixxd size. In particular this holds when $u \to F'(u)$ is bounded and smooth. When $F(u) \geq -\text{const}(1+u^2)$ is smooth but may be large positive for large u, approximating F by bounded functions, we can still construct solutions in $\mathcal{L}^1_\infty$ by a compactness argument. The solution turns out to be unique when $F \in C^2$ and $F'(u) = \underline{O}(|u|^{p-1})$ where $p=(n-1)/(n-3)$ when n>3 and $p<\infty$ when n=3. But there are no examples of non-uniqueness when this condition fails and there are still solutions. For the details of this including a better result when n=4, see the lectures by Walter Strauss.

CHAPTER 3 CAUCHY'S PROBLEM FOR SECOND ORDER HYPERBOLIC OPERATORS WITH VARIABLE COEFFICIENTS

Using the energy inequality for wave operators with variable coefficients we shall get a rather complete theory of Cauchy's problem for such operators. The chapter ends with a section on the dangers for scattering theory of abandoning hyperbolicity.

3.1 The energy inequality. Let

$$P = \sum a^{jk}(x)\partial_j\partial_k + \sum a_j(x)\partial_j + a_o ,$$

defined in an open part of R^n, have C^∞ coefficients and a real principal part such that the corresponding quadratic form has Lorentz signature $+-\ldots-$. We shall write the coordinates $x=(x^1,\ldots,x^n)$ with upper indices and put $\partial_j=\partial/\partial x^j$. To every x, consider the cotangent plane T^*_x spanned by differential forms $\xi=\sum\xi_j dx^j$ with the inner product

$$(\xi,\eta) = \sum a^{jk}(x)\xi_j\xi_k$$

and a dual tangent plane T_x spanned by differential operators $Q=\sum b^j\partial_j$ with the inner product

$$(Q,Q)=\sum a_{jk}(x)b^j b^k .$$

Here the matrices (a_{jk}) and (a^{jk}) are inverses of each other.

When the principal part of a constant coefficient differential operator P_o is in $\mathrm{Hyp}^o(\vartheta,m)$, i.e. has order m and is strongly hyperbolic with respect to ϑ, write $P_o \in \mathrm{hyp}^o(\vartheta,m)$.

Let $\vartheta(x) = \sum \vartheta_j(x)dx^j$ be a continuous differential form such that $(\vartheta,\vartheta)>0$ for all x. Then $P_x \in \mathrm{hyp}^o(\vartheta_x,2)$ for all x, where and index x means that the coefficients of P and ϑ are frozen at the point x. Let $\Gamma_x = \Gamma(P_x,\vartheta_x)$: $(\xi,\xi)>0$, $(\xi,\vartheta_x)>0$ be the corresponding hyperbolicity cone in the cotangent plane at x and $K_x = K(P_x,\vartheta_x)$: $\langle Q,\xi\rangle=\sum b^j\xi_j \geqq 0$ for all ξ in Γ_x, be the associated propagation cone in the tangent plane at x. A hypersurface $s(x)=0$ where s is real and smooth is said to be space-like at x if $ds(x) \in \pm\Gamma_x$.

The energy identity (2.7) transformed to variable coefficients reads

$$(1)\qquad QuPv + QvPu = \sum \partial^k T_k(Q,u,v) + R(u,v)$$

where

$$Q = \sum b^k\partial_k$$

has real coefficients, R is a bilinear form in the derivatives of order <2 of u and v,

$$\partial^k = a^{kj}\partial_j , \quad b_k = \sum a_{kj}b^j$$

and

$$T_k(Q,u,v) = Qu\partial_k v + \partial_k uQv - b_k(\partial u,\partial v) .$$

It suffices to prove this when Q and $P = \sum a^{jk}\partial_j\partial_k$ have constant coefficients and in this case the verification is immediate. Integrating (1) with $v=\bar{u}$ over a region in R^n, we get an integral over its boundary S,

$$\int T_\nu(Q,u,\bar{u})\, dS$$

where

$$T_\nu(Q,u,\bar{u}) = \sum \nu_k a^{kj} T_j(Q,u,\bar{u}) = 2 \operatorname{Re} Qu\overline{Nu} - (Q,N)(\partial u,\partial\bar{u})$$

with ν the exterior normal of S and $N=\sum a^{kj}\nu_k\partial_j$ the corresponding derivative. Using a simple fact of Lorentz geometry, namely that

$$C \to 2(A,C)(B,C)-(A,B)(C,C)$$

is a positive definite quadratic form when A and B are both time-like and in the same cone, it follows that

$$(2) \qquad u \to T_\nu(Q,u,\bar{u})$$

is a positive definite hermitian form on the space-like part of S provided Q_x is in the interior of K_x and ν in Γ_x for all x in question.

We shall now consider P in a band X: $0 \leq x_1 < T$ and assume that all P_x form a compact part of $hyp^o(\vartheta,2)$ with $\vartheta=(1,0,\ldots,0)$ and that all the derivatives of all the coefficients of P are bounded. Let hyp(X,2) be this class.

Let V be a block, the intersection of X and a backward cone

$$((y_2-x_2)^2+\ldots+(y_n-x_n)^2)^{1/2} \leq c_o(y_1-x_1)$$

with its vertex y outside X so that $y_1>T$. The number $c_o>0$ is taken so small that all parts S_t, M_t of the boundary of $V_t = V \cap (0 \leq x_1 \leq t)$ are uniformy space-like with respect to P, i.e. the forms (2) with $Q=\partial/\partial x^1$ are uniformly positive on them. Then, from Green's formula follows an energy inquality for C^2 functions u, namely

$$(3) \qquad |D^1u,S_t|^2 \leq c|D^1u,S_o|^2 + c|D^1u,V_t|^2 + c|Pu,V_t|^2$$

with c>0 independent of t and the block. We shall rewrite it using a weighted norm,

$$|v,V_t|^2_\tau = \int_{V_t} |v(x)|^2 e^{-2\tau x^1} dx$$

where $\tau \geq 1/2$ and analogously for $|D^1v,V_t|_\tau$. We then have

Lemma 3.1 If u is a C^2 function, then

$$\tau|D^1u,V_t|^2_\tau \leq c|D^1u,S_o|^2 + c|D^1u,V_t|^2_\tau + c|Pu,V_t|^2_\tau$$

with c>0 independent if t and the block V.

Proof. Multiplying (3) with t replaced by s by $e^{-2\tau s}$ and integrating with respect to s from 0 to t gives the desired result. The details are left to the reader.

From the lemma follows in a trivial way

Uniqueness theorem Suppose that, close to a block V_t,

$$u \in C^2, \quad Pu(x) = \underline{O}(|Du(x)|+|u(x)|) \text{ and } u(x)=\underline{O}(x_1{}^2).$$

Then u=0 in the block.

Note. We have now reverted to lower indices for the coordinates.

Proof. The lemma shows that

$$\tau|D^1u,V_t|_\tau^2 \leq c|D^1u,V_t|_\tau^2$$

for some number c. Taking $\tau > c$, the theorem follows.

Emission. To distinguish between the past and the future we shall now introduce oriented half-open bands $X = I \times R^{n-1}$ and their duals $X' = I' \times R^{n-1}$ where I and I' are intervals of the x_1-axis and $I': t_1 < x_1 \leq t_2$ when $I: t_1 \leq x_1 < t_2$ and conversely. Put $\varepsilon(X)=+1$ in the first case and $\varepsilon(X)=-1$ in the second case and put $\vartheta(X)=(\varepsilon(X),0,\ldots,0)$. When $P \in \mathrm{hyp}(X, \vartheta(X))$, consider parts Y of X of the form $s(x) \geq 0$ when $s \in C^2$ is real and $ds(x) \in \Gamma(P_x, \vartheta(X))$ for all x. The intersection of all such Y containing a given subset Z of X will be called the emission of Z in the $\vartheta(X)$ direction, $Em(Z)=Em(P,\vartheta(X),Z)$. The uniqueness theorem can also be stated as

(4) $\qquad u \in C^2 \Rightarrow \operatorname{supp} u \subset Em(P,\vartheta(X),Z)$

where

$$Z = (\operatorname{supp} D^1u \cap \partial X) \cup \operatorname{supp} Pu .$$

The emission of a point Z=y, denoted by $K(P,\vartheta,y)$, will be called the propagation cone of P from y in the ϑ direction. Its tangent at y is the propagation cone $K_y = K(P_y,\vartheta)$ in the tangent plane at y. (Figure 5).

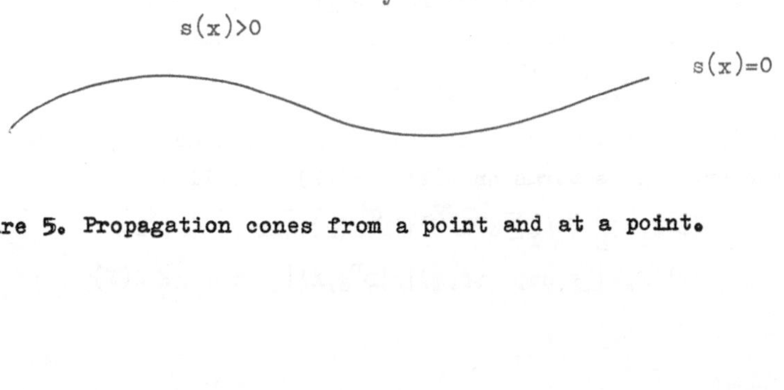

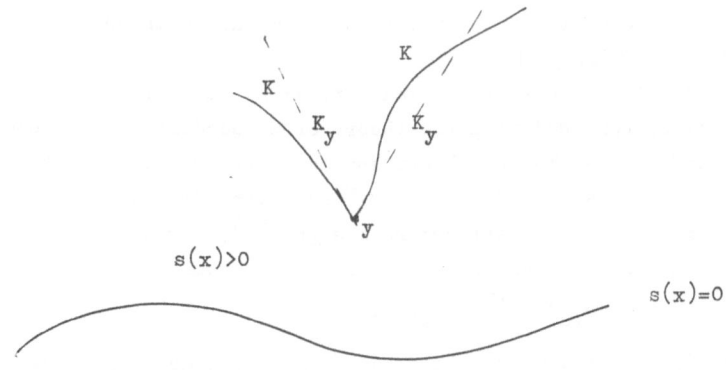

Figure 5. Propagation cones from a point and at a point.

3.2 <u>Inverses of hyperbolic operators on a band</u>. Let X be a half-open band and X' its dual. Let $\mathcal{E}(X)$ be the space of C^∞ functions from X with compact supports equipped with the Schwartz topology. By definition, a distribution on X is a continuous anti-linear function from $\mathcal{E}(X)$. A locally integrable function f(x) from X gives rise to the distribution

$$g \to (f,g) = \int_X f(x)\overline{g(x)}dx$$

and we shall use this notation, including the integral also for general $f \in \mathcal{E}'(X)$, the space of distributions on X. The derivative $\partial^\alpha f(x) = \partial(X)^\alpha f(x)$ is defined by the formula

$$(\partial^\alpha f,g) = (-1)^{|\alpha|}(f,\partial^\alpha g).$$

Away from ∂X, $\partial^\alpha f$ is the ordinary distribution derivative but, if $f \in C^1$, an integration by parts gives

$$(\partial_1(X)f,g) = -\int_X \partial_1 f(x)\overline{g(x)}dx + \int_{\partial X} f(x)\overline{g(x)}dx_2 \dots dx_n$$

with ∂X suitable oriented. Hence, if $\alpha_1 > 0$, the derivative $\partial(X)^\alpha f$ of a smooth function differs from the classical derivative $\partial^\alpha f$ by a term at the boundary. Note that if $f \in \mathcal{E}(X')$, then f=0 close to ∂X and not such terms appear. The pair $\mathcal{E}(X')$, $\mathcal{E}'(X)$ corresponds to the pair $\mathcal{D}, \mathcal{D}'$ of ordinary distribution theory in open subsets of R^n.

We can now formulate our principal result.

<u>Big Theorem</u>. Let X be a half-open band and let $P \in hyp(X,2)$. Then

$$P\colon \mathcal{E}(X') \to \mathcal{E}(X') \text{ and } \mathcal{E}'(X) \to \mathcal{E}'(X)$$

are linear homeomorphisms such that, in both cases,

$$\text{supp } u \subset Em(P, \partial(X), \text{supp } Pu).$$

Note. This result extends to strongly hyperbolic operators of higher order, $P \in hyp(X,m)$, and to systems. The proofs are similar to those given below. They are given in detail in |13|.

The proof combines the energy inequality with some simple functional analysis including a piece borrowed from the theory of pseudodifferential operators. Its main ingredient is a scale of Hilbert spaces to be described next.

Let S(X') be the space of complex C^∞ functions from X' vanishing close to ∂X whose derivatives of all orders are $\underline{O}(|x|^{-k})$ for all k as $x \to \infty$. Let S'(X) be the dual of S(X) via an extension of the duality

$$(f,g) = \int_X f(x)\overline{g(x)}dx$$

on $S(X') \times S(X)$. In particular, S(X') is a dense subspace of S'(X). With $\tau > 0$ a real parameter, define norms on S(X') S'(X) as follows

(5) $$|D^r f,X|_\tau^2 = \int_X \sum e^{-2\varepsilon\tau x_1}|D^\alpha f(x)|^2 dx\ , \quad |\alpha| \leq r,$$

(6) $$|D^{-r}f,X|_\tau = \sup |(f,g)|/|D^r g,X'| \quad \text{for } g \in S(X)$$

where, as above, $\varepsilon=\varepsilon(X)$ is $1(-1)$ when $T=T(X)$ is closed from below (above). For r of arbitrary sign, let $H^r_\tau(X)$ be the closure of $S(X')$ with respect to the norm $|D^r f,X|_\tau$. Then, by the definition above, $H^{-r}_\tau(X)$ is the dual of $H^r_\tau(X')$ when $r\geq 0$. But the latter space is effectively a Hilbert space (its inner product is obtained from (5) in the natural way) and hence reflexive. This shows that, for all r,

$$(7) \qquad H^{-r}_\tau(X') \text{ is the dual of } H^r_\tau(X)$$

and that (6) holds for r of arbitrary sign. Of course, $H^r_\tau(X)\subset S'(X)$ for all r.

It follows from (5) and (6) that the functions $r\to|D^r f,X|_\tau$ are not decreasing. Hence the spaces $H^r_\tau(X)$ decrease as r increases,

$$\ldots H^{-2}_\tau(X) \subset H^{-1}_\tau(X)\subset H^0_\tau(X)\subset H^1_\tau(X) \subset H^2_\tau(X)\ldots \;.$$

By (5), $|D^r D_j f,X|_\tau \leq |D^{r+1}f,X|_\tau$ for all derivatives D_j when $r\geq 0$ and hence, by (6)

$$|D^{-r}D_j f,X|_\tau = \sup|(f,D_j g)|/|D^r g,X'|_\tau \leq$$
$$\leq \sup|(f,D_j g)|/|D^{r-1}D_j g,X'|_\tau \leq |D^{1-r}f,X|_\tau$$

when $r<0$. Hence D_j induces continuous maps from $H^r_\tau(X)$ to $H^{r-1}_\tau(X)$ for all r and this proves that all maps

$$(8) \qquad D^\alpha : H^r_\tau(X) \to H^{r-|\alpha|}_\tau(X)$$

are continuous.

Next, consider the pseudodifferential operator

$$\Lambda_0 = (1+D_2^2+\ldots+D_n^2)^{1/2} = \mathcal{F}^{-1}(1+\xi_2^2+\ldots+\xi_n^2)^{1/2}\mathcal{F}$$

where $\mathcal{F}$ is the Fourier transform with respect to $x_2,\ldots,x_n$. We shall prove

<u>Lemma</u> 3.2 The operator

$$\Lambda = \Lambda(X) = \varepsilon(X)\partial_1 + \Lambda_0$$

induces linear isomorphisms

$$(9) \qquad S(X') \to S(X'), \quad S'(X)\to S'(X), \quad H^r_\tau(X) \to H^{r-1}_\tau(X)$$

for all r and if $b\in C^\infty(\overline{X})$ has bounded derivatives of all orders, then

$$(10) \qquad b\Lambda^s - \Lambda^s b : H^r(X) \to H^{r-s+1}(X)$$

is continuous for all integral r and s.

Proof. Consider the operator

$$\Lambda_{\mathcal{F}} = \varepsilon\partial_1 + \mathcal{F}\Lambda_0\mathcal{F}^{-1} = \varepsilon\partial_1 + (1+\xi_2^2+\ldots+\xi_n^2)^{1/2}$$

and assume that $T(X)$ is an interval $t_1\leq x_1<t_2$, making $\varepsilon=\varepsilon(X)=1$. Then $\Lambda_{\mathcal{F}}$ has the inverse map

$$f \to \int_{t_1}^{x_1} e^{-t(1+\xi_2^2+\ldots+\xi_n^2)^{1/2}} f(t,\xi_2,\ldots,\xi_n)\,dt$$

which is obviously linear and continuous from S(X') to S(X'). Since $\mathcal{F}$ is a linear isomorphism of S(X'), this proves that $\Lambda_{\mathcal{F}}$ and $\Lambda = \mathcal{F}^{-1}\Lambda_{\mathcal{F}}\mathcal{F}$ are linear isomorphisms of the same space. In particular, $\Lambda(X')$ is an isomorphism of S(X) and hence its adjoint is an isomofphism of S'(X).

To proceed, consider, for f in S(X'), the norm square

$$|\Lambda f,X|_\tau^2 = \int_X e^{-2\tau x_1}(|\partial_1 f(x)|^2+|\Lambda_0 f(x)|^2+\partial_1(f(x)\Lambda_0\overline{f(x)}+\Lambda_0 f(x)\overline{f(x)}))dx=$$
$$=|D^1 f,X|_\tau^2 + 2\int_{\partial X'} e^{-2\tau t} 2\mathrm{Re}\,\Lambda_0 f(x)\overline{f(x)}dx_2\ldots dx_n + 2\tau\int_X e^{-2\tau x_1}\mathrm{Re}\,\Lambda_0 f(x)\overline{f(x)}dx.$$

Since, putting $\xi'=(\xi_2,\ldots,\xi_n)$,

$$\int \Lambda_0 f(x)\overline{f(x)}\,dx_2\ldots dx_n = (2\pi)^{1-n}\int (1+|\xi'|^2)^{1/2}|\mathcal{F}f(x_1,\xi')|^2 d\xi' \geq 0$$

for all x_1 in T(X), this proves that $|\Lambda f,X|_\tau \geq |D^1 f,X|_\tau$. But, by the triangle inequality,

$$2^{-1}|\Lambda f,X|_\tau^2 \leq |D_1 f,X|_\tau^2 + |\Lambda_0 f,X|_\tau^2 = |D^1 f,X|_\tau^2 .$$

Hence $f \to |\Lambda f,X|_\tau$ and $f \to |D^1 f,X|_\tau$ are equivalent norms on S(X'). An iteration using the fact that Λ is an isomorphism then proves that

$$f \to |D^r\Lambda f,X| \quad\text{and}\quad f \to |D^{r+1}f,X|$$

are also equivalent norms when $r \geq 0$. Hence the maps

$$\Lambda(X)\colon H_\tau^r(X) \to H_\tau^{r-1}(X)$$

and, by duality, the maps

$$\Lambda(X')\colon H_\tau^{1-r}(X') \to H_\tau^{-r}(X')$$

are isomorphisms when $r \geq 0$. Changing X to X', this proves the last part of (9).

By a standard result of the theory of pseudodifferential operators, there are constants c such that, putting $[\Lambda_0,a] = \Lambda_0 a - a\Lambda_0$,

$$|D^r[\Lambda_0,a]\,h,R^{n-1}| \leq c|D^r h,R^{n-1}|$$

when $h \in S(R^{n-1})$, $r \geq 0$ and all the derivatives of a are bounded. Here

$$|D^r h,R^{n-1}|^2 = \int \sum |D^\alpha h(x)|^2 dx_2\ldots dx_n,\quad |\alpha| \leq r.$$

Hence, obviously, the commutator $[\Lambda,b] = (\partial b(x)/\partial x_1) + [\Lambda_0,b]$ induces continuous maps $H^r(X) \to H^r(X)$ for all $r \geq 0$. Hence its adjoint $\bar{b}\Lambda(X') - \Lambda(X')\bar{b}$ induces continuous maps $H_\tau^r(X') \to H_\tau^r(X')$ for all $r \leq 0$ but then $[\Lambda,b]$ itself induces continuous maps $H_\tau^r(X) \to H_\tau^r(X)$ for all values of r. Hence (10) holds when s=1. Since

$$b\Lambda^{s+1} - \Lambda^{s+1}b = (b\Lambda^s - \Lambda^s b)\Lambda + \Lambda^s(b\Lambda - \Lambda b)$$

it then holds by induction for all $s \geq 0$ and also for all $s<0$. This proves the lemma.

Dropping the auxiliary paramoter τ and putting $H^r(X) = H_0^r(X)$, we have the following basic result

<u>Theorem</u> 3.1 Under the hypotheses of the Big Theorem, $PS(X')$ is dense in every $H^s(X)$ and there are constants c depending on s, P and X such that

$$(11) \qquad |D^{1+s}f,X| \leq c|D^sPf,X|$$

for all f in $S'(X)$.

Proof. When s=0, (11) follows from Lemma 3.1. More precisely, we have

$$\tau |D^1f,X|^2_\tau \leq c|D^1f,X|^2_\tau + c|Pf,X|^2_\tau .$$

Replacing f by $\Lambda^s f$ and introducing the commutator $[P,\Lambda^s] = P\Lambda^s - \Lambda^s P$, we get from this that

$$\tau |D^1\Lambda^s f,X|^2_\tau \leq c|\Lambda^s Pf,X|^2_\tau + |D^1\Lambda^s f,X|^2_\tau + c|[P,\Lambda^s]f,X|^2_\tau$$

for all s and $f \in S(X')$. By Lemma 3.2, the last term is majorized by

$$c|D^{1+s}f,X|^2_\tau .$$

Making τ large enough, this proves (11) for all s. Since $P: H^{s+2}(X) \to H^s(X)$ is continuous, (11) holds when $f \in H^{s+2}(X)$. In particular, since s is arbitrary,

$$f \in \text{some } H^s(X), Pf = 0 \Rightarrow f = 0$$

so that, passing to the adjoint P^* and the dual band X',

$$g \in H^{-s}(X'), \quad P^*g = 0 \Rightarrow g = 0.$$

Since $P^*g=0$ means that $(PS(X'),g)=0$, it follows that $PS(X')$ is dense in $H^s(X)$ and this for all s. Hence, by continuity, (11) holds when $Pf \in H^s(X)$. Since the inequality is trivially true when the right side is infinite, this finishes the proof.

<u>Proof of the Big Theorem.</u> Combining the uniqueness theorem with the **previous** one shows that $P: \mathcal{E}(X') \to \mathcal{E}(X')$ and hence also its adjoint $P^*: \mathcal{E}'(X) \to \mathcal{E}'(X)$ are linear isomorphisms. The statement about emission then follows from (4).

<u>Note</u>. The Big Theorem contains the uniqueness theorem (and generalizes it). In fact, suppose that $u \in C^2$ and that $Pu=0$ in a region $Y \subset X$ and that $D^1u=0$ at $\partial Y \cap \partial X$. Then, if $f \in \mathcal{E}(X)$ vanishes outside Y an integration by parts shows that

$$(Pu,f)=(u,P^*f)= 0,$$

i.e. $Pu=0$ in Y in the distributional sense on X. Hence, by the Big Theorem, $u=0$ outside $Em(Pu)$, in particular $u=0$ close to ∂Y. (Figure 6)

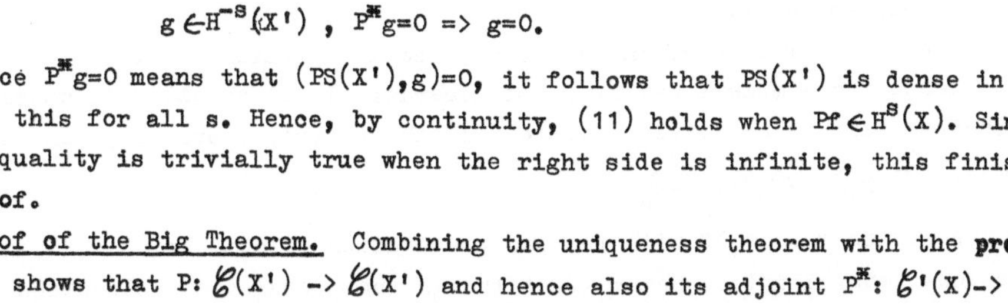

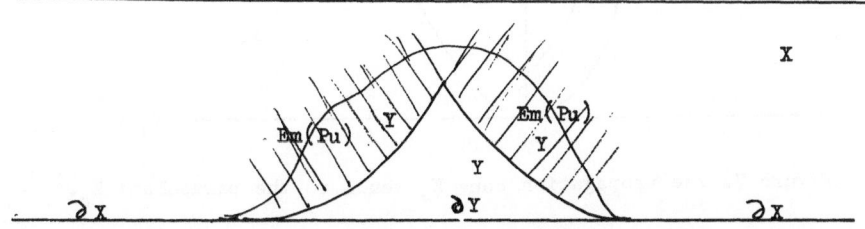

Figure 6. Uniqueness from the Big Theorem.

3.3 The dangers of infinite and vanishing propagation velocities. That an operator

$$P = \sum a_{jk}\partial_j\partial_k$$

with constant coefficients is strongly hyperbolic with respect to the time variable $x_1 = x\vartheta$ means precisely that

$$P(\xi) = a_{11}(\xi_1 - c_1(\xi))(\xi_1 - c_2(\xi))$$

where c_1 and c_2, only depending on $\xi' = (\xi_2, \ldots, \xi_n)$, are real and different when ξ' is real and $\neq 0$. Provided we measure distances in $x_2, \ldots, x_n$-space by the Euclidean metric form $x_2^2 + \ldots + x_n^2$, the number c_k is the propagation velocity of a plane wave,

$$x \to f(x\xi),\ x_1 = c_k(\xi),\ \xi_2^2 + \ldots + \xi_n^2 = 1\ ,$$

travelling in the direction ξ', normal to the hyperplanes $x_2\xi_2 + \ldots + x_n\xi_n = \text{const}$. When P is hyperbolic with variable coefficients,

$$P = \sum a_{jk}(x)\partial_j\partial_k + \sum a_k(x)\partial_k + a_0\ ,$$

the solutions u of $Pu=0$ describe wave propagation in an inhomogeneous medium with two propagation velocities, $c_1(x,\xi)$, $c_2(x,\xi)$ defined as above with respect to the principal part of P.

The principal part of a second order hyperbolic operator has Lorentz signature. Hence the simplest way it can degenerate is to an elliptic operator whose principal part is negative definite. To take a typical example, consider

$$(12)\qquad P = \partial_1 x_1 \partial_1 - \partial_2^2 - \ldots - \partial_n^2\ ,$$

strongly hyperbolic when $x_1 > 0$ with the two propagation velocities $\pm\ x_1^{1/2}$ in all directions and elliptic when $x_1 < 0$. Although the propagation velocities are infinite when $x_1 = 0$, a ray of light from $x=0$ is just a parabola $x_1 = t$, $x_k = 2t^{1/2}a_k$, $\sum a_k^2 = 1$. The future propagation cone from $(\varepsilon, 0, \ldots, 0)$ is

$$K_\varepsilon : x_1 \geq \varepsilon,\ x_2^2 + \ldots + x_n^2 \leq 2(x_1 - \varepsilon)$$

and tends to a paraboloid when $\varepsilon \to 0$. (Figure 7).

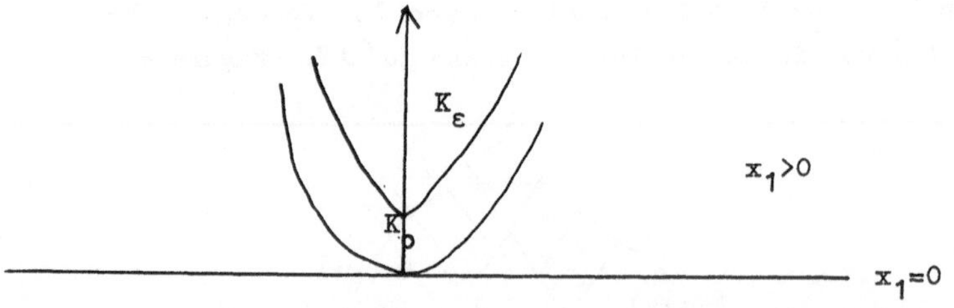

Figure 7. The propagation cone K_ε tends to the paraboloid K_0.

We shall see that, for the operator (12), there is no uniqueness for weak solutions of Cauchy's problem with data on $x_1=0$. We shall construct a weak solution $u\neq 0$ of $Pu=0$, vanishing when $x_1<0$, whose first derivatives are locally square integrable. In addition, the supports of the distributions $x_1 \to u(x)$ are uniformly bounded when x_1 is bounded from above.

Let $w(x_2,\dots,x_n)\in C_0^\infty$ be real and let $v=v_\varepsilon$ solve Cauchy's problem

$$x_1>\varepsilon \Rightarrow Pv=0,\quad x_1=\varepsilon \Rightarrow v=0,\ \partial_1 v = w/\varepsilon.$$

The energy identity, in this case

$$2\partial_1 vPv = \partial_1 F(v) - 2\partial_2(\partial_1 v\partial_2 v) - \dots - 2\partial_n(\partial_1 v\partial_n v) - (\partial_1 v)^2$$

where

$$F(v) = x_1(\partial_1 v)^2+\dots+(\partial_n v)^2$$

shows that

$$\int_{x_1=t} F(v)dx_2\dots dx_n + \int_{\varepsilon<x_1<t} (\partial_1 v)^2 dx = \int w(x_2,\dots,x_n)^2 dx_2\dots dx_n.$$

The same equality with v,w replaced by $\partial^\alpha v$, $\partial^\alpha w$ holds for all space derivatives ∂^α ($\alpha_1=0$). Extending v by zero when $x_1<\varepsilon$, integrations by parts show that

$$\int v(x)Pf(x)dx = -\varepsilon^{1/2}\int_{x_1=\varepsilon} f(x)w(x)dx_2\dots dx_n$$

for all $f\in C_0^\infty(R^n)$. Letting $\varepsilon\to 0$ and chosing a subsequence where all $\partial^\alpha v$ converge weakly, locally in L^2, the sequence will actually converge strongly to a limit u with the desired properties. In fact, the emission of of supp w, viz. supp $w + K_0$ (see Figure 7), has a compact intersection with every hyperplane x_1=const.

<u>Note</u>. There is a similar phenomenon for one space variable. The first order hyperbolic operator in two variables, $P=2\sqrt{t}\partial_t + \partial_x$ has the propagation velocity $1/2\sqrt{t}$ and, for every distribution f in one variable,

$$u=\theta(t)f(x-\sqrt{t})$$

is a weak solution of $Pu=0$ in all of R^2.

Consider now another way of passing from hyperbolic to elliptic, namely through zero propagation velocity. Take for instance the operator

$$P=\partial_1^2 + x_1(\partial_2^2+\dots+\partial_n^2) \tag{13}$$

which is elliptic when $x_1>0$, strongly hyperbolic when $x_1<0$ and degenerate with 0 propagation velocity when $x_1=0$. Then no solution $u\neq 0$ of $Pu=0$ when $x_1<0$, weak or otherwise, which vanishes in an open set bounded partly by an open subset A of the hyperplane $x_1=0$ can be extended to a solution across the hyperplane $x_1=0$. This is a consequence of Holmgren's uniqueness theorem (see Hörmander |6|). It says that if such an extension exists, it must vanish close to A when $x_1>0$. But then, by the ellipticity, the solution vanishes when $x_1>0$

so that, by Holmgren's theorem again, it vanishes everywhere (Figure 8)

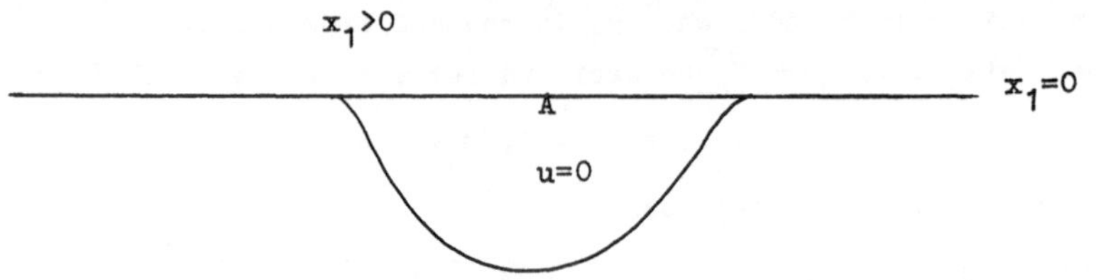

Figure 8. No solution u of Pu=0 (P according to (13)) in the whole space can vanish in the region indicated without vanishing identically.

Example. When a=b=1, the system in two variables t and x,

$$m^2u+v_t+w_x=0,\ u_t-av=0,\ u_x+bw=0$$

where $m\neq 0$ and the indices denote differentiation, is invariant under linear coordinate transformations and **simultaneous** contravariant transformations of v and w. When a and b are real and constant, the system is hyperbolic when ab>0. In fact, its characteristic determinant is $a\xi^2-b\tau^2-abm^2$, ξ and τ corresponding to $\partial/\partial x$ and $\partial/\partial t$. Violating the hyperbolicity condition can have drastic effects. E.g., when a=1, b=t, the system is equivalent to the equation

$$t(mu+u_{tt})-u_{xx}=0$$

with $v=u_t$, $w=u_x/t$. As done above, one can prove that there are weak solutions $u,v,w\neq 0$ which vanish identically when t<0. On the other hand, when a=-t, b=1, we get the following equation for v,

$$-m^2tv+v_{tt}+tv_{xx}=0.$$

By virtue of Holmgren's uniqueness theorem, no solution of this equation such that, e.g., the support of v is bounded in the x-direction when t<0, can be extended to a solution when t>0.

Note. A current recipe for scattering in physics is the following one. There is a hyperbolic equation or a system with constant coefficients whose solutions are called free waves. Introducing a compact perturbation of the coefficients of the **system**, one expects the perturbed **system** to have solutions which, for large negative and positive time behave like free waves, the incoming and outgoing waves. According to the preceding example one has better not introduce perturbations which destroy the hyperbolicity.

CHAPTER 4 MICROLOCAL CALCULUS. PROPAGATION OF SINGULARITIES

The general idea of microlocal calculus is to study the singularities of a distribution u at a point x by looking at the Fourier transforms of its localizations $\varphi(y)u(y)$ where $\varphi \in C_0^\infty$ equals 1 close to x. We shall review some of the basic results of this calculus including a simple version of the important propagation of singularities theorem. Microlocal calculus also exists in a hyperfunction version invented by Mikio Sato. For recent results in the subject, see |11|. The material here is from Hörmander |7| and Duistermaat-Hörmander |2|.

4.1 Singular support and singular spectrum. The support of a distribution u(x) in R^n, supp u, is the smallest closed set outside of which u vanishes. Similarly, the singular support of u, sing supp u, is the smallest closed set outside of which u is a C^∞ function. This set will now be analyzed further.

Let v be a distribution with compact support and

$$\hat{v}(\xi) = \mathcal{F}v(\xi) = \int e^{-ix\xi}\, v(x)dx$$

its Fourier transform. It is well known that v is a C^∞ function if and only if $\hat{v}$ is rapidly decreasing in the sense that

$$(1) \qquad |\hat{v}(\xi)| \leq C_N(1+|\xi|)^{-N}, \quad N=1,2,\ldots \quad .$$

Let $\sum(v)$, called the high frequency set of v, be the set of directions for which (1) does not hold. This means that an $\eta \neq 0$ is outside $\sum(v)$ precisely when (1) holds for all ξ in some conical neighborhood of η. This makes the high frequency set $\sum(v)$ a closed cone (or, if we want, a closed subset of the (n-1)-sphere) which is empty if and only if v is a C^∞ function. We can also say, somewhat loosely, that $\sum(v)$ is the set of high frequencies needed to put together v modulo a C^∞ function.

Multiplication by a smooth function can only decrease the high frequency set. We have

Lemma 4.1 If v is a distribution with compact support, then

$$\varphi \in C_0^\infty \Rightarrow \sum(\varphi v) \subset \sum(v) .$$

Proof. If $\hat{v}$ is rapidly decreasing in an open cone Γ, then for some N,

$$|\widehat{\varphi v}(\xi)| \leq \int_\Gamma |\hat{\varphi}(\xi-\eta)\hat{v}(\eta)|d\eta + C_N \int_{C\Gamma} |\hat{\varphi}(\xi-\eta)|(1+|\eta|)^N d\eta.$$

Splitting the first integral into one where $|\eta|>|\xi|/2$ and one where $|\eta|<|\xi|/2$, we see that it is a rapidly decreasing function of ξ. Since $|\eta| \leq |\xi-\eta|+|\xi|$, the second integral is at most

$$\text{const } (1+|\xi|)^N \int_{C\Gamma} |\hat{\varphi}(\xi-\eta)|(1+|\xi-\eta|)^N d\eta$$

where the integral is rapidly decreasing when $|\xi-\eta|>\varepsilon|\xi|$ for some $\varepsilon>0$ and this is true in every closed subset of Γ. This completes the proof.

According to the previous lemma, it makes sense to define the high frequency set $S_x(u)$ of a distribution u at a point x as the intersection of the high frequency sets of localizations of u. We put, by definition,

$$S_x(u) = \bigcap \sum(\varphi u) \quad \text{for } \varphi \in C_0^\infty \quad \text{and } \varphi(x) \neq 0.$$

In particular, $S_x(u) \neq \emptyset$ if and only if x is in the singular support of u. We can also assert that, in the sense of inclusion, $\lim\sup \sum(\varphi u) = S_x(u)$ as the supports of $\varphi \in C_0^\infty$ tend to x. In fact, given an η outside $S_x(u)$, there is a φ_η in C_0^∞ with $\varphi_\eta(x) \neq 0$ such that $\sum(\varphi_\eta u)$ does not contain η. Hence there is a product ψ of such functions such that $\sum(\psi u) \subset V$, a given open cone containing $S_x(u)$. But then, if $\varphi \in C_0^\infty$ vanishes close enough to x, then $\varphi/\psi \in C_0^\infty$ and hence $\sum(\varphi u) = \sum(\psi(\varphi/\psi)u) \subset V$.

We are now ready for the basic concept of microlocal analysis, the singular spectrum of a distribution.

__Definition__. When u is a distribution in an open part A of R^n, the closed set

$$S(u) = \text{sing spec } u = ((x,\xi) \in A \times R^n \setminus 0 \; ; \; \xi \in S_x(u)) \; ,$$

is called the singular spectrum of u.

Note. The high frequency set at x, $S_x(u)$, is then the fiber over x in the singular spectrum $S(u)$ and the projection of $S(u)$ onto x-space is the singular support of u. The word 'microlocal' refers to this resolution of the singular support.

__Example__. The singular supports of the distributions in one variable, $u=\delta(x)$, $(x\pm i0)^{-1}$ all consist of the origin and the fibers there of their singular spectra are, respectively, R, the positive axis R_+ and the negative axis R_-. In fact, their Fourier transforms are 1, $\theta(\xi)$, $\theta(-\xi)$ where θ is the Heaviside function and if $\varphi \in C_0^\infty$, then $(\widehat{\varphi u})(\xi)$ equals $\varphi(0)$ or tends to $\varphi(0)$ as $\xi \to \infty$ and $-\infty$ respectively. Differentiations and integrations do not change the singular spectra of these distributions. Note that $\delta(x)$ and the principal value of $1/x$, namely $((x+i0)^{-1}+(x-i0)^{-1})/2$, have the same singular spectrum.

__Example__. Let $E(x)$ be a fundamental solution of a differential operator P with constant coefficients. Since $P(D)E(x)=\delta(x)$ and differentiations do not increase the singular spectrum (Lemma 4.3 below), the fiber of $S(E)$ over $x=0$ is at least that of $S(\delta)$, i.e. all of $R^n \setminus 0$ (generalization of the preceding example). When P is hypoelliptic, i.e. every fundamental solution of P is a C^∞ function outside the origin, this is the entire singular spectrum of E.

Products and convolutions. The singular spectra of products and convolutions behave in a simple way. Let, e.g., u and v be distributions in one variable with compact supports whose high frequency sets are the positive axis. Then, for any ξ, the function

$$\eta \to \hat{u}(\xi-\eta)\hat{v}(\eta)$$

is small when $\eta \to +\infty$ since the first factor is small there and small also when $\eta \to -\infty$ since the second factor is small there. Hence the convolution integral

$$(\hat{u} * \hat{v})(\xi) = \int \hat{u}(\xi-\eta)\hat{v}(\eta)d\eta$$

is absolutely convergent. Since it is also of at most polynomial growth in ξ, the formula

$$(uv)(x) = (2\pi)^{-1}\int e^{ix\xi}\,(u*v)(\xi)d\xi$$

defines a product with, presumably, good natural properties. Localizing and passing to several variables, this observation can be made into a proof of

Lemma 4.2 There is a well defined product **uv** of two distributions u and v provided $S_x(u)+S_x(v)$ does not contain the origin for any x and one has

$$S_x(uv) \subset (S_x(u)+S_x(v)) \cap S_x(u) \cap S_x(v)$$

for all x.

For convolutions there is a similar result.

Lemma 4.3 When u is a distribution and P a differential operator with smooth coefficients, then

$$S(Pu) \subset S(u).$$

When u and v are distributions in R^n and one of them has compact support, then

$$S(u*v) \subset ((x+y,\xi) \; ; \; (x,\xi) \in S(u) \text{ and } (y,\xi) \in S(v)).$$

Note. Letting

$$S^{\xi}(u) = (x; \; (x,\xi) \in S(u))$$

denote the fiber of S(u) under ξ, the previous formula becomes, simply,

$$S^{\xi}(u*v) \subset S^{\xi}(u)+S^{\xi}(v).$$

Proof. When (x,ξ) is outside S(u), choose $\varphi \in C_0^{\infty}$ with $\varphi = 1$ close to x. Then $\varphi Pu = P\varphi u$ close to x and hence, by Lemma 4.1 and the obvious fact that differentiations do not increase high frequency sets, we have $S_x(Pu) \subset \Sigma(P\varphi u) \subset \Sigma(\varphi u)$ so that (x,ξ) is not in S(Pu). To prove the second part, consider the convolution

$$(u*v)(x) = \int u(x-y)u(y)dy$$

and assume that

ξ is not in $S_{x-y}(u) \cap S_y(v)$ for any y.

We have to show that ξ is not in $S_x(u*v)$. This will be done using the obvious fact that $\Sigma(f*g)\subset\Sigma(f)\cap\Sigma(g)$ for any two distributions of compact support. Taking v of compact support, we can first write $u*v$ as a sum of convolutions $u*w$ where w has so small support that $S_{x-y}(u)\cap\Sigma(w)$ does not contain ξ when y is in the support of w. It suffices then to prove that, under this hypothesis, ξ is not in $S_x(u*w)$. We can now find a $\varphi\in C_0^\infty$ such that $\varphi=1$ close to the set x - supp w while $\Sigma(\varphi u*w)\subset\Sigma(\varphi u)\cap\Sigma(w)$ still does not contain ξ. It follows that ξ is not in $S_x(u*w)\subset\Sigma(\varphi u*w)$.

Note. The concept of the singular spectrum S(u) of a distribution u extends to distributions on a manifold X. Their singular spectra are parts of the cotangent bundle $T^*(X)$ minus its zero section.

4.2 Certain parametrices. Let P(D), $D=\partial/i\partial x$, be a differential operator with constant coefficients. Under certain hypotheses we are going to construct parametrices of P of the form

$$(3)\qquad E(x) = (2\pi)^{-n}\int_{|\xi|>R} e^{ix\zeta}P(\zeta)^{-1}\,d\zeta \qquad ,\zeta=\xi-iv(\xi),$$

with $\xi\to v(\xi)$ a bounded real smooth vector field and

$$(4)\qquad |\xi|>R \Rightarrow |P(\zeta)| > \text{const} > 0.$$

Then E is a well-defined distribution and

$$P(D)E(x) = (2\pi)^{-n}\int_{|\xi|>R} e^{ix\zeta}\,d\zeta = \delta(x) + H(x)$$

where

$$H(x) = (2\pi)^{-n}\int_{|\xi|<R} e^{ix\zeta}\,d\zeta$$

is an entire function. In this sense, E is a parametrix, i.e. an approximate fundamental solution.

We now suppose that that the characteristic polynomial $P(\xi)$ has a real principal part, $a(\xi)$, and that it is of principal type, i.e.

$$\xi \text{ real and } \neq 0 \Rightarrow \operatorname{grad} a(\xi)\neq 0.$$

In particular, the real conical hypersurface $A: a(\xi)=0$ is non-singular outside the origin. Note that any strongly hyperbolic operator P is in this class and that, if $P\in \mathrm{hyp}(\vartheta)$, we can take R=0 and, in (3), $v(\xi)=t\vartheta$ for some large t, getting a fundamental solution and not only a parametrix.

Next, let V be the set of C^∞ vector fields $\xi\to v(\xi)$, mapping $R^n\setminus 0$ into R^n and homogeneous of degree zero, such that

$$a(\xi)=0,\ \xi\neq 0 \Rightarrow v(\xi).\operatorname{grad} a(\xi)\neq 0.$$

To construct such fields, consider functions v_η from the sphere $|\xi|=1$ to R^n with small supports around η such that, when $a(\eta)=0$,

$$\varepsilon(\eta)v_\eta(\xi).\text{grad } a(\xi) > 0$$

close to η and ≥ 0 for all ξ. Here the sign function $\varepsilon(\eta)$ is given and equals +1 or -1 close to every component of $A\setminus 0$. Adding a finite number of such functions, suitably chosen, and making them homogeneous of degree zero produces a vector field v in V such that

$$(5) \qquad \xi \in A\setminus 0 \Rightarrow \varepsilon(\xi)v(\xi).\text{grad } a(\xi) > 0.$$

It is obvious that any two v in V with the same sign function $\varepsilon(\xi)=\varepsilon_v(\xi)$ are linearly homotopic in V. When P has degree m and is hyperbolic, $A\setminus 0$ has m components (sheets) giving rise to 2^m different possible sign functions.

In order to construct parametrices, let B be a compact part of V. Then, if v is in B, ξ is large and P has degree m,

$$P(\xi-isv(\xi)) = a(\xi) - isv(\xi).\text{grad } a(\xi) + \underline{O}(|\xi|^{m-1})$$

so that, for some $c>0$,

$$|P(\xi-isv(\xi))| \geq 2c \min(s,|\xi|)|\xi|^{m-1} - \underline{O}(|\xi|^{m-1})$$

when $s>0$. Hence there is an $R>0$ such that

$$(6) \qquad |\xi|>R,\ s\geq 1 \Rightarrow |P(\xi-isv(\xi))| \geq c\min(s,|\xi|)|\xi|^{m-1}.$$

In particular, taking $s=1$, we have (4) and a parametrix $E=E_v$. Different choices of such v which are homotopic within V will only change E by an entire function. To see this, consider the smeared version of (3),

$$\int E(x)f(x)dx = (2\pi)^{-n}\int_{|\xi|>R} F(-\zeta)P(\zeta)^{-1}d\zeta$$

where F is the Fourier-Laplace transform of $f\in C_0^\infty$. The integrand on the right, the product of an analytic function and $d\zeta = d\zeta_1 \ldots d\zeta_n$, is a closed differential form and $\zeta \to F(-\zeta)$ is a rapidly decreasing function when $\operatorname{Im}\zeta$ is bounded. Hence, by Stokes's theorem and (6), changes of v, i.e. changes of the chain of integration only leaves an integral over $|\xi|=R$. Hence the difference between two right sides of (3) with different v's is an integral over $|\xi|=R$ which is an entire function.

The advantage of the parametrices just introduced is that their singular spectra are very simple and well adapted to a proof to follow of the important propagation of singularities theorem.

<u>Theorem</u> 4.1 Under the above hypothesis about P, the singular spectrum of the parametrix $E=E_v$ is given by the formula

$$a(\eta)\neq 0 \Rightarrow S^\eta(E)=0,\quad a(\eta)=0 \Rightarrow S^\eta(E)=R(\eta)$$

where $S^\eta(E)$ denotes the fiber of S(E) under η and $R(\eta)$ is the half-ray

$$(x;\ x=\varepsilon(\eta)s \text{ grad } a(\eta),\ s\geq 0,\ \varepsilon=\varepsilon_v)$$

Note. In terms of fibers over x-space, $S_0(E)=R^n\setminus 0$ and, if $x\neq 0$, $S_x(E)$ is empty or a half-ray $(s\eta;\ s\geq 0)$, the latter case occurring precisely when x is on the half-ray $R(\eta)$.

Proof. Consider first the polynomials $t\to P(\zeta+t\eta)$ and let $P_\eta(\zeta)$ be the leading term so that, if P is of degree m,

$$(7)\qquad P(\zeta+t\eta) = t^{m-k}(P_\eta(\zeta) + \underline{O}(t^{-1}))$$

where k=1 when $a(\eta)=0$ and k=0 when $a(\eta)\neq 0$. In other words, $P_\eta(\zeta)=a(\eta)$ when $a(\eta)\neq 0$ and

$$P_\eta(\zeta) = \operatorname{grad} a(\eta).\zeta \ +b(\eta)$$

when $a(\eta)=0$. Here $b(\zeta)$ is homogeneous of degree m-1 and $P(\zeta)=a(\zeta)+b(\zeta)$ + lower terms.

In order to determine the singular spectrum of $E(x)$ we shall study the high frequency sets of the distributions $E(x)f(x)$ where $f\in C_0^\infty$. We then have to determine the growth for large t of functions

$$(8)\qquad t\to \int e^{-i\eta tx}E(x)f(x)dx \ = (2\pi)^{-n}\int_{|\xi+t\eta|>R} e^{i\zeta x}F(-\zeta)P(\zeta+t\eta)^{-1}d\zeta$$

where F is the Fourier-Laplace transform of f, $|\eta|=1$ and

$$\zeta=\xi-iv(\xi+t\eta) = \xi-iv(\eta+t^{-1}\xi) .$$

Letting $t\to\infty$, the formulas (6) and (7) and dominated convergence proves that

$$(9)\qquad t^{k-m}\int e^{-it\eta x}E(x)f(x)dx \to (2\pi)^{-n}\int_{R^n} e^{i\zeta x}F(-\zeta)P_\eta(\zeta)^{-1}d\zeta$$

where now $\zeta=\xi-iv(\eta)$. When $a(\eta)\neq 0$, the right side is simply $f(0)/a(\eta)$. When $a(\eta)=0$, a passage to the limit in (6) with ξ replaced by $\xi+t\eta$ proves that

$$(10)\qquad s\geq 1 \Rightarrow |P_\eta(\xi-isv(\eta))| \geq cs .$$

Hence the right side of (9) equals

$$\int E_\eta(x)f(x)dx$$

where

$$E_\eta(x) = (2\pi)^{-n}\int e^{i\zeta x}P_\eta(\zeta)^{-1}d$$

is a fundamental solution of the first order hyperbolic differential operator $P_\eta(D)$. Its hyperbolicity cone $\Gamma(P_\eta,v(\eta))$ with respect to $v(\eta)$ is the half-space $\varepsilon(\eta)\xi.\operatorname{grad} a(\eta) >0$. Hence the support of $E_\eta(x)$ is contained in and, in this simple case, actually equal to the corresponding propagation cone which is just the half-ray $R(\eta)$ (Theorem 1.2). All this shows that S(E) is not less than the set described in the theorem.

To prove the opposite inclusion note that, changing the region of integration in the last integral of (8) to, e.g., $|\xi|\leq t^{1/2}$ just amounts to adding a function of t which is rapidly decreasing uniformly with respect to η. Hence it suffices to consider

(11) $$t \to (2\pi)^{-n} \int_{|\xi|<t^{1/2}} e^{i\zeta x} F(-\zeta) P(\zeta+t\eta)^{-1} d\zeta .$$

First, let $a(\theta)=0$, $|\theta|=1$, keep η close to θ, choose v so that $v(\eta)=v(\theta)$ there and write

(12) $$P(\zeta+t\eta) = t^m(P_\eta(t^{-1},\zeta) - t^{-2}Q_\eta(t^{-1},\zeta))$$

where

$$P_\eta(t^{-1},\zeta) = a(\eta) + t^{-1}(\text{grad } a(\eta).\zeta + b(\eta)),$$

making $Q_\eta(t^{-1},\zeta)$ a polynomial in η, t^{-1}, ζ. Inserting (12) into (11) and employing the identity

$$(A-B)^{-1} = \sum_0^{N-1} A^{-k-1}B^k + A^{-N}B^N(A-B)^{-1} ,$$

we get a sum of terms

(13) $$(2\pi)^{-n}\, t^{-m-2k} \int_{|\xi|<t^{1/2}} P_\eta(t^{-1},\zeta)^{-k-1} Q_\eta(t^{-1},\zeta)^k F(-\zeta) d\zeta$$

and a term

(14) $$(2\pi)^{-n} t^{-m-2N} \int_{|\xi|<t^{1/2}} P_\eta(t^{-1},\zeta)^{-N} Q_\eta(t^{-1},\zeta)^N P(\zeta+t\eta)^{-1} F(-\zeta) d\zeta ,$$

which, if N is large enough, is of arbitrarily small growth in t, uniformly with respect to η. In the terms (13), $\zeta = \xi - iv(\eta + t^{-1}\xi) = \xi - iv(\theta)$ when t is large enough and η close to θ. Keeping this ζ for all ξ and extending the region of integration to all of R^n does not change the growth for large t of the resulting terms, namely

(15) $$(2\pi)^{-n}\, t^{-m-2k} \int P_\eta(t^{-t},\zeta)^{-k-1} Q_\eta(t^{-1},\zeta)^k F(-\zeta) d\zeta =$$
$$= t^{-m-2k} \int Q_\eta(t^{-1},D) E_\eta^{(k+1)}(t^{-1},x) f(x) dx$$

where

$$E_\eta^{(k+1)}(t^{-1},x) = (2\pi)^{-n} \int e^{ix\zeta} P_\eta(t^{-1},\zeta)^{-k-1} d\zeta$$

is a fundamental solution of the differential operator $P_\eta^{k+1}(t^{-1},D)$ which, precisely as $E_\eta(x)$, vanishes outside the half-ray $R(\eta)$. Hence, if $f=0$ close to $R(\theta)$, all the terms (15) vanish when η is close to θ. Hence, if $a(\theta)=0$, $S^\theta(E) \subset R(\theta)$. If $a(\theta)\neq 0$, we can just repeat the same arguments with (12) replaced by

$$P(\zeta+t\eta) = t^m(P_\eta(\zeta) - t^{-1}Q_\eta(t^{-1},\zeta))$$

where now η is close to θ so that $P_\eta(\zeta)=a(\eta)$ is just a constant. In (13) and (14) we shall then have the factors t^{-m-k} and t^{-m-N} but our arguments work as before. The fundamental solutions appearing in the modified formula (15) are now equal to $a(\eta)^{-k-1}\delta(x)$ so that the corresponding terms all vanish when $f=0$ close to $x=0$. This finishes the proof of the theorem.

4.3 Propagation of singularities. Let

$$P(x,D) = \sum a_\alpha(x)D^\alpha, \quad |\alpha| \leq m,$$

with smooth coefficients be of principal type with real principal part, i.e. the principal part

$$a(x,D) = \sum a_\alpha(x)D^\alpha, \quad |\alpha| = m,$$

has real coefficients and $a_\xi(x,\xi) = \partial a(x,\xi)/\partial\xi \neq 0$ when ξ is real and $\neq 0$. A bicharacteristic of P is, by definition, a solution $x=x(t)$, $\xi=\xi(t)$ of Hamilton's equations, ($a_x(x,\xi) = \partial a(x,\xi)/\partial x$),

$$dx = a_\xi(x,\xi)dt, \quad d\xi = -a_x(x,\xi)dt$$

for which $a(x,\xi)=0$. (Note that $da(x,\xi)=0$ for every solution). When $a(x,\xi)=a(\xi)$ has constant coefficients, $a_x=0$ and the bicharacteristics are straight lines

(16) $$x = a_\eta(\eta)t + y, \quad a(\eta)=0 .$$

The following theorem (Hörmander |8|), proved here only for constant coefficients, is a basic ingredient of microlocal theorey.

Theorem 4.2 Let P be a smooth differential operator of principal type and with real principal part in an open subset Ω of R^n, let u be a distribution in Ω and γ an open connected bicharacteristic over Ω. Then, if $S(Pu)$ does not contain γ, either $\gamma \subset S(u)$ or $\gamma \cap S(u) = \emptyset$.

Note. In other words, $S(u) \setminus S(Pu)$ is a union of maximal connected bicharacteristics.

Proof (for constant coefficients). Let $(y,\eta) \in S(u) \cap \gamma$. In particular, γ is some curve (16) with t running over some interval containing the origin. Let $f \in C_0^\infty$ be 1 close to y. Then

$$Pfu = v + fPu$$

where $v=Pfu-fPu$ vanishes outside supp df, which we can imagine as a ring around y (see Figure 9). Let E be one of the parametrices of P of Theorem 4.1. Then, modulo entire functions,

$$fu \equiv E*v + E*fPu$$

so that, by (2)

$$S^\eta(fu)_\gamma \subset (S^\eta(v) + S^\eta(E))_\gamma \cup (S^\eta(Pu) + S^\eta(E))_\gamma$$

where a subscript γ denotes intersection with γ. Note that, by Theorem 4.1 $S^\eta(E)$ is a half-ray from the origin with the direction $\pm a_\eta(\eta)$ where the sign is at our disposal. In particular, the formula above implies

$$S^\eta(fu)_\gamma \subset (S^\eta(v)_\gamma + S^\eta(E)) \cup (S^\eta(Pu)_\gamma + S^\eta(E)).$$

If $S^{\eta}(Pu)_{\gamma}$ is empty so is the last parenthesis and we can disregard it. Let I be an interval outside y of the projection $\pi\gamma$ of γ into x-space (see Figure 9) and choose f such that

$$\text{supp } df \cap \pi\gamma \subset I$$

and E such that the direction of the half-ray $S^{\eta}(E)$ points from I to y as indicated by the arrow of the picture. Then

$$S^{\eta}(v)_{\gamma} + S^{\eta}(E)$$

is contained in an interval J on $\pi\gamma$ on the other side of y. This contradiction proves that if $S(Pu)\cap\gamma$ is empty and γ has some point in common with S(u), then all of γ is in S(u).

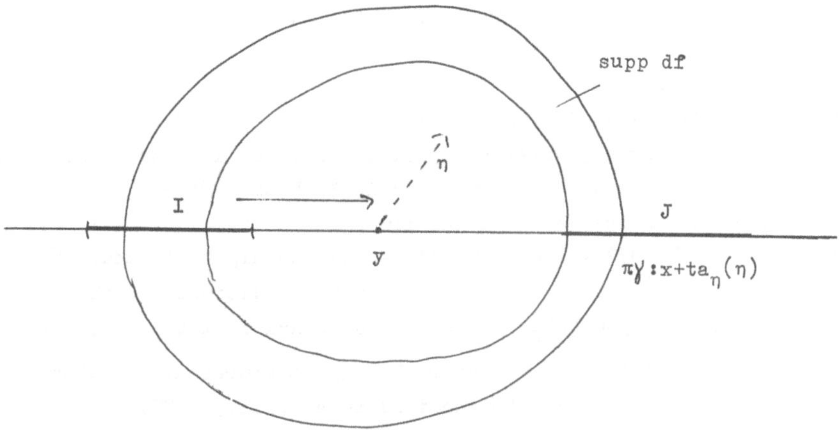

Figure 9. To the proof of the propagation of singularities theorem.

References

|1| Atiyah, M.F., Bott, R., Gårding L. Lacunas for hyperbolic differential operators with constant coefficients. Acta Math. 124(1970)109-189.

|2| Duistermaat J.J. and Hörmander L. Fourier integral operators II. Acta Math 128(1972)183-269.

|3| Gårding L. Math. Rev. 31.2 (1966) p. 3722. Review of T. Kakita Hyperbolic convolution operators Can. J. Math. 17(1965)559.582. See also L. Ehrenpreis Solutions of some problems of division Part V. Hyperbolic operators Am. J. Math. 84(1962)324-328.

|4| Gårding L. Local hyperbolicity.Israel J. Math. 13(1972)65-81.

|5| Gårding L. Support functions of plurisubharmonic functions. J. Math. and Mech. 17(1967)225-240

|6| Hörmander L. Linear Partial Differential Operators. Springer 1963.

|7| Hörmander L. Fourier Integral Operators. Acta Math. 127(1971)79-183.

|8| Hörmander L. Linear differential operators. Proc. Int. Congress of Math. Nice 1970. 121-133.

|9| Larsson E. Generalized hyperbolicity. Arkiv f. Mat. 7.2(1967)11-32

|10| Lax A. On Cauchy's problem for partial differential equations with multiple characteristics. Comm. Pure and Appl. Math. 10(1957)390-398.

|11| Publications of the Research Institute for Mathematical Sviences (Kyoto) vol 12, suppl. Feb 1977 (Proc. of OJI seminar on algebraic analysis 1976).

|12| Svensson S.Leif. Necessary and sufficient conditions for the hyperbolicity of polynomials with hyperbolic principal part. Ark f. Mat 8(17)(1969)145-162

|13| Gårding L. Introduction to hyperbolicity. Lectures at the CIME school "Hyperbolicity", Cortona 1976 (Liguori Editore S.r.l., 1977)

METHOD OF CHARACTERISTICS IN THE EXTERNAL FIELD PROBLEM

or

HOW TO RECOGNIZE AN ACAUSAL EQUATION WHEN YOU SEE ONE

Daniel Zwanziger
Physics Department, New York University
New York, New York 10003

I. Introduction

In these lectures, the method of characteristics will be used to obtain a non-perturbative expansion of the retarded Green's function, and to reveal acausal propagation or improperly posed initial value problems. As particular results, we show that avoidance of the latter pathologies uniquely resolves the operator ordering ambiguity (due to non-commutativity of generally covariant derivatives) of a vector particle in an external gravitational field, and we rederive the acausal propagation of Rarita-Schwinger waves minimally coupled to an external electromagnetic potential.

Following the introduction (this section), we will briefly review the method of characteristics (Sec. II), and apply it in succession to a Klein-Gordon particle in an external scalar field (Sec. III), a Dirac particle minimally coupled to an external electromagnetic potential (Sec. IV), a vector or Proca particle in an external gravitational field described by Einstein's general relativity (Sec. V) and a Rarita-Schwinger particle minimally coupled to an external electromagnetic potential (Sec. VI), concluding with a brief perspective on the higher spin problem (Sec. VII).

The unified method by which these various cases are treated is to construct the fundamental retarded solution (also known as the retarded Green's function or retarded propagator or retarded commutator) in the neighborhood of its singular surface or surfaces. For the Klein-Gordon and Dirac equations, the fundamental retarded solution is represented explicitly and non-perturbatively as a series in powers of the distance from the light cone, following the method of Hadamard.[1] For matrix equations, namely the Dirac, Proca and Rarita-Schwinger equations, the resolution is achieved by multiplying the original matrix differential operator by another appropriate matrix differential operator, which we may call a "resolving divisor," a generalization of the Klein-Gordon divisor. It is so chosen that, in the product, the matrix of highest derivatives is a scalar matrix. This allows a unified treatment of non-singular systems (Dirac) and of singular systems (Proca and Rarita-Schwinger). Use of the resolving divisor has the advantage, over the previous determinental method,[2,3] of revealing the true multiplicity of the characteristics. In the Rarita-Schwinger case, for example, the determinental method gives $(n^2)^3[n^2 + (\frac{2e}{3m})F^d\cdot n)^2]$, where $(F^d\cdot n)^\kappa = \frac{1}{2}\epsilon^{\kappa\lambda\mu\nu}F_{\mu\nu}n_\lambda$ and n^μ is the normal to the characteristic surface. But with the resolving divisor, the leading polynomial is $n^2[n^2 + (\frac{2e}{3m^2}F^d\cdot n)^2]$ which shows simple characteristics. This may offer the basis for an existence theorem.

The results presented here provide a sample of what is known about acausal phenomena in the external field problem. We take this occasion to mention some other results and where further references may be found, without attempting a systematic survey of the literature. Following the original program of Fierz and Pauli,[4] difficulties with the charged spin-3/2 quantized field began to appear in the early 1940's[5] and later.[6] It was realized,[2] using the Rarita-Schwinger formalism,[7] that they were due to acausal characteristics of the corresponding classical field equation, and that this feature and other pathologies are a general phenomenon which afflict other higher spin equations in interaction.[3,8] In particular it was shown that the Proca equation[9] for a vector field representing a massive spin-one particle, though well behaved when coupled minimally and with arbitrary anomalous magnetic moment, is badly behaved if it has an anomalous electric quadrupole moment, or is coupled to an external symmetric tensor field.[3] Existence theorems have been established for the Proca equation.[10,11] Acausal behavior has also been established for a spin-two particle in an external electromagnetic field.[12] Non-linear systems may also be studied by the method of characteristics[3] and this has been exploited to yield restrictions on the possible interactions among massive vector mesons.[13] It should not be thought that acausal propagation only afflicts unphysical quantities, as for example in the Coulomb gauge where observable quantities remain causal despite apparently instantaneous propagation. For it has been shown, in the case of the Proca equation in an external tensor field, that the commutator of the energy density with itself is non-vanishing at space-like separation.[14]

A systematic discussion of general spin equations and of the relation of the retarded fundamental solution and the homogeneous quantized equation has been presented by A.S. Wightman,[15,16] and related material by R. Seiler.[17] Acausal phenomena have been studied by R. Guertin and T. Wilson[18] in the Hamiltonian formalism in which, at the cost of manifest invariance, constraints are eliminated and only dynamical degrees of freedom remain. For an alternative approach based on the Bhabha equation see R. Krajcik and M. Nieto.[19] Other difficulties with higher spin have been exhibited by A.S. Wightman.[20,21] For the Proca equation in an external gravitational field we make use of the method developed by De Witt and Brehme,[22] following Hadamard,[1] and extended by Hobbs,[23] who calculated the retarded Lienard-Wiechert potential in curved space-time. An elementary exposition of the method of characteristics is presented by Courant and Hilbert,[24] and modern results by L. Garding.[25]

II. Method of Characteristics

We briefly outline the method of characteristics.[24]

Suppose the solution $\phi(x)$ of an equation of the form

$$A^{\mu\nu}(x)\partial_\mu\partial_\nu\phi(x) + B^\mu(x)\partial_\mu\phi(x) + C(x)\phi(x) = 0 \qquad (2.1)$$

is singular on some surface $S(x) = 0$ with leading singularity $\psi(S(x))$. We call $S(x)=0$ a characteristic surface. Here $\psi(S)$ is some singular function (e.g. $\psi(S) = \delta^{(n)}(S)$ or $\theta(S)$ etc.) whose singular support is at $S=0$,

$$\phi(x) = H(x)\psi(S(x)) + \text{less singular terms} \qquad (2.2)$$

and $H(x)$ is a smooth function. On substituting this expression for $\phi(x)$ into Eq. (2.1), the leading singularity on the left hand side is found to be

$$A^{\mu\nu}(x)\partial_\mu S(x)\partial_\nu S(x)H(x)\psi''(S(x)) \qquad (2.3)$$

This will vanish if $S(x)$ satisfies

$$\tfrac{1}{2}A^{\mu\nu}(x)\partial_\mu S(x)\partial_\nu S(x) = 0. \qquad (2.4)$$

(More generally, the leading singularity vanishes if

$$\tfrac{1}{2}A^{\mu\nu}(x)\partial_\mu S(x)\partial_\nu S(x) = CS(x) \qquad (2.4a)$$

where C is a constant or a function of x.) This is a first order partial differential equation which determines $S(x)$ as follows. We write $x = x^\mu = (t,\underset{\sim}{x})$, and suppose that at $t = t_0$, $\phi(t_0,\underset{\sim}{x})$ is singular on some given surface in 3-space defined by $S_0(\underset{\sim}{x}) = 0$. Then the characteristic surface $S(x) = S(t,\underset{\sim}{x}) = 0$ is provided by the solution of Eq.(2.4) [or (2.4a)] with initial condition $S(t_0,\underset{\sim}{x}) = S_0(\underset{\sim}{x})$.

In addition we may determine how the strength of the singularity $H(x)$ propagates along $S(x)$ by equating to zero the coefficient of $\psi'(S)$ on the left hand side of Eq. (2.1) (after substitution of Eq.(2.2) and (2.4)), namely

$$2\partial_\mu S(x)A^{\mu\nu}(x)\partial_\nu H(x) + A^{\mu\nu}(x)\partial_\mu\partial_\nu S(x)H(x) + B^\mu(x)\partial_\mu S(x)H(x) = 0 \qquad (2.5)$$

which is valid for x on $S(x) = 0$. (In the following sections we will see that the "less singular terms" of Eq.(2.2) contribute in the next order). We will solve this equation by the method of ray or bi-characteristic, assuming $S(x)$ is given. The first order partial differential equation (2.4) may be regarded as the Hamilton-Jacobi equation corresponding to a Hamiltonian system $H(p,x) = \tfrac{1}{2}A^{\mu\nu}(x)p_\mu p_\nu$ with trajectories, or rays $x(\tau)$ given by $\dot{x}^\mu = \partial/\partial p^\mu H(p,x)$, with $p_\mu = \partial_\mu S(x)$. (Actually $S(x)$ itself may be found from the complete solution of the Hamiltonian system). Given $S(x)$, the ray $x(\tau)$ satisfies

$$\dot{x}^\mu = \frac{dx^\mu}{d\tau} = A^{\mu\nu}(x)\partial_\nu S(x). \qquad (2.6)$$

We have $\dot{x}^\mu\partial_\mu S(x) = 0$ by Eq.(2.4), so the ray $x^\mu(\tau)$ lies on the surface $S(x)$, and all points on $S(x)$, (a 3-dimensional manifold) may be parametrized by τ and 2 other parameters, say $\hat{x}$, which parametrize the points on $S_0(\underset{\sim}{x}) = 0$, (a 2-dimensional manifold).

$$x^\mu = x^\mu(\tau,\hat{x}) \text{ for } x \text{ on } S(x) = 0. \tag{2.7}$$

Similarly we write $H(x) = H(\tau,\hat{x})$ for x on $S(x) = 0$ and we have

$$\partial_\mu S(x)A^{\mu\nu}(x)\partial_\nu H(x) = \frac{dx^\nu}{d\tau}(\tau,\hat{x})\partial_\nu H(x) = \frac{\partial}{\partial\tau} H(\tau,x). \tag{2.8}$$

Thus Eq.(2.5) becomes

$$2\frac{\partial H}{\partial\tau}(\tau,x) + A^{\mu\nu}(x)\partial_\mu\partial_\nu S(x)H(\tau,x) + B^\mu(x)\partial_\mu S(x)H(\tau,x) = 0\ , \tag{2.9}$$

where $x = x(\tau,\hat{x})$. We now turn to some applications of this method.

III. Klein-Gordon Particle in an External Scalar Field

The Lagrangian density

$$L = \tfrac{1}{2}\partial_\mu\phi\partial^\mu\phi - \tfrac{1}{2}m^2\phi^2 - \tfrac{1}{2}S(x)\phi^2 + \eta(x)\phi \tag{3.1}$$

corresponds to the equation of motion

$$\partial^2\phi + m^2\phi + S(x)\phi = \eta(x). \tag{3.2}$$

We have introduced the probe field $\eta(x)$ as a technique for studying characteristic surfaces, which may be defined as the surfaces along which a singular response to a small localized disturbance $\eta(x)$ propagates. In particular with

$$\eta(x) = \delta^4(x-y) , \tag{3.3}$$

which allows the disturbance to be located at an arbitrary space-time point y, the solution, if there is one, which vanishes for $x^0<y^0$ is called the fundamental retarded solution $D^R(x,y)$,

$$[\partial^2 + m^2 + S(x)]D^R(x,y) = \delta^4(x,y) \tag{3.4}$$

$$D^R(x,y) = 0 \qquad x^0<y^0 \tag{3.5}$$

(If the equation were non-linear we would replace $\eta(x)$ by $g\,\overset{4}{\delta}(x-y)$ and study the dependence of ϕ to first order in g. Since the equation is linear, g may be factored out). There is an analogous definition of the advanced fundamental solution D^R, and one may show $D^A(x,y) = D^R(y,x)$ (exercise). The definition of hyperbolicity is that $D^R(x,y)$ exist. If it does not exist one may show that the Cauchy initial value problem is not well posed,[1,25] for an initial surface x^0 = const. Einsteinian causality states that $D^R(x,y)$ also vanish at space-like separations,

$$D^R(x,y) = 0 \qquad x^2<0. \tag{3.6}$$

These fundamental solutions are important in quantum field theory because the field commutator is given by [15,16,21]

$$i[\phi(x),\phi(y)] = D(x,y) \tag{3.7}$$

$$D(x,y) = D^R(x,y) - D^A(x,y) \tag{3.8}$$

so our definition implies the vanishing of the commutator at space-like separations, which is sometimes called microcausality. The retarded fundamental solution also governs the time evolution of the one particle theory. (For a Dirac particle $\langle \underset{\sim}{x}|e^{-iHt}|\underset{\sim}{x}'\rangle\theta(t) = iS^R(x,t;\underset{\sim}{x}',o)\gamma^0$.) The boundary of the support of $D^R(x,y)$ is, by definition, a characteristic surface, so the requirement of causality is that this characteristic lie inside or on the light cone centered at y.

We now investigate how the presence of the external field $S(x)$ affects $D^R(x,y)$ in the neighborhood of its characteristic surface. Without loss of generality we set y=0 and write $D^R(x) \equiv D^R(x,o)$. Since the highest derivatives in Eq.(3.4) constitute the wave operator ∂^2, we try, for the leading singularity of $D^R(x)$, the retarded fundamental solution of the wave equation

$$D_0^R(x) = (2\pi)^{-1}\theta(x^0)\delta(x^2) = (4\pi r)^{-1}\delta(t-r) \tag{3.9}$$

which satisfies

$$\partial^2 D_0^R(x) = \delta^4(x). \tag{3.10}$$

The Hamilton-Jacobi equation, (2.4), corresponding to the Klein-Gordon equation is

$$(\partial S)^2 = 0 , \tag{3.11}$$

which is satisfied by $S(x) = t-r$, which is the argument of the δ-function in Eq. (3.9). Thus the characteristic surface is

$$S(x) = t-r = 0 , \tag{3.12}$$

the equation for the future light cone. The ray equation (2.6) reads

$$\frac{dx^\mu}{d\tau} = g^{\mu\nu}\partial_\nu S(x) = (1,\hat{x}) \tag{3.13}$$

with solution $t = \tau$, $\underset{\sim}{x} = \tau\hat{x}$, or, writing r for τ,

$$t = r, \quad \underset{\sim}{x} = r\hat{x}. \tag{3.14}$$

The equation of propagation of the leading singularity, Eq.(2.9) with $B_\mu = 0$, $A^{\mu\nu}(x) = g^{\mu\nu}$ and

$$\partial_\mu\partial_\nu S = \begin{pmatrix} 0 & 0 \\ 0 & -\frac{\delta_{ij}}{r} + \frac{x_i x_j}{r^3} \end{pmatrix}$$

reads

$$2\frac{\partial H(r,\hat{x})}{\partial r} + \frac{2}{r} H(r,\hat{x}) = 0 \tag{3.15}$$

with solution

$$H(r,\hat{x}) = \frac{1}{r} f(\hat{x}). \tag{3.16}$$

In order to agree with the singularity of $D_0^R(x)$ in the neighborhood of the origin, which correctly produces the inhomogeneous term $\delta^4(x)$, we take $f(\hat{x}) = (4\pi)^{-1}$. This fixes the leading singularity of $D^R(x)$,

$$D^R(x) = \frac{1}{4\pi r}\,\delta(t-r) + \text{less singular terms}. \tag{3.17}$$

Having determined the leading singularity, the general solution may be found[1] by expanding in powers of $t-r$,

$$D^R(x) = (4\pi r)^{-1}\delta(t-r) + \theta(t-r)\sum_{n=0}^{\infty} G_n(\underset{\sim}{x})(t-r)^n \tag{3.18}$$

$$S(x) = \sum_{n=0}^{\infty} S_n(\underset{\sim}{x})(t-r)^n \tag{3.19}$$

where

$$S_n(\underset{\sim}{x}) = \frac{1}{n!}\left(\frac{\partial}{\partial t}\right)^n S(t,\underset{\sim}{x})\Big|_{t=r} . \tag{3.20}$$

Upon inserting this expansion into $(\partial^2 + m^2 + S)D^R(x) = 0$, which holds for $t>0$, we verify that the coefficients of $\delta''(t-r)$ and $\delta'(t-r)$ vanish. The coefficient of $\delta(t-r)$ must also vanish, which gives

$$2[\frac{\partial G_o(\underset{\sim}{x})}{\partial r} + \frac{1}{r} G_o(\underset{\sim}{x})] + \frac{1}{4\pi r}[m^2 + S_o(\underset{\sim}{x})] = 0 , \qquad (3.21)$$

as does the coefficient of $\theta(t-r)(t-r)^n$,

$$2(n+1) \frac{\partial G_{n+1}}{\partial r}(\underset{\sim}{x}) + \frac{1}{r}G_{n+1}(\underset{\sim}{x}) + (-\underset{\sim}{\nabla}^2 + m^2)G_n(\underset{\sim}{x}) + \sum_{m=0}^{n} S_m(\underset{\sim}{x})G_{n-m}(\underset{\sim}{x}) = 0. \qquad (3.22)$$

These are ordinary differential equations with solutions

$$G_o(\underset{\sim}{x}) = -\frac{1}{8\pi r}\int_o^r [m^2 + S_o(\underset{\sim}{x}')]dr' + \frac{f_o(x)}{r} \qquad (3.23a)$$

$$G_{n+1}(\underset{\sim}{x}) = \frac{-1}{2(n+1)r}\int_o^r r' [(-\underset{\sim}{\nabla}'^2 + m^2)G_n(\underset{\sim}{x}') + \sum_{n=0}^{m} S_m(\underset{\sim}{x}')G_{n-m}(\underset{\sim}{x}')]dr' + \frac{f_n(x)}{r} , \qquad (3.23b)$$

where $\underset{\sim}{x}' = r'x$. Because the integrals vanish linearly in r, the first term in the above equations is regular. But the second term, $f_n(\hat{x})/r$, which arises as an integration constant, has a $1/r$ singularity, and produces the singular term $\delta(\underset{\sim}{x})$ when acted on by the wave operater, $\partial^2 1/r = 4\pi\delta(\underset{\sim}{x})$. This violates the homogeneous equation at $r=0$, $t>0$, and hence we have $f_n(\hat{x}) = 0$. On restoring the y dependence which was suppressed, we obtain with $\underset{\sim}{x} \to \underset{\sim}{x}-\underset{\sim}{y}$, and with new integration variable s, $r' = |\underset{\sim}{x}-\underset{\sim}{y}|s$,

$$G_o(\underset{\sim}{x}-\underset{\sim}{y},y) = -(8\pi)^{-1}\int_o^1 [m^2 + S_o(y^o + rs, \underset{\sim}{y} + (\underset{\sim}{x}-\underset{\sim}{y})s)]ds \qquad (3.24a)$$

$$G_{n+1}(\underset{\sim}{x}-\underset{\sim}{y},y) = -(2n+2)^{-1}r\int_o^1 s [(-\underset{\sim}{\nabla}^2+m^2)G_n((\underset{\sim}{x}-\underset{\sim}{y})s,y)$$

$$+ \sum_{n=0}^{m} S_m(y^o + rs, \underset{\sim}{y}+(\underset{\sim}{x}-\underset{\sim}{y})s)G_{n-m}((\underset{\sim}{x}-\underset{\sim}{y})s,y)]ds \qquad (3.24b)$$

$$D^R(x,y) = (4\pi r)^{-1}\delta(t-r) + \theta(t-r)\sum_{n=0}^{\infty} G_n(\underset{\sim}{x}-\underset{\sim}{y},y)(t-r)^n \qquad (3.25)$$

where $t=y^o-x^o$, $r = |\underset{\sim}{x}-\underset{\sim}{y}|$.

This expansion of the retarded propagator may be useful if there is a preferred coordinate frame, for example, if the external field $S(x)$ is time-independent in some frame. $S(t,\underset{\sim}{x}) = S(\underset{\sim}{x})$. However, in general, a manifestly covariant expression would be preferable. Instead of rearranging the above series, it is more convenient to pose the covariant expansion, with $z \equiv x-y$

$$D^R(x,y) = \theta(z^o)[\frac{1}{2\pi}\delta(z^2) + \theta(z^2)\sum_{n=0}^{\infty}\frac{1}{n!}(\frac{1}{4}z^2)^n G_n(x,y)]. \qquad (3.26a)$$

Making use of

$$\partial_\mu [(x^2)^n\theta(x^2)] = \delta_{n,o}2x_\mu\delta(x^2) + 2nx_\mu(x^2)^{n-1}\theta(x^2)$$

$$\partial^2 [(x^2)^n\theta(x^2)] = 4\delta_{n,o}\delta(x^2) + 4n(n+1)(x^2)^{n+1}\theta(x^2)$$

we obtain, with $\Sigma(x) \equiv m^2 + S(x)$,

$$[\partial^2 + \Sigma(x)]D^R(x,y) = \delta^4(z) + \theta(z^o)\delta(z^2)B(x,y) + \theta(z^o)\theta(z^2)\sum_{n=0}^{\infty}\frac{1}{n!}(\frac{z^2}{4})^n C_n(x,y)$$

where

$$B(x,y) = (2\pi)^{-1}\Sigma(x) + 4[G_o(x,y) + (x-y)^n\partial_\mu G_o(x,y)]$$

$$C_n(x,y) = [\partial^2 + \Sigma(x)]G_n(x,y) + [(n+2)G_{n+1}(x,y) + (x-y)^\mu \partial_\mu G_{n+1}(x,y)] ,$$

where the derivatives act on the first argument of the G's. Thus a formal solution of the equation for $D^R(x,y)$ is obtained if $B = C_n = 0$. This will be true if G_n is given by

$$G_n(x,y) = -\int_0^1 s^n H_n(y + (x-y)s,y)ds \qquad (3.26b)$$

where

$$H_n(x,y) = [\partial^2 + \Sigma(x)]G_{n-1}(x,y) \qquad n>0 \qquad (3.26c)$$

$$H_0(x,y) = (8\pi)^{-1}\Sigma(x) , \qquad (3.26d)$$

and the wave operator acts on the x variable. To verify that B and C_n vanish, observe that

$$\begin{aligned}(x-y)^\mu \partial_\mu G_n(x,y) &= -\int_0^1 s^{n+1}\frac{\partial}{\partial s}H_n(y + (x-y)s,y)ds \\ &= -H_n(x,y) + (n+1)\int_0^1 s^n H_n(y + (x-y)s,y)ds \\ &= -(\partial^2 + \Sigma(x))G_{n-1}(x,y) - (n+1)G_n(x,y).\end{aligned}$$

For the case $S(x)=0$, $\Sigma(x)=m^2$, one obtains $G_n = [8\pi(n+1)!]^{-1}(-m^2)^n$, and Eqs.(8.26a-d) reduce to the Bessel series.

The advantage of the expansions (3.24) - (3.25) or (3.26a-d) is their simplicity. They only involve one-dimensional integration whereas conventional perturbation theory (Born series) requires 4-dimensional integration. We postpone until the next section a discussion of the validity of these formal solutions, which are asymptotic away from the light cone.

If the characteristic surface is unchanged by the perturbation, the latter is classified as "weak." If, in addition, the leading singularity is the same as in the free case, we may call it "very weak." Thus the scalar external field is a very weak perturbation of the Klein-Gordon equation.

IV. Minimally Coupled Dirac Equation

We pose

$$L = \bar{\psi}(i\not{\partial} + e\not{A}-m)\psi + \bar{\eta}\psi + \bar{\psi}\eta \qquad (4.1)$$

corresponding to

$$(i\not{\partial} + e\not{A}-m)\psi = \eta \qquad (4.2)$$

Here $\eta(x)$ is a spinor probe and $A_\mu(x)$ is a given external electromagnetic potential. Our conventions are $\not{\partial} = \gamma^\mu\partial_\mu$, $\not{A} = \gamma^\mu A_\mu$, $\{\gamma^\mu,\gamma^\nu\} = 2g^{\mu\nu}$, $\bar{\psi} = \psi^\dagger\gamma^o$. The retarded fundamental solution $S^R_{\alpha\beta}(x,y)$ satisfies Eq.(4.2) with $\eta_\alpha(x) = \delta_{\alpha\beta}\delta^4(x-y)$ corresponding to a source at y with spinor component β. The spinor indices α,β=1,2,3,4 will be suppressed and we write

$$(i\not{\partial} + e\not{A}-m)S^R(x,y) = \delta^4(\kappa,y) \qquad (4.3a)$$

$$S^R(x,y) = 0 \qquad x^o<y^o \qquad (4.3b)$$

Our strategy in solving a matrix equation is to reduce it to another equation in which the matrix of highest derivatives is a scalar matrix. This will be done by multiplying the original matrix differential operator by a resolving divisor. For this purpose we pose

$$S^R(x,y) = -(i\not{\partial} + e\not{A}+m)\Delta^R(x,y) , \qquad (4.4)$$

which gives as the equation for $\Delta^R(x,y)$

$$[(\partial-ieA)^2 + m^2 - ie\sigma F]\Delta^R(x,y) = \delta^4(x,y) \qquad (4.5a)$$

$$\Delta^R(x,y) = 0 \qquad x^o<y^o \qquad (4.5b)$$

where $\sigma\cdot F \equiv \frac{1}{4}[\gamma^\mu,\gamma^\nu]F_{\mu\nu}$. The resolving divisor $-(i\not{\partial} + e\not{A}+m)$ would be the so-called Klein-Gordon divisor for e=0. The matrix of highest derivatives is a scalar matrix, as is the right hand side, and the leading operator is again the wave operator.

We temporarily suppress the y variable and pose

$$\Delta^R(x) = H(\underset{\sim}{x})\,\delta(t-r) + \text{less singular terms.} \qquad (4.6)$$

Here $H(\underset{\sim}{x})$ is a scalar, so the matrix character of Δ^R only enters the non-leading term. Thus again we have $S(x)$ = t-r, and Eq.(2.9) reads, with τ replaced by r,

$$2\frac{\partial H}{\partial r}(r,\hat{x}) + \frac{2}{r}H(r,\hat{x})-2ie[A^o(r,\hat{x})-\hat{x}\cdot\underset{\sim}{A}(r,\hat{x})]H(r,x) = 0. \qquad (4.7a)$$

This is completed with the initial condition

$$\lim_{r\to o} 4\pi r H(r,\hat{x}) = 1 , \qquad (4.7b)$$

so the inhomogeneous term $\delta^4(x)$ results at the origin. The solution is immediate,

$$H(\underset{\sim}{x}) = (4\pi r)^{-1}\exp[ieI(\underset{\sim}{x})] \qquad (4.8a)$$

$$I(\underset{\sim}{x}) = \int_o^r[A^o(x')-\underset{\sim}{x}\cdot\underset{\sim}{A}(x')]dr' , \qquad (4.8b)$$

where x' = (r,rx). The last integral is the line integral $\int A_\mu dx^\mu$ along the ray x of the light cone.

Having determined the leading singularity of Δ^R, we do not pause to obtain the non-covariant expansion, the analog of Eqs.(3.18) and (3.23), but instead we construct the analog of the covariant expansion (3.26a-d). For this purpose we write

$$\Delta^R(x,y) = \exp[ieI(x,y)]\Delta_I{}^R(x,y) \tag{4.9}$$

where

$$I(x,y) = \int_0^1 A_\mu[y + (x-y)s](x-y)^\mu ds \tag{4.10}$$

The factor $\exp[ieI(x,y)]$ makes $\Delta_I{}^R(x,y)$ a gauge invariant quantity and takes most of the sting out of the first order derivative term $-2ieA^\mu\partial_\mu$, as we shall see. Note first that

$$\begin{aligned}\frac{\partial}{\partial x^\mu} I(x,y) &= \int_0^1 ds\{A_\mu[y + (x-y)s] + (x-y)^\nu \frac{\partial}{\partial x^\mu}A_\nu[y + (x-y)s]\\ &= \int_0^1 ds[\frac{\partial}{\partial s}\{sA_\mu[y + (x-y)s] - s\frac{\partial}{\partial s}A_\mu[y + (x-y)s] + s(x-y)^\nu\partial_\mu A_\nu[y+(x-y)s]]\end{aligned}$$

so

$$\frac{\partial}{\partial x^\mu} I(x,y) = A_\mu(x) - A_\mu{}^I(x,y). \tag{4.11}$$

where

$$A_\mu{}^I(x,y) = -\int_0^1 ds\ s(x-y)^\nu F_{\mu\nu}[y+(x-y)s] \tag{4.12}$$

This gives

$$[\frac{\partial}{\partial x^\mu} -ieA_\mu(x)]\Delta^R(x,y) = \exp[-ieI(x,y)]\,[\frac{\partial}{\partial x^\mu} -ieA_\mu{}^I(x,y)]\Delta_I{}^R(x,y)\ , \tag{4.13}$$

and $\Delta_I{}^R(x,y)$ is the retarded solution of

$$\{[\frac{\partial}{\partial x^\mu} -ieA_\mu{}^I(x,y)]^2-m^2-ie\sigma\cdot F(x)\}\Delta_R{}^I(x,y) = \delta^4(x-y). \tag{4.14}$$

Note that A^I is expressed in terms of the electromagnetic field F by Eq.(4.12), so $\Delta_R{}^I$ satisfies a gauge invariant equation.

We now pose, with $z \equiv x-y$

$$\Delta_I{}^R(x,y) = \theta(z^0)\{(2\pi)^{-1}\delta(z^2)+\theta(z^2)\sum_{n=0}^{\infty}\frac{1}{n!}(\frac{z^2}{4})^n G_n(x,y)\quad , \tag{4.15a}$$

where $G_n(x,y)$ are matrix coefficients to be determined. Note that $(x-y)^\mu A_\mu{}^I(x,y)=0$, by Eq.(4.12), which greatly simplifies the equations for G_n since terms of the form $A_I{}^\mu(x,y)\partial_\mu[(x-y)^2]$ vanish. In fact, on substituting the expansion (4.15) into the equation for $\Delta_R{}^I(x,y)$, we obtain a formal solution, provided the G_n satisfy

$$(n+1)G_n(x,y) + (x-y)^\mu\partial_\mu G_n(x,y) + H_n(x,y) = 0,$$

where

$$H_n(x,y) = \{[\frac{\partial}{\partial x^\mu} -ieA_\mu{}^I(x,y)]^2+m^2-ie\sigma\cdot F(x)\}G_{n-1}(x,y) \tag{4.15b}$$

and

$$G_{n-1}(x,y) = (2\pi)^{-1}. \tag{4.15c}$$

The solution to this equation is given by

$$G_n(x,y) = -\int_0^1 ds\ s^n H_n[y + (x-y)s,y]\ , \tag{4.15d}$$

as is shown below Eq.(3.26b). Finally, we recall that the desired retarded propagator for the Dirac field is given by

$$S^R(x,y) = -[i\partial\!\!\!/_x + eA\!\!\!/(x) + m]\{\exp[ieI(x,y)]\Delta_R{}^I(x,y)\}, \tag{4.15e}$$

where I(x,y) and $A^I(x,y)$ are defined in Eqs.(4.10) and (4.12).

Let us now briefly consider whether the formal solutions, Eqs.(4.15), (3.18) or (3.26) actually converge to a solution of the given problem. In general such expansions converge only if the external potential is analytic. Examples of the converse are easily found. Suppose, for example, that the support of the external potential S(x) or $A_\mu(x)$ lies entirely inside the future light cone AyB with vertex at y, but nowhere touching it, as shown in Figure 1. The true retarded fundamental solution

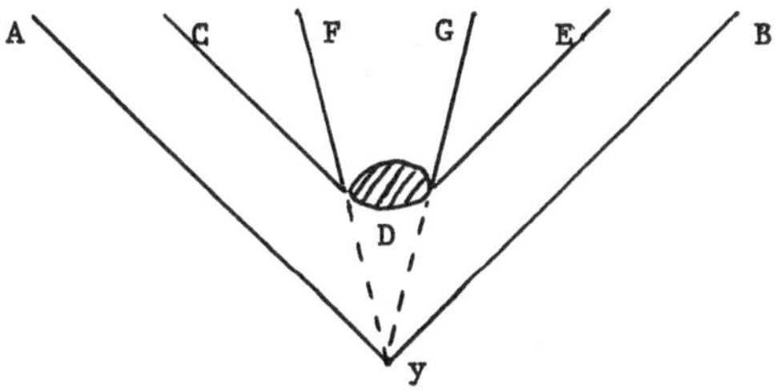

Fig. 1. The shaded region is the support of the external field.

will be influenced by the external field in the region CDE, the zone of influence of the external field, which is the union of future light cones with vertex in the support of the external field. On the other hand, the non-covariant expansion (3.18) about the light cone AyB will not reflect the presence of the external field at all, since the true solution coincides with the free solution outside of CDE. The covariant expansion (3.26) or (4.15) will be influenced by the external field only in the region FDG which consists of the cone of straight lines with vertex at y and passing through the support of the external field, (less the part between the vertex y and the support of the external field). Thus neither expansion converges to the true solution in this case. However Hadamard has shown[26] that if the external field is analytic, the expansions do converge to the true solution, and any external field may be approximated by a sequence of analytic functions. In the neighborhood of the light cone, the formal expansions are accurate.

Note that the coefficient of the leading singularity, Eq.(4.8) is influenced by the external field, although the singular surface is the same as the wave equation. This is true both for the Dirac and the Klein-Gordon equations with minimal electromagnetic coupling. We call this situation "moderately weak."

We make two remarks:

1) The expansion in powers of t-r (or $(x-y)^2$) is not perturbative. The coefficients of $(t-r)^n\theta(t-r)$ given by quadrature are exact to all orders in the electric charge e.

2) Causality in no way follows from the relativistic covariance of the equation. If the Klein-Gordon field ϕ were coupled to an external symmetric tensor field $A^{\mu\nu}(x)$, by $\partial_\mu[A^{\mu\nu}(x)\partial_\nu\phi(x)]$, the characteristic surface would not be the Minkowski light cone $x^2=0$. Causality and hyperbolicity are a feature of the Klein-Gordon and Dirac equations when coupling does not involve the highest derivatives. However, as we shall see in the following sections, this is not necessarily true for higher spin, in which the matrix of highest derivatives is a singular matrix.

V. Proca Equation in an External Gravitational Field

We make a brief digression on higher spin equations. (See also the review by Prof. R. Seiler in this volume). Beyond spin ½, the number of components of the fields grows much more rapidly than the spin degree of freedom: 4 components for a vector field to describe spin-one with 3 degrees of freedom, 16 for the Rarita-Schwinger field to describe spin 3/2 etc. For this reason it is necessary that the fields be subject to constraints, namely conditions that hold at a given time. In order to obtain consistency between the equations of motion in the presence of interaction and the constraints, which must be preserved in time, Fierz and Pauli[4] in 1939 made the ingenious proposal that constraints be a consequence of the Lagrangian equations. This may be achieved if the matrix of highest derivatives is singular. For in this case it possesses a null left eigenvector, on contraction with which, a lower order equation results. The simplest example of this is the free Proca lagrangian to describe spin-one:

$$L = -\tfrac{1}{4}G^{\mu\nu}G_{\mu\nu} + \tfrac{1}{2}m^2\phi^\mu\phi_\mu - \eta^\mu\phi_\mu \tag{5.1}$$

when $G_{\mu\nu} \equiv \partial_\mu\phi_\nu - \partial_\nu\phi_\mu$, corresponding to the lagrangian equation of motion

$$\partial^2\phi_\lambda - \partial_\lambda\partial^\mu\phi_\mu + m^2\phi_\lambda = \eta_\lambda \tag{5.2}$$

On taking the divergence of this equation, we find, in the homogeneous case, $\partial\cdot\phi=0$.

The method of characteristics achieves its full power in the case of coupling to a given external gravitational field described by a metric tensor $g_{\mu\nu}(x)$. The partial derivatives ∂_μ are replaced by covariant derivatives[27] ∇_μ. However there is a factor ordering problem because of the non-commutativity of the ∇_μ,

$$[\nabla_\mu,\nabla_\nu]\phi_\lambda = R_\lambda{}^\kappa{}_{\mu\nu}\phi_\kappa , \tag{5.3}$$

which gives, upon contraction,

$$[\nabla^\mu,\nabla_\nu]\phi_\mu = R_\nu{}^\kappa\phi_\kappa \tag{5.4}$$

where $R_{\lambda}{}^{\kappa}{}_{\mu\nu}$ is the Riemann curvature tensor and $R_{\mu\nu} = R_{\nu\mu}$ is the Ricci tensor $R_\nu{}^\kappa = g^{\lambda\mu}R_\lambda{}^\kappa{}_{\mu\nu}$. Thus, to the prescription $\partial_\mu \longrightarrow \nabla_\mu$, we must add an arbitrary multiple of $R_\nu{}^\kappa\phi_\kappa$ to obtain the Proca equation in the presence of an external gravitational field,

$$g^{\mu\nu}\nabla_\mu\nabla_\nu\phi_\lambda - \nabla_\lambda\nabla^\mu\phi_\mu + m^2\phi_\lambda + cR_\lambda{}^\mu\phi_\mu = \eta_\lambda . \tag{5.5}$$

To understand this from the Lagrangian point of view, observe that the possible term in the free lagrangian $c(\partial^\mu\phi_\nu\partial^\nu\phi_\mu - \partial^\nu\phi_\nu\partial^\mu\phi_\mu)$ is a divergence (and thus does not affect the free equations of motion) but the corresponding term of the generally covariant lagrangian is not.

On setting $\eta_\lambda(x) = g_\lambda^\sigma\delta^4(x,y)$, so there is a source for each spin component at an arbitrary space-time point y, we obtain the equation for the fundamental retarded so lution

$D_\lambda^\sigma(x,y)$ (we suppress the superscript R),

$$[g_\kappa{}^\lambda\nabla^\mu\nabla_\mu - \nabla_\kappa\nabla^\lambda + m^2g_\kappa{}^\lambda + cR_\kappa{}^\lambda]D_\lambda{}^\sigma(x,y) = g_\kappa{}^\sigma\delta^4(x,y) \tag{5.6a}$$

Here $D_\lambda{}^\sigma(x,y)$ is a bilocal quantity, with λ a vector index at x and σ a vector index at y, and the operator in brackets acts at x. In general relativity the geodesics issuing from a point y may cross (caustic), but we seek the retarded fundamental solution $D_\lambda{}^\sigma(x,y)$ for x in some sufficiently small neighborhood $\mathcal{U}_y$ of y so that this does not occur. For the retarded boundary condition we require the vanishing of $D_\lambda{}^\sigma(x,y)$ for $x\varepsilon\mathcal{U}_y$ in the past with respect to (written <) a space-like surface Σ_y through y dividing $\mathcal{U}_y$

$$D_\sigma{}^\lambda(x,y) = 0 \quad x\varepsilon\mathcal{U}_y \quad x<\Sigma_y. \tag{5.6b}$$

(Note that we avoid the issue of whether the past and future cones are globally disjoint, and our discussion deals only with what may be called "local causality.") For causality we require that $D_\lambda{}^\sigma(x,y)$ vanish unless the unique geodesic path from y to x is future time-like or light-like. (Because a geodesic is obtained by parallel displacement, its tangent vector is always either time-like or space-like or null).

Our basic strategy is, as before, to reduce this to an equation where the matrix of highest derivatives is a scalar matrix. As a first step in this direction, we factor out of D an operator which, in the absence of interaction, would be the Klein-Gordon divisor of the Proca equation. Put

$$D_\lambda{}^\sigma(x,y) = [g_\lambda{}^\rho + m^{-2}\nabla_\lambda\nabla^\rho]S_\rho{}^\sigma(x,y). \tag{5.7}$$

In the resulting equation for S, the terms involving 4∇'s are $m^{-2}[\nabla^\mu\nabla_\mu\nabla_\kappa-\nabla_\kappa\nabla^\mu\nabla_\mu]$ $\nabla^\rho S_\rho{}^\sigma(x,y)$. Since $\nabla^\rho S_\rho{}^\sigma(x,y)$ is a scalar in x, this term is $m^{-2}[\nabla^\mu\nabla_\kappa-\nabla_\kappa\nabla^\mu]$ $\nabla_\mu\nabla^\rho S_\rho{}^\sigma(x,y)$, which, by Eq.(5.4) is $m^{-2}R_\kappa{}^\mu\nabla_\mu\nabla^\rho S_\rho{}^\sigma(x,y)$. We thus obtain, as the equation for S,

$$[g_\kappa{}^\rho\nabla_\mu\nabla^\mu + c'R_\kappa{}^\lambda\nabla_\lambda\nabla^\rho + m^2g_\kappa{}^\rho + cR_\kappa{}^\rho]S_\rho{}^\sigma(x,y) = g_\kappa{}^\sigma\delta^4(x,y), \tag{5.8}$$

where

$$c' = m^{-2}(c+1). \tag{5.9}$$

Notice that because of the interaction the Klein-Gordon divisor has not given a scalar matrix as the matrix of highest derivatives. There is the additional term $c'R_\kappa{}^\lambda\nabla_\lambda\nabla^\rho$ which depends on the external field $R_\kappa{}^\lambda(x)$, which may affect the propagation. Nevertheless there is a gain over the original equation (5.6a) since the matrix of highest derivatives is not singular. (This matrix may be defined by replacing the differential operator ∂_μ by the vector $n_\mu = \partial_\mu S$, where $S=0$ is the equation of a characteristic surface).

Consider first the case where $c'=0$. All derivatives are contained in the scalar matrix operator $g_\kappa{}^\rho\nabla^\mu\nabla_\mu$, the matrix of highest derivatives being $g_\kappa{}^\rho g^{\mu\nu}(x)\partial_\mu\partial_\nu$. In this case the local cone of propagation coincides with the local light cone, defined by $g^{\mu\nu}(x)n_\mu n_\nu=0$, and the propagation is obviously causal. One may verify using Hamilton-Jacobi theory (exercise) that the leading singularity of $S_\lambda{}^\mu(x,y)$ is

$$\delta(\sigma(x,y))g_\lambda{}^\mu. \tag{5.10}$$

Here $\sigma(x,y)$ is the scalar function of x and y, introduced by Hadamard[1] and employed by

De Witt and Brehm[22] and Hobbs[23] which is the generalization of $\frac{1}{2}(x-y)^2$ to curved space, defined as follows. Let $z^\mu(x,y,\lambda)$ designate the unique geodesic path from y to x with affine parameter λ, $0 \leq \lambda \leq 1$, so $z^\mu(x,y,o) = y^\mu$ and $z^\mu(x,y,1) = x^\mu$. Put $\dot{z}^\mu = \partial z^\mu/\partial\lambda$, then

$$\sigma(x,y) = \tfrac{1}{2}\int_o^1 \dot{z}^\mu g_{\mu\nu}(z)\dot{z}^\nu d\lambda \tag{5.11}$$

[To verify (5.10) show, by Hamilton-Jacobi theory, that $\sigma(x,y)$ satisfies

$$\tfrac{1}{2}\partial_\mu\sigma(x,y)g^{\mu\nu}(x)\partial_\nu\sigma(x,y) = \sigma(x,y) \tag{5.12}$$

and use Eq.(2.4a)]. The support of $S_\mu{}^\nu(x,y)$ and hence of $D_\mu{}^\nu(x,y)$ is inside and on the cone of future null geodesics defined by $\sigma(x,y)=0$.

We now consider the nature of propagation for $c'\neq 0$. To lighten the notation, we rewrite Eq.(5.8) in an obvious matrix notation

$$[I\nabla\cdot\nabla + c'R\cdot\nabla\nabla + m^2I + cR]S(x,y) = I\delta^4(x,y). \tag{5.13}$$

and pose

$$S(x,y) = [I(\nabla\cdot\nabla + \nabla\cdot c'R\cdot\nabla) - c'R\cdot\nabla\nabla]E(x,y) \tag{5.14}$$

The resulting equation for E reads

$$[I\nabla\cdot\nabla(\nabla\cdot\nabla + \nabla\cdot c'R\cdot\nabla) + \text{lower order terms}]\,E(x,y) = I\delta^4(x,y). \tag{5.15}$$

Here the matrix of highest derivatives is a scalar matrix, as desired, and thus, a resolving divisor for the Proca equation with $c'\neq 0$ is

$$(I + m^{-2}\nabla\nabla)[I(\nabla\cdot\nabla + \nabla\cdot c'R\cdot\nabla) - c'R\cdot\nabla\nabla],$$

as is seen by combining Eqs.(5.7) and (5.14). The next term in Eq.(5.15) is

$$\{c'R\cdot\nabla[\nabla,\nabla\cdot c'R\cdot\nabla] + [c'R\cdot\nabla\nabla,\nabla\cdot\nabla]\}E(x,y),$$

which is of lower order because of the commutators.

The propagation is controlled by the leading scalar operator in Eq.(5.15) because it is strongly hyperbolic if its factors are.[25] If one repeats the reasoning of Sec.II for this equation, one obtains the Hamilton-Jacobi equation for the characteristic surface $S=0$, with $n_\mu \equiv \partial_\mu S$,

$$[n_\mu g^{\mu\nu}n_\nu][n_\kappa(g^{\kappa\lambda} + c'R^{\kappa\lambda})n_\lambda] = 0. \tag{5.16}$$

The first factor defines an ordinary ray which propagates along the null geodines (light rays). The second factor defines an extraordinary ray whose propagation is determined by the "extraordinary metric," $h^{\mu\nu} = g^{\mu\nu} + c'R^{\mu\nu}$, which depends on the Ricci tensor.

It is clear that if one considers a completely arbitrary Ricci tensor, then, in general, the propagation for $c'\neq 0$ will be acausal, or worse, the equation may cease to be hyperbolic (ill posed initial value problem). However we restrict our consideration to the case where the Ricci tensor is specified by Einstein's field equation $R^{\mu\nu} - \frac{1}{2}g^{\mu\nu}R_\lambda^\lambda = 2T^{\mu\nu}$, (in convenient units) in terms of the stress-energy tensor $T^{\mu\nu}$ of familiar physical field. Suppose, in particular the source of the external gravitational field is an external electromagnetic field $F_{\mu\nu}$,

$$T^{\mu\nu} = F^{\mu\lambda}F_{\lambda}{}^{\nu} - \tfrac{1}{4}g^{\mu\nu}F^{\kappa\lambda}F_{\lambda\kappa}.$$

(The analysis and conclusion are similar if the source is a perfect fluid). From $T^{\mu}_{\mu} = 0$ we have $R^{\mu}_{\mu} = 0$, and thus $R^{\mu\nu} = 2T^{\mu\nu}$, which gives as the "extraordinary metric tensor" $h^{\mu\nu} = g^{\mu\nu} + 2c'T^{\mu\nu}$

$$h^{\mu\nu} = g^{\mu\nu}(1 - \tfrac{1}{2}c'F^{\kappa\lambda}F_{\lambda\kappa}) + 2c'F^{\mu\lambda}F_{\lambda}{}^{\nu}. \tag{5.17}$$

The normals $n_{\mu} = \partial_{\mu}S$ to the characteristic surface $S(x)=0$ are determined by

$$n_{\mu}h^{\mu\nu}n_{\nu} = 0 \tag{5.18}$$

We fix our attention on a particular point x and choose a coordinate system such that $g^{\mu\nu}$ is Minkowskian at that point $g^{\mu\nu} = \mathrm{diag}(1,-1,-1,-1)$. Suppose further, for simplicity, there is no electric field, $\underset{\sim}{E}=0$, but that there is a non-vanishing magnetic field $\underset{\sim}{B}$. (Because $T^{\mu\nu}$ is invariant under $\underset{\sim}{E} \longrightarrow \underset{\sim}{E}\cos\theta + \underset{\sim}{B}\sin\theta$, $\underset{\sim}{B} \longrightarrow -\underset{\sim}{E}\sin\theta + \underset{\sim}{B}\cos\theta$, and because our equations are covariant, this describes a more general situation). Then with $F_{\mu\nu} = (\underset{\sim}{E},\underset{\sim}{B})$, Eq.(5.18) reads

$$(n^{o})^2[1 + c'\underset{\sim}{B}^2] - \underset{\sim}{n}^2[1 - c'\underset{\sim}{B}^2] - 2c'(\underset{\sim}{n}\cdot\underset{\sim}{B})^2 = 0 , \tag{5.19}$$

with solution

$$(n^{o})^2 = \underset{\sim}{n}^2\,\frac{1-c'\underset{\sim}{B}^2}{1+c'\underset{\sim}{B}^2} + 2c'\,\frac{(\underset{\sim}{n}\cdot\underset{\sim}{B})^2}{1+c'\underset{\sim}{B}^2} . \tag{5.20}$$

Consider now the propagation direction $\underset{\sim}{n}$ perpendicular to the magnetic field, $\underset{\sim}{n}\cdot\underset{\sim}{B}=0$, so

$$(n^{o})^2 = \underset{\sim}{n}^2\,\frac{1-c'\underset{\sim}{B}^2}{1+c'\underset{\sim}{B}^2} . \tag{5.21}$$

For a "weak" magnetic field, defined by $-1<c'\underset{\sim}{B}^2<1$, n^{o} is real, as it must be for a hyperbolic system, with maximum velocity of propagation

$$v_{max} = \frac{n^{o}}{|\underset{\sim}{n}|} = \frac{1-c'\underset{\sim}{B}^2}{1+c'\underset{\sim}{B}^2} . \tag{5.22}$$

Thus for $c'>0$ and weak field, the extraordinary ray has $v_{max}<1$ which is subluminal and thus causal. But for $c'<0$ and weak field, v_{max} exceeds the speed of light (one) and the propagation is acausal. On the other hand for a sufficiently strong electric field $|c'\underset{\sim}{B}^2|>1$, n^{o} is pure imaginary. In this case the system is not hyperbolic, no retarded solution exists, and the Cauchy initial value problem is not well posed. At the transition values $c'\underset{\sim}{B}^2 = 1$, the maximum velocity of the extraordinary ray drops to zero, whereas for $c'\underset{\sim}{B}^2 = -1$ the extraordinary ray attains infinite velocity. Since we require causal propagation for weak magnetic fields $c'<0$ is definitely excluded. On the other hand, unless some principle of maximum field strength is adopted, $c'>0$ is also excluded.

Thus the requirement that the Proca equation be causal and allow a well posed initial value problem in an arbitrary external gravitational field fixes uniquely $c'=0$ or, by Eq.(5.9), $c=-1$ which resolves the ambiguity of the minimal prescription $\partial_{\mu} \longrightarrow \nabla_{\mu}$. Thus the Proca Eq.(5.5) in a gravitational field reads

$$\nabla^{\mu}\nabla_{\mu}\phi_{\lambda} - \nabla_{\lambda}\nabla^{\mu}\phi_{\mu} + m^2\phi_{\lambda} - R_{\lambda}{}^{\mu}\phi_{\mu} = \eta_{\lambda} , \tag{5.23a}$$

or, by Eq.(5.4)

$$\nabla^{\mu}G_{\mu\lambda} + m^2\phi_{\lambda} = \eta_{\lambda} \tag{5.23b}$$

where

$$G_{\mu\lambda} = \nabla_{\mu}\phi_{\lambda} - \nabla_{\lambda}\phi_{\mu} . \tag{5.23c}$$

Gravitational coupling of a scalar field deserves to be called "strong" because it changes the characteristic surface from the Minkowski null cone to another one given by $\sigma(x,y)=0$ determined by a more general metric $g^{\mu\nu}(x)$ through Eqs.(5.10)-(5.12). This is also the case for the Proca equation with the particular gravitational coupling given in Eqs.(5.23). For other values of the constant c in Eq.(5.5), where the coupling term $cR_{\mu}{}^{\nu}\phi_{\nu}$ appears, corresponding to a different ordering of the covariant derivatives, the coupling may be called "too strong" since it leads to acausal propagation or non-hyperbolicity. The term $cR_{\mu}{}^{\nu}\phi_{\nu}$ is a gravitational quadrupole coupling. Thus we may say anomolous gravitational quadrupole coupling of the Proca particle is "too strong."

Finally we recall that the minimal electromagnetic coupling prescription is also ambiguous for the Proca equation. For the prescription $\partial_{\mu} \longrightarrow D_{\mu} = \partial_{\mu} - ieA_{\mu}$ leads to an arbitrary magnetic moment coupling $i\lambda F_{\mu}{}^{\nu}\phi_{\nu}$ because of the non-commutativity of the D_{μ}: $[D_{\mu},D_{\nu}]=-ieF_{\mu\nu}$, precisely analogous to the gravitational term $cR_{\mu}{}^{\nu}\phi_{\nu}$. However the requirements of hyperbolicity and causality are not as restrictive for electromagnetic as for gravitational coupling, becuase the Proca equation remains causal and hyperbolic[3,10] for any value of λ.

VI. Rarita-Schwinger Equation with Minimal Electromagnetic Coupling

Returning to flat space we consider the fundamental retarded solution of the Rarita-Schwinger equation

$$\not\pi g_\kappa^{\ \lambda} - \gamma_\kappa \pi^\lambda - \pi_\kappa \gamma^\lambda + \gamma_\kappa \not\pi \gamma^\lambda - m(g_\kappa^{\ \lambda} - \gamma_\kappa \gamma^\lambda) D_\lambda^{\ \mu}(x,y) = g_\kappa^{\ \mu} \delta^4(x-y) \tag{6.1}$$

where a matrix notation for Dirac spinor indices is employed (all quantities here are 4x4 Dirac matrices, including $D_\lambda^{\ \mu}(x,y)$ for each vector index λ and μ) and

$$\pi_\kappa = i \frac{\partial}{\partial x^\kappa} + eA_\kappa(x) \ , \tag{6.2}$$

with $[\pi_\kappa, \pi_\lambda] = ieF_{\kappa\lambda}(x)$. We introduce a matrix notation for the vector indices also, and rewrite Eq.(6.1) in the form

$$[\not\pi - \gamma\pi - \pi\gamma + \gamma\not\pi\gamma - m(1-\gamma\gamma)] D(x,y) = \delta^4(x,y) \ . \tag{6.3}$$

We pose

$$D(x,y) = [1 - \frac{1}{3}\gamma\gamma - \frac{1}{3m}\pi\gamma + \frac{1}{3m}\gamma\pi - \frac{2}{3m^2}\pi\pi] E(x,y) \ , \tag{6.4}$$

which gives

$$[\not\pi + \frac{2}{3}\frac{ie}{m^2} F^d \cdot \gamma\gamma^5\pi - m + \frac{ie}{3m} F^d \cdot \gamma\gamma^5\gamma] E(x,y) = \delta^4(x,y) \tag{6.5}$$

where $\gamma^5 = \gamma_0\gamma_1\gamma_2\gamma_3$, $(\gamma^5)^2 = -1$ and $F^d_{\kappa\lambda} \equiv \frac{1}{2}\varepsilon_{\kappa\lambda}^{\ \ \mu\nu} F_{\mu\nu}$, so $(F^d \cdot \gamma)_\kappa = F^d_{\ \kappa}{}^{\lambda}\gamma_\lambda$ etc. In the free case the last equation would read $(i\not\partial - m)E(x,y) = \delta^4(x,y)$, so the operator in bracket in Eq.(6.4) corresponds to the "Dirac divisor" of the Rarita-Schwinger operator. To obtain an operator whose leading terms would be the wave operator in the free case, we pose

$$E(x,y) = [\not\pi - \frac{2}{3}\frac{ie}{m^2} F^d \cdot \gamma\gamma^5\pi] G(x,y) \tag{6.6}$$

and obtain

$$[X + \text{l.o.}] G(x,y) = \delta^4(x,y) \tag{6.7}$$

where

$$X \equiv \pi^2 - \frac{4}{3}\frac{ie}{m^2} F^d \cdot \pi\pi\gamma^5 + \frac{4}{9}\frac{e^2}{m^4} F^d \cdot \gamma\pi \cdot F^d \cdot \gamma\pi \tag{6.8}$$

and "l.o." means a term which is lower order in derivatives than the leading term which is represented explicitly. On squaring X one finds

$$X^2 = [2\pi^2 + \frac{4}{9}\frac{e^2}{m^4}(\pi \cdot F^d)^2] X - \pi^2 [\pi^2 + \frac{4}{9}\frac{e^2}{m^4}(\pi \cdot F^d)^2] \tag{6.9}$$

and hence, on posing

$$G(x,y) = [-X + 2\pi^2 + \frac{4}{9}\frac{e^2}{m^4}(\pi \cdot F^d)^2] H(x,y) \tag{6.10}$$

we have

$$\{\pi^2 [\pi^2 + \frac{4}{9}\frac{e^2}{m^4}(\pi \cdot F^d)^2] + \text{l.o.}\} H(x,y) = \delta^4(x,y) \tag{6.11}$$

The analysis is now similar to the preceeding section. There is a cone of ordinary rays corresponding to the factor π^2, and a cone of extraordinary rays, with normals to the cone $n_\mu = \partial_\mu S$ satisfying

$$n^2 + \frac{4}{9}\frac{e^2}{m^4}(n\cdot F^d)^2 = 0. \tag{6.12}$$

Suppose, for simplicity that the magnetic field $\mathbf{B}$ vanishes, with electric field $\mathbf{E}$ non-vanishing. (For the opposite assumption see Ref. 2). We have

$$(n^o)^2 - \mathbf{n} - (\frac{2e}{3m^2}\mathbf{n}\times\mathbf{E})^2 = 0.$$

Consider a propagation direction $\mathbf{n}$ perpendicular to $\mathbf{E}$,

$$(n^o)^2 = \mathbf{n}^2[1 + (\frac{2e}{3m^2}\mathbf{E})^2] \ . \tag{6.13}$$

Observe that n^o is real, as it must be for a hyperbolic system, but the propagation is acausal,

$$v_{max} = \frac{n^o}{|\mathbf{n}|} = \quad 1 + (\frac{2e}{3m^2}\mathbf{E})^2 > 1. \tag{6.14}$$

This coupling is thus, again, too strong to be causal.

VII. Perspective

The acausality of the electrically coupled Rarita-Schwinger equation for spin-3/2 and the spin-two equation has been known for some time.[2,12] Since then no alternative equation for spin-3/2 or spin-2 has been exhibited which remains causal when electrically coupled. It appears a difficult task to find one starting from the various possible free particle equations, unless one had in hand a method which guaranties a priori that the coupled equation will remain causal. The difficulty is compounded if one also requires that the equation remain causal (and hyperbolic!) when coupled to an external gravitational field,[29] because, as we have seen for the Proca equation, this condition is more restrictive than electromagnetic coupling. The difficulty, as our examples show, is that when the matrix of highest derivatives is singular, as it must be to imply constraints, lower order terms, which contain coupling determine the maximum velocity of propagation and the propriety of the initial value problem.

If the difficulty is insuperable, and no one has overcome it, then we will have to abandon the Fierz-Pauli program of writing a free Lagrangian for a particle of arbitrary but unique mass and spin which may then be coupled in various ways.

A different mechanism for eliminating redundant components is gauge invariance, whereby equivalence classes of solutions and of initial data correspond to the same physical situation. Many physically interesting theories fall into this category: electrodynamics, general relativity, Yang-Mills theory and the Lagrangian formulation of fluid dynamics.[30] A striking success in this direction is the demonstration by Deser and Zumino[31] of the causality of supergravitational coupling of spin-3/2 and spin-2 fields.

The above perspective is mere prejudice on the part of the author. For what we call "the fundamental problem of higher spin equations" remains wide open: either produce a higher spin equation for a particle with unique mass and spin that may be coupled electrically (with arbitrary charge[29]) and gravitationally, or prove that none exists.

Acknowledgement: These lectures benifitted greatly from many instructive conversations with other members of the Ettore Majorana Summer School of Mathematical Physics, particularly Profs. L. Garding, R. Seiler and A.S. Wightman. I am particularly grateful for essential aid to my good friend and collaborator Giorgio Velo, whose own modesty and generous shouldering of administrative responsibilities kept him from lecturing himself.

References

1. Jacques Hadamard, Lectures on Canchy's Problem in Linear Partial Differential Equations, Yale University Press, New Haven, Connecticut (1923), reprinted by Dover Publications, New York (1952).

2. G. Velo and D. Zwanziger, Phys. Rev. 186, 1337 (1969).

3. G. Velo and D. Zwanziger, Phys. Rev. 188, 2218 (1969).

4. M. Fierz and W. Pauli, Proc. Roy. Soc. (London) A173, 211 (1939).

5. S. Kusaka, J. Weinberg (unpublished).

6. K. Johnson and E.C.G. Sudarshan, Ann. Phys. (N.Y.) 13, 126 (1961).

7. W. Rarita and J. Schwinger, Phys. Rev. 60, 61 (1941).

8. G. Velo and D. Zwanziger in "Troubles in the External Field Problem for Invariant Wave Equations," (A.S. Wightman, M.D. Cin, G.J. Iverson and A. Perlmutter, Eds.), p.8, Gordon and Breach, New York, (1971).

9. A. Proca, Compt. Rend. 202, 1420 (1936).

10. G. Velo, Communications in Mathematical Physics 43, 171 (1975).

11. G. Velo, Annales de l'Institut Henri Poincaré 22, 249 (1975).

12. G. Velo, Nuclear Physics B 43 389 (1972).

13. G. Velo, Nuclear Physics B 65, 427 (1973).

14. T. Darkhosh, New York University dissertation in the department of physics, June 1976.

15. A.S. Wightman, "Relativistic Wave Equations as Singular Hyperbolic Systems," pp.441-477 in "Partial Differential Equations," Proc. of Symposia in Pure Math. Vol. XXIII Amer. Math. Soc. (1973).

16. A.S. Wightman, this volume. Further references may be found here.

17. R. Seiler, this volume and references found there.

18. R. Guertin and T. Wilson, Annals of Physics 104, 427 (1977), Phys. Rev. D15, 1518 (1977). Many further references may be found here.

19. R. Krajcik and M. Nieto, Phys. Rev. D13, 924 (1976). An extensive bibliography may be found here.

20. A.S. Wightman in "Troubles in the External Field Problem for Invariant Wave Equations," p.1, Gordon and Breach, New York, (1971).

21. A.S. Wightman in "Studies in Mathematical Physics," (E.H. Lieb, B. Simon and A.S. Wightman, Eds.), Princeton University Press, Princeton (1976).

22. B. DeWitt and R. Brehme, Annals of Physics (N.Y.) 9, 220 (1960).

23. J. Hobbs, Annals of Physics (N.Y.) 47, 141 (1968).

24. R. Courant and D. Hilbert, Methods of Mathematical Physics, Vol. 2, pp. 590, 596, 618, 619.

25. L. Garding, this volume.

26. Ref.1, paragraphs 193-196.

27. See, for example, C. Misner, K. Thorne and J. Wheeler, Gravitation, W.H. Freeman & Co., San Francisco (1973).

28. The contents of this section were communicated to Prof. Wightman in a letter dated Jan. 1972. I am grateful to Prof. Wightman for remembering this otherwise forgotten manuscript and suggesting that it be made available to a wider audience.

29. See, however, the intriguing result of J. Madore, Phys. Letts. 55B, 217 (1975) who shows, in the linear approximation, that the acausal modes of the minimally coupled Rarita-Schwinger field are eliminated if gravitational coupling is in-

cluded and the coupling constants are related by $m=e/\sqrt{3G}$, where m and e are mass and electric charge, and G is the gravitational coupling constant.

30. Because of the non-uniqueness of the potentials, see, for example, F. Bretherton, J. Fluid Mech. 44, 19 (1970).

31. S. Deser and B. Zumino, Phys. Rev. Lett. 38, 1433 (1977).

Particles with Spin S $\leq$ 1 in an External Field

R. Seiler
Institut für theoretische Physik, Freie Universität Berlin
1000 Berlin 33, Arnimallee 3, Germany

The main purpose of the lectures will be to summarize results on particles with spin zero, one half and one in external fields. We start with a brief historical account. A second section is devoted to the canonical formalism of particles with spin zero and a third one to scattering theory for particles with spin s $\leq$ 1. In an appendix we summarize assumptions on free relativistic wave equations and fix the notation. In particular the Cauchy problem for the Fierz-Pauli spin $^3/_2$ equation is solved and the S-matrix for a Dirac electron in an external field of arbitrary size is constructed.

Lecture 1, History and Main Results

The interpretation of relativistic wave equations had a most interesting historical development. The first example - Maxwell's equation (Maxwell (1873)) - was an equation for a classical force field. Later Schrödinger (1926) attempted to use the equation, now called the Klein-Gordon equation (Klein (1926), Gordon (1926)), for the interpretation of the hydrogen spectrum; he had in mind a one particle theory for an election in the external field of the nucleus. This point of view had some serious shortcomings. One - the lack of the spin degeneracy of discrete energy eigenvalues - was overcome by Dirac's (1928) famous equation for a point particle with spin s = 1/2. However, as Dirac noticed already, the problem of splitting consistently the solutions with support on the in the forward from those in the backward lightcone remained the same as with the Klein-Gordon equation. This problem was made explicit by Klein (1929). The Klein paradox manifests itself in the case of the Klein-Gordon equation through an indefinite energy norm and imaginary eigenvalues of the Hamiltonian H. We come back to this question in the second lecture.

Dirac's equation for spin 1/2 could not be interpreted as a classical field equation because of the indefinite field energy. The problems with particle and field interpretation were overcome only after Dirac proposed the whole theory and the equation was put into the context of a many-body quantum field theory.

In the thirties equations for particles with arbitrary spin were considered. From the beginning it was clear that interest in relativistic wave equations is only well motivated if interaction is included. The equations proposed by Dirac (1936) for particles with higher spin were shown to be inconsistent if minimal coupling is introduced (this result was stated by Fierz (1939) and proved by Fierz and Pauli (1939); an updated version is included in Wightman's lecture notes).

In 1936 Pauli argued that quantization of a spin zero field according to Fermi-Dirac is inconsistent with locality and relativistic covariance. The general result, the theorem on spin and statistic for free fields, was given by Fierz (1939) and in a slightly generalized form by Pauli (1940). We will reproduce the argument using the notation of the appendix on free relativistic wave equations.

The solutions of $(\not{D}-m)\psi = 0$ can be parametrized in terms of the Fourier components $a_\varepsilon(p,s)$, $\varepsilon=\pm, p\in H_m, p^0>0, s=1\ldots n$, (67). Boson respectively Fermion quantization is introduced by demanding $a_\varepsilon(p,s)$ to be Fock space operators with vacuum Ω and the commutation (+) respectively anticommutation (-) relations

$$[a_\varepsilon(p,s), a_{\varepsilon'}(p',s')]_\pm = \frac{2\omega(p)}{m} \delta_{\varepsilon\varepsilon'} \delta^{(3)}(p - p'). \qquad (1)$$

All other (anti-) commutators shall vanish. Let Ψ now be the quantized field defined by (1) and (67). Then the commutator respectively anticommutator can be computed; if we use identities (69) for the projectors $\Lambda_{\pm}(p)$ the result is

$$[\psi(x),\psi^{+}(y)]_{\pm} = \frac{1}{(2\pi)^3} \int d\mu(p) \left(e^{-ipx}\Lambda_{+}(p) \pm \sigma e^{ipx}\Lambda_{-}(p)\right)$$

$$= \frac{1}{(2\pi)^3} \int d\mu(p) \left(e^{-ipx}\left(\frac{m+\not p}{2m}\right)\left(\frac{\not p}{m}\right)^{n-2} \pm \sigma e^{ipx}\left(\frac{m-\not p}{m}\right)\left(\frac{-\not p}{m}\right)^{n-2}\right)$$

$$= \frac{m+\not\partial}{2m}\left(\frac{\not\partial}{m}\right)^{n-2}\frac{1}{(2\pi)^3}\int d\mu(p)\left(e^{-ipx} \pm \sigma e^{ipx}\right).$$

This distribution has its support in the lightcone only if $\pm\sigma = -1$. Hence, the requirement of causality implies the well-known relation of spin and statistics.

In the fourties and fifties perturbative quantum field theory developed. The external field problem for electrons and positrons was solved (Feynman (1949), Mathews and Salam (1953), Schwinger (1954)). The renormalization program could be carried through quite easily for this particular case.

In the late sixties Wightman proposed the investigation of <u>stability of relativistic wave equations</u>. The aim was to characterize the equations giving rise to theories consistent with generally accepted principles (Wightman (1968)). One year later Velo and Zwanziger (1969) showed that the propagation cone of several well accepted wave equations depends on the external field and the type of coupling used. This discovery was preceded by an early note on the non-causality of the commutator of two observables in the Fierz-Pauli spin 3/2 theory with minimally coupled electromagnetic field (J. Weinberg (1943), S. Kusaka and J. Weinberg). A systematic investigation of all possible couplings to external fields for the most important wave equations of the form (65) with spin $s \leqslant 1$ produced the Velo-Zwanziger phenomenon even for the spin zero case (Wightman (1971)).

Existence of solutions of the Cauchy problem for the wave equations considered was more or less taken for granted in the early discussion of the Velo-Zwanziger phenomenon. Justification, however, turned out to be a difficult problem and has only partially been achieved. There are two main sources of difficulties. The first one is the vanishing of det γ^o for all relativistic wave equations for one mass and spin with the notable exception of the Dirac equation. For that reason (65) cannot be written in a Schrödinger type form $i\partial_t \psi = H\psi$. The second one comes from the fact that free relativistic wave equations typically give rise to hyperbolic systems. However, if an external field is added hyperbolicity is difficult to prove (see Svenson's theorem in the lectures by Garding). I should like to mention just two examples already discussed by Velo and Zwanziger where they have found non-causal propagation. The first case is the Procca spin one equation with symmetric tensor coupling - or the corresponding Petiau-Duffin-Kemmer equation. The equation is of the form

$$(\not{p} - m + B(x))\, \psi\, (x) = 0 \qquad (2\ )$$

We assume B to be smooth and localized in space and time, $B\in C_o^{\infty}(R^4;C^N)$. It can be shown that the Cauchy problem has a solution provided the external field is sufficiently small (Velo (1975), Minkowski and Seiler (1971)). There is, however, no fundamental solution for the operator $\not{p}$-m+B(x) if the external symmetric tensor field is large. This is demonstrated by Garding in his lectures. The second case is the Fierz-Pauli spin 3/2 equation with minimal coupling to the electromagnetic field $A_\mu(x)$; μ=0,1,2,3. The wave equation is again of the form (65); the partial differential operator L can be expressed in terms of Dirac matrices and $\pi = p + A$,

$$L(\pi) = (\not{\pi}-m)g_\mu{}^\nu - (\gamma_\mu\pi^\nu + \pi_\mu\gamma^\nu) + \gamma_\mu(\not{\pi}+m)\gamma^\nu.$$

Applying results of Leray and Ohya (1964) we will demonstrate that the Cauchy problem has a unique solution in a test function space of the Gevrey class.

The Gevrey test function space $\gamma_p^{(\alpha)}(\Omega)$ with indices $\alpha \geqslant 1$, p positive integer or infinity, consists of all functions $f\in C^\infty(\Omega)$ with $\sup_n (1+|n|)^{-\alpha}||D^n f||_p^{1/|n|}<\infty$. $\Omega\subset R^4$ denotes either the strip Y, $0<x^0<|Y|$ or the plane S_t, $x^0=t$. The Gevrey functions have the following important properties:

i) for α=1, p=∞ they are analytic; for α>1, however, they are non-quasi analytic, i.e. there exists a decomposition of unity by elements of $\gamma_\infty^{(\alpha)}(\Omega)$ with support arbitrarily small.

ii) $\gamma_\infty^{(\alpha)}(\Omega)$ is an algebra and $\gamma_p^{(\alpha)}(\Omega)$ a module over $\gamma_\infty^{(\alpha)}(\Omega)$.

As a first step towards the solution of the Cauchy problem we multiply $L(\pi)$ by the operator $k(\pi)$, the Klein-Gordon divisor K divided by $(m+\not{p})$,

$K(p) = (m+\not{p})k(p)$,

where p is replaced by π. The result is

$$M_L(\pi) = k(\pi)L(\pi) = (\not{\pi}-m)g_\mu{}^\nu + \lambda(\pi_\mu + \tfrac{1}{2}m\gamma_\mu)\gamma^5.\gamma.F^\nu$$

$$M_R(\pi) = L(\pi)k(\pi) = (\not{\pi}-m)g_\mu{}^\nu + \lambda\tilde{F}_\mu.\gamma.\gamma^5(\pi^\nu + \tfrac{1}{2}m\gamma^\nu).$$

Here we use the Dirac γ-matrices (appendix, remark 9). λ is a constant, $3m^2\lambda = 2i$, and $\tilde{F}$ the dual field strength tensor $F_\mu{}^\nu = \tfrac{1}{2}\varepsilon_\mu{}^{\nu k}{}_\lambda F_k{}^\lambda$, $\varepsilon^{0123}= 1$. The Cauchy problem for L can be solved if the corresponding one for M_R, M_L is solved. This will be shown below. The latter problem is simpler because the coefficient matrix of the time derivative in the M's is nonsingular.

Now we come to the second step where we apply the results of Leray and Ohya on linear hyperbolic systems:

Theorem 1: Suppose $A_\mu \in \gamma_\infty^{(\alpha)}(Y)$, $F(x)$ is uniformly bounded in x by a constant sufficiently small, $u_o \in \gamma_2^{(\alpha)}(S_o)$, $v \in \gamma_2^{(\alpha)}(Y)$ and $1<\alpha<2$. Then the Cauchy problem $Mu = v$ with data $u(x^o=0) = u_o$ and inhomogeneity v has a unique solution $u \in \gamma_2^{(\alpha)}(Y)$. Here M is either M_L or M_R. Furthermore we assume $A_\mu=0$ in a neighbourhood of $x^o=0$.

The proof is simply a verification that the assumptions of theorem 1 imply those of the theorems by Leray and Ohya. We will sketch the main ideas. The regularity assumptions of Leray and Ohya are immediately transcribed into the regularity assumptions of the theorem using the result already mentioned that $\gamma_\infty^{(\alpha)}(\Omega)$ is an algebra and $\gamma_2^{(\alpha)}(\Omega)$ a module over this algebra. The main assumption concerns the principal part $a_{pr}(x,p)$ of $a(k,p) = \det M$, where the determinant of a matrix with non-commuting entries is defined by

$$\det M = \sum_\pi \operatorname{sign}(\pi)\, M_{1,\pi(1)} \cdots M_{16,\pi(16)}.$$

The principal part can be computed to be

$$a_{pr}(k,p) = (p^2)^7 \,\{p^2 - \lambda^2 (p\tilde{F})^2\}.$$

The theorem of Leray and Ohya can be applied if a_{pr} factorizes into regular hyperbolic terms ($g(k,p)$ is regular hyperbolic if the roots p_o^i, $i = 1 \dots k$, of $g(k,p) = 0$ are real, finite, distinct and if every limit polynomial $\lim_{|x|\to\infty} g(x,p)$ has the same property). This is obviously correct if the external field strength F is sufficiently small.

The knowledge on M can be used to solve the Cauchy problem for L. If u is a solution of the Cauchy problem for M_R with data u_o then $\psi = k(\pi)u$ is a solution of the Cauchy problem for L with data $\psi_o = (k(\pi)u)(x^o=0)$. The question is now whether all Cauchy data $\psi_o \in \gamma_2^{(\alpha)}(S_o)$ are of this form. This can easily be shown if we use the additional assumption stated at the end of the above theorem: $A(x) = 0$ for $\varepsilon > x^o \geqslant 0$.
It suffices then to choose $u_o=(2m)^{-1}\psi_o$ so that $\psi_o = (k(\pi)u)(x^o=0)$ holds. Hence, under the above mentioned conditions the Cauchy problem for L has a solution in $\gamma_2^{(\alpha)}(Y)$ for data in $\gamma_2^{(\alpha)}(S_o)$. Uniqueness follows from uniqueness of the Cauchy problem for M_L.

More recently non-perturbative results were found on quantum fields with external interaction. This will be the subject of the remaining two lectures. The next one is devoted to the problem of the time evolution operator for spin zero (Schroer, Seiler, and Swieca (1970), Hochstenbach (1976)) and spin one half (Bongaarts (1970), Seiler (1972)). In the last lecture results on the existence of the S-matrix will be presented (Bellissard (1975,1976), Wightman (1973)).

Lecture 2, Canonical Theory for Spin 0 and 1/2

In this lecture we consider quantization of the Klein-Gordon and Dirac equation with minimal electromagnetic and scalar external fields. The external fields are supposed to be test functions in $\mathcal{S}(R^4)$. We follow mainly the articles by Hochstenbach (1976) and Seiler (1972).

Quantization proceeds in two steps. Firstly, classical phase space with time evolution is described. Secondly, the quantization map from phase space to the boson respectively fermion field algebra of linear operators on physical Hilbert space is defined. Time evolution on phase space induces via the quantization map an automorphism of the field algebra. The last and most fundamental question arises whether this automorphism is unitarily implementable.

The lecture is divided into two main parts. In the first one we discuss the Klein-Gordon in the second one the Dirac equation. In each part we discuss first the classical field aspects, the structure of phase space and time evolution. Then we apply general results on the unitary implementability of automorphism of boson respectively fermion algebras (Shale (1962), Shale and Stinespring (1965)). Finally the explicit construction of the unitary operator implementing the automorphism or rather of the vacuum for the theory with interaction is described, thereby reproducing the results of Shale and Stinespring.

1. The Klein-Gordon equation with minimally coupled electromagnetic and scalar external fields can be rewritten in two component form,

$$i\partial_t \Phi = H\,\Phi,\quad H = \begin{pmatrix} 0 & 1 \\ \vec{\pi}^2+m^2 & 0 \end{pmatrix} - A_0 \qquad (3)$$

$$\pi = p+A,\quad \Phi = \begin{pmatrix} \phi \\ \psi \end{pmatrix},\quad m = m_0+m_1,\quad m_0 \in R.$$

ϕ is the classical field and ψ the conjugate momentum. We adopt the follwoing general assumption: the external fields are test functions, more precisely if we talk about the time dependent case we assume $A \in \mathcal{S}(R^4,R^4)$, $m_1 \in \mathcal{S}(R^4,R)$ otherwise $A \in \mathcal{S}(R^3,R^4)$, $m_1 \in \mathcal{S}(R^3,R)$. This class of external fields is denoted by $\mathcal{J}$.

The Hamiltonian with no interaction H_0 is selfadjoint on the domain $D(H_0) = H^2(R^3) \oplus H^1(R^3)$ in $\mathcal{H}_E = H^1(R^3) \oplus H^0(R^3)$; $H^k(R^3)$ denotes the Sobolev space with norm $||f||_{2,k} = ||(1+\vec{p}^2)^{k/2} f||_2$. The norm and scalar product in $\mathcal{H}_E$ will be denoted by $||\ \ ||_E$ and $(\ ,\ )_E$, respectively. $\mathcal{H}_E$ is continuously injected - by identification - in a Hilbert space $\mathcal{H}_B$, the one body phase space, with the scalar product $(\Phi,\Phi)_B = (\Phi,|H_0|^{-1}\Phi)_E$. The symplectic structure is given by the imaginary part of the scalar product $(\ ,\ )_B$.

The structure of phase space is quite interesting and a short elucidation might be helpful for understanding quantization. The following model shows already all the essential properties: Let $L = C \oplus C$ be the complex linear vector space considered as a vector space over R with scalar product $(\Phi,\Phi') = \mathrm{Re}\,(\bar{\phi}\,\phi'+\bar{\psi}\psi')$; $\Phi,\Phi' \in L$. Consider furthermore the matrix operator $H = \begin{pmatrix} 0 & 1 \\ 1 & 0 \end{pmatrix}$ on L. It has the spectral decomposition $H = p_+ - p_-$, $2p_\pm = \begin{pmatrix} 1 & \pm 1 \\ \pm 1 & 1 \end{pmatrix}$. There are two natural complex structures on L, the first one inherited from $C \oplus C$ and denoted by i and the second one associated to H and defined by $j=i(p_+ - p_-)$. Let L_i and L_j denote the respective complexifications. The scalar product on L_i is the ordinary unitary one and the one on L_j is defined by

$$(\Phi_1,\Phi_2)_j = (\Phi_1,\Phi_2) + i(j\Phi_1,\Phi_2). \tag{4}$$

The imaginary part $i\sigma(\Phi_1,\Phi_2)$ of (4) can be expressed in terms of the i-sesquilinear form

$$q(\Phi_1,\Phi_2) = (\bar{\phi}_1\psi_2 + \bar{\psi}_1\phi_2),$$

using the identity

$$q(\Phi_1,\Phi_2) = i(j\Phi_1,\Phi_2) \tag{5}$$

There are several Lagrangian subspaces $\mathcal{L} \subset L$ with respect to the symplectic structure σ giving rise to complex conjugations K, $K = 1 \oplus (-1)$ on $L = \mathcal{L} + j\mathcal{L}$.

The complex linear transformation d

$$d : \Phi \to \Phi' = p_+\Phi + \overline{p_-\Phi}$$

maps L_i unitarily onto L_j: d intertwines the two complex structures on L.

The time evolution $\Phi(t) = (\exp -itH)\Phi$ gives rise to the relation $(d\Phi)(t) = (\exp -it1)(d\Phi)$. The generators of the two groups have different spectrum, in particular only the second one has positive spectrum.

We return now to the one body phase space $\mathcal{H}_B$ of the spin zero theory. There, too, is in addition to i a second complex structure $J = i(P_+ - P_-)$. $P_\pm$ denotes the spectral projectors of H_0 projecting on positive respectively negative energy states in $\mathcal{H}_B$. The complex space $\mathcal{H}_{B,J}$ - complexification of $\mathcal{H}_B$ with J - is unitarily mapped onto $\mathcal{H}_p = L^2(R^3,d\mu) \oplus L^2(R^3,d\mu)$ by D

$$D : \Phi \to A = \binom{a}{b},\ \phi(x) = \left(\frac{1}{2\pi}\right)^{3/2} \int d\mu(e^{-ipx}a(p)+e^{ipx}\overline{b(p)}),\ \psi(x) = \left(\frac{1}{2\pi}\right)^{3/2} \int d\mu\omega(e^{-ipx}a(p)-e^{ipx}\overline{b(p)}).$$

$\mathcal{H}_p$ will be called the particle-antiparticle phase space. On $\mathcal{H}_p$, $\mathcal{H}_B$ respectively, real bilinear forms (called charges) Q_p, Q_B are defined, given by

$$Q_B(\Phi_1,\Phi_2) = \int d^3x\,(\bar{\phi}_1\psi_2 + \bar{\psi}_1\phi_2) = Q_p(D\Phi_1,D\Phi_2),$$

$$Q_p(A_1,A_2) = \int d\mu(\bar{a}_1 a_2 - b_1\bar{b}_2). \tag{6}$$

The symplectic structure on $\mathcal{H}_p$,

$$\sigma(A_1,A_2) = \mathrm{Im}\,(A_1,A_2) \tag{7}$$

can be expressed in terms of the charge due to the identity

$$Q_p(A_1,A_2) = i(J\Phi_1,\Phi_2). \tag{8}$$

The Hamiltonian H defined by (3) is the sum of a selfadjoint operator H_o and a bounded, in general non-symmetric, perturbation. The domain of definition of H is $D(H_o)$. For any interaction in $\mathfrak{J}$ the equation of motion (3) leads to a propagator B (t,s) with the characteristic properties:

i) B maps $D(H_o)$ into itself and is uniformly bounded for (s,t) in a compact set in R^2.
ii) B is strongly continuous in both variables.
iii) $B(r,s)B(s,t) = B(r,t)$
iv) $B(t,t) = 1$.

We will use the notation $B(t) = B(t,0)$. Existence of a propagator for any interaction in $\mathfrak{J}$ follows from standard results (see e.g. Reed and Simon (1975) Vol II, Theorems X.69 and 70). For our special case it is convenient to use the interaction picture because $H_1(t) = e^{iH_ot}(H(t)-H_o)e^{-iH_ot}$ is bounded and strongly continuous in t. The equation of motion formulated in $\mathcal{H}_p$ is

$$i\partial_t A(t) = K(t)A(t)$$

$$K(t) = \begin{pmatrix} 1 & 0 \\ 0 & -1 \end{pmatrix} \tilde{H}_1(t). \qquad \tilde{H}_1(t) = DH_1(t)D^{-1}$$

K(t) is real linear; the propagator P(t) is given in terms of a norm convergent Dyson series

$$P(t) = \sum_{n=0}^{\infty} R_n(t),\ R_o = 1,\ R_n(t) = -i\int_0^t dt'K(t')R_{n-1}(t'). \tag{9}$$

P(t) can be easily checked to be an isometry with respect to charge,

$$Q_p(P(t)A_1,P(t)A_2) = Q_p(A_1,A_2). \tag{10}$$

In the case of a time independent external field the propagator B on $\mathcal{H}_B$ is an abelien group, $B(t_1)B(t_2) = B(t_1+t_2)$, with generator H. The analysis of H is simplified if one makes use of the second natural sesquilinear form on $\mathcal{H}_B$, the classical field energy,

$$\begin{aligned} E[\Phi] &= \int d^3x(\phi(\vec{\pi}^2+m^2)\phi+\bar{\psi}\psi - A_o(\bar{\phi}\psi+\bar{\psi}\phi)) \\ &= \int d^3x(\bar{\phi}(\vec{\pi}^2+m^2-A_o^2)\phi + |\psi-A_o\phi|^2). \end{aligned} \tag{11}$$

The classical field energy is positive definite for external fields in $\mathcal{J}$ subject to the supplementary condition $m^2-A_o^2 \geqslant \varepsilon > 0$. The subclass of those interactions is called $\mathcal{K}$. It is readily seen that there exist constants c_1,c_2 so that $c_1E[\Phi] \leqslant (\Phi,\Phi)_E \leqslant c_2E|\Phi|$, hence the quadratic forms are topologically equivalent. In this case H is selfadjoint on $\mathcal{H}_E$ with scalar product E and $(\Phi,B(t)\Phi)$ is uniformly bounded in t. If the external field does not belong to $\mathcal{K}$ B(t) is expected to blow up in time. This is the way the Klein paradox manifests itself in this context. In fact this claim can be supported by the analysis of explicit examples (Snyder and Weinberg (1940), for more information on spectral properties of H we refer to Narnhofer (1974), Weder (1976) and Nenciu (1976).

We come now to the second step of quantization, the definition of the field algebra, the automorphism generated by time evolution on classical phase space and the question of unitary implementability. Let $\mathcal{F}$ be the Fock space over $\mathcal{H}_p$ with vacuum Ω and particle operators $\underline{a}(p)$, $\underline{b}(p)$ generating the operator C^* algebra $\mathcal{A}$. (They are, of course, symbolic quantities and we should really talk about the exponentiated form of the smeared-out selfadjoint linear combinations.) The real linear quantization map R is defined by

$$R(A) = \sqrt{2}\ \mathrm{Im}\ (A,\underline{A}),\quad A\in \mathcal{H}_p,\quad \underline{A} = \begin{pmatrix}\underline{a}\\ \underline{b}\end{pmatrix}\ , \tag{12}$$

where the complex conjugate operator means the adjoint. The canonical commutation relation on finite particle states $\mathcal{F}_o$ holds,

$$[R(A_1),R(A_2)] \quad = i\mathrm{Im}\ (A_1,A_2). \tag{13}$$

The right-hand side of (12) and (13) can be expressed in terms of the charge (6). It follows now clearly from (10):

<u>Theorem 2:</u> For any interaction in $\mathcal{J}$ the time evolution operator P(t) defined by (6) generates a canonical automorphism α_t, on the boson field algebra $\mathcal{A}$,

$$\alpha_t(R(A)) = R(P(t)A). \tag{14}$$

A necessary and sufficient condition for existence of a unitary operator on Fock space implementing α_t is the following condition (Shale (1962)):

<u>Theorem 3:</u> Let P be a symplectic (i.e. P preserves σ, (7)) real linear mapping defined on $\mathcal{H}_p$. The automorphism generated by P can be unitarily implemented on Fock space if and only if $(P^TP)^{1/2}-1$ is Hilbert-Schmidt (H.S.)

We wish to translate this condition into one on the off-diagonal part of P only.

<u>Corollary 1:</u> Let P be defined by (9) for an interaction in $\mathcal{J}$,

$$P = \begin{pmatrix} P_{++} & P_{+-} \\ P_{-+} & P_{--}\end{pmatrix} .$$

The Shale condition is equivalent to the condition P_{+-} and P_{-+} are Hilbert-Schmidt. The proof is divided into 5 steps:

1. The Shale condition is equivalent to $P^T P - 1$ is H.S.:
This follows from the identity

$$P^T P - 1 = ((P^T P)^{1/2} - 1)((P^T P)^{1/2} + 1)$$

and the fact that $(P^T P) + 1$ has a bounded inverse.

2. Next we translate the conditions on P into conditions on a new operator S,

$$S = \begin{pmatrix} 1 & 0 \\ 0 & C \end{pmatrix} P \begin{pmatrix} 1 & 0 \\ 0 & C \end{pmatrix} . \qquad (15)$$

C is complex conjugation and reflection on $L^2(R^3, d\mu)$ $(cb)(p) = \overline{b(-p)}$. Of course P_{+-}, P_{-+} is H.S. whenever S_{+-}, S_{-+} is, and vice versa. The advantage of working with S is, that S is complex linear on $\mathcal{H}_p$ (because P is complex linear with respect to the complex structure $i\begin{pmatrix} 1 & 0 \\ 0 & -1 \end{pmatrix}$ and pseudo-unitary with respect to the sesquilinear form associated to the charge, i.e. let $g = \begin{pmatrix} 1 & 0 \\ 0 & -1 \end{pmatrix}$, then

$$SgS^* g = 1 \qquad (16)$$

$$gS^* gS = 1 \qquad (17)$$

3. If S_{+-}, S_{-+} is HS then $S^* S - 1$ is HS too: This is a consequence of the following identity resulting from (17)

$$S^* S - 1 = 2 \begin{pmatrix} S^*_{-+} & S^*_{++} \\ S^*_{--} & S^*_{+-} \end{pmatrix} \begin{pmatrix} S_{-+} & 0 \\ 0 & S_{+-} \end{pmatrix} . \qquad (18)$$

4. To prove the converse we first show that S_{++} and S_{--} have a bounded inverse: From (16) and (17) follows

$$S^*_{++} S_{++} = 1 + S^*_{-+} S_{-+} \geqslant 1, \; S^*_{++} S_{++} = 1 + S^*_{+-} S_{+-} \geqslant 1. \qquad (19)$$

The first inequality implies $\ker S_{++} = 0$, the second $(\text{range } S_{++})^{\perp} = 0$.

5. The pseudo-unitarity relations imply

$$\begin{pmatrix} S^*_{-+} & S^*_{++} \\ S^*_{--} & S^*_{+-} \end{pmatrix} = \begin{pmatrix} S^*_{++} & 0 \\ 0 & S^*_{--} \end{pmatrix} \begin{pmatrix} M & 1 \\ 1 & M^* \end{pmatrix} \qquad (20)$$

$M = S^{*-1}_{++} S^*_{-+}$. The first factor on the right-hand side of (20) is invertable as we have seen in the last paragraph.

6. The second factor on the right-hand side of (20) has a bounded inverse: Due to the closed graph theorem it is sufficient to show that $\begin{pmatrix} M & 1 \\ 1 & M^* \end{pmatrix}$ and its adjoint have vanishing kernels. Let $x = \begin{pmatrix} x_1 \\ x_2 \end{pmatrix}$ be an element of the kernel, then $Mx_1 = -x_2$, $M^*x_2 = -x_1$. This gives $M^*Mx_1 = x_1$. The unitarity equations lead to

$$M^*M = S_{--}^{*-1} S_{+-}^{*} S_{+-} S_{--}^{-1} = S_{--}^{*-1} (S_{--}^{*} S_{--} - 1) S_{--}^{-1}$$

$$= 1 - S_{++}^{*-1} S_{--}^{-1} .$$

Hence, x_1 is an element of ker $S_{++}^{*-1} S_{--}^{-1}$ and, therefore, vanishes. But then also $x_2 = 0$. The proof of the adjoint case is similar, concluding the proof of the corollary.

For applications it is important to note that the HS condition on P_{+-}, P_{-+} can be reduced to the corresponding one for the Born term:

Theorem 4: For any interaction in $\mathcal{J}$ a necessary and sufficient condition for α (14) to be unitarily implementable for $t \in I = [o,T]$ is $R_{1+-}(t) = -i \int_o^t ds K_{+-}(t)$ is Hilbert-Schmidt and continuous in this norm for $t \in I$.

Remark: The sufficiency was shown by Bongaarts (1970) and Schroer, Seiler and Swieca (1970); Hochstenbach (1976) showed the necessity.

Proof: Let us first obtain an estimate on the HS norm of $R_{n+-}(t)$, $\rho_n(t) = ||R_{n+-}(t)||_2$ in terms of $k = \sup_I ||K(t)||$ and $\rho = \sup_I \rho_1(t)$.

Definition (8) of R_n implies the inequality

$$\rho_{n+1}(t) \leq k \int_o^t ds\, \rho_n(s) + r_n \rho \qquad (21)$$

where r_n is defined by $r_n = \sup_I ||R_n(t)||$. By iteration it follows from (21)

$$\rho_{n+1}(t) \leq \rho\left(\frac{(kt)^n}{n!} + \ldots + \frac{(kt)^{n-\ell}}{(n-\ell)!} r_\ell + \ldots + r_n\right)$$

Furthermore, it follows again from (21) that

$$r_n \leq ktr_{n-1} \leq \ldots \leq \frac{(kt)^n}{n!}$$

Combining the two last unequalities we end up with

$$\rho_{n+1}(t) \leq \rho(n+1) \frac{(kt)^n}{n!}$$

Hence, the convergence in the HS norm of

$$P_{+-}(t) = \sum_n R_{n+-}(t)$$

is proved. For the proof of the converse statement, namely that $P_{+-}(t)$ HS implies

R_{1+-} HS, we refer to Hochstenbach (1976).

Application of the theorem leads to the following result:

Corollary 2: For any interaction in $\mathcal{J}$ with $\vec{A} = 0$, α is unitary implementable. For time independent interactions in $\mathcal{J}$ the condition is also necessary.

Proof: The argument will be given for the time independent case. (The more general case is reduced to this case by Fourier transformation in the time variable.) The Hilbert-Schmidt norm of the Born term $R_1(t)$ can be computed; the result is

$$||\int_o^t K_{+-}(t')dt'||^2_{HS} = \text{const.}\int d\mu(p)d\mu(p')$$

$$\left|\frac{(\omega(\vec{p})-\omega(\vec{p}'))A_o(p-p')-\vec{A}(\overrightarrow{p-p'})[\overrightarrow{(p+p')}+2\vec{A}(\overrightarrow{p-p'})]+m_1^2(\overrightarrow{p-p'})}{\omega(p)+\omega(p')}\right|^2 \sin^2(\frac{\omega(\vec{p})+\omega(\vec{p}')}{2})t$$

The integral exists for $\vec{A} = 0$ and for this case only. Furthermore it is continuous in t for $\vec{A} = 0$. This proves the corollary.

Now we come to an alternative approach to the problem of unitary implementability of a canonical automorphism α. It relies essentially on the explicit construction of the unitary transformation implementing α. Idea and methods go back to Friedrichs and are based on the following observation: Let a, b be two commuting boson operators with vacuum Ω; let V be the Bogoliubov-Valatin transformation:

$$V(\alpha) : \begin{pmatrix} a \\ b^* \end{pmatrix} \quad \begin{pmatrix} a(\alpha) \\ b^*(\alpha) \end{pmatrix} = V(\alpha) \begin{pmatrix} a \\ b^* \end{pmatrix}, \quad V(\alpha) = \begin{pmatrix} \mathrm{Ch}\alpha & \mathrm{Sh}\alpha \\ \mathrm{Sh}\alpha & \mathrm{Ch}\alpha \end{pmatrix}$$

The unitary transformation $W(\alpha) = \exp \alpha(A-A^*)$, $A=ba$, implements V,

$$V(\alpha) \begin{pmatrix} a \\ b^* \end{pmatrix} = W(\alpha) \begin{pmatrix} a \\ b^* \end{pmatrix} W^{-1}(\alpha).$$

The new vacuum $\Omega(\alpha)$, $a(\alpha)\Omega(\alpha) = b(\alpha)\Omega(\alpha) = 0$, is explicitly given by

$$\Omega(\alpha) = \sqrt{1-\mathrm{Tgh}^2\alpha} \quad \exp(-\mathrm{Tgh}\alpha A^*)\Omega,$$

unique up to a phase and analytic in α.

The Lie algebra generated by A and A^* has 3 dimensions. As a basis one can choose A, A^* and $[A,A^*]$. With that information the following formula for the normal form of W can be derived by standard differential equation methods,

$$W(\alpha) = \sqrt{1-\mathrm{Tgh}^2\alpha} : \exp \{\mathrm{Tgh}\alpha(A-A^*) - 2 \frac{\mathrm{Sh}^2\alpha/_2}{\mathrm{Ch}\alpha} [A,A^*]\} : \tag{22}$$

The automorphism α on the boson algebra $\mathcal{A}$ can be formulated in terms of S (15) and the particle operators $\underline{a}(p)$ and $\underline{b}^*(-p)$,

$$\alpha_t : \begin{pmatrix} \underline{a}(p) \\ \underline{b}^*(-p) \end{pmatrix} \rightarrow \begin{pmatrix} \underline{a}(t,p) \\ \underline{b}^*(t,-p) \end{pmatrix} = \int d\mu(p')\, S(p,p') \begin{pmatrix} \underline{a}(p) \\ \underline{b}^*(-p) \end{pmatrix} \tag{23}$$

Clearly α_t is unitarily implementable if and only if there exists a new vacuum $\Omega(t)$ in Fock space. We will prove the result:

Theorem 5: For any interaction in $\mathcal{J}$ with S_{+-} Hilbert-Schmidt the operator $L=S_{++}^{-1}S_{+-}$ is HS too; furthermore there exists a vacuum $\Omega(t)$,

$$\Omega(t) = (\det(1-L^*L))^{1/2} \exp - \int d\mu(p)d\mu(p')\underline{a}^*(p)L(p,p')\underline{b}^*(p')\Omega. \tag{24}$$

$\Omega(t)$ is unique up to a phase and analytic in the interactions (i.e. if $A \rightarrow z_1 A$ and $m_1 \rightarrow z_2 m_1$, then $\Omega(t,z_1,z_2)$ is analytic in z_1 and z_2).

Proof: As we have shown previously (step 4 in the proof of corollary 1), S_{++}^{-1} is a well-defined bounded operator; therefore $L = S_{++}^{-1}S_{+-}$ is HS and has the representation $L = \Sigma\lambda_i f_i \otimes g_i$, $\{f_i\}$, $\{g_\ell\}$ orthonormal basis of $L^2(R^3,d\mu)$, $\Sigma\lambda_i^2<\infty$. Now we argue that the right-hand side of (24) is a well-defined element of Fock space. The determinant is given by the infinite product $\prod_{i=1}^{\infty}(1-\lambda_i^2)$. For $\lambda_i < 1$ this is finite if and only if $\Sigma\lambda_i^2 < \infty$ (v. Neumann (1938)). But the eigenvalues of L are strictly smaller than one as we have seen before (step 5 in the proof of corollary 1). The exponential term in (24) in terms of $a_i^* = \int d\mu(p)f_i(p)\underline{a}(p)$, $b_i^* = \int d\mu(p)\overline{g_i(p)}\underline{b}^*(-p)$ is $\exp - \sum_{i=1}^{\infty} \lambda_i a_i^* b_i^* \Omega$. Since the operators $a_i^* b_i^*$ commute for different indices the problem is reduced to the case of two modes only, a case we have discussed before. We will not give the remainder of the proof of theorem because it is straight forward by realizing that uniqueness up to a phase of $\Omega(t)$ can be translated into uniqueness up to a phase of the Fock space vacuum Ω, provided S_{++} is invertible and has the total Hilbert space as its range. The latter follows because $\ker S_{++} = 0$ holds due to the relation $S_{++}S_{++}^* \geq 1$, (19). With this we finish the section on the Klein Gordon equation.

2. The discussion of the Dirac equation will closely follow the one of the Klein-Gordon equation. For analogous quantities we will use the same symbols as in the former case. The equation of motion can be written in the form (3),

$$i\partial_t \psi = H\,\psi, \quad H = \gamma^o(\vec{\pi}\vec{\gamma}+m) - A_o. \tag{25}$$

Always assuming the external fields to be in $\mathcal{J}$, H is selfadjoint with domain $D(H_o) \subset \mathcal{H}_B = \mathbb{C}^4 \otimes L^2(R^3)$. $\mathcal{H}_B$ stands for the spin ½ one body phase space. Scalar product of $\mathcal{H}_B$ is associated to the quadratic form $Q_B[\psi]$ by polarization.

The complex Hilbert space $\mathcal{H}_B$ carries an additional complex structure $J = i(P_+ - P_-)$ where $P_\pm$ denotes the spectral projectors of H_o for positive and negative spectrum, $H_o = \omega(P_+ - P_-)$. The complexification $\mathcal{H}_{B,J}$ is unitarily mapped onto the one particle phase space $\mathcal{H}_p = \mathbb{C}^2 \otimes L^2(R^3,d\mu) \otimes \mathbb{C}^2 \otimes L^2(R^3,d\mu)$. The mapping is explicitly given by

$$D : \psi \to A = \begin{pmatrix} a(p,s) \\ b(p',s') \end{pmatrix} ; \; s,s' = \pm 1.$$

$$\psi(x) = \left(\frac{1}{2\pi}\right)^{3/2} \sum_s \int d\mu(p) \, (e^{-ipx} u(p,s) a(p,s) + e^{ipx} v(p,s) \overline{b(p,s)}).$$

The charge gives rise to a bilinear form on $\mathcal{H}_p$,

$$Q_p(A_1,A_2) = \sum_s \int d\mu(p) \, (\overline{a_1(p,s)} a_2(p,s) + b_1(p,s) \overline{b_2(p,s)}).$$

As in the spin zero case there is a propagator P for the equation of motion (25) where K(t) is defined analogously,

$$K(t) = (1 \otimes (-1)) \, D H_1(t) D^{-1} \, , \; H_1(t) = e^{iH_o t}(H-H_o)e^{-iH_o t} \qquad (26)$$

$H_1(t)$ is the interaction part of H in the interaction picture. As previously P is given by a norm convergent Dyson series and is norm continuous in the arguments.

We come now to the second step of quantization, the definition of the field algebra, the automorphism generated by the time evolution on $\mathcal{H}_p$ and the question of unitary implementability. Let $\mathcal{F}$ be the Fock space over $\mathcal{H}_p$ with vacuum Ω and one particle operators $\underline{a}(p,s), \underline{b}(p,s)$, $s = \pm 1$. The real linear quantization map R is defined by

$$R(A) = \sqrt{2} \; \mathrm{Re}(A,\underline{A}), \; A \in \mathcal{H}_p, \; \underline{A} = \begin{pmatrix} \underline{a}(p,s) \\ \underline{b}(p,s) \end{pmatrix}$$

It is readily seen that the canonical anticommutation relations hold,

$$[R(A_1), R(A_2)]_+ = \mathrm{Re}(A_1,A_2).$$

Time evolution generates a canonical automorphism α_t and theorem 2 holds with the only change that $\mathcal{A}$ now denotes the fermion algebra.

Time evolution can alternatively be expressed in terms of the unitary transformation S(t),

$$S(t) = (1 \otimes C) P(t) (1 \otimes C), \qquad (27)$$

acting on $\underline{a}(p,s)$ and $\underline{b}^*(p,s)$:

$$\alpha_t : \begin{pmatrix} \underline{a} \\ \underline{b}^* \end{pmatrix} \to \begin{pmatrix} \underline{a}(t) \\ \underline{b}(t) \end{pmatrix} = S(t) \begin{pmatrix} \underline{a} \\ \underline{b}^* \end{pmatrix} \qquad (28)$$

The question of unitary implementability is answered by

Theorem 6 (Shale and Stinespring (1965)). Let S be the complex linear transformation defined by (27) and (9) for an external field in $\mathcal{J}$. Then the automorphism α_t (28) is unitarily implementable on the Fock space $\mathcal{F}$ if and only if S_{+-} and S_{-+} are Hilbert-Schmidt.

Remark: The original version of the theorem is slightly more general. There only the real linear structure of the vector space $\mathcal{H}_p$ is used and the criterion - corresponding to the one given in the theorem - is that iP - Pi is a real Hilbert-Schmidt operator.

The question of implementability can again be formulated in terms of the Born term only. Statement and proof are identical with the one for spin zero case, theorem 4.

Application of theorem 4 leads to the result almost identical with the spin zero case:
Corollary 3: For any interaction in $\mathcal{J}$ with $\vec{A} = m_1 = 0$ the automorphism α_t is unitarily implementable.

We have not investigated the necessity of the conditions in the corollary. However, it is clear that a generic interaction in I will not pass the test of theorem 4 because the integral

$$||\int_o^t dt' K_{+-}(t')||^2_{HS} = \int d\mu(p) d\mu(q) \mathrm{Tr}(\not{p}-m)\gamma^o(-\not{A}(\vec{p}-\vec{q})+m_1(\vec{p}-\vec{q}))(\not{q}+m)(-\not{A}(\vec{p}-\vec{q}) + +m_1(\vec{p}-\vec{q}))\gamma^o \frac{\sin^2 \frac{1}{2}(\omega(\vec{p})+\omega(\vec{q}))t}{(\omega(\vec{p})+\omega(\vec{q}))^2} \quad (29)$$

will not converge. On the other hand, the integral in (29) is finite under the assumptions of the corollary.

In the fermion case there is an alternative approach to the problem of unitary implementability , too, as in the previously discussed spin zero case. However, there is a slight complication compared to the former case as the following model shows:

Let a and b be two anticommuting fermion operators with vacuum Ω and V the Bogobubov-Valatin transformation defined by

$$V(\alpha) : \begin{pmatrix} a \\ b^* \end{pmatrix} \quad \begin{pmatrix} a(\alpha) \\ b^*(\alpha) \end{pmatrix} = V(\alpha) \begin{pmatrix} a \\ b^* \end{pmatrix}, \quad V(\alpha) = \begin{pmatrix} \cos\alpha & \sin\alpha \\ -\sin\alpha & \cos\alpha \end{pmatrix}$$

V can be implemented by the unitary transformation $W(\alpha) = \exp\alpha(A-A^*)$, $A = ba$. The explicit formula for the new vacuum $\Omega(\alpha)$ depends on whether $\cos\alpha \neq 0$ or not,

$$\Omega(\alpha) = \begin{cases} (1+tg^2\alpha)^{-1/2} \exp(-tg\alpha A^*)\Omega & , \cos\alpha \neq 0 \\ \pm A^*\Omega & , \cos\alpha = 0 . \end{cases} \quad (30)$$

$\Omega(\alpha)$ is analytic in α for an appropriate choice of the sign. The exceptional case $\cos\alpha = 0$ is called a strong Bogolinbov-Valatin transformation (Labonté (1974)).

As mentioned previously the automorphism α_t generated by the unitary transformation

S on $\mathcal{H}_p$ is implementable if and only if there is a new vacuum $\Omega(t)$.

The main assumption in the following theorem is the Hilbert-Schmidt property of S_{+-}, S_{-+},

$S_{+-} = \Sigma\, \mu_k\, f_k \otimes g_k$.

As the previous example shows the unitarity equations do not imply the μ_k's to be smaller than one as it was in spin zero case. It will be necessary to split $\mathcal{H}_p$ into a subspace $\mathcal{H}'_p$ and the orthogonal complement $\mathcal{H}''_p$ where $\mathcal{H}'_p$ is the finite dimensional subspace spanned by the g_k's with $\mu_k^2 = 1$. Hence, $\mathcal{H}'_p$ is the eigenspace of $S_{+-}S_{+-}$ with eigenvalue one; for convenience we reorder the g's so that

$$\mathcal{H}'_p = \langle g_1, \ldots, g_N \rangle. \tag{31}$$

With this notation we are ready to state the result on the existence and explicit form of the new vacuum (Friedrichs (1951), Labonté (1974)):

Theorem 7: Let S be the unitary propagator (27) for an interaction in $\mathcal{J}$. Suppose S_{+-}, S_{-+} to be Hilbert-Schmidt and define the particle-antiparticle operators by

$$a_i^* = \Sigma \int d\mu(p)\; f_i(p,s) a^*(p,s)$$

$$b_k^* = \Sigma \int d\mu(p)\; \overline{g_k(p,s)} b^*(-p,s).$$

Then the operator $L = S_{++}^{-1}\, S''_{+-}$ is Hilbert-Schmidt, where S''_{+-} denotes the restriction of S_{+-} to the subspace $\mathcal{H}''_p = \mathcal{H}'^{\perp}_p$.

The new vacuum $\Omega(t)$ is given by the formula

$$\Omega(t) = \pm(\det(1+L^*L))^{-\frac{1}{2}} \exp - \int d\mu(p) d\mu(p') a^*(p,s) L(p,p',s,s') b^*(-p',s') \prod_{\ell=1}^{N} A_\ell^* \,\Omega \tag{32}$$

where $A_\ell = a_\ell b_\ell$. The vacuum is unique up to a phase and the sign in (32) has to be determined by continuity as in the two mode case (30).

The only part where the proof of this theorem differs essentially from the corresponding one for the Klein-Gordon equation (theorem 5) is in the demonstration that L is Hilbert-Schmidt on $\mathcal{H}''_p$. We will focus our attention only on this point. Due to the unitarity of S the inequalities hold

$S_{++}^* S_{++} \leq 1, \; S_{+-}^* S_{+-} \leq 1.$

Hence, S_{++} is a bounded linear operator with norm not larger than one. Let ϕ be in the range of S''_{+-} with $\varepsilon = \max_k (1-\mu_k^2)^{-1}$ we have

$$||\phi||^2 \leq \varepsilon \sum_k (1-\mu_k^2)|(\phi,g_k)|^2 = \varepsilon\,(\phi,(1-S_{+-}^* S_{+-})\phi) = \varepsilon ||S_{++}\phi||^2.$$

Hence, S_{++} has a bounded inverse on the range S''_{+-}, proving that L is HS on $\mathcal{L}''_p$.

To finish this chapter we add some additional remarks:

1. In the spin zero and spin one half case there are not only expressions for the new vacuum but also for the unitary transformation implementing Bogoliubov-Valatin transformations of the boson respectively fermion algebra (Friedrichs (1951), Berezin (1965), Labonté (1974), Ruijsenaars (1977)).

2. The unitary operators implementing the automorphism can be chosen to be strongly continuous in the time variable. In the abstract setting and for time independent external fields this follows from a theorem by Kallman (1971). In the constructive version and for general interactions in $\mathcal{J}$ it follows from the fact that the new vacuum can be chosen to be a continuous function of time. This follows from the explicit formula (24) and (32) and the continuity properties of the propagator.

3. On the properties of the self-adjoint generator, the energy operator, the following is known: In the case of spin zero with interactions in $\mathcal{J}$ the spectrum is bounded from below (by zero) if $m^2-A_o^2 \geqslant 0$, Hochstenbach (1976). If this inequality does not hold the spectrum is expected to be unbounded from below (Klein Paradox). According to Schroer and Swieca (1970) such a theory might still have a physically respectable interpretation. A similar statement holds for the spin one half case. The spectrum is bounded below if $0 \notin \sigma(H)$, where H is given by (25). (Weinless (1969)).

4. The quantum fields for interactions in $\mathcal{J}$ are local, i.e. all commutators of ϕ,π and their adjoints vanish for space like arguments.

Lecture 3, Scattering Theory

In this last lecture we will present results on scattering for particles with spin $s \leq 1$ in external fields. The general frame is again a relativistic wave equation with interaction B,

$$(\not{p} - m + B)\, \psi = 0. \tag{33}$$

B is a matrix valued test function, $B \in \mathcal{S} = \mathcal{S}(R^4, \mathbb{C}^N \times \mathbb{C}^N)$. The Cauchy problem with data ψ^{out}, respectively ψ^{in}, is summarized by the Yang-Feldman equations,

$$\psi(x) = \psi^{in}(x) - \int dy S_R(x-y)B(y)\psi(y) \tag{34}$$

$$\psi(x) = \psi^{out}(x) - \int dy S_A(x-y)B(y)\psi(y). \tag{35}$$

S_R and S_A are the retarded and advanced fundamental solution of the free equation $(\not{p}-m)\psi^{in}=0$. The out-field can be expressed in terms of the in-field and the Capri-Wightman operators T_R and T_A (see lecture notes of the course by Wightman (this volume) and Capri (1969)),

$$\psi^{out}(f) = \psi^{in}(T_R^{-1}T_A f). \tag{36}$$

This equation defines a canonical automorphism α, the scattering automorphism of the field algebra of the in-field. The problem is now the following: under what conditions can α be implemented in Fock space of ψ^{in}.

It will turn out that α is implementable in many cases where we have found non-existence of a unitary operator implementing time evolution. This striking fact will be first explained for the case of the Dirac spin 1/2 theory. After that we prove the implementability for the case of arbitrary relativistic wave equations with small and so-called nice interactions.

The result of unitary implementability for the Dirac spin 1/2 case with arbitrary interaction in $\mathcal{S}$ has been announced by Wightman (private communication). It is a generalization of a result of Bellissard (1976) who proved it for sufficiently small interactions. We will give two alternative proofs. The result for the general case of small, nice interactions is due to Bellissard (1976).

In our first discussion of the Dirac spin 1/2 theory we will use the same formalism as in the last lecture. The aim is to show clearly the difference between the problem of implementing the time evolution automorphism α_t and the scattering automorphism α. In order to simplify matters, we suppose that the interaction vanishes outside a finite interval $[0,T]$. The equation of motion in the interaction picture is given by

$$i\partial_t \psi = H_1(t)\psi \tag{37}$$

where $H_1(t)$ is an operator on the one-body space $\mathcal{H}_B = L^2(R^3) \otimes \mathbb{C}^4$ defined as follows:

$$H_1(t) = e^{iH_0 t} V e^{-iH_0 t}, \quad V = -\gamma_0 B. \tag{38}$$

There is a propagator for (37) defined by the norm convergent Dyson series

$$U(t) = \sum_{n\geq 0} U_n(t), \; U_n(t) = -i \int_{-\infty}^{t} dt' H_1(t') U_{n-1}(t'). \tag{38}$$

According to the theorem by Shale and Stinespring (lecture 2) the scattering automorphism α is implementable if and only if the operator $U = U(\infty)$ satisfies the following condition: $P_+ U P_-$ is Hilbert-Schmidt. Recall that P_+ and P_- denote the spectral projectors of H_0 for positive and negative spectrum. In the following we will use the notation $U_{++} = P_+ U P_+$, etc.

To show the mechanism of the proof of unitary implementability we consider first the kernel of the Born term,

$$U_{1,+-}(p_1,p_2) = -i \int_{-\infty}^{+\infty} e^{i(\omega(p_1)+\omega(p_2))t} \, V_{+-}(t,p_1-p_2)dt \; .$$

Since V vanishes for large $|t|$, the integral can be transformed by n partial integrations, $n \in Z_+$,

$$U_{1,+-}(p_1,p_2) = -i \int_{-\infty}^{+\infty} dt \; e^{i(\omega(p_1)+\omega(p_2))t} \left(\Gamma^n \left(i\frac{d}{dt} \right)^n V(t,p_1-p_2) \right)_{+-} \tag{39}$$

Γ^n denotes n successive application of Friedrichs Γ operator (Friedrichs (1960))

$$(\Gamma A)(p,q) = \begin{pmatrix} \dfrac{A_{++}(p,q)}{\omega(p)-\omega(q)} & \dfrac{A_{+-}(p,q)}{\omega(p)+\omega(q)} \\[2ex] -\dfrac{A_{-+}(p,q)}{\omega(p)+\omega(q)} & -\dfrac{A_{--}(p,q)}{\omega(p)-\omega(q)} \end{pmatrix} \tag{40}$$

If we choose $n\geq 2$ in (39) the integrand is a Hilbert-Schmidt kernel, with a HS norm uniformly bounded in t vanishing for large $|t|$. Hence $U_{1,+-}$ is Hilbert-Schmidt due to the standard

<u>Lemma 1:</u> Let $B(t)$ be a HS operator for all $t \in [a,b]$, $B(t)$ strongly continuous for $t \in [a,b]$, and $||B(t)||_{HS} \leq f(t)$ for some integrable $f(t)$. Then $A = \int_a^b dt\, B(t)$ is H.S. and $||A||_{HS} \leq \int_a^b dt \; ||B(t)||_{HS}$. For a proof see e.g. Boongaarts (1970).

Next we come to the n-th order term in (38). U_{n+-} can be written as a sum of n terms,

$$U_{n+-} = (-i)^n \sum_{k=1}^{n} U_{nk}$$

$$U_{nk} = \int_{-\infty}^{\infty} dt_1 \int_{-\infty}^{t_1} dt_2 \int_{-\infty}^{t_{n-1}} dt_n \, (H_1(t_1)\ldots H_1(t_{k-1}))_{++} H_{1+-}(t_k) H_{1,--}(t_{k+1})\ldots H_{1,--}(t_n).$$

Each term U_{nk} can be rewritten in the following form

$$U_{nk} = \int_{-\infty}^{+\infty} dt_k \, A_{++}(t_k)\, H_{1+-}(t_k)\, A_{--}(t_k) \tag{41}$$

$$A_{++}(t) = \int_{t}^{\infty} dt_{k-1} \int_{t_2}^{\infty} dt_1 (H_1(t_1)\ldots H_1(t_{k-1}))_{++}$$

$$A_{--}(t) = \int_{-\infty}^{t} dt_{k+1} \ldots \int_{-\infty}^{t_{n-1}} dt_n \, H_{1,--}(t_{k+1}) \ldots H_{1,--}(t_n)$$

The essential point in the representation (41) is that the integration over t_k is done last and the range of integration is the interval $(-\infty, +\infty)$. Furthermore, A_{++}, A_{--} and their derivatives are uniformly bounded and norm continuous in t_k.

Before we estimate the H.S. norm of U_{nk} we should like to make some remarks on the Friedrichs Γ operator. Γ has the characteristic property

$$[H_o, \Gamma(H_1(t))] = H_1(t). \tag{42}$$

From this follow the two identities,

$$\begin{aligned} H_1(t) &= -i\Gamma(H_1'(t)) + i\Gamma(\dot{H}_1(t)) \\ &= -\Gamma^2(H_1''(t)) + 2\Gamma^2(\dot{H}_1'(t)) - \Gamma^2(\ddot{H}_1(t)), \end{aligned} \tag{43}$$

where prime denotes total time derivative and dote the time derivative acting on V only (see definition of H_1 (38)). The two identities can be used to transform U_{nk} by partial integration as for the Born term.

If we insert the expression (43) for $H_1(t)$ into (41), it is easily seen that the contribution from the last term $\Gamma^2(\ddot{H}_1(t))$ gives rise to a HS operator. The second term gives

$$\begin{aligned} &2 \int_{-\infty}^{+\infty} dt \, A_{++}(t)\Gamma^2(\dot{H}_1'(t))A_{--}(t) = \\ &= -2 \int dt \, A_{++}'(t)\Gamma^2(H_1(t))A_{--}(t) - 2 \int dt \, A_{++}(t)\Gamma^2(\dot{H}_1(t))\, A_{--}'(t), \end{aligned} \tag{44}$$

and the first one

$$-\int dt\, A_{++}(t)\Gamma^{2}(H_1''(t))\, A_{--}(t) =$$

$$= -\int dt\, A_{++}''(t)\Gamma^{2}(H_1(t))A_{--}(t) -2\int dt A_{++}'(t)\Gamma^{2}(H_1(t))A_{--}'(t) \quad (45)$$

$$-\int dt\, A_{++}(t)\Gamma^{2}(H_1(t))A_{--}''(t).$$

The Hilbert-Schmidt norm of U_{nk} can now be estimated. Let α be defined by

$$\alpha = \sup_t (||H_1(t)||,||H_1'(t)||),$$

then one gets the bounds

$$||A_{++}^{(s)}(t)|| \le 2\,(\alpha T)^{k-1}\ \mathrm{Max}\,(\frac{1}{(k-3)!},\ 1)$$

$$||A_{--}^{(s)}(t)|| \le 2\,(\alpha T)^{n-k}\ \mathrm{Max}\,(\frac{1}{(n-k-2)!},\ 1),\ s = 0,1,2.$$

The superscript (s) denotes time derivatives and the factorial is defined via the Gamma function. The second term (44) can be estimated by

$$||2\int_{-\infty}^{+\infty} dt\, A_{++}\Gamma^{2}(\dot{H}_1')A_{--}||_{HS} \le c(n,k)\int_{-\infty}^{+\infty} dt||\Gamma^{2}(\dot{H}_1)||_{HS} \quad (46)$$

and the first one (45) by

$$||\int dt\, A_{++}\Gamma^{2}(H_1'')A_{--}||_{HS} \le c(n,k)\int dt||\Gamma^{2}(H_1)||_{HS} \quad (47)$$

where we used the abbreviation

$$c(n,k) = 2^{3}(\alpha T)^{n-1}\ \{\mathrm{Max}\,(\frac{1}{(k-3)!},1)\}\cdot\{\mathrm{Max}\,(\frac{1}{(n-k-2)!},1)\}.$$

Due to the inequality

$$\sum_{k=1}^{n} c(n,k) \le 2^{3}(\alpha T)^{n-1}n \qquad \text{for} \quad n \le 4$$

$$\le 2^{3}(\alpha T)^{n-1}\ \frac{2^{n-2}}{(n-3)!} \qquad \text{for } n > 4$$

we get the estimate

$$||U_{n+-}||_{HS} \le 2^{4}(\alpha T)^{n-1}\ n\ \tau \qquad \text{for } n \le 4 \quad (48)$$

$$\le 2^{4}(\alpha T)^{n-1}\ \frac{2^{n-2}}{(n-3)!}\ \tau \qquad \text{for } n > 4$$

where we introduced the notation

$$\tau = \text{Max}\ (\int_{-\infty}^{+\infty} dt \| \Gamma^2(H_1(t)) \|_{HS},\ \| \int_{-\infty}^{+\infty} dt \| \Gamma^2(\dot{H}_1(t)) \|_{HS}.$$

Hence, the Dyson series (38) for U_{+-} converges in the Hilbert-Schmidt norm. This proves

Theorem 8: Consider the Dirac spin 1/2 case with an interaction B. We assume that B is in $\mathcal{S}(R^4, \mathbb{C}^4 \times \mathbb{C}^4)$ and supp $B \subset (a,b) \times R^3$ for some a,b finite. Then the off-diagonal part U_{+-} of the propagator defined by (38) is Hilbert-Schmidt.

There is an alternative formalism useful for a discussion of this theorem based on ideas of Mathews and Salam (1953), Wightman (1968, 1969), Capri (1969), and Bellissard (1976). The method is particularly good for a discussion of a perturbative approach and for the corresponding Euclidean problem. We start from the two equations

$$S_R(x,y;B) = S_R(x-y) - \int dz\ S_R(x-z)B(z)S_R(z,y;B) \tag{49}$$

$$\psi^{out}(x) = \psi^{in}(x) - \int dz\ S(x-y)B(y)S_R(y,z;B)B(z)\psi^{in}(z) \tag{50}$$

This last equation defines the scattering automorphism α in terms of the retarded fundamental solution $S_R(x,y;B)$ of (33). We discuss first the solutions of the integral equation (49) and later check whether the scattering automorphism can be implemented.

One can try to solve (49) by iteration. This can, in fact, be done if B is sufficiently small as we will demonstrate below. It gives an expansion of $S_R(B)$ in terms of B. Equation (49) is an integral equation for a distribution with an inhomogeneous part $S_R(x-y)$ which is again a distribution. On the other hand, we do not necessarily need $S_R(x,y;B)$ for defining α. For instance, it is enough to know $B(x)S_R(x,y;B)$. It turns out that it is even better to discuss the integral equation

$$T(x,y;B) = T(x,y) - \int dz\ T(x,z)T(z,y;B) \tag{51}$$

where we introduced the notation,

$$T(x,y) = A^{-1}(x)B(x)S_R(x-y)A(x) \tag{52}$$

$$T(x,y;B) = A^{-1}(x)B(x)S_R(x,y;B)A(x) \tag{53}$$

$$A(x) = (1 + x_o^2 + \vec{x}^2)^{-L}. \tag{54}$$

L is large and chosen so that $\partial^\mu \hat{A}(p)$ is uniformly bounded in p and square integrable for $|\mu| < \mu_o$. μ_o will be chosen appropriately. Furthermore, A(x) is by definition in O_M.

Clearly the integral equation (51) is as good for our purpose as (49). Next we will show that T(x,y) is a kernel of a bounded operator on $L^2(R^4) \times C^4$. This is easily done if we introduce a method of Bellissard's (1975) and define the following Banach space of kernels: Consider a function $K \in C_o^\infty (R^4 x R^4)$ and the norm

$$||K||_{m,\alpha} = \sum_{|\mu|,|\nu| \leq m} \sup_{p,q} (1+(p^o-q^o)^2)^\alpha \ (1+(\vec{p}-\vec{q})^2)^\alpha |\partial_p^\mu \partial_q^\nu K(p,q)|.$$

The closure of all K's with respect to this norm is denoted by $\mathcal{N}(m,\alpha)$. The following two statements hold:

Lemma 2: If $K \in \mathcal{N}(m,\alpha)$, $\alpha > 3/2$, $m > 0$, the operator K defined on $L^2(R^4) \otimes \mathbb{C}^4$ is bounded.

Lemma 3: For $m \geq 1$ and $\alpha > 0$ the inequality holds:
$||KS_R L||_{m,\alpha} \leq c(m,\alpha).||K||_{m,\alpha}||L||_{m,\alpha}$, where S_R is the retarded fundamental solution of the free Dirac equation and c a constant depending on m and α.

For the proof of the first statement we consider the inequality

$|\int dq K(p,q) f(g)| \leq ||K||_{m,\alpha} \int d^4q (1+(p^o-q^o)^2)^{-\alpha} \ (1+(\vec{p}-\vec{q})^2)^{-\alpha} \ |f(q)|$. Due to Young's inequality the L^2 norm of Kf is finite if $(1+(p^o)^2)^{-\alpha} \ (1+p^2)^{-\alpha}$ is an L^1 function. This proves lemma 2. For the proof of the second statement we refer to Bellissard (1975).

Now we are ready to state the

Theorem 9: If B is a test function, $B \in \mathcal{S}(R^4, \mathbb{C}^4 \times \mathbb{C}^4)$, then the operator T defined by (52) is bounded by $\tau_1 = \inf_{\substack{m \geq 1 \\ \alpha > 0}} c(m,\alpha)||\widehat{A^{-1}B}||_{m,\alpha}||\hat{A}||_{m,\alpha} < \infty$. $\widehat{A^{-1}B}$ and $\hat{A}$ denote the kernels $\widehat{A^{-1}B}(p-q)$ and $\hat{A}(p-q)$.

Proof: We show first that $\hat{B}(p-q)$ is in $\mathcal{N}(m,\alpha)$ for any $m \geq 0$, $\alpha > 0$ whenever B is a test function: For any m and α there is a constant so that

$$\sum_{|\mu| \leq m} |\partial^\mu \hat{B}(p)| \leq \text{const.} \ (1+(p^o)^2+\vec{p}^2)^{-2\alpha} \quad . \tag{55}$$

Due to the inequality $(1+(p^o)^2+\vec{p}^2)^2 \geq (1+(p^o)^2)(1+\vec{p}^2)$ the right-hand side of (55) is bounded by $(1+(p^o)^2)^{-\alpha} \ (1+\vec{p}^2)^{-\alpha}$. Now the statement of the theorem follows from the definition and lemma 2 and 3,

$$||\hat{T}||_{m,\alpha} \leq c(m,\alpha)||\widehat{A^{-1}B}||_{m,\alpha}||\hat{A}||_{m,\alpha}, \ m > 1, \ \alpha > 0.$$

The right-hand side is finite for an appropriate choice of m and α. This proves the theorem. Similarly one gets the

Corollary 4: Under the same assumptions as in the preceding theorem the inequality holds,

$$||T^n||_{m,\alpha} \leqslant c^n(m,\alpha)||\widehat{A^{-1}B}||_{m,\alpha}||\hat{B}||^{n-1}_{m,\alpha}||\hat{A}||_{m,\alpha}, \quad m \geqslant 1. \tag{56}$$

The right-hand side of (56) is finite for appropriately chosen m,α, for instance $m = \alpha = 10$. Furthermore, the Neumann series of (51) and interaction λB converges in $\mathcal{N}(m,\alpha)$ and in the uniform norm topology of bounded operators on $L^2(R^4)$ for λ in a neighbourhood of zero.

From this point on our discussion of (51) bifurcates in two branches. First we consider the consequences of convergence of the Neumann series in $\mathcal{N}(m,\alpha)$ (lemma 4) and after that we investigate the convergence in the uniform norm topology of bounded operators.

Consider a kernel $K(p,q) \in \mathcal{N}(m,\alpha)$, $m \geqslant 1$. The restriction of K to the forward and backward mass shell defines four new kernels on $L^2(R^3,d\mu) \otimes C^4$

$$(\hat{K}_{\varepsilon\varepsilon'}f)(\vec{p}) = \int\limits_{\substack{q^o=\omega(\vec{q})\\ p^o=\omega(\vec{p})}} K(\varepsilon p,\varepsilon' q)f(\vec{q})d(\vec{q}) \ . \quad \varepsilon = \pm,\ \varepsilon' = \pm. \tag{57}$$

Now we can state

Lemma 4: For $m \geqslant 1$, $\alpha > 1$, the two operators $\hat{K}_{+-}$, $\hat{K}_{-+}$ defined by (57) are Hilbert-Schmidt.

Proof: We show the statement for the first case. From the definition of the norm we get the inequality

$$\int d\mu(\vec{p})d\mu(\vec{q})|K(p,-q)|^2 \leqslant ||K||^2_{m,\alpha} \int d\mu(\vec{p})d\mu(\vec{q})(1+(\omega(\vec{p})+\omega(\vec{q}))^2)^{-2\alpha}(1+(\vec{p}+\vec{q})^2)^{-2\alpha}$$

$$\leqslant ||K||^2_{m,\alpha}\left(\int d\mu(\vec{p})(1+\vec{p}^2)^{-2\alpha}\right)^2 .$$

The integral converges if $\alpha > 1$. This proves the lemma.

Now we apply those results to prove that the scattering automorphism is implementable on the Fock space over the one-particle space. The one-body phase space $\mathcal{H}_B$ is $L^2(R^3,d^3x) \otimes \mathbb{C}^4$. By Fourier transformation this space is unitarily mapped onto

$\mathcal{H}_+ \oplus \mathcal{H}_-$, $\mathcal{H}_\pm \subset L^2(R^3,d\mu(\vec{p})/2\omega(\vec{p})) \otimes \mathbb{C}^4$

$$\psi(x) = \left(\frac{1}{2\pi}\right)^{3/2} \int d\mu(\vec{p})(e^{-ipx}\psi_+(\vec{p}) + e^{ipx}\psi_-(\vec{p})).$$

In terms of $\psi_\pm(\vec{p})$ equation (50) can be rewritten

$$\begin{aligned}\psi^{out}_\varepsilon(\vec{p}_1) &= \psi^{in}_\varepsilon(\vec{p}_1)+\varepsilon K(\varepsilon p_1) \sum_{\varepsilon'} \int B(\varepsilon p_1-\varepsilon' p_2)\psi^{in}_{\varepsilon'}(\varepsilon' p_2)d\mu(\vec{p}_2) + \\ &+ \varepsilon K(\varepsilon p_1) \sum_{\varepsilon'} \int d\mu(p_2)d\mu(p_3)d\mu(p_4)\hat{B}(\varepsilon p_1-p_2)\hat{S}_R(p_2,p_3,B)\hat{B}(p_3-\varepsilon' p_4)\psi^{in}_{\varepsilon'}(p_4).\end{aligned} \tag{58}$$

As we mentioned already previously the Shale-Stinespring test of implementability is the Hilbert-Schmidt property of the off-diagonal part in (58). Substituting the kernel T for S_R according to (53) it follows easily from lemma 4 and convergence of the Neumann series in $\mathcal{N}(m,\alpha)$ that α is implementable for any interaction λB with λ in a neighbourhood of zero.

Now we come to the second branch of the discussion of equation (51) and prove
Theorem 10: Suppose that B satisfies the same conditions as in theorem 8. Then the Neumann series converges in the uniform norm topology of bounded operators.
Proof: The core of the proof is the following well-known a priori estimate (see Garding's course)
Lemma 5: Let f be in $C_o^\infty(R^4)$ and define

$$||f||^2_{m,s_o,s_1} = \sum_{|\mu|\leqslant m} ||\partial^\mu f||^2_{o,s_o,s_1}$$

$$||f||^2_{o,s_o,s_1} = \int_{s_o}^{s_1} ds \int d^3x |f(s,\vec{x})|^2 .$$

Then for any $f \in C_o^\infty$ with supp f in the interior of $[s_o,s_1] \times R^3$,

$$||\Delta_r f||_{m+1,s_o,s_1} \leqslant \int_{s_o}^{s_1} ds ||f||_{m,s_o,s} \leqslant |s_1 - s_o| \; ||f||_{m,s_o,s_1}, \quad m \geqslant 0. \tag{59}$$

Of course, (59) then holds for the completion of all f's denoted by $\mathcal{H}_{m,s_o,s_1}$.

For functions in $C_o^\infty(R^4,\mathbb{C}^4)$ and S_R instead of Δ_r we get

$$||S_R f||_{m,s_o,s_1} \leqslant \text{const.} \int_{s_o}^{s_1} ds ||f||_{m,s_o,s}, \quad m \geqslant 0. \tag{60}$$

Suppose now that B satisfies the hypothesis of the theorem. Then we get for any $f \in \mathcal{H}_{m,s_o,s_1}$, $s_o,s_1 \in R$,

$$||Tf||_{m,s_o,s_1} \leqslant ||Tf||_{m,a,b} ,$$

where a and b are as in theorem 8 and by iteration of (59) and (60),

$$||T^n f||_{m,s_o,s_1} \leqslant K^n \frac{|a-b|^n}{n!} ||f||_{m,s_o,s_1} , \quad m \geqslant 0, \; K \text{ constant}. \tag{61}$$

Since this inequality holds for any real s_o,s_1, the statement of the theorem follows.

In order to complete this second proof of theorem 8 we would have to show the HS property of the off-diagonal operator defined in (58). For this the argument given previously for the inequalities (47), (48), and (46) has to be translated into this language. We will not do that.

The case of relativistic wave equations with arbitrary spin cannot be treated for arbitrary external fields. Let us define the class of <u>nice interactions</u> (Bellissard (1976)). $B \in \mathcal{S}(R^4, \mathbb{C}^N \times \mathbb{C}^N)$ is called a nice interaction if there are partial differential operators A_r, A_l, B_r, B_l with coefficients in $\mathcal{S}(R^4, \mathbb{C}^N \times \mathbb{C}^N)$ so that

i) $$(\not p - m + B)(K + B_r) = (p^2 - m^2)1 + A_r \tag{62}$$

$$(K + B_l)(\not p - m + B) = (p^2 - m^2)1 + A_l \tag{63}$$

ii) A_r A_l are at most of first order.

iii) The coefficients of A_r, A_l, B_r, B_l converge to zero in $\mathcal{S}$ if B does. The motivation of this concept is the following result:

<u>Theorem 11:</u> Consider a relativistic wave equation of the form (65) with a Harish-Chandra number $n \leq 3$ (see appendix, equation (66)),

$\not p^{n-2}(\not p \not p - p^2) = 0.$

Then the minimal coupling to an external field $p \to p+A$ is a nice interaction. For a proof we refer to Velo and Zwanziger (1971) and Bellissard and Seiler (1972)

<u>Remark:</u> Examples of relativistic wave equations with $n \leq 3$ are the Petiau-Duffin-Kemmer spin zero and spin one equation and the Dirac spin 1/2 equation. The Fierz-Pauli spin 3/2 equations are not of this type (n=4).

The important property of nice interactions is the following identity: Let $\Delta_R(x,y,A_r)$ be the retarded fundamental solution of the operator which appears on the right-hand side of (62). Then

$$S_R(x,y;B) = (K + B_r)\,\Delta_R(x,y,A_r)$$

is a retarded fundamental solution of (43).

In order to get a convenient form of $S_R(x,y,B)$ in terms of $\Delta_R(x,y,A_r)$ the following formal manipulation is useful which will be justified in retrospect: For $h \in \mathcal{S}(R^4, \mathbb{C}^N)$,

$$\begin{aligned}(1 + A_r\Delta_R)h &= (p^2 - m^2 + A_r)\,\Delta_R h \\ &= (\not p - m + B)(K + B_r)\Delta_R h \\ (\not p - m + B)h &= (1 + BS_R)(\not p - m)h\end{aligned}$$

where Δ_R and S_R denote the free retarded propagators. From the two last identities one gets

$$(1 + BS_R)^{-1}B = (1 + (\not p - m)B_r\Delta_R)(1 + A_r\Delta_R)^{-1}B. \tag{64}$$

Now we are prepaired to prove the

Theorem 12 Bellissard (1976)): Consider the wave equation (48) with a nice interaction λB, $B \in \mathcal{S}(R^4,\mathbb{C}^N \times \mathbb{C}^N)$, $\lambda \in C$. Then the sandwitched retarded fundamental solution $B(x)S_R(x,y,B)B(y)$ can be defined as follows: Let

$$B(x)S_R(x,y,B)B(y) = \int(1+BS_R)^{-1}(x,z)B(z)S_R(z-y)B(y)dz$$

where $(1+BS_R)^{-1}B$ is defined by (64). For any interaction λB the right-hand side is in a Banach space of kernels $\mathcal{N}(m,\alpha,p)$ to be defined below for a $p > 0$ and $\forall m > 1$, $\alpha > 0$, provided λ is in a neighbourhood of zero.

Remark: The Banach space $\mathcal{N}(m,\alpha,p)$ mentioned above is a straightforward generalization of $\mathcal{N}(m,\alpha)$. $\mathcal{N}(m,\alpha,M)$ is the closure of all $C_o^\infty(R^4xR^4,\mathbb{C})$ functions with respect to the norm

$$||K||_{m,\alpha,M} = \sum_{|\mu|,|\nu|\leq m} \sup_{p,q} \frac{(1+(p^o-q^o)^2)^\alpha(1+(\vec{p}-\vec{q})^2)^\alpha}{(1+|p|^2)^M(1+|q|^2)^M} |\partial_p^\mu\partial_q^\nu K(p,q)|$$

where $|p|$ denotes the Euclidean length. The importance of those spaces is that a partial differential operator P of degree 2M in ∂_x with coefficients in $\mathcal{S}$ maps $\mathcal{N}(m,\alpha)$ continuously into $\mathcal{N}(m,\alpha,M)$.

Proof of theorem 12: According to a slight generalization of lemma 3 the Neumann series of $(1+A_r(\lambda)\Delta_R)^{-1}(\lambda B)$ converges in $\mathcal{N}(m,\alpha)$, $m \geqslant 1$, $\alpha > 0$, for λ in a neighbourhood of zero. Furthermore, the partial differential operator $(1+(\not{p}-m)B_r\Delta_R)$ maps the result into a new space $\mathcal{N}(m,\alpha,P)$ where P has to be chosen according to the degree of B_r.

Finally we note that lemma 4 generalizes to kernels in $\mathcal{N}(m,\alpha,P)$ if α is replaces by α-M. Hence, we get the

Corollary 5: Under the assumptions of theorem 12 the scattering automorphism α for an arbitrary spin equation with nice interaction λB defined by (49) and (50) is implementable if λ is in a neighbourhood of zero.

We terminate this lecture with a few remarks:

1. Theorem 8 is proved and formulated for external fields with compact support in time. It is readily seen that the proof can be adapted to the case $B \in \mathcal{S}(R^4,\mathbb{C}^4 x\mathbb{C}^4)$ at least if we use the formalism presented at the beginning of the lecture.
2. The crucial fact, why the formalism using Dyson's expansion for U (38) respectively the a priori bound (61) works for arbitrary size of the external field is the term $^1/n!$ in (61) or equivalently the time ordering in (38). Such a factor is missing in the Banach space approach of Bellissard.
3. The Euclidean version of integral equation (49) can be treated by Fredholm technics at least for the Dirac case. The integral equation for the operator

$K(p_1,p_3) = \int d^4p_2\ \hat{B}(p_1-p_2)\ S\ (p_2,p_3,B)$

where S is the kernel of the resolvent of the elliptic operator ($\not{p}$+m+B) and $K_o(p,q) =$ $= B(p-q)(\not{q}+m)^{-1}$, $K(p_1,p_3) = K_o(p_1,p_3) - \int d^4p_2 K_o(p_1,p_2)K(p_2,p_3)$.

Obviously K_o is a bounded operator on $L^2(R^4)$. A simple power counting argument proves that K_o^3 is Hilbert-Schmidt. Therefore the general machinery (see e.g. Ruston (1948)) can be applied.

Application of such technics to the Minkowski space version of integral equation (49) is made difficult by the fact that Δ_R improves differentiability only by one order whereas its Euclidean counterpart does better , namely two (see the "Little Theorem" in Garding's course).

Appendix on Free Relàtivistic Wave Equations

We summarize assumptions generally used in the context of relativistic wave equations for particles with one mass m and spin s.

A1. The relativistic wave equations are of the form

$$(\not p - m)\psi(x) = 0, \quad p_\mu = i\partial_\mu, \qquad (65)$$

where $\not p = \Sigma\, i\partial_\mu \gamma^n$ $(= -\Sigma\, \partial_\mu \beta^n)$. The γ and β-matrices are endomorphismes of $\mathbb{C}^N$ for a particular N and ψ is a distribution, $\psi \in \mathcal{S}'(R^4, C^N)$.

A2. There exists a tensor representation S of SL(2,C) so that $S(A)\not b S(A)^{-1} = \not{\Lambda(A)b}$ where Λ is the representation of SL(2,C) by the one component of the Lorentz group L^1. Due to a theorem by Garding (1943) (see also the lecture notes by Wightman) we can assume with no restriction of generality that every irreducible constituent of S induces a one valued respectively two valued representation of L^1.

A3. On $\mathbb{C}^N$ there exists a non-degenerate, S-invariant hermitian scalar product $[\ ,\ ]$. It is related to the unitary scalar product by the matrix η, $[\ ,\ ] = (\ ,\eta\ )$.

A4. The matrices γ^μ are $[\ ,\ ]$-symmetric.

A5. The scalar product $[\ ,\ ]$ is positive definite on $P\mathbb{C}^N$, where P denotes the eigenprojector of γ^o for the eigenvalue 1. This eigenvalue is assumed to be the only strictly positive eigenvalue of γ^o.

A6. There exists an endomorphism C of $\mathbb{C}^N$ so that $-\bar\gamma^\mu = C\gamma^\mu C^{-1}$, $\overline{S(A)} = CS(A)C^{-1}$ and the map $\sigma = C^+C$, $C^+ = \eta^{-1} C^* \eta$, is assumed to be 1 or -1 according to whether S induces a one or two valued representation of L^1.

We will not give a detailed discussion of these assumptions. There is an abundant literature on this subject. Instead, we add a few remarks:

1. The general case of relativistic wave equations is discussed in Wightman's lectures. A famous example which does not fit into our restricted class of equations is the Petiau-Duffin-Kemmer version of the wave equation $\Box\,\phi = 0$.

2. Covariance (68) and the spectral condition on γ^o (A5) implies the support properties of the solutions of (65),

$$\operatorname{supp} \hat\psi \subset H_m = \{p \in R^4 \mid p^2 = m^2\}.$$

3. The antilinear mapping of $\mathcal{S}'(R^4, C^N)$ - charge conjugation - defined by

$$\mathcal{C} : \psi \to C^{-1}\bar\psi$$

leaves the set of solutions invariant. It maps positive frequency solutions covariantly onto negative ones.

4. There is a Klein-Gordon divisor K(p)

$$K(p) = (\not{p}+m) + \frac{1}{m}\sum_{k=0}^{n-3}\left(\frac{\not{p}}{m}\right)^k (\not{p}\not{p}-p^2), \qquad (66)$$

with the defining property $K(p)(\not{p}-m) = p^2-m^2$. Existence follows from Harish-Chandra's (1947) relation: $(\gamma^o)^{n-2}((\gamma^o)^2-1) = 0$, for an appropriate choice of n (Speer (1969), Glass (1971)). We will call n the Harish-Chandra number of the particular wave equation.

5. The solutions of (65) can be parametrized in terms of plane waves. Choose a basis $u_\pm(p,s)$ of $E_\pm(p) = \ker(\pm\not{p}-m)$. In particular we denote by $u_+(s)$ the basis in $\ker(\gamma^o-1)$. The choice can be made such that $u_\pm(p,s) = S(p)\, u_\pm(s)$,

$$u_-(s) = C^{-1}\overline{u_+(s)},\ [u_\varepsilon(p,s),u_{\varepsilon'}\, p,s')] = \delta_{\varepsilon\varepsilon'}\delta_{ss'},\ \varepsilon = \pm,\ s = 1....n.$$

$S(p)$ is a boost for the momentum $p\in H_m$. The parametrization is now given by

$$\psi(x) = \left(\frac{1}{2\pi}\right)^{3/2} \sum_s \int d\mu(p)\, (a_+(p,s)e^{-ipx}u_+(p,s) + \overline{a_-(p,s)}e^{ipx}u_-(p,s)) \qquad (67)$$

$$d\mu(p) = (2\omega(p))^{-1}d^3p,\ \omega(p) = \sqrt{p^2+m^2}$$

We will also use the notation a,u,b,v for a_+,u_+,a_-,u_-.

6. There exists a scalar Lagrange density, $\mathcal{L} = \psi^+(\not{\partial}-m)\psi$, and a vector current $j^\mu = \psi^+\gamma^\mu\psi$ so that the corresponding charge defines a scalar product,

$$(\psi,\psi) = \int d^3x j_o(x) = \sum_s \int d\mu(p)\, (|a_+(p,s)|^2-\sigma|a_-(p,s)|^2). \qquad (68)$$

The scalar product is positive on positive frequency solutions of (65) ($a_-(p,s) = 0$). This follows notably from (A 5).

7. The projectors $\Lambda_\pm(p)$ onto $E_\pm(p)$ can be expressed in terms of the u's or the γ's,

$$\Lambda_\varepsilon(p) = \varepsilon^\sigma \sum_s u_\varepsilon(p,s) \otimes u_\varepsilon^+(p,s),\quad \varepsilon = \pm$$
$$= \left(\frac{\varepsilon\not{p}}{m}\right)^{n-2} \frac{m+\varepsilon\not{p}}{2m}. \qquad (69)$$

The first identity is based on the covariant normalization of the u's, $u_\varepsilon^+(p,s)\gamma^o u_\varepsilon(p,s') = \varepsilon\sigma^{1/2(1-\varepsilon)}\omega(p)/m\,\delta_{ss'}$, following from (A.2) and the definition. The second one follows from the Harish-Chandra relation (remark 4).

8. The linear space of positive frequency solutions can be normed by means of the charge (68). The transformation $\psi(x) \to S(A)\psi(\Lambda^{-1}(A)(x-a))$ is a unitary representation of the inhomogeneous Poincaré group iSL(2C). The mass of the irreducible constituents is m; the spin is determined by detailed properties of the γ-matrices.

9. In the special case of the Dirac equation the γ-matrices generate a Clifford algebra: $\{\gamma^\mu,\gamma^\nu\} = 2g^{\mu\nu}$, where g is the diagonal matrix (1,-1,-1,-1). Furthermore, the γ-matrices anticommute with $\gamma^5=\gamma^0\gamma^1\gamma^2\gamma^3$ and can be chosen so that $\gamma^\mu=\gamma_\mu^*$, $\mu=0,1,2,3$.

References

J. Bellissard and R. Seiler (1972), Lettere al Nuovo Cimento 5, 221.

J. Bellissard (1976), Commun. math. Phys. 46, 53.

J. Bellissard (1975), Commun. math. Phys. 41, 235.

F.A. Berezin (1965), Methods of Second Quantization, english transl. Academic Press, New York and London, 1966.

P.J.M. Bongaarts (1970), Ann. Phys. 56, 108.

A. Capri (1969), J. Math. Physics 10, 575.

A. Capri (1967), thesis, Princeton University, Princeton N.J. (unpublished).

P.A.M. Dirac (1928), Proc. Roy. Soc. 117, 610 and 118, 351.

P.A.M. Dirac (1936), Proc. Roy. Soc. A 155, 447.

R.P. Feynman (1949), Phys. Rev. 76, 749.

M. Fierz and W. Pauli (1939), Proc. Roy. Soc. A 173, 211.

M. Fierz (1939), Helv. Phys. Acta 12, 3.

K.O. Friedrichs (1951-1953), Commun. Pure Appl. Math., collectively reissued by Interscience, New York (1953).

L. Garding (1943), Kungl. Fysiografiska Sällskapets i Lund Forhandlingar 13, 229.

A.S. Glass (1971), Commun. math. Phys. 23, 176.

W. Gordon (1926), Z. Phys. 40, 117.

Harish-Chandra (1947), Phys. Rev. 71, 793.

W.J.M.A. Hochstenbach (1976), Commun. math. Phys. 51, 211.

R.R. Kallman (1971), Journal of Functional Analysis 7, 43.

O. Klein (1926), Z. Phys. 37, 895.

O. Klein (1929), Z. Phys. 53, 157.

S. Kusaka and J. Weinberg, unpublished manuscript, Department of Physics, University of California, Berkeley.

G. Labonté (1974), Commun. math. Phys. 36, 59.

J. Leray and Y. Ohya (1964), Systèmes linéaires hyperboliques non-stricts, Colloque CBM, Louvrain; reprinted in Battelle Rencontres on Hyperbolic Equations and Waves (1968).

J.C. Maxwell (1873), A Treatise on Electricity and Magnetism.

P. Minkowski and R. Seiler (1971), Phys. Rev. D4, 359.

H. Narnhofer (1974), Acta Physica Austriaca 40, 306.

G. Nenciu (1976), Commun. math. Phys. 48, 235.

J.V. Neumann (1938), Comp. Math. 6, 1.

W. Pauli (1936), Ann. de l'Institut H. Poincaré 6, 137.

W. Pauli (1940), Phys. Rev. 58, 716.

M. Reed and B. Simon (1975), Fourier Analysis and Self-Adjointness; Academic Press, New York, San Franziski, London.

S.N.M. Ruijsenaars (1977), Commun. math. Phys. 52, 267.

A.F. Ruston (1951), Proc. London Math. Soc. 53, 109.

A. Salam and P.T. Matthews (1953), Phys. Rev. 90, 690

B. Schroer, R. Seiler and A. Swieca (1970), Phys. Rev. D2, 2927.

B. Schroer and A. Swieca (1970), Phys. Rev. D2, 2938.

E. Schrödinger (1926), Quantisierung als Eigenwertproblem (vierte Mitteilung), Annalen d. Physik 81, 109.

J. Schwinger (1954), Phys. Rev. 93, 615.

R. Seiler (1972), Commun. math. Phys. 25, 127.

D. Shale (1962), Transaction of American math. Soc. 103, 149.

D. Shale and W.F. Stinespring (1965), J. Math. Mech. 14, 315.

H. Snyder and J. Weinberg (1940), Phys. Rev. 57, 307.

E.R. Speer (1969), Ann. of Math. Studies No. 62, Princeton Univ. Press., N.J., Univ. of Tokyo Press, Tokyo.

G. Velo and D. Zwanziger (1969), Phys. Rev. 186, 1337 and 188, 2218.

G. Velo and D. Zwanziger (1971), in lectures from the Coral Gables Conference on Fundamental Interactions at High Energy; G.J. Iverson, A. Perlmutter, Editors; Gordon and Breach Science Publishers.

G. Velo (1975), Ann. Inst. Henri Poincaré 22, 249.

R. Weder (1977), Helv. Phys. Acta 50, 105.

J. Weinberg (1943), thesis, University of California.

M. Weinless (1969), Jour. Funct. Anal. 4, 350.

A. Wightman (1968), "Symmetry Principles at High Energy, Fifth Coral Gables Conference", A. Perlmutter, C.A. Hurst and B. Kursunoglu, Eds. W.A. Benjamin, New York 1968, page 291.

A.S. Wightman (1971), Proc. Symp. in Pure Math., Vol. 23, Berkeley 1971, Providence Rhode Island.

NONLINEAR INVARIANT WAVE EQUATIONS

Walter A. Strauss*
Brown University
Providence, R. I. 02912 USA

The contrast between the linear and nonlinear theory is striking. You can see this already for the simplest kind of nonlinearity for the simplest kind of global problem, the Cauchy problem. (This is the problem of solving the equation with given initial conditions at $t = 0$, say.) To solve it locally is easy (cf. Courant-Hilbert), but as soon as the solution is continued far away from $t = 0$, nonlinear phenomena appear. These include: blow-up in a finite time, inability to prove uniqueness of the solution, existence of solitary waves (traveling waves, solitons, instantons), scattering (existence of in and out states) without spatial or temporal inhomogeneities in the equation, and (for more complicated nonlinearities) shock waves (spontaneous appearance of discontinuities). We will not discuss the last phenomenon here.

The nonlinear theory is in good shape for only a couple of simple examples. General hyperbolic equations are out of reach. Our main example will be the nonlinear Klein-Gordon equation

$$\text{(NLKG)} \qquad u_{tt} - \Delta u + m^2 u + F(u) = 0$$

$m \geq 0$, $x \in \mathbb{R}^n$, n = space dimension, $F(0) = F'(0) = 0$. The behavior of the solution is very sensitive to the nature of the nonlinear term $F(u)$. For instance, for $m^2u + F(u) = +u+u^3$, all solutions are asymptotically free. For $+u-u^3$, some solutions are asymptotically free, others are stationary and still others blow up in a finite time. For $-u+u^3$, some solutions are stationary and no solutions blow up in a finite time. There are seven chapters.

I. The Cauchy problem: existence, uniqueness, regularity, boundedness, blow-up.

II. Invariance and conservation laws. We consider the Euclidean equation $\Delta u = F(u)$, its invariance under the conformal group, and the method of multipliers to obtain conservation laws and "energy" identities. By changing one variable $x_0 \to it$, we apply these results to NLKG.

*Supported by NSF Grant MCS75-08827.

III. Solitons. We first prove the existence of solitary waves for a large class of nonlinearities F for NLKG. Then we specialize to the sine-Gordon equation, $F(u) = \sin u$, and outline the theory of solitons and the inverse scattering method.

IV. Basic ideas in nonlinear scattering. This is discussed in the framework of a group of nonlinear transformations $U(t) = \exp(itH)$ and is applied to NLKG. For NLKG for certain F there is weak scattering (Theorem 4.2), and for other F there is no scattering (Theorem 4.3).

V. The wave operators and low-energy scattering. In the general framework of Chapter IV, there are strong scattering of small ("low-energy") inputs, existence of the wave operators, and an inverse scattering result.

VI. Asymptotic completeness. For NLKG for $F(u) = u^3$ and $n = 3$, there is strong scattering. The scattering operator S is a well-defined nonlinear covariant operator defined on a Banach space $\mathcal{F}$ which is dense in the usual Hilbert space. This uses Chapter II.

VII. Other invariant wave equations. We briefly discuss recent results on (A) the classical coupled Maxwell-Dirac equations, (B) Maxwell, Dirac and Klein-Gordon equations with other interaction terms, and (C) the nonlinear Schrödinger equation.

In spite of its physical interest, there was practically no global mathematical theory of classical nonlinear wave equations until 1961. In that year K. Jörgens published the first, and in some ways still the best, existence theorem. The guiding force behind the further development of the theory, especially the scattering theory, has been I. E. Segal. In a series of papers, he put Jörgens' theorem in a general context [1963a], proved the existence of weak solutions for strong nonlinearities [1963b] and proved the existence of the wave operators [1966] and of the low-energy scattering operator [1968]. And it has been his direct influence which stimulated almost all the other results discussed in these notes except Chapter III. The concept of the soliton, on the other hand, developed from problems in fluid mechanics and from the Fermi-Pasta-Ulam problem.

The classical scattering theory of Chapter VI is the basis of a constructive approach to quantum field theory: see Bałaban, Rączka et.al. [1975, 1976].

Note: Formulas are numbered separately for each chapter. Formula (x.y) means formula (y) in Chapter x. Formula (y) refers to the current chapter.

I. EXISTENCE

We consider NLKG where n is the space dimension, m the mass, and F the interaction term. Famous examples are $F(u) = u^3$ and $F(u) = \sin u$. For simplicity we assume $u = u(x,t)$ is real-valued. Let $G(u) = \int_0^u F(v)dv$. By a solution of NLKG we mean a solution in the sense of distributions defined on the whole of space and time.

The most basic fact about NLKG is the conservation of energy. Multiply NLKG by u_t to get

$$(1) \qquad \partial_t\{\tfrac{1}{2} u_t^2 + \tfrac{1}{2}|\nabla u|^2 + \tfrac{1}{2}m^2u^2 + G(u)\} + \nabla_x\cdot\{-u_t\nabla_x u\} = 0$$

where $\partial_t = \partial/\partial t$ and $\nabla_x = \nabla$ is the spatial gradient. Integrate over space to get

$$E = E[u] = \int\{\tfrac{1}{2}u_t^2 + \tfrac{1}{2}|\nabla u|^2 + \tfrac{1}{2}m^2u^2 + G(u)\}d^nx = \text{constant.}$$

In this chapter we consider the existence, uniqueness and regularity properties of the solutions of the Cauchy problem: NLKG with initial conditions $u(x,0)$ and $u_t(x,0)$. We shall first state all the main facts and then sketch the proofs.

Theorem 1.1. (Existence of Weak Solutions). *Assume* F *is a continuous function such that* $uF(u)$ *and* $G(u)$ *are bounded below. If the initial conditions have finite energy, then there exists at least one solution of NLKG of finite energy with those initial solutions. If the initial conditions only have locally finite energy (finite energy on bounded sets), there is a solution of locally finite energy.* (Actually it suffices to assume that $uF(u)$ and $G(u)$ are bounded below by $-c(1+u^2)$ where c is a constant.)

Presumably the solution is unique, but that is an open problem. We know how to prove uniqueness only under growth restrictions on $F(u)$ as $|u| \to \infty$.

Theorem 1.2. (Uniqueness and Regularity). *Let* $n = 3$. *Let* F *be a* C^1 *function (locally Lipschitz is enough) such that* $G(u) \geq 0$ *for all* u *and*

$$(2) \qquad |F(u)| = O(1 + |u|^{1-\varepsilon}G(u)^{-\varepsilon+2/3}) \quad \text{as} \quad |u| \to \infty$$

for some $\varepsilon > 0$. *Then there exists a solution of [locally] finite energy which is bounded in bounded regions of space-time, for any initial conditions satisfying the same sort of properties. This solution is unique. It is as smooth as* F *and the initial conditions permit; in particular, it is* C^∞ *if* F *and the initial data are.*

Example 1. $F(u) = \sin u$, $G(u) = 1-\cos u \geq 0$. This F is bounded and C^∞. Therefore there is a unique C^∞ solution for arbitrary C^∞

initial data.

Example 2. $F(u) = |u|^{p-1}u$ ($= u^p$ in case p is odd) where $p > 1$. Then $G(u) = |u|^{p+1}/(p+1) \geq 0$. (2) is satisfied if and only if $p < 1+(p+1)2/3$. That is, $p < 5$. So if $p < 5$ we get unique, smooth solutions. If $p \geq 5$, Theorem 2 is inapplicable and we only know there exist solutions as in Theorem 1.

Numerical computations indicate that the solutions are bounded (and hence unique) even for $p \geq 5$. In fact as p increases for fixed initial data, the amplitude seems to decrease and the number of oscillations to increase. See Strauss and Vazquez [1977].

Other dimensions. If $n = 1$, the bound (2) on F is not needed in Theorem 2. If $n = 2$, any power $p < \infty$ is okay. If $n \geq 4$, it is okay for $p \leq 1 + 2/(n-2)$. It ought to be okay for $p < 1 + 4/(n-2)$, but this has not been proved.

Theorem 1.3. (Boundedness). *Let* $n = 3$ *and let* F *be a* C^2 *function such that* $G(u) \geq 0$ *and*

$$|F(u)| = O(|u|^{1-\varepsilon}G(u)^{-\varepsilon+2/3}) \quad \text{as } |u| \to \infty$$

for some $\varepsilon > 0$. *Then the solution is uniformly bounded in all space-time.*

In the preceding theorems we have assumed a lower bound on G. Without such an assumption we are in trouble. We can see this already in the o.d.e. $u_{tt} + u^3 = 0$, all of whose solutions are bounded. For the o.d.e. $u_{tt} - u^3 = 0$, the kinetic and potential energy terms ($u_t^2/2$ and $-u^4/4$) can both become enormous and still add up to a fixed energy E; its solutions may blow up in a finite time. We present two results of this type for NLKG.

Theorem 1.4. (Blow-up). *Let* n *be arbitrary.*

(a) Let $F \leq 0$, $F' \leq 0$, $F'' \leq 0$ *and*

$$\int^{\infty} |G(s)|^{-1/2} ds < \infty \quad \text{for large } s. \tag{3}$$

Then there exist nice (C_c^{∞}) *initial data for which the solution blows up in a finite time.*

(b) Let $sF(s) \leq (2+\varepsilon)G(s)$ *for all* s *for some* $\varepsilon > 0$. *If there exist initial data with energy* $E < 0$ *and* $\int uu_t dx > 0$ *at* $t = 0$, *the solution with those data blows up in a finite time.*

For instance, if $F(u) = -|u|^{p-1}u$, there are solutions which blow up in a finite time.

Proof of Theorem 2. The theorem is due to Jörgens [1961]. Shortly thereafter, a more elegant proof was found by Segal [1963a] which however only applies to powers $p \leq 3$ $(p \leq 1 + 2/(n-2))$. For the sake of simplicity we shall sketch the proof for the case $F(u) = u^3$, $n = 3$, $m = 1$. An excellent reference is Reed [1976].

The key to the proof is the Sobolev embedding theorem $H^1 \subset L^{2n/(n-2)} = L^6$ for $n = 3$ and the corresponding Sobolev inequality

$$(4) \qquad (\int u^6 dx)^{1/6} \leq c(\int |\nabla u|^2 dx)^{1/2}$$

valid for functions $u \in L^6$ for which $\nabla u \in L^2$. (See, for instance, Friedman [1969].) This implies that the mapping $u \to u^3$ is locally Lipschitz from H^1 into L^2, namely

$$(5) \qquad ||u^3 - v^3||_2 \leq c(||\nabla u||_2^2 + ||\nabla v||_2^2)||\nabla(u-v)||_2$$

where $|| \; ||_p$ is the L^p norm. Denote

$$||u||_e^2 = \int (u_t^2 + |\nabla u|^2 + u^2) dx.$$

We define approximate solutions to $Lu + F(u) = u_{tt} - \Delta u + u + F(u) = 0$ in the same way as the standard Picard method for ordinary differential equations. Let $u^{(-1)}(x,t) = 0$. For $m \geq 0$ let $u^{(m)}$ be the unique solution of the <u>linear</u> problem

$$Lu^{(m)} + F(u^{(m-1)}) = 0$$
$$u^{(m)}(x,0) = \phi(x), \quad u_t^{(m)}(x,0) = \psi(x).$$

Then $L(u^{(m+1)} - u^{(m)}) + F(u^{(m)}) - F(u^{(m-1)}) = 0$. Multiply this by $(u^{(m+1)} - u^{(m)})_t$ and integrate to get

$$||u^{(m+1)} - u^{(m)}||_e^2(t) \leq 4\{\int_0^t ||F(u^{(m)}) - F(u^{(m-1)})||_2(s) ds\}^2$$
$$\leq c\{\int_0^t ||u^{(m)} - u^{(m-1)}||_2 (||u^{(m)}||_2^2 + ||u^{(m-1)}||_2^2) ds\}^2$$

by (5). It follows that

$$E_m(t) \equiv ||u^{(m+1)} - u^{(m)}||_e^2(t) \leq k \int_0^t E_{m-1}(s) ds$$

for $0 \leq t \leq T$ with k depending on T. This is easily solved:

$$E_m(t) \leq (kt)^m/m! \sup_s E_0(s).$$

Hence $u^{(m)}$ is a Cauchy sequence in the energy norm. It is easy to see the limit is a solution.

Uniqueness: Suppose u and v both are solutions. Let $w = u-v$. Then $Lw + (u^3-v^3) = 0$. Multiply by w_t to get

$$\frac{1}{2}||w(T)||_e^2 = -\int_0^T\int (u^3-v^3)w_t\,dxdt$$
$$\leq \int_0^T ||u^3-v^3||_2 ||w_t||_2 dt \leq c\int_0^T ||w||_e^2\, dt.$$

This implies $w = 0$.

The regularity is proved by multiplying the approximate equation by Δu_t, $\Delta^2 u_t$ and so on to obtain L^2 estimates of higher derivatives. The local boundedness is proved as in Theorem 3 (see below).

We also have the local energy inequality. It comes from integrating (1) over the interior of any light cone and using the divergence theorem. We get (see sketch)

$$\int_T e(u)dx - \int_B e(u)dx + \int_K (e(u)-u_t u_r)dSdt/\sqrt{2} = 0.$$

where $e(u) = u_t^2/2 + |\nabla u|^2/2 + m^2u^2/2 + G(u)$ is the energy density and $r = |x|$. But

$$e(u) - u_t u_r = \frac{1}{2}(u_t-u_r)^2 + \frac{1}{2}(|\nabla u|^2-u_r^2) + \frac{1}{2}m^2u^2 + G(u)$$

has four non-negative terms if $G \geq 0$. Hence

(6) $$\int_T e(u)dx \leq \int_B e(u)dx$$ and

(7) $$\int_K (\tfrac{1}{2}m^2u^2+G(u))dx \leq \sqrt{2}\, E[u].$$

Now if the initial data in Theorem 2 are only of locally finite energy, we "cut them off" for large $|x|$ and solve the resulting problem. By (6), this solution will be independent of the cut-off in a large piece of light cone. Hence it trivially converges to a limit which is the desired solution.

Proof of Theorem 1. For F = a power, this theorem was proved by Segal [1963b] and Lions [1964]. In the present form it is due to Strauss [1970]. The difficulty is that $F(u)$ grows rapidly as $u \to \infty$, so that Sobolev's inequality is inapplicable. The idea of the proof is to cut $F(u)$ down for large u: find a sequence of functions $F_j \to F$, each F_j not too large at infinity. For each j we can solve the equation (uniquely!) by Theorem 2:

(8) $$u_{tt}^{(j)} - \Delta u^{(j)} + u^{(j)} + F_j(u^{(j)}) = 0.$$

The usual energy estimate gives

$$\frac{1}{2}||u^{(j)}(t)||_e^2 + \int G_j(u^{(j)})dx = E[u^{(j)}] \leq \text{const.}$$

Since G_j are uniformly bounded below, $u^{(j)}$ have uniformly bounded energy. Now we use some standard compactness theorems (due to Rellich): $u^{(j)}$ has a subsequence $u^{(k)}$ which converges weakly in the energy norm and also almost everywhere (a.e.) to some function u. Hence $G_k(u^{(k)})$ converges a.e., but for u to be a solution of the differential equation, we need $G_k(u^k) \to G(u)$ in the sense of distributions. This depends on a convergence theorem and on the assumed lower bound on $uF(u)$. Once this is done, each of the terms in (8) converges as $j \to \infty$ in the sense of distributions. By the weak limits, the solution satisfies the inequality

$$\int (\tfrac{1}{2}u_t^2 + \tfrac{1}{2}|\nabla u|^2 + \tfrac{1}{2}u^2 + G(u))dx \leq E.$$

The details are given in Strauss [1970] and Reed [1976]. If the initial data are only of locally finite energy, we approximate them by data of compact support, say. The passage to the limit works exactly as before and one gets in the limit

$$\int_T e(u)dx \leq \int_B e(u)dx$$

for any B and T as in the sketch above.

<u>Proof of Theorem 3</u>. We sketch the proof for $u_{tt} - \Delta u + u + u^3 = 0$. We represent the solution in terms of the initial data and u^3 by its retarded Green's function:

(9) $$u(x,t) = u_0(x,t) - \int_0^t\int R(x-y,t-\tau)u^3(y,\tau)dyd\tau,$$

where

(10) $$R(x,t) = \frac{1}{4\pi}\,\frac{\delta(|x|-t)}{t} + \frac{\theta}{4\pi}\,J_1((t^2-|x|^2)^{1/2})(t^2-|x|^2)^{-1/2}$$

where J_1 is the Bessel function, where $\theta = \theta(t^2-r^2)$ **is** 1 inside the light cone and 0 outside it, and where u_0 is a free solution (of the linear Klein-Gordon equation). It is well-known that u_0 is bounded. In order to estimate the integrals, we use the local energy inequality (7). The first integral in (9) is

$$\int_0^t \int_{|x-y|=t-\tau} |u(y,\tau)|^3 dS\,\frac{d\tau}{t-\tau}\,.$$

The part of this from 0 to t-1 less than a multiple of E[u] by (7) since $2u^3 \leq u^2 + u^4$. Similarly (7) can be used to estimate the Bessel term in (9). We are left with the integral

$$\int_{t-1}^{t} \leq c\left(\iint |u|^5 dS d\tau\right)^{3/5}\left(\iint \left(\frac{1}{t-\tau}\right)^{5/2} dS d\tau\right)^{2/5}$$

$$\leq c(\max|u|)^{3/5}\left(\iint u^4 dS d\tau\right)^{3/5}.$$

Hence $\max |u(t)| \leq c(1 + \max_{[t-1,t]} |u|^{3/5})$.

Hence u is bounded. This proof is from Morawetz and Strauss [1972], Lemma 6.

<u>Proof of Theorem 4</u>. Part (a) is due to Keller [1957] if $n \leq 3$ and Glassey [1973b] if $n \geq 4$. Part (b) is due to Levine [1974]. Both blow up proofs reduce the problem to an ordinary differential inequality whose solutions obviously blow up. Note that the assumption of (b) implies (3).

We first give Keller's proof in the simple case $n = 1$, $m = 0$. Let $v(t)$ be the (space-independent) solution of $v_{tt} + F(v) = 0$, $v(0) = \alpha$, $v_t(0) = \beta$, α and β being constants. We can solve for v explicitly:

$$v_t^2/2 + G(v) = \beta^2/2 + G(\alpha) = E$$

$$v_t = \pm(2E - 2G(v))^{1/2}$$

$$t = \int_\alpha^v (2E - 2G(v))^{-1/2} dw$$

for $t \geq 0$, assuming $E > 0$, $\beta \geq 0$. By assumption there exist large α and β so that the last integral converges to a finite value T as $v \to \infty$. Therefore $v(t) \to \infty$ as $t \to T$.

Now let $u(x,t)$ be any solution of NLKG such that $u(x,0) \geq \alpha$ and $u_t(x,0) \geq \beta$ in the cone $|x| \leq T-t$. In that cone we use the retarded Green's function (which is non-negative for $n \leq 3$) to express $u(x,t)$ in terms of the initial data and the interaction term $F(u)$. By a comparison argument, $u(x,t) \geq v(t)$ in the cone $|x| \leq T-t$. Since v blows up, so does u.

The proof of (b) is identical in any dimension. It is based on the "convexity method". Multiply NLKG by u and integrate to get

$$I'' + \int(-u_t^2 + |\nabla u|^2 + uF(u))dx = 0$$

where $I = I(t) = \int u^2/2\ dx$. Let $\alpha = \varepsilon/4 > 0$. From the equation above, subtract $(2+4\alpha)$ times the energy E to get

$$I'' - (2+2\alpha)\int u_t^2 dx - 2\alpha\int|\nabla u|^2 dx$$
$$- \int[(2+4\alpha)G(u) - uF(u)]dx = -(2+4\alpha)E.$$

Hence by assumption,

$$I'' > (2+2\alpha)\int u_t^2\ dx.$$

By Schwarz' inequality,

$$II'' > (1+\alpha)\int u_t^2 dx \int u^2 dx \geq (1+\alpha)(I')^2.$$

Furthermore, $I'(0) > 0$ by assumption. Hence $J = I^{-\alpha}$ satisfies $J''(t) < 0$ and $J(0) > 0$, $J'(0) < 0$. Hence $J(t) \leq J(0) + tJ'(0)$. Hence $J(T) = 0$ for some $T > 0$. Hence $\int u^2 dx \to 0$ as $t \nearrow T$. This proof shows that <u>if</u> a solution exists up to time T, then it blows up at T.

II. INVARIANCE AND CONSERVATION LAWS

In most of this chapter we consider the Euclidean equation

(1) $$\Delta u = F(u(x)) \qquad (x \in \mathbb{R}^N)$$

where F is a real function such that $F(0) = 0$ and $u(x)$ is a smooth real function going to zero as $|x| \to \infty$.

The simplest identity can be obtained by multiplying (1) by u,

(2) $$0 = (-\Delta u+F(u))u = \nabla\cdot(-\nabla u\ u) + |\nabla u|^2 + uF(u)$$

so that

(3) $$0 = \int (|\nabla u|^2 + uF(u))dx.$$

Another simple device, if u and v are two solutions of (1), is to multiply the u-equation by v, the v-equation by u, and subtract:

(4) $$0 = \nabla\cdot(-\nabla u\ v + u\ \nabla v) + F(u)v - uF(v).$$
$$0 = \int (F(u)v - uF(v))dx.$$

This is a familiar procedure for eigenvalue problems.

It is good to bear in mind the effect of scale changes. If $u(x)$ is a solution of (1), then $v(x) = \alpha u(\lambda x)$ satisfies $\Delta v = \alpha\lambda^2 F(v/\alpha)$. For instance, if equation (1) is $\Delta u = cu^p + du^q$, then α and λ may be chosen so that $\Delta v = \pm v^p \pm v^q$.

Equation (1) can be written variationally as $\delta E[u] = 0$, where

$$E[u] = \int\{\tfrac{1}{2}|\nabla u|^2 + G(u)\}dx$$

is the energy and $G(u) = \int_0^u F(v)dv$. This can be expressed formally as follows. Let T_ε be a smooth family of transformations such that $T_0 = I$. Let $M = dT_\varepsilon/d\varepsilon$ at $\varepsilon = 0$. For any function $u = u(x)$,

$$\frac{d}{d\varepsilon}\Big|_{\varepsilon=0} E[T_\varepsilon u] = (E'(u),Mu) = (-\Delta u+F(u),Mu).$$

Here M stands for "multiplier". If u is a solution of (1), this expression vanishes. This illustrates the general principle of Noether [1918] that if a one-parameter family of transformations leaves a variational problem invariant, the solution satisfies a conservation law. In our case it means that the product $(-\Delta u+F(u))(Mu)$ is a divergence.

It is well-known that the Laplace operator is invariant under the *conformal group* $\mathscr{L}$, the group of transformations on $\mathbb{R}^N$ which preserve angles. If $N \geq 3$, this group consists of four types of transformations: translations, rotations, dilation and inversions. The total dimension of $\mathscr{L}$ is therefore $N(N-1)/2 + 2N + 1$ (= 15 if $N = 4$). On the other hand, equation (1) is invariant only under the Galilean group but not under the whole conformal group, with the exception of one particular F. We propose to exploit this fact, looking separately at the various generators of $\mathscr{L}$.

The translation T_ε: $u(x) \to u(x+\varepsilon a)$, where a is a constant vector, has $M = a\cdot\nabla$ as its infinitesimal generator. Writing $(-\Delta u+F(u))(Mu)$ as a divergence, we get the conservation law

$$0 = \nabla\cdot\{-(a\cdot\nabla u)\nabla u + a(|\nabla u|^2/2 + G(u))\}.$$

We get N independent laws by choosing a as the unit vector in the coordinate direction x_k:

$$(5) \qquad 0 = \{-u_k^2 + \tfrac{1}{2}|\nabla u|^2 + G(u)\}_k + \sum_{j\neq k}\{-u_j u_k\}_j$$

where subscripts denote partial derivatives.

The rotations give the $N(N-1)/2$ multipliers $x_k u_j - x_j u_k$ for $j \neq k$ and the conservation laws

$$(6) \qquad 0 = \nabla\cdot\{(-x_k u_j + x_j u_k)\nabla u\} + \{x_k(|\nabla u|^2/2 + G(u))\}_j - \{x_j(|\nabla u|^2/2 + G(u))\}_k.$$

The dilation $u \to u_\lambda$ leaves the Dirichlet integral invariant, where $u_\lambda(x) = \lambda^m u(\lambda x)$. To find the correct value of m, we calculate $\nabla u_\lambda(x) = \lambda^{m+1}(\nabla u)(\lambda x)$ and

$$E[u_\lambda] = \int\{\tfrac{1}{2}\lambda^{2m+2}|(\nabla u)(\lambda x)|^2 + G(\lambda^m u(\lambda x)\}dx$$
$$= \int\{\tfrac{1}{2}\lambda^{2m+2-N}|\nabla u(y)|^2 + \lambda^{-N}G(\lambda^m u(y))\}dy$$

where $y = \lambda x$, $dy = \lambda^N dx$. The first term is invariant if $2m+2-N = 0$ or $m = (N-2)/2$. For this choice of m,

$$(7) \qquad 0 = \frac{d}{d\lambda}E[u_\lambda]\bigg|_{\lambda=1} = \int\{-NG(u) + muF(u)\}dy.$$

The multiplier is

$$Mu = \frac{d}{d\lambda}\lambda^m u(\lambda x)\bigg|_{\lambda=1} = x\cdot\nabla u + mu.$$

The conservation law is

$$(8) \qquad 0 = \frac{N-2}{2}\, uF(u) - NG(u) + \nabla\cdot\{(x\cdot\nabla u)\nabla u + \tfrac{1}{2}x|\nabla u|^2 + \frac{N-2}{2}\, u\nabla u + xG(u)\}.$$

Equations (3) and (7) provide some non-trivial information about possible solutions of (1). We have

$$\int |\nabla u|^2 dx = -\int uF(u)dx = \frac{-2N}{N-2}\int G(u)dx$$

(if $N \neq 2$). Therefore

$$E[u] = \frac{1}{N}\int |\nabla u|^2 dx \geq 0.$$

Theorem 2.1. *If* u *is a solution of (1), smooth and zero at infinity, then the energy is positive (except if* $u \equiv 0$*). There can be no solution of (1) if any one of the following five functions is positive (for* $s \neq 0$*):*

$$sF(s),\ G(s),\ H(s),\ -H(s),\ K(s)$$

where we assume $N \neq 1$, $H(s) = (N-2)sF(s) - 2NG(s)$, *and* $K(s) = sF(s) - 2G(s)$. (*Assume* $N \geq 3$ *for the statement about* $K(s)$.)

We have proved this theorem except for the statement about $K(s)$ which will be proved later. The one-dimensional case $(N = 1)$ is truly exceptional since it permits solutions even if $G \geq 0$. The theorem is due in part to Derrick [1964] and in part to Strauss [1977a].

We remark that, in a different context, the above considerations give us the Virial Theorem. If we suppose that F and G depend on x as well as u, then (7) becomes

$$0 = \int\{-NG(x,u) + \frac{N-2}{2}\, uF(x,u) - r\,\frac{\partial G}{\partial r}(x,u)\}dx, \quad r = |x|.$$

We have seen above that the nonlinear equation is not invariant under the transformation $u \to u_\lambda$. However, it is invariant in the special case

$$-NG(u) + \frac{N-2}{2}\, uF(u) = 0, \quad G' = F.$$

That is, $G(u) = \text{const}\ u^{2N/(N-2)}$. In this case, our variational problem is equivalent to finding the best Sobolev constant $||\phi||_{2N/(N-2)} \leq \text{const}\ ||\nabla\phi||_2$. See Strauss [1977a].

The inversion $V: x \to x/x\cdot x$ is the fourth kind of conformal transformation. It leaves the unit sphere $|x|^2 = 1$ invariant and $V^2 = I$. If we let $v(x) = |x|^{2-N}u(x|x|^{-2})$, a calculation shows that $\int|\nabla v(x)|^2 dx = \int|\nabla u(y)|^2 dy$. An N-parameter family of inversions is given by $y = V_a(x)$ where

$$y/|y|^2 = x/|x|^2 + a \qquad (a \in \mathbb{R}^N).$$

That is, $V(y) = T_a V(x)$ where T_a is translation. So we may write $V_a = VT_aV$ or

$$y = V_a(x) = \frac{x + a|x|^2}{1 + 2a\cdot x + |a|^2|x|^2} .$$

These inversions give us N rather complicated conservation laws. The multipliers are essentially

$$\frac{\partial}{\partial\varepsilon} u(V_{\varepsilon a}(x))\Big|_{\varepsilon=0} = |x|^2 a\cdot\nabla u - 2(a\cdot x)(x\cdot\nabla u).$$

We find it simpler to make the calculations using multipliers systematically.

Method of Multipliers. This method provides some useful identities which are not conservation laws. It also gives an independent derivation of the infinitesimal generators of the conformal group.

Theorem 2.2. *Let* $M = \Sigma \ell_i(x)\partial_i + p(x)$ *where* $\partial_i = \partial/\partial x_i$. *Let* $q = -2^{-1}\ \Sigma\partial\ell_i/\partial x_i + p$. *Then (for any* C^2 *function* $u(x)$) *we have the identity:*

$$\begin{aligned}(-\Delta u+F(u))(Mu) &= \Sigma(\partial\ell_i/\partial x_j)(\partial_i u)(\partial_j u) + q|\nabla u|^2 \\ &\quad - \tfrac{1}{2}\Delta p\ u^2 + puF(u) - (\nabla\cdot\ell)G(u) \\ &\quad + \nabla\cdot\{-\nabla u\ Mu + \ell(|\nabla u|^2/2 + G(u)) - \nabla p\ u^2/2\}\end{aligned}$$

where $\ell = (\ell_1,\ldots,\ell_N)$.

Proof. Simply carry out the divergence in the last term and cancel terms. There is a rationale behind the calculation which brings out its underlying structure. Let us write

$$-\Delta = -\nabla^2 = (-\mathrm{div})(\mathrm{grad}) = B^*B$$

and break M into its antisymmetric and symmetric parts:

$$M = M_a + M_s, \quad M_a = \tfrac{1}{2}(M-M^*), \quad M_s = \tfrac{1}{2}(M+M^*).$$

Then

$$(9) \qquad (-\Delta u,Mu) = (Bu,BMu) = (Bu,[B,M_a]u) + (Bu,BM_s,u)$$

because $(Bu,M_aBu) = (v,M_av) = \frac{1}{2}(v,Mv) - \frac{1}{2}(Mv,v) = 0$. Let us denote $\ell_{ij} = \partial\ell_i/\partial x_j$, $\ell_{ijk} = \partial^2\ell_i/\partial x_j\partial x_k$ and so on. Then formally

$$M^* = \Sigma(-\partial_i)(\ell_i) + p = -\Sigma\ell_i\partial_i - \Sigma\ell_{ii} + p$$

$$M_a = \Sigma\ell_i\partial_i + \frac{1}{2}\Sigma\ell_{ii}$$

$$M_s = -\frac{1}{2}\Sigma\ell_{ii} + p = q$$

$$[B,M_a] = (\partial_j)(\Sigma\ell_i\partial_i + \frac{1}{2}\Sigma\ell_{ii}) - (\Sigma\ell_i\partial_i + \frac{1}{2}\Sigma\ell_{ii})(\partial_j)$$

$$= \sum_i (\ell_{ij}\partial_i + \frac{1}{2}\ell_{iij})$$

Therefore (9) takes the form

$$(10) \quad (-\Delta u, Mu) = \sum_j \int \left[\sum_i(\ell_{ij}\partial_i u + \frac{1}{2}\ell_{iij}u) + \partial_j(qu)\right][\partial_j u]dx$$

$$= \int\left[\sum_{i,j}\ell_{ij}\partial_i u\partial_j u + q|\nabla u|^2 - \frac{1}{2}\Delta p\ u^2\right]dx.$$

This is the same as the identity in the theorem without the nonlinear term and with the divergence integrated out.

<u>The radial derivative</u>. An important multiplier which is not invariant is $\partial u/\partial r$, modified so that it becomes antisymmetric. Thus, with $r = |x|$,

$$\ell_i = \frac{x_i}{r}, \qquad \sum_{i=1}^N \ell_{ii} = \frac{N-1}{r}, \qquad q = 0$$

and the multiplier is

$$Mu = \frac{\partial u}{\partial r} + \frac{N-1}{2r}\ u.$$

Upon integration, Theorem 2.2 reduces to the following identity.

$$(11) \quad (-\Delta u+F(u),Mu) = \int(|\nabla u|^2 - u_r^2)\frac{dx}{r} + \frac{(N-1)(N-3)}{4}\int u^2\ \frac{dx}{r^3}$$

$$+ \frac{N-1}{2}\int(uF(u)-2G(u))\frac{dx}{r}.$$

The reason this identity is useful, even though M is not invariant, is that each term on the right may be positive for $N \geq 3$. One word of caution, however: the singularity at $r = 0$ could spoil everything. In fact, the worst singularity comes from the very last term in Theorem 2.2.,

$$\nabla\cdot(-\frac{1}{2}\nabla p\ u^2) = -\ \frac{N-1}{4}\ \nabla\cdot(\frac{x}{r^3}\ u^2).$$

This term integrates to zero only if $N \geq 4$. If $N = 1$ or 2 it diverges. If $N = 3$, it integrates to $2\pi u^2(0)$. Thus (11) is valid for $N \geq 4$ and, with the additional positive (!) term $2\pi u^2(0)$ on the right side, it is also valid for $N = 3$. The statement about $K(s)$ in Theorem 2.1 follows from (11).

Theorem 2.3. *If* $N \geq 3$, *every first-order differential operator* M *for which* $(-\Delta u, Mu) = 0$ *for all functions* u *comes from a combination of translations, rotations, dilation and inversions. They form a* $(N^2+3N+2)/2$ *dimensional space.*

Proof. This gives a separate justification of our earlier considerations. If by now you are tired of this sort of analysis, skip the proof. From (10) we see that necessary and sufficient conditions for M to be invariant are

$$\ell_{ij} + \ell_{ji} = 0 \quad \text{for} \quad i \neq j, \tag{12}$$

$$\ell_{ii} + q = 0 \quad \text{for all} \quad i, \tag{13}$$

$$\Delta p = 0. \tag{14}$$

We shall solve (12) and (13). They are $N(N+1)/2$ equations for $N+1$ unknowns.

Our notation is $\ell_{ijk} = \partial^2\ell_i/\partial x_j\partial x_k$, $q_i = \partial q/\partial x_i$, and so on. Let i,j,k be distinct indices. Then

$$q_{ii} = -\ell_{jjii} = -\ell_{jiij} = +\ell_{ijij} = +\ell_{iijj} = -q_{jj},$$

$$q_{ii} = -q_{jj} = +q_{kk} = -q_{ii},$$

Hence $q_{ii} = 0$ for each i, which shows that (14) is redundant. Next,

$$\ell_{ijk} = -\ell_{jik} = -\ell_{jki} = +\ell_{kji} = +\ell_{kij} = -\ell_{ikj} = -\ell_{ijk},$$

$$\ell_{ijk} = 0,$$

$$q_{ij} = -\ell_{kkij} = -\ell_{kijk} = +\ell_{ikjk} = 0.$$

Thus $\nabla q_i = 0$, $q_i = \text{constant} = -\alpha_i$. So

$$\ell_{jj} = -q = \sum_{i=1}^{N} \alpha_i x_i + \beta.$$

Define $f_j(x)$ by the equation

$$\ell_j(x) = x_j \sum_{i=1}^{N} \alpha_i x_i - \frac{1}{2}\alpha_j \sum_{i=1}^{N} x_i^2 + \beta x_j + f_j(x).$$

Then $f_{jj} = 0$ and $f_{jk} + f_{kj} = 0$ for $j \neq k$ by (12). It follows that $f_j(x) = \sum_k \gamma_{jk}x_k + \delta_j$ where $\gamma_{jk} = -\gamma_{kj}$ and δ_j are constants. Hence

$$\ell_j(x) = x_j \sum_i \alpha_i x_i - \frac{1}{2}\alpha_j \sum_i x_i^2 + \beta x_j + \sum_i \gamma_{ji}x_i + \delta_j.$$

That is,

$$\ell = (\alpha\cdot x)x - \tfrac{1}{2}\alpha|x|^2 + \beta x + \Gamma x + \delta$$

where ℓ, α and δ are vectors and Γ is a skew matrix. There are N alphas, one beta, $N(N-1)/2$ gammas and N deltas.

From (13) and (10) the multiplier is $M = \ell\cdot\nabla + (1-N/2)q$ and Theorem 2.2 simplifies to:

$$(15)\qquad (-\Delta u+F(u))(Mu) = (1-N/2)quF(u) + NqG(u)$$
$$+ \sum_{j=1}^{N} \frac{\partial}{\partial x_j}\left\{- \frac{\partial u}{\partial x_j} Mu + \ell_j(\tfrac{1}{2}|\nabla u|^2+G(u)) + \frac{2-N}{4}\alpha_j u^2\right\}.$$

Choosing $\delta_k = 1$ and the other coefficients zero, we get the translational identity (5). Choosing $\gamma_{jk} = 1 = -\gamma_{kj}$ and the others zero, we get the rotational identity (6). Choosing $\beta = 1$ and the others zero, we get the dilational identity (8). Finally, let us choose $\alpha = 2a$. Then the multiplier is

$$Mu = 2(a\cdot x)(x\cdot\nabla u) - |x|^2(a\cdot\nabla u) + (N-2)(a\cdot x)u.$$

If $\alpha_k = 2$ and $\alpha_j = 0$ for $j \neq k$, then

$$(16)\qquad Mu = (x_k^2 - \sum_{i\neq k} x_i^2)\partial_k u + 2x_k \sum_{j\neq k} x_j\partial_j u + (N-2)x_k u$$

We note from (15) that $(-\Delta u + F(u),Mu) = 0$ only if $q = 0$ (the translational and rotational identities) or $G(u) = \text{const } u^{2N/(N-2)}$.

For the proof of Theorem 2.3 the notes of Tartar [1976] have been most helpful.

<u>The Relativistic Case</u>. Take the nonlinear Klein-Gordon equation

$$\text{(NLKG)}\qquad u_{tt} - \Delta_x u + m^2u + F(u) = 0, \quad x \in \mathbb{R}^N.$$

We can transfer each of the Euclidean identities by making the following changes:

$$N = n+1,$$
$$x \to (x_1,x_2,\ldots,x_n,it), \quad x_N = x_{n+1} = it,$$
$$F(u) \to m^2u + F(u).$$

Thus there are $(N^2+3N+2)/2$ identities which immediately follow from their Euclidean counterparts. Here they are, after integration over space coordinates only.

From the multiplier $u_t = \partial_t u$, we get the <u>energy</u>

$$\int e(u)dx = \int(\tfrac{1}{2}u_t^2 + \tfrac{1}{2}|\nabla u|^2 + \tfrac{1}{2}m^2u^2 + G(u))dx = \text{constant}.$$

From the multiplier $u_k = \partial_k u$, we get the momenta

$$\int u_t u_k \, dx = \text{constant}.$$

We get the angular momenta from the multipliers $x_k u_t + t u_k$ and $x_k u_j - x_j u_k$:

$$\int (x_k e(u) + t u_k u_t) dx = \text{const.}$$

and

$$\int (x_k u_j - x_j u_k) u_t dx = \text{const.}$$

The next two identities are due to Morawetz [1975]: From the multiplier $Mu = tu_t + ru_r + \frac{n-1}{2} u$, where $r = |x|$ is the spatial radius, we get the dilational identity

$$0 = \frac{d}{dt} \int (te(u) + ru_r u_t + \frac{n-1}{2} uu_t) dx + \frac{1}{2} \int H(u) dx$$

where

$$H(u) = (n-1)uF(u) - 2(n+1)G(u) - 2m^2 u^2.$$

Finally we get the inversional or conformal identities. From the multiplier $(k = N = n+1$ in (16))

$$Mu = (t^2+r^2)u_t + 2rtu_r + (n-1)tu,$$

we get the identity

$$(17) \quad 0 = \frac{d}{dt} \int [(t^2+r^2)e(u) + 2rtu_r u_t + (n-1)tuu_t - \frac{n-1}{2} u^2] dx + t \int H(u) dx.$$

From the multiplier (see (16))

$$Mu = tx_k u_t + \frac{1}{2}(t^2 + 2x_k^2 - r^2) u_k + x_k \sum_{j \neq k} x_j u_j + \frac{n-1}{2} x_k u,$$

we get the identity

$$0 = \frac{d}{dt} \int [tx_k e(u) + \frac{1}{2}(t^2 + 2x_k^2 - r^2) u_k u_t + x_k \sum_{j \neq k} x_j u_j u_t + \frac{n-1}{2} x_k u_t u] dx + \frac{1}{2} \int x_k H(u) dx.$$

Another identity due to Morawetz [1968] is obtained using the spatial radial derivative as the multiplier. Thus

$$Mu = \frac{\partial u}{\partial r} + \frac{n-1}{2r} u, \quad r = |x|.$$

In analogy to (11), we get

$$(18) \qquad 0 = \frac{d}{dt}\int u_t\left(u_r + \frac{n-1}{2r}u\right)dx + \int\left(|\nabla u|^2 - u_r^2\right)\frac{dx}{r}$$
$$+ \frac{(n-1)(n-3)}{4}\int u^2 \frac{dx}{r^3} + \frac{n-1}{2}\int (uF(u)-2G(u))\frac{dx}{r}$$

for $n \geq 3$, with the extra term $2\pi u^2(0,t)$ in case $n = 3$.

In analogy to (4) we have, for any pair of solutions, the familiar identity

$$(19) \qquad \frac{d}{dt}\int (uv_t - u_t v)dx = \int (F(u)v - uF(v))dx.$$

In analogy to (2) we have (in the complex case)

$$0 = \frac{d}{dt}\int u_t\bar{u}dx + \int\left(|\nabla u|^2 - |u_t|^2 + m^2|u|^2 + \bar{u}F(u)\right)dx.$$

Its imaginary part gives the conservation of <u>charge</u> in case $\bar{u}F(u)$ is real.

III. SOLITONS

A. Solitary Waves. In scattering theory an important role is naturally played by waves $u(x)$ which are stationary in some Lorentz frame. In another frame they look like $u(x-ct)$ where c is a constant vector. We could also consider $\exp(i\omega t)u(x)$, as in the linear theory. More generally we define a solitary wave as a solution $u(x,t)$ of a wave equation whose maximum amplitude at time t, $\max_x |u(x,t)|$, does not tend to zero as $|t| \to \infty$, although the energy is finite. For instance for NLKG, $u(x-ct)$ is a solitary wave if it approaches a zero of $G(u)$ fast enough as $|x| \to \infty$. For a solution of the form $u(x-ct)$ or $\exp(i\omega t)u(x)$, NLKG reduces to an elliptic equation of the form (2.1) studied in Chapter II. Following Strauss [1977a], we shall show that solutions of (2.1) are not difficult to find. We assume $F(0) = 0$. Here $F(u)$ is allowed to include the term m^2u. We recall from Theorem 2.1 that any solution has positive energy.

In one dimension it is easy to find explicit solutions which vanish at infinity. For example, the equation $-u_{xx} + m^2u - \lambda u^p = 0$ has the solution

$$u(x) = a\ \mathrm{sech}^{2/(p-1)}(bx),\quad a = ((p+1)m^2/2\lambda)^{1/p-1},\quad b = (p-1)m/2.$$

Theorem 3.1. *Under the following conditions, there exists a constant* $\lambda > 0$ *and a nonnegative solution* $u \not\equiv 0$ *of the equation*

$$-\Delta u + m^2u + F_1(u) = \lambda F_2(u)$$

for which $\int |\nabla u|^2 dx < \infty$, $\int G_1(u(x))dx < \infty$, *and* $u(x)$ *decays exponentially as* $|x| \to \infty$. *Here* $G_1 = \int F_1$, $G_2 = \int F_2$. *The conditions are* $m > 0$ *and*

(1) $F_1(s) \geq 0$, $F_2(s) > 0$ *for* $s > 0$

(2) $F_i(s) = o(s)$ *as* $s \to 0$ $(i = 1,2)$

(3) $F_2(s) = o(s^\ell + F_1(s))$ *and* $= O(s^\ell + G_1(s)/s)$ *as* $s \to \infty$

where $\ell = 1+4/(n-2)$ *if* $n \geq 3$ ($\ell < \infty$ *if* $n = 2$; *drop (3) entirely if* $n = 1$).

Example 1. $-\Delta u + u = |u|^{q-1}u$ with $q > 1$, $n \geq 3$. We have scaled out the coefficients to be 1. Here $F(s) = s - |s|^{q-1}s$, $G(s) = s^2/2 - |s|^{q+1}/(q+1)$. Let $\alpha^{-1} = 2^{-1} - (q+1)^{-1}$. By Theorem 2.1, there is no non-trivial solution if $((n-2)/2)sF(s) - nG(s) = -s^2 + (1-\alpha^{-1}n)|s|^{q+1}$ is of one sign; that is, $\alpha \leq n$ or $q \geq 1 + 4/(n-2)$. So assume $1 < q < 1 + 4/(n-2)$. Any solution must satisfy the identities of Theorem 2.1 which reduce to:

$$\alpha(n-2)\int|\nabla u|^2dx = \frac{n\alpha}{\alpha-n}\int|u|^2dx = n\int|u|^{q+1}dx.$$

Theorem 3.1 asserts the existence of a solution, where $F_1(s) = 0$, $F_2(s) = |s|^{q-1}s$, mass $m = 1$, and a scale change is used to make $\lambda = 1$. More sophisticated methods show that this solution is the first one of an infinite sequence of distinct solutions.

Example 2. $-\Delta u + (m^2-\omega^2)u + |u|^{p-1}u - \lambda|u|^{q-1}u = 0$ where $x \in \mathbb{R}^n$, $m^2 - \omega^2 > 0$ and p and q are distinct numbers larger than 1. We distinguish four cases.

Case A: $q < p$. Theorem 3.1 asserts the existence of a non-trivial solution for some $\lambda > 0$. Note that

$$G(s) = \frac{1}{2}(m^2 - \omega^2)s^2 + \frac{1}{p+1}|s|^{p+1} - \frac{\lambda}{q+1}|s|^{q+1}$$

is bounded below. There is a number λ_* so that, for $\lambda \leq \lambda_*$, $G(s)$ is non-negative and the only solution is the trivial one, according to Theorem 2.1.

According to Theorem 2.1, the (integrated) energy is necessarily positive, even though the function $G(s)$ is not allowed to be positive if a non-trivial solution is to exist. On the other hand, for a standing wave solution $\exp(i\omega t)u(x)$ of the NLKG equation, the energy density $\frac{1}{2}|\nabla u|^2 + \frac{1}{2}\omega^2u^2 + G(u)$ may be positive. This is the case if $\omega > 0$ and λ is slightly larger than λ_*.

Anderson [1971] has computed these solutions in the case $n = 3$, $p = 5$, $q = 3$. His most interesting result is that the positive solution appears to be stable with respect to perturbations of the initial data of NLKG in the cases when the energy density is positive. For this choice of p, q and n, the inequalities

$$\lambda s^4 = (2s)(\tfrac{1}{2}\lambda s^3) \leq \tfrac{1}{2}(2s)^2 + \tfrac{1}{2}(\tfrac{1}{2}\lambda s^3)^2 = 2s^2 + \tfrac{1}{8}\lambda^2 s^6,$$

$$G(s) = \tfrac{1}{2}s^2 + \tfrac{1}{6}s^6 - \tfrac{\lambda}{4}s^4 \geq (\tfrac{1}{6} - \tfrac{1}{32}\lambda^2)s^6$$

show that $\lambda_* = 4/\sqrt{3}$, if we normalize to make $m^2 - \omega^2 = 1$.

Case B: $p < q < 1 + 4/(n-2)$. Theorem 3.1 is again applicable. We can prove the existence of an infinite sequence of non-trivial solutions for each $\lambda > 0$.

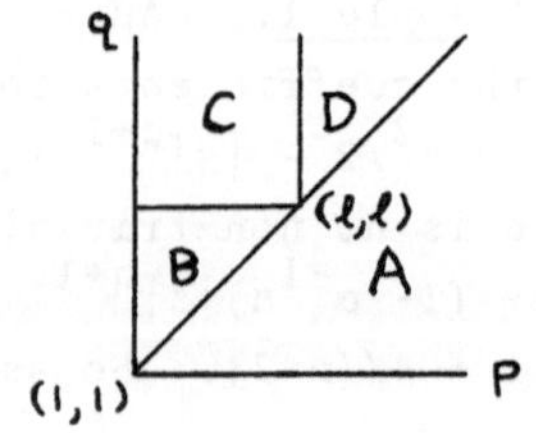

Case C: $p \leq 1 + 4/(n-2) \leq q$. Let $\alpha^{-1} = 2^{-1} + (q+1)^{-1}$ and $\beta^{-1} = 2^{-1} + (p+1)^{-1}$. Then $\alpha \leq n \leq \beta$ and

$$\frac{n-2}{2} sF(s) - nG(s) = -s^2 - (1 - \frac{n}{\beta})|s|^{p+1} + (1 - \frac{n}{\alpha})\lambda|s|^{q+1}.$$

By Theorem 2.1, there can be no non-trivial solution.

Case D: $1 + 4/(n-2) < p < q$. This case remains open: we do not know whether or not there exists a non-trivial solution.

Proof of Theorem 3.1. We sketch the proof in the case of Example 2, for $n \geq 3$ and $q < \max(p, 1+4/(n-2))$:

$$-\Delta u + u + |u|^{p-1}u = \lambda|u|^{q-1}u.$$

For the general case, see Strauss [1977a]. Let $F_1(u) = |u|^{p-1}u$, $F_2(u) = |u|^{q-1}u$. We use the direct method of the calculus of variations. Let

$$L = \inf \int (\tfrac{1}{2}(|\nabla u|^2 + u^2) + G_1(u))dx = \inf \mathscr{L}(u)$$

subject to the constraints

$$\int(|\nabla u|^2 + u^2)dx < \infty, \quad u = u(|x|),$$
$$\mathscr{M}(u) = \int G_2(u)dx = 1.$$

Obviously $0 < L < \infty$. Choose any minimizing sequence:

$$\mathscr{L}(u_\nu) \to L, \quad \mathscr{M}(u_\nu) = 1.$$

Because each term in $\mathscr{L}(u_\nu)$ is positive, each term is bounded. By Sobolev's inequality (see (1.4)), $\int|u_\nu|^{\ell+1}dx$ is also bounded. Standard compactness theorems as in the proof of Theorem 1.1 imply that there exists a subsequence, still denoted u_ν, which converges weakly to a limit u and $u_\nu(x) \to u(x)$ almost everywhere. We want to show $\mathscr{M}(u) = 1$. This follows from the assumption that the u_ν depend only on $|x|$, that

$$q+1 < \max(p+1, 2+4/(n-2))$$

and a little measure theory (see Strauss [1970]). From the weak convergence and Fatou's lemma, it follows that $\mathscr{L}(u) \leq \liminf \mathscr{L}(u_\nu) \leq L$. Hence $\mathscr{L}(u) = L$ and $\mathscr{M}(u) = 1$. Thus u attains the minimum. By the method of Lagrange multipliers, u satisfies the differential equation for some λ.

Now $u^+(x) = \max\{u(x),0\}$ is also a solution of the minimum problem, if we define $F_i(s) = 0$ for $s \leq 0$. This shows that there exists a non-negative solution. Finally, the fact that the mass is positive together with linear spectral theory implies the exponential decay.

B. Solitons. Solitary waves have been known in classical physics ever since Scott-Russell [1844], riding on horseback alongside a canal, observed a wave in the form of "a large solitary elevation, a rounded, smooth and well-defined heap of water, which continued its course along the channel apparently without change of form or diminution of speed." But it was long considered a rather unimportant curiosity for it was generally supposed that if two solitary waves were initially launched on a collision course, the nonlinear interaction would completely destroy their integrity. With the advent of the computer it became possible to test this assumption. The first such test was performed by Perring and Skyrme [1962] for what is now known as the sine-Gordon equation

$$\text{(SG)} \qquad \phi_{tt} - \phi_{xx} + \sin\phi = 0.$$

They were amazed to find that the solitary waves emerged from the collision having exactly the same shapes and velocities with which they entered! Independently, Zabusky and Kruskal [1965] found the same behavior for the Korteweg-deVries equation. They coined the term "soliton" for these remarkable solitary waves. Since then, several other particular examples of nonlinear equations with solitons have been discovered. See Scott et al.[1973] for a survey and references. Every one of these examples has space dimension one. No examples of solitons (in the proper, narrow sense) have yet been found for a scalar equation in higher dimensions. (June 1977 update: an example has been reported in Rome by Zakharov.)

We will discuss only the sine-Gordon equation (SG), which is invariant under Lorentz transformations, including translations in space and time, as well as the symmetries

$$\phi \to -\phi \quad \text{and} \quad \phi \to \phi + 2\pi n$$

(n an integer). It has the constant solutions $\phi = 2\pi n$. If one regards ϕ as an angle, multiples of 2π are identified, and there is only one constant solution, the vacuum. The energy is

$$E = \int \{\tfrac{1}{2}\phi_t^2 + \tfrac{1}{2}\phi_x^2 + 1 - \cos\phi\}dx.$$

A good reference on (SG) is Rajaraman [1975].

In the notation of Chapter I, $G(\phi) = 1 - \cos\phi$ vanishes for $\phi = 2\pi n$ (n = integer). Thus the asymptotic values of solitary waves must be multiples of 2π. A stationary solution $\phi(x)$ would satisfy the equation $\phi_{xx} = \sin\phi$, which is easily integrated. Assuming $\phi_x \to 0$ and $\phi \to 2\pi n$ as $x \to \pm\infty$, the solution is

$$\phi(x) = 4\tan^{-1}(\exp(\pm x)).$$

Transforming to another Lorentz frame, we get

$$s_\pm(x,t) = 4\tan^{-1}\exp[\pm(x-ct)(1-c^2)^{-1/2}]$$

for $|c| < 1$. The picture of s_+

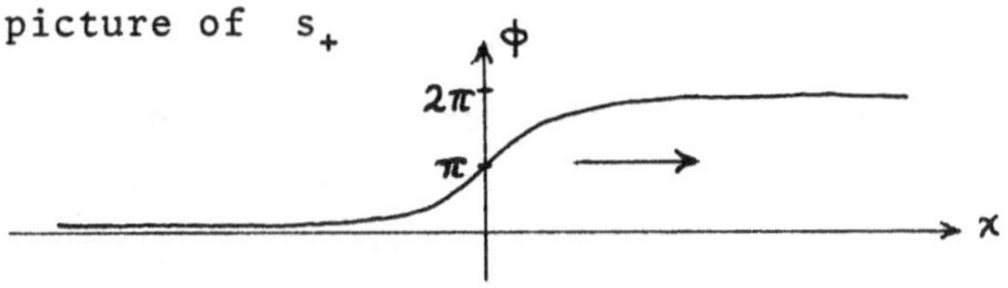

moves at speed $|c|$ to the right ($c > 0$) or left ($c < 0$) and

$$s_\pm(+\infty,t) - s_\pm(-\infty,t) = \pm 2\pi.$$

These are the *solitons*.

What is the asymptotic behavior as $t \to \pm\infty$ of the *general* solution of (SG)? The answer can be seen clearly in computer-generated motion pictures. It has a discrete part composed of a finite number of solitons and antisolitons and a continuous, or transient, part whose amplitude goes to zero as $t \to \pm\infty$. The discrete part has been analyzed mathematically in great detail but the proof that the transient part actually goes to zero has so far eluded rigorous proof.

It is possible to explicitly write down formulas for those solutions which have no continuous part at all. For instance, there is the *soliton-antisoliton* solution

$$s_{+-}(x,t) = 4\tan^{-1}\left[\frac{\sinh(ct/\sqrt{1-c^2})}{c\cosh(x/\sqrt{1-c^2}}\right].$$

An explicit computation shows that

$$s_{+-}(x,t) \sim s_+(x,-(t+\Delta/2)) + s_-(x,t+\Delta/2) \qquad \text{as} \quad t \to -\infty$$

$$s_{+-}(x,t) \sim s_+(x,-(t-\Delta/2)) + s_-(x,t-\Delta/2) \qquad \text{as} \quad t \to +\infty,$$

where $\Delta/2 = \sqrt{1-c^2}\log c/c < 0$. Thus s_{+-} consists of a soliton and an antisoliton coming towards each other, momentarily annihilating one another at time $t = 0$, emerging unscathed and separating from each other with only a time delay Δ. For formulas for the multisoliton (purely discrete) solution, see Appendix A of Scott et al. [1973].

The general solution of (SG) has a similar behavior. As $t \to \infty$, it breaks up into a finite number of solitons (or antisolitons) arranged so that the fastest soliton is at the front and the slowest is at the rear. As $t \to -\infty$, the arrangement is reversed. There is an interaction region where the shape of the solution looks complicated, but in the passage from time $-\infty$ to time $+\infty$, the solitons are completely unchanged in shape and speed. Only a time delay remains as evidence of the interaction. The formula for this delay shows that they are the same as if the solitons collided only pairwise. Thus we could say that triple collisions have no effect at all.

C. The Inverse Scattering Method. Why does such a nonlinear equation possess solutions which interact almost like linear waves? The answer is that there is indeed a linear problem lurking in the background. For convenience, let us write the sine-Gordon equation in characteristic coordinates

$$X = \frac{x+t}{2}, \qquad T = \frac{x-t}{2}, \tag{4}$$
$$\frac{\partial^2\phi}{\partial X \partial T} = \sin\phi.$$

Consider the *linear* eigenvalue problem

$$\begin{cases} \dfrac{\partial v_1}{\partial X} + i\zeta v_1 = q v_2 \\ \dfrac{\partial v_2}{\partial X} - i\zeta v_2 = -q v_1 \end{cases} \tag{5}$$

where ζ is the eigenvalue, $q = q(X,T) = -\frac{1}{2}\partial\phi/\partial X$ plays the role of the potential, and T plays the role of a parameter. Note that q depends on ϕ and hence on T. The magic is that the eigenvalues ζ are independent of T.

In fact, assume the following T-dependence of the eigenfunctions

$$\frac{\partial}{\partial T}\begin{bmatrix} v_1 \\ v_2 \end{bmatrix} = \frac{i}{4\zeta}\begin{bmatrix} \cos\phi & \sin\phi \\ \sin\phi & -\cos\phi \end{bmatrix}\begin{bmatrix} v_1 \\ v_2 \end{bmatrix}. \tag{6}$$

Differentiate (5) with respect to T and (6) with respect to X. Use (4) and the definition of q. We find that ζ is independent of T.

By studying the linear problem (5) we can gather a great deal of information about (SG). In fact we can entirely reconstruct ϕ! This is done by the inverse scattering technique which has been worked out for (SG) by Ablowitz et al. [1974].

Define the functions χ and ψ as the solutions of (4) with the asymptotic forms

$$\chi \to \begin{pmatrix}1\\0\end{pmatrix} e^{-i\zeta X} \quad \text{as} \quad X \to -\infty, \ \mathrm{Im}\,\zeta \geq 0,$$
$$\psi \to \begin{pmatrix}0\\1\end{pmatrix} e^{i\zeta X} \quad \text{as} \quad X \to +\infty, \ \mathrm{Im}\,\zeta \geq 0.$$

Thus χ is small at $-\infty$ and ψ at $+\infty$. Let

$$v = \begin{pmatrix} v_1(X,\zeta) \\ v_2(X,\zeta) \end{pmatrix}, \qquad v^\dagger(x,\zeta) = \begin{pmatrix} \bar{v}_2(X,\bar{\zeta}) \\ -\bar{v}_1(X,\bar{\zeta}) \end{pmatrix}$$

where the bar denotes complex conjugate. If v is a solution of (5) so is $v^\dagger$. All solutions are linear combinations of ψ and $\psi^\dagger$. In particular,

$$\chi = a(\zeta)\psi^\dagger + b(\zeta)\psi, \quad \zeta = \xi + i\eta.$$

The coefficient $a(\zeta)$, originally defined for $\zeta = \xi$ real, can be continued analytically into the upper half-plane $\mathrm{Im}\,\zeta > 0$. The zeros ζ_j $(j = 1,\ldots,N)$ of $a(\zeta)$ in the upper half-plane are the discrete eigenvalues of (5). At these values we have $\chi(X,\zeta_j) = c_j\psi(X,\zeta_j)$. From (6) the T-dependence of the coefficients is found to be

$$a(\xi) = a_0(\xi), \qquad b(\xi) = b_0(\xi)\exp(-iT/2\xi),$$
$$c_j = c_{j0}\exp(-iT/2\zeta_j).$$

The classical scattering data for (5) is $\{a(\xi), b(\xi), c_j, \xi_j\}$. The inverse scattering problem is to reconstruct the potential from them. This is done by the Gelfand-Levitan-Marchenko equation (see Marchenko [1955]) as follows. Let

$$B(X) = \frac{1}{2\pi}\int_{-\infty}^{\infty} \frac{b(\xi)}{a(\xi)} e^{i\xi X} d\xi - i\sum_{j=1}^{N} c_j e^{i\zeta_j X}.$$

Solve the GLM integral equation

$$K(X,y) = \bar{B}(X+y) - \int_X^\infty \int_X^\infty \bar{B}(y+z)B(w+z)K(X,w)\,dz\,dw.$$

Then $q(X) = -2K(X,X)$. The great advantage of (GLM) is that it is linear.

In our case q depends on the parameter T and so do b, c_j, B and K. This allows a complete solution of the sine-Gordon equation: given "initial" data $\phi(X,0)$, calculate the scattering data at $T = 0$, hence the scattering data at any T, then solve the inverse problem to get $\phi(X,T)$.

Unfortunately, an explicit solution of (GLM) is impossible in general. However, some particular solutions can be calculated, those corresponding to the discrete spectrum.

If ζ is a discrete eigenvalue, so is $-\bar{\zeta}$. (Conjugate the equation (5).) If $b(\xi) = 0$, the function $B(X)$ is a finite sum and (GLM) can be solved explicitly. The solution turns out to be

$$\frac{1}{4}\left(\frac{\partial\phi}{\partial X}\right)^2 = \frac{d^2}{dX^2} \log \det(I + AA^*)$$

where

$$A_{\ell m} = \frac{\sqrt{c_\ell \bar{c}_m}}{\zeta_\ell - \bar{\zeta}_m} e^{i(\zeta_\ell - \bar{\zeta}_m)x} .$$

This gives us the *multisolitons*. For instance, a single soliton corresponds to a single purely imaginary eigenvalue $\zeta = i\eta$; it is

$$\phi = 4 \tan^{-1}(\exp \pm \{(\eta+1/4\eta)(x-x_0) + (\eta-1/4\eta)t\}).$$

Corresponding to a pair ζ, $-\bar{\zeta}$ of complex eigenvalues with $\mathrm{Re}(\zeta) \neq 0$ is the multisoliton solution

$$\phi = 4 \tan^{-1}\left[\frac{\eta \cos\{\xi\eta(t-t_0) - (4-\nu)x\}}{\xi \cosh\{\eta\nu(x-x_0) - (4-\nu)t\}}\right],$$

where $\nu = 2 + (2|\zeta|^2)^{-1}$. This solution is particularly interesting in case $|\zeta| = \frac{1}{2}$, $\nu = 4$, in which case it reduces to a *breather* solution:

$$\phi = 4 \tan^{-1}\left[\frac{\eta \cos\{\xi\eta(t-t_0)\}}{\xi \cosh\{4\eta(x-x_0)\}}\right].$$

A breather has a fixed location where it oscillates in time, like a standing wave.

We have seen how the solitons are related to a linear eigenvalue problem. Another characteristic feature is the existence of a *Bäcklund transformation* which carries solutions of (4) into solutions of (4). It is $\phi \to \psi$ where

$$\psi_X = \phi_X + 2a \sin\left(\frac{\psi+\phi}{2}\right), \quad \psi_T = -\phi_T + \frac{2}{a} \sin\left(\frac{\psi-\phi}{2}\right)$$

where a is a constant. Now if we think of a as small and expand ψ in powers of a,

$$\psi \sim \sum_{j=0}^{\infty} \psi_j(X,T)a^j \quad \text{as } a \to 0,$$

the coefficients ψ_j can be determined from ϕ. Of course

$$\left(\tfrac{1}{2}\phi_X^2\right)_T + (\cos\phi - 1)_X = 0$$

and the same for ψ. If we substitute the series for ψ into this, and equate like powers of a, we get an infinite sequence of distinct *conservation laws*. For instance, the second one is

$$[2\phi_{TTTX}\phi_X + 4\phi_{TTX}\phi_{TX} + \phi_T^2\phi_{TX}\phi_X]_T + [\cdots]_X = 0.$$

The existence of infinitely many exact conservation laws is another striking phenomenon related to the existence of the solitons themselves.

IV. BASIC IDEAS IN NONLINEAR SCATTERING

The idea of a scattering state is that the interaction has no effect asymptotically far in the future and deep in the past. A typical result in mathematical scattering theory is the following one.

Theorem 4.1. *For every "nice" solution of the equation*

$$u_{tt} - \Delta u + m^2 u + \lambda u^3 = 0$$

$x \in \mathbb{R}^3$, $\lambda > 0$, $m > 0$, *there exists a pair of "free" solutions* u_+ *and* u_- *of the linear Klein-Gordon equation* $v_{tt} - \Delta v + m^2 v = 0$ *such that* $u(t) - u_\pm(t) \to 0$ *in the energy norm as* $t \to \pm\infty$. *The scattering operator* $S: \phi_- \to u_+$ *is a well-defined nonlinear operator on a certain Banach space of free solutions.*

We will build up the theory gradually and finally prove this theorem in Chapter VI. The theory in Chapter IV and V is a development of that of Segal [1966, 1968] and a slight modification of the presentation in Strauss [1974]. The latter article presents some different examples from these lecture notes. The material is also presented in Reed and Simon [1977]. The functional analytic framework is as follows.

Let X be a Hilbert space with norm $|\ |_2$. Let $U(t)$ be a family of (nonlinear) operators on X. We may think of a state $f \in X$ evolving into the state $U(t)f$ with the passage of time t. We assume

$$(1) \qquad U(t)U(s) = U(t+s), \quad U(0) = I.$$

Here $-\infty < t < \infty$. It follows that these operators are invertible and $U(t)^{-1} = U(-t)$. Assume a reference system $U_0(t)$ is also given satisfying (1). Assume $U_0(t)$ is a unitary linear operator for each t. Given a state f, we look for states f_+ and f_- such that

$$|U(t)f - U_0(t)f_\pm|_2 \to 0 \quad \text{as} \quad t \to \pm\infty.$$

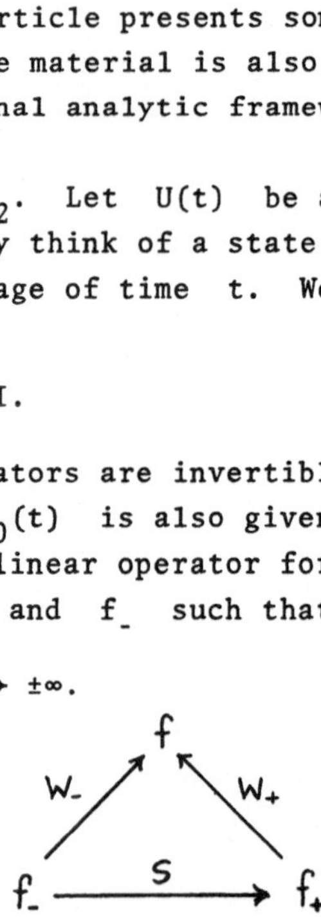

f_+ and f_- are necessarily unique because $U_0(t)$ is unitary. We define the scattering operator as the map $f_- \to f_+$. We may also think of it as taking $U_0(t)f_-$ into $U_0(t)f_+$. The wave operators are defined as $W_\pm: f_\pm \to f$ or $U_0(t)f_\pm \to U_0(t)f$. They act from free states to interacting states. They are called complete if $\text{Range}(W_-) = \text{Range}(W_+)$. If they are complete and W_+ is one-one, we can form $S = W_+^{-1}W_-$. It is standard to denote $U(t) = \exp(itH)$ and $U_0(t) = \exp(itH_0)$ but we will not use this notation.

The <u>intertwining relations</u> are

(2) $$W_\pm U_0(T) = U(T)W_\pm \quad \text{and} \quad SU_0(T) = U_0(T)S.$$

<u>Proof</u>: $U(t)W_+f_+ \sim U_0(t)f_+$ as $t \to +\infty$. Replacing t by $t+T$, $U(t)U(T)W_+f_+ \sim U_0(t)U_0(T)f_+$ as $t \to +\infty$. Hence the first relation in (2). Similarly for W_-. Letting $f = W_+f_+ = W_-f_-$, we have $U(t)U(T)f \sim U_0(t)U_0(T)f_\pm$ as $t \to \pm\infty$, hence the last relation in (2).

The usefulness of (2) in linear scattering theory is well-known. Often W_+, W_- and S can be shown to be unitary operators and then (2) says that the free and interacting groups are unitarily equivalent. Also S commutes with the free group and hence with its generator; hence S is diagonalized in the spectral representation of the free group.

If S is nonlinear and $U_0(t) = \exp(itH_0)$ is linear then the last relation in (2) can be differentiated formally to yield the relation

$$H_0Sg = S'(g)H_0g.$$

We now fix the precise assumptions to be used later.

<u>Hypothesis I</u>: X is a Hilbert space. $U_0(t)$ is a unitary group of operators on X. P is an operator from a domain $D(P) \subset X$ into X such that $P0 = 0$.

<u>Definition</u>: A <u>perturbed solution</u> is a function $u(t)$ with values in $D(P)$ for which $Pu(t)$ is continuous in X which satisfies the equation

(3) $$\frac{d}{dt}[U_0(-t)u(t)] = U_0(-t)Pu(t).$$

The idea is that $u(t) = U(t)f$, $P = iH - iH_0$ is the perturbation operator and (3) is the familiar equation

$$\frac{d}{dt}\left[e^{-itH_0}\, e^{itH}\, f\right] = e^{-itH_0}\, Pe^{itH}\, f.$$

Thus (3) is the equation $\frac{du}{dt} = iH_0u + Pu$ written in the Heisenberg picture.

It is convenient to write (3) in integral form:

(4) $$u(t) = U_0(t-T)u(T) + \int_T^t U_0(t-s)Pu(s)ds$$

If we let $T \to \pm\infty$, we may write (with $u_\pm(t) = U_0(t)f_\pm$):

(5) $$u_\pm(t) = u(t) + \int_t^{\pm\infty} U_0(t-s)Pu(s)ds$$

(the Yang-Feldman equations). Subtracting the + and - equations,

(6) $$u_+(t) - u_-(t) = \int_{-\infty}^{\infty} U_0(t-s)Pu(s)ds.$$

Hypothesis II: In addition to the Hilbert norm $|\ |_2$ there are two other norms $|\ |_3$ and $|\ |_1$ defined on X, except that we allow them to be $+\infty$ or zero for some $0 \neq f \in X$. We assume

$$|U_0(t)f|_3 \leq c|t|^{-d}|f|_1 \quad \text{for} \quad |t| \geq t_0$$

where c, d and t_0 are positive constants. d is called the "decay rate".

This means that the free group decays in a certain norm $|\ |_3$ even though it is constant in the Hilbert norm. In the following discussion we will always assume I and II. Each of the following theorems assumes some additional condition. The first result is an observation of Segal [1966] and Matsumura [1976], which says that if P is "strong enough", the scattering operator exists in a weak sense.

Hypothesis: Assume $\{f \in X: |f|_1 < \infty\}$ is dense in X and there is a constant r such that

(7) $$|(Pf,g)| \leq b(f)|g|_2^{1-r}|g|_3^r .$$

Theorem 4.2. *Let* u(t) *be a perturbed solution such that* b(u(t)) *is bounded. If* $r > 1/d$, *then* $U_0(-t)u(t)$ *converges weakly in* X *as* $t \to \pm\infty$.

Proof. We must find f_+ and f_- in X such that as $t \to \pm\infty$

$$(U_0(-t)u(t),h) \to (f_\pm,h) \quad \text{for all} \quad h \in X.$$

It suffices to take $|h|_1 < \infty$, which is a dense set. By (5) it suffices to show

$$\int_{-\infty}^{\infty} |(U_0(-t)Pu(t),h)|dt < \infty.$$

Now the integrand is

$$|(Pu(t),U_0(t)h)| \leq b|h|_2^{1-r}|U_0(t)h|_3^r \leq bc^r|t|^{-dr}|h|_1^r|h|_2^{1-r}$$

for $|t| \geq t_0$. This is integrable since $dr > 1$. Q.E.D.

The trouble with Theorem 4.2 is that the weak limits may not be unique. If we want strong limits, the next result shows how careful we must be. It is analogous to the situation for the Coulomb potential. The method is due to Glassey [1973a].

Hypothesis: Let $B(\ ,\ )$ be an invariant bounded bilinear form on X; that is, $|B(f,g)| \leq c|f|_2|g|_2$ and $B(U_0 t)f, U_0(t)g) = B(f,g)$ for all $f,g \in X$. Assume

$$(8) \qquad |B(Pf-Pg,g)| \leq b|g|_3^{p-1}|f-g|_2$$

where $1 < p < 1 + 1/d$ and b depends boundedly on $|f|_2 + |g|_2$.

Theorem 4.3. *Let* $|h|_1 < \infty$ *and* $B(PU_0(t)h,U_0(t)h) \geq c_0 t^{-d(p-1)}$ *for* $t \geq t_0$ *where* $c_0 > 0$. *Let* $u(t)$ *be any perturbed solution. Then* $|u(t)-U_0(t)h|_2$ *does NOT go to zero as* $t \to +\infty$.

Proof. On the contrary, suppose $|u(t)-U_0(t)h|_2 \to 0$. We have

$$\frac{d}{dt} B(U_0(-t)u(t),h) = B(U_0(-t)Pu(t),h) = B(Pu(t),U_0(t)h).$$

Therefore $\int_0^T B(Pu(t),U_0(t)h)dt$ has a limit as $T \to +\infty$. On the other hand, we have by assumption

$$|B(Pu(t)-PU_0(t)h,U_0(t)h)| \leq b|U_0(t)h|_3^{p-1}|u(t)-U_0(t)h|_2.$$

Since $|u(t)|_2 \to |h|_2$, b is bounded. The last factor goes to zero as $t \to +\infty$. By Hypothesis II, the expression on the right is $o(t^{-d(p-1)})$. So by the assumption of this theorem,

$$B(Pu(t),U_0(t)h) \geq \tfrac{1}{2}c_0 t^{-d(p-1)}.$$

Since $d(p-1) < 1$, this is not integrable. This is a contradiction. Q.E.D.

The following simple criterion for the existence of S (in the sense of the norm) is well-known in the linear theory and forms the basis of our later work.

Lemma 4.4. *Let* $u(t)$ *be a perturbed solution such that* $\int_{-\infty}^{\infty}|Pu(t)|_2 dt < \infty$. *Then there exist unique* $f_\pm \in X$ *such that*

$$|u(t)-U_0(t)f_\pm|_2 \to 0 \quad \textit{as} \quad t \to \pm\infty.$$

Proof. Define f_+ by

$$f_+ = U_0(-t)u(t) + \int_t^{+\infty} U_0(-s)Pu(s)ds.$$

Differentiate with respect to t. By (3) we get zero. So f_+ is independent of t. Multiply f_+ by $U_0(t)$ and take norms to get

$$|U_0(t)f_+ - u(t)|_2 \le \int_t^{+\infty} |U_0(-s)Pu(s)|_2 dt = \int_t^{\infty} |Pu(s)|_2 ds.$$

By assumption this expression tends to zero as $t \to +\infty$. Similarly at $-\infty$. Q.E.D.

Application to NLKG. Write it as

$$\text{(NLKG)} \qquad \phi_{tt} - \Delta\phi + m^2\phi + F(\phi) = 0 . \qquad (x \in \mathbb{R}^n)$$

For definiteness, take $F(\phi) = \lambda|\phi|^{p-1}\phi$ with $m > 0$, $\lambda > 0$, $p > 1$. Allow ϕ to be complex. By §I we know the existence of weak solutions and, if p is small enough, their uniqueness, regularity and boundedness.

To fit (NLKG) into the Hilbert space context, we write it as a pair of equations:

$$u = \begin{bmatrix} u_1 \\ u_2 \end{bmatrix} = \begin{bmatrix} \phi \\ \phi_t \end{bmatrix}, \quad \frac{du}{dt} = iH_0 u + Pu$$

where

$$iH_0 = \begin{bmatrix} 0 & I \\ \Delta - m^2 & 0 \end{bmatrix}, \quad P = \begin{bmatrix} 0 & 0 \\ -F & 0 \end{bmatrix}.$$

The free equation corresponds to the case $P = 0$. The basic norm is the energy norm; thus we define

$$|u|_2 = \{\int (|\nabla u_1|^2 + m^2|u_1|^2 + |u_2|^2) dx\}^{1/2}.$$

The space X is the Hilbert space of all pairs of functions for which $|u|_2$ is finite. Then

$$|Pu|_2^2 = \int |F(\phi)|^2 dx = \lambda^2 \int |\phi|^{2p} dx \le \lambda^2 \int |\phi|^2 dx \cdot \sup_x |\phi|^{2p-2}.$$

Hence Lemma 4.4 is applicable if ϕ *is a solution such that*

$$\text{(9)} \qquad \int_{-\infty}^{\infty} \sup_x |\phi(x,t)|^{p-1} dt < \infty.$$

This illustrates how natural the L^∞ norm is in nonlinear problems.

Next we show that *Theorem 4.2* (weak scattering) *is applicable if* $p > 1 + 2/n$ *and* $n \ge 3$. For this purpose we define $|u|_3 = \sup_x |u_2(x)|$. Write $f = [f_1, f_2]$, $g = [g_1, g_2]$ as we wrote $u = [u_1, u_2]$. Then

$$|(Pf,g)| \le \lambda \int |f_1|^p |g_2| dx.$$

If $p \ge 2$, (7) is valid with $r = 1$ and

$$b(f) = \text{const} \int |f_1|^p dx \le \text{const} \int (|f_1|^2 + |f_1|^{p+1}) dx.$$

For any solution of (NLKG), $b(u(t))$ is bounded by a multiple of the energy. If $1+2/n < p < 2$ $(n \geq 3)$, we have

$$|(Pf,g)| \leq \lambda\left(\int |f_1|^2 dx\right)^{p/2}\left(\int |g_2|^2 dx\right)^{(2-p)/2} \sup_x |g_2(x)|^{p-1}$$

so that (7) is valid with $r = p-1$. We claim Hypothesis II is satisfied with $d = n/2$ and some norm $|\ |_1$. This is just a precise statement of a well-known fact about solutions of the free KG equation. We omit the definition of $|\ |_1$ which is a bit complicated and not important here.

Next we show how *Theorem 4.3* (nonexistence of strong scattering) *is applicable to NLKG if* $1 < p \leq 1+2/n$ $(1 < p \leq 2$ if $n = 1)$. We define

$$B(f,g) = \int (f_2\bar{g}_1 - f_1\bar{g}_2)dx. \tag{10}$$

By equation (2.19) this bilinear form is invariant for free solutions $(F = 0)$. We define $|f|_3 = \sup_x |f_1(x)|$, a choice different from that above. Now

$$B(Pf-Pg,g) = \lambda\int (|f_1|^{p-1}f_1 - |g_1|^{p-1}g_1)\bar{g}_1 dx.$$

By Schwarz's inequality, (8) holds with

$$b = \text{const}(|f|_2 + |g|_2)^{p-1}|g|_2^{2-p}.$$

As above, Hypothesis II is satisfied with $d = n/2$. Finally, let $\psi = U_0(t)h$ be a free solution. We know that

$$B(P\psi,\psi) = \lambda\int |\psi_1|^{p+1}dx = O(t^{-(p-1)n/2}).$$

For Theorem 4.3 we need $t^{(p-1)n/2}\int|\psi_1|^{p+1}dx$ bounded away from zero as $t \to +\infty$. This can be seen by Fourier transformation (see Strauss [1974], Example 3.4).

To summarize, we have shown the following. Let $\phi(x,t)$ be a solution of NLKG which has its energy bounded for all t and is continuous in t with values in the space of functions of finite energy. If $n \geq 3$ and $p > 1 + 2/n$, then $U_0(-t)u(t)$ converges weakly in X as $t \to \pm\infty$. If $1 < p < 1+2/n$ $(p \leq 2$ if $n = 1)$, then $U_0(-t)u(t)$ does not converge strongly in X to any non-zero element $h \in X$ with $|h|_1 < \infty$.

V. THE WAVE OPERATORS AND LOW-ENERGY SCATTERING

In this chapter, we show that the wave operators $W_\pm : f_\pm \to f$ exist, that the scattering operator $S: f_- \to f \to f_+$ exists for small f_- ("low energy"), and that the interaction can often be recovered from S (the inverse scattering problem). The first three theorems are essentially due to Segal.

We continue to assume Hypotheses I and II. We also assume throughout this chapter

Hypothesis III: There is a constant c such that

$$|f|_3 \leq c|f|_2 \quad \text{for all} \quad f \in X.$$

In each of the following theorems we assume a slightly different condition that the interaction term P is of sufficiently high degree.

The first theorem asserts "low energy waves are asymptotically free".

Hypothesis 5.1. There exist constants $b > 0$, $\delta > 0$ and $q \geq 1$, $dq > 1$ such that

$$|Pf|_1 + |Pf|_2 \leq b|f|_3^q \quad \text{if} \quad |f|_2 \leq \delta.$$

In case $q = 1$ we also assume $b = b(|f|_2) \to 0$ as $|f|_2 \to 0$.

Theorem 5.1. *Let* $u(t)$ *be a perturbed solution. If* $|u(0)|_1 + |u(0)|_2$ *is sufficiently small, then there exist* f_+ *and* f_- *in* X *such that* $|u(t)-U_0(t)f_\pm|_2 \to 0$ *as* $t \to \pm\infty$.

Proof. For convenience we denote various positive constants by c. Denote $f = u(0)$, and

$$m(t) = \sup_{0\leq s\leq t} [(1+s)^d|u(s)|_3 + |u(s)|_2].$$

We claim that $m(t) < \delta$ for all t. By assumption, this is true when $t = 0$, and hence for small t. In any interval in which $m(t) < \delta$, we have the following estimates. From (4.4) with $T = 0$,

$$\begin{aligned}
|u(t)|_2 &\leq |f|_2 + \int_0^t |Pu(s)|_2 \, ds \\
&\leq |f|_2 + \int_0^t b|u(s)|_3^q \, ds \\
&\leq |f|_2 + b\int_0^t (1+s)^{-dq} m(s)^q \, ds \\
&\leq |f|_2 + cb\, m(t)^q
\end{aligned}$$

We note that, for any g, $|U_0(t)g|_3 \leq c|U_0(t)g|_2 = c|g|_2$, so that

$$|U_0(t)g|_3 \leq \min\{c|g|_2, ct^{-d}|g|_1\} \leq c(1+t)^{-d}(|g|_1 + |g|_2).$$

Again from (4.4)

$$|u(t)|_3 \leq |U_0(t)f|_3 + \int_0^t |U_0(t-s)\,Pu(s)|_3 ds$$
$$\leq c(1+t)^{-d}(|f|_1 + |f|_2) + \int_0^t c(1+t-s)^{-d}(|Pu(s)|_1 + |Pu(s)|_2)ds.$$

The integral is less than

$$cb\ m(t)^q \int_0^t (1+t-s)^{-d}(1+s)^{-dq}ds,$$

which in turn is less than $c(1+t)^{-d}$ because $dq > 1$ and $dq \geq d$.

Putting these estimates together, we obtain the inequality

$$m(t) \leq c_1 + cb\ m(t)^q,$$

where $m(0) \leq c(|f|_1 + |f|_2) = c_1$. This is of the form $m \leq c_1 + \varepsilon(m)m$, where $\varepsilon(m) \to 0$ as $m \to 0$. Thus $m \leq 2c_1$ for all time if $2\varepsilon(2c_1) \leq 1$, say. This proves the claim.

As we showed above, $|Pu(t)|_2 \leq bm(t)^q(1+t)^{-dq}$. Since $m(t)$ is bounded and $dq > 1$, $|Pu(t)|_2$ is integrable as $t \to +\infty$, so that we may apply Lemma 4.4. Q.E.D.

Definition. If $v(t)$ is an X-valued continuous function of t, let

$$N(v) = \sup_{-\infty<t<\infty} \{|v(t)|_2 + (1+|t|)^d |v(t)|_3\}.$$

Let $\Sigma = \{f \in X \mid N(U_0(\cdot)f) < \infty\}$. Σ is a Banach space with the norm $|f|_\Sigma = N(U_0(\cdot)f)$. By Hypothesis II, $|\ |_\Sigma$ is a weaker norm than $|\ |_1 + |\ |_2$. The space Σ will be our space of scattering states.

Our next result is analogous to Theorem 5.1 but with initial conditions at $-\infty$. It says that inputs of small Σ-norm are indeed scattering states. First we need a slightly stronger hypothesis.

Hypothesis 5.2. There exists $q \geq 1$ with $dq > 1$ and there exists b depending boundedly on $|f|_2 + |g|_2$ such that

$$|Pf-Pg|_1 + |Pf-Pg|_2 \leq b(|f|_3 + |g|_3)^{q-1}|f-g|_3$$
$$+ b(|f|_3 + |g|_3)^q |f-g|_2$$

for all $f,g \in X$. In case $q = 1$ we assume $b \to 0$ as $|f|_2 + |g|_2 \to 0$.

Theorem 5.2. (Low-Energy Scattering). *If* $f_- \in \Sigma$, $|f_-|_\Sigma < \eta$ *with* η *sufficiently small, there exists a unique perturbed solution* $u(t)$ *such that*

$$|u(t)-U_0(t)f_-|_2 \to 0 \quad \textit{as} \quad t \to -\infty, \qquad N(u) \leq 2|f_-|_\Sigma.$$

Furthermore, there exists a unique $f_+ \in \Sigma$ *such that*

$$|u(t)-U_0(t)f_+|_2 \to 0 \quad \textit{as} \quad t \to +\infty, \qquad |f_+|_\Sigma \leq 2N(u).$$

The scattering operator $f_- \to f_+$ *is one-one and continuous from* $\{f_- \in \Sigma \mid |f_-|_\Sigma < \eta\}$ *into* Σ.

Proof. Let $f_- \in \Sigma$. Let $u_-(t) = U_0(t)f_-$. Equation (4.5) may be written as

$$u = u_- + \mathcal{P}u, \qquad \mathcal{P}u(t) = \int_{-\infty}^{t} U_0(t-s)Pu(s)ds.$$

We claim that $N(\mathcal{P}u-\mathcal{P}v) \leq \beta N(u-v)$, where $\beta = b'\cdot(N(u)+N(v))^{q-1}$ and b' has the same properties as b. This is proved in a manner similar to the estimates of the preceding theorem. As in that proof,

$$|\mathcal{P}u(t)|_2 \leq \int_{-\infty}^{t} b'|u(s)|_3^q \leq b'N(u)^q \int_{-\infty}^{t} (1+|s|)^{-dq}ds$$

$$\leq cb'N(u)^q,$$

$$|\mathcal{P}u(t)|_3 \leq b'N(u)^q \int_{-\infty}^{t} (1+|t-s|)^{-d}(1+|s|)^{-dq}ds$$

$$\leq b'N(u)^q c|t|^{-d}.$$

These estimates imply $N(\mathcal{P}u) \leq b'N(u)^q$. To prove the claim, we use the analogous estimates applied to the difference $\mathcal{P}u - \mathcal{P}v$.

By assumption, $\beta \to 0$ as $N(u) + N(v) \to 0$. Hence $\beta \leq 1/2$ if $N(u) + N(v)$ is sufficiently small. Define a sequence of approximate solutions as

$$u_0 = u_-, \qquad u_{n+1} = u_- + \mathcal{P}u_n.$$

Then $N(u_{n+1}) \leq N(u_-) + \frac{1}{2}N(u_n)$ so long as $N(u_n)$ is small enough. By induction $N(u_n) \leq 2N(u_-)$ so long as $N(u_-)$ is small enough. Since $u_{n+1} - u_n = \mathcal{P}u_n - \mathcal{P}u_{n-1}$, we also have

$$N(u_{n+1} - u_n) \leq 2^{-n} N(u_1 - u_0)$$

if $N(u_-)$ is small enough. Thus $\{u_n\}$ is a Cauchy sequence in the norm N. Its limit u satisfies $u = u_- + \mathscr{P}u$. This proves the first statement of the theorem. It follows that $|u(t) - u_-(t)|_2 \to 0$. By Lemma 4.4, f_+ exists. The same kind of estimate as above shows that $|f_+|_\Sigma \leq 2N(u)$.

Uniqueness: if u and v are two such solutions then $u = u_- + \mathscr{P}u$, $v = u_- + \mathscr{P}v$, so that $N(u-v) = N(\mathscr{P}u - \mathscr{P}v) \leq N(u-v)/2$, whence $u-v = 0$. Continuity: given another small incoming state g_-, let $v_-(t) = U_0(t)g_-$, $v(t)$ the corresponding perturbed solution and $v_+(t) = U_0(t)g_+$ the outgoing state. Then we may use equation (4.5) to write

$$u_+(t) - v_+(t) = u(t)-v(t) + \int_t^{+\infty} U_0(t-s)(Pu(s)-Pv(s))ds.$$

Just as in the above claim, $N(u_+-v_+) \leq N(u-v) + \beta N(u-v)$. But, as we showed above,

$$N(u-v) \leq N(u_--v_-) + N(\mathscr{P}u - \mathscr{P}v)$$
$$\leq N(u_--v_-) + \beta N(u-v) .$$

So we have

$$N(u_+-v_+) \leq (1+\beta)(1-\beta)^{-1}N(u_--v_-)$$

for η sufficiently small. So S is Lipschitz-continuous. Q.E.D.

The next result is the existence of the wave operators $W_\pm$. It could be termed the "Cauchy problem at $\pm\infty$". No smallness assumption is needed for the initial data. We state the result for W_-. There are two possible hypotheses.

Hypothesis A. There exist $q > 1 + 1/d$ and b depending boundedly on $|f|_2 + |g|_2$ such that

$$|Pf-Pg|_2 \leq b(|f|_3 + |g|_3)^{q-1}|f-g|_3 + b(|f|_3 + |g|_3)^q|f-g|_2.$$

Hypothesis B. There exists $q > 1$, $q > 1/d$ and b depending boundedly on $|f|_2 + |g|_2$ such that

$$|Pf-Pg|_1 + |Pf-Pg|_2 \leq b(|f|_3 + |g|_3)^{q-1}|f-g|_3$$
$$+ b(|f|_3 + |g|_3)^q|f-g|_2.$$

(Thus B is the same as Hypothesis 5.2 except $q \neq 1$; A does not require a bound in the $|\ |_1$ norm but q must be larger.)

Theorem 5.3. (Existence of W_-). *Assume either* A *or* B. *If* $f_- \in \Sigma$ *there exists a finite time* T *and a unique perturbed solution* $u(t)$ *in the time interval* $-\infty < t \leq T$ *such that* $|u(t)-U_0(t)f_-|_2 \to 0$ *as* $t \to -\infty$ *and*

$$N_T(u) = \sup_{-\infty<t\leq T} (|u(t)|_2 + (1+|t|)^d|u(t)|_3) < \infty.$$

Proof. N_T is a norm for functions defined for $-\infty < t \leq T$. Define $\mathscr{P}u(t)$ as in the proof of Theorem 5.2 except that $-\infty < t \leq T$. We claim that

$$(1) \qquad N_T(\mathscr{P}u-\mathscr{P}v) \leq \varepsilon(T)\beta N_T(u-v)$$

where $\varepsilon(T) \to 0$ as $T \to -\infty$ and β depends boundedly on $N_T(u) + N_T(v)$. Assuming (1) for the moment, it follows that, for each $k > 0$, there exists T such that $\mathscr{P}$ is a contraction mapping on $\{u \mid N_T(u) \leq k\}$. We define $u_0 = u_- = U_0(\cdot)f_-$, $u_{n+1} = u_- + \mathscr{P}u_n$. Then there is a T depending on $N_T(u_-)$ so that $\{u_n\}$ is a Cauchy sequence in the N_T norm. The limit is our desired solution. Its uniqueness also follows from (1).

To prove (1) we estimate as before:

$$c|\mathscr{P}u(t)|_3 \leq |\mathscr{P}u(t)|_2 \leq \int_{-\infty}^t |Pu(s)|_2 ds$$

$$\leq \beta \int_{-\infty}^t |u(s)|_3^q \, ds \leq \beta|t|^{1-dq}N_T(u)^q.$$

So under Hypothesis A we can say

$$N_T(\mathscr{P}u) \leq \beta'|T|^{1-dq+d} \to 0 \quad \text{as} \quad T \to -\infty.$$

Applying the same estimate to $\mathscr{P}u-\mathscr{P}v$, we get (1).

Under Hypothesis B we estimate $|\mathscr{P}u(t)|_3$ as in the preceding theorems:

$$|\mathscr{P}u(t)|_3 \leq \beta\int_{-\infty}^t (1+t-s)^{-d}(1+|s|)^{-dq}ds.$$

Since $dq > \max\{d,1\}$, the last integral is $o(|t|^{-d})$ and we get (1) once again. Q.E.D.

Next we consider the Inverse Scattering Problem. There are many formulations; we state it as the determination of the interaction operator P from the scattering operator S.

We assume Hypothesis 5.2. Therefore S exists locally. In addition we assume P is homogeneous:

$$P(\varepsilon f) = \varepsilon^p Pf, \quad \text{where } p > 1,\ p \geq q,\ \varepsilon > 0$$

for any $f \in X$ with $|f|_2$ and $|\varepsilon f|_2$ small. By Hypothesis 5.2 this implies $|Pf|_1 \leq c|f|_2^{p-q}|f|_3^q$.

We also postulate a bounded bilinear form $B(f,g)$ on X which is invariant under the free group. (Cf. Theorem 4.3.) Then we have

Theorem 5.4. *The scattering operator* S *defined in Theorem 5.2 determines the integral*

$$\int_{-\infty}^{\infty} \{B(P\phi,\psi) + B(\phi,P\psi)\}dt$$

for arbitrary free solutions ϕ *and* ψ *of finite* N*-norm.*

Proof. Let $u(t)$ and $v(t)$ be perturbed solutions. Denote $\bar{u}(t) = U_0(-t)u(t)$ and $\bar{v}(t) = U_0(-t)v(t)$. Assume that $f_\pm = \lim \bar{u}(t)$ and $g_\pm = \lim \bar{v}(t)$ exist in X as $t \to \pm\infty$. Then

$$\frac{d}{dt} B(\bar{u},\bar{v}) = B(\frac{d\bar{u}}{dt}, \bar{v}) + B(\bar{u}, \frac{d\bar{v}}{dt})$$

$$= B(Pu,v) + B(u,Pv)$$

from the definition of a perturbed solution. Integrating this, we get

$$B(f_+,g_+) - B(f_-,g_-) = \int_{-\infty}^{\infty} \{B(Pu,v) + B(u,Pv)\}dt.$$

The right side converges because of the left side. We now look at the leading term in the expansion of the right side for small f_- and g_-.

Let $h \in \Sigma$ and $\phi(t) = U_0(t)h$. We choose $f_- = \varepsilon h$. For small enough ε, Theorem 5.2 is applicable to f_-. Thus $N(u) \leq 2\varepsilon|h|_\Sigma$ and

$$N(u-\varepsilon\phi) = N(\mathscr{P}u) \leq \beta N(u)^p = O(\varepsilon^p)$$

by hypothesis and by the estimates in the proof of Theorem 5.2. Furthermore by Hypothesis 5.2,

$$|Pu-P(\varepsilon\phi)|_2 \leq b(|u|_3+|\varepsilon\phi|_3)^{q-1}|\mathscr{P}u|_3 + b(|u|_3+|\varepsilon\phi|_3)^q|\mathscr{P}u|_2$$

for each t. Hence as $\varepsilon \to 0$ we certainly have

$$\sup_t\ (1+|t|)^{dq}|Pu(t) - P(\varepsilon\phi(t))|_2 = o(\varepsilon^p).$$

We also let $k \in \Sigma$ and $\psi(t) = U_0(t)k$. The same estimates are valid for the perturbed solution $v(t)$ with incoming data $g_- = \varepsilon k$. Hence the expression

$$\sup_t\ (1+|t|)^{dq}[B(Pu,v) - B(P(\varepsilon\phi),\varepsilon\psi)]$$

is $o(\varepsilon^{p+1})$. Thus

$$\int_{-\infty}^{\infty} B(Pu,v)dt = \varepsilon^{p+1} \int_{-\infty}^{\infty} B(\mathcal{P}\phi,\psi)dt + o(\varepsilon^{p+1}).$$

Since $f_+ = S(f_-) = S(\varepsilon h)$ and $g_+ = S(g_-) = S(\varepsilon k)$, this means that

$$\text{(2)} \qquad \lim_{\varepsilon\to 0} \varepsilon^{-p-1} \{B(S(\varepsilon h),S(\varepsilon k)) - B(\varepsilon h,\varepsilon g)\}$$
$$= \int_{-\infty}^{\infty} \{B(P\phi,\psi) + B(\phi,P\psi)\}dt.$$

The left side is determined entirely by S. Q.E.D.

Application to NLKG. The set-up is the same as at the end of Chapter 4. In order to avoid complications, we assume space dimension $= n = 3$. We choose $|u|_3 = \sup_x |u_1(x)|$. Hypothesis III is false with the choice of $|u|_2$ in Chapter 4. Here we define (with $u =$ the pair $[u_1,u_2]$):

$$|u|_2 = \{\int[|(m^2-\Delta)u_1|^2 + |\nabla u_2|^2 + m^2|u_2|^2]dx\}^{1/2}.$$

Since $m > 0$, this is equivalent to the Sobolev norm H^2 in u_1 and H^1 in u_2. The Sobolev embedding theorem states that $H^2 \subset L^\infty$. This is just another way of saying that Hypothesis III holds. It is easy to check that the norm $|\ |_2$ is conserved for free solutions; this implies that $U_0(t)$ is unitary on the Hilbert space X with this norm (Hypothesis I).

As mentioned in Chapter 4, Hypothesis II is valid with $d = n/2 = 3/2$. There the exact definition of the norm $|\ |_1$ was not important. Now we need it to be as small as possible. The best choice is

$$|u|_1 = \sum_{|\alpha|\leq 3} |\partial^\alpha u_1(x)|dx + \sum_{|\beta|\leq 2} |\partial^\beta u_2(x)|dx$$

where ∂^α is any partial derivative of order ≤ 3 and ∂^β of order ≤ 2. Hypothesis II is a statement about the uniform decay of free solutions. It is proved by explicitly expressing $U_0(t)$ in terms of the Green's function (1.10), integrating by parts and estimating carefully. (See Morawetz and Strauss [1972] Appendix B, Reed [1976] page 98, or von Wahl [1971].)

Hypotheses 5.1 and 5.2 are valid if $p \geq 3$. We choose $q = p-2$. Indeed, Hypothesis 5.1 reduces to

$$\sum_{|\beta|\leq 2}\int|\partial^\beta(\phi^p)|dx + \{\int[|\nabla(\phi^p)|^2 + m^2|\phi^p|^2]dx\}^{1/2} \leq b \sup_x |\phi(x)|^{p-2}.$$

For simplicity, let us use a subscript x to denote each derivative. Then the main terms are

$$\int |\phi^{p-1}\phi_{xx}|dx, \quad \int |\phi^{p-2}\phi_x^2|dx, \quad \int |\phi^{p-1}\phi_x|^2 dx.$$

These are, respectively, less than

$$||\phi||_\infty^{p-2}||\phi||_2||\phi_{xx}||_2, \quad ||\phi||_\infty^{p-2}||\phi_x||_2^2, \quad ||\phi||_\infty^{p-1}||\phi_x||_2$$

where we denote $||\phi||_\infty = \sup_x |\phi(x)|$ and $||\phi||_2 = \{\int|\phi(x)|^2 dx\}^{1/2}$. Thus Hypothesis 5.1 is valid with $b = \text{const } |u|_2^2$. Hypothesis 5.2 is proved in exactly the same way except we must estimate differences; we leave it as an exercise.

Therefore *the scattering operator exists locally in the sense of Theorem 5.2 provided* $p \geq 3$. Note that the coupling constant λ may be positive or negative. The small number η in Theorem 5.2 depends on λ. In fact, the correct condition in this case is that $|\lambda||f_-|_\Sigma^{p-1}$ should be small. This follows from the homogeneity (cf. the first page in Chapter 2): if u is a solution with coupling constant λ then αu is a solution with coupling constant $\lambda\alpha^{1-p}$.

If S is considered as depending on the complex number λ as well as the input $f_- \in \Sigma$, it is not difficult to show that S is an analytic function on the domain $\{(\lambda,f_-) \mid |\lambda||f_-|_\Sigma^{p-2} < \eta\}$ for η sufficiently small, provided the nonlinear term $F(u)$ is analytic in u (for instance $F(u) = \lambda u^p$, p an integer ≥ 3). See Rączka and Strauss [1977].

We next apply Theorem 5.4 in case $F(u) = \lambda|u|^{p-1}u$, $p \geq 3$. We claim that S *determines the coupling constant* λ. We choose the standard bilinear form at the end of Chapter 4. We choose $\psi = 2\phi$. Thus S determines

$$\lambda(2-2^p) \iint |\phi|^p dx\, dt$$

for all free solutions ϕ. Just choose a single $\phi \neq 0$ and λ is determined. In a similar way a quite general interaction term $F(x,u)$ can be determined. See Morawetz and Strauss [1973], Strauss [1974] or Morawetz [1975].

Finally, *the wave operators exist if* $p > 2+2/n = 8/3$. To show this, we verify Hypothesis A with the choice $q = p-1$ (different from the previous choice). We need to show that

$$\{\int [|\nabla(\phi^p)|^2 + m^2|\phi^p|^2]dx\}^{1/2} \leq b \sup_x |\phi(x)|^{p-1}.$$

This is easy to verify with b depending on the energy of ϕ.

VI. ASYMPTOTIC COMPLETENESS

Our main goal is Theorem 4.1, which is due to Morawetz and Strauss [1972]. It is the case $n = 3$, $p = 3$. From Theorem 5.3 we know the wave operators exist: given a free solution $u_-(t)$, there exists a perturbed solution $u(t)$ defined for $-\infty < t \leq T$. By Theorem 1.2, $u(t)$ exists for all time as a smooth solution and by Theorem 1.3 it is uniformly bounded. We do not know, however, whether $u_+(t)$ exists unless we are willing to assume $N(u_-)$ is small. In this chapter we allow arbitrarily large u_-. By Lemma 4.4 it suffices to prove (4.9):

$$\int^{+\infty} \sup_x |u(x,t)|^2 dt < \infty.$$

What must be proved is the asymptotic decay of the solutions of the perturbed equation.

In this chapter we go back to the notation u for a solution of NLKG. (Note, however, that in Chapters IV and V, u denoted the pair $[\phi,\phi_t]$ where ϕ was a solution of NLKG.)

Perhaps surprisingly, *the* $m = 0$ *case* is easier. This is because of its conformal invariance. Take the equation

$$u_{tt} - \Delta u + |u|^{p-1}u = 0, \quad x \in \mathbb{R}^n.$$

The coupling constant is one since this can be achieved by scaling. We reach into Chapter 2 for identity (2.17). After some algebraic manipulations (see Strauss [1968]), it can be written as

$$(1) \quad \frac{d}{dt}\int \left[\frac{1}{2}(r^2+t^2)(|\nabla v|^2+v_t^2) + 2rtv_rv_t + \frac{(n-1)(n-3)}{8}\,\frac{r^2+t^2}{r^2}\,v^2\right] dx$$

$$+ \frac{d}{dt}\int (r^2+t^2)\,\frac{|u|^{p+1}}{p+1}\,dx + t((n-1)p-(n+3))\int \frac{|u|^{p+1}}{p+1}\,dx = 0.$$

where $v = r^{(n-1)/2}u(x,t)$. If we assume $n \geq 3$ and $p \geq 1+4/(n-1)$, then each of these integrals is positive. In particular

$$(2) \qquad \int |u|^{p+1}dx = O(t^{-2}).$$

If $n = 3$ and $p = 3$, this can be combined with the basic integral equation (1.9) (equation (4.4) in abstract form) to show the uniform decay. Indeed the Green's function $\mathscr{R}(x,t) = \delta(r-t)/4\pi t$ satisfies the estimate

$$(3) \qquad \left|\int \mathscr{R}(x-y,t)\psi(y)dy\right| \leq \frac{c}{t}\int |\nabla\psi(y)|dy,$$

which is the analogue of Hypothesis II of Chapter 4 for the case

$m = 0$. Thus from (1.9)

$$|u(x,t)| \leq |u_0(x,t)| + \int_0^t \frac{c}{t-s} \left(\int |\nabla u|^2 dy\right)^{1/2} \left(\int |u|^4 dy\right)^{1/2} ds$$

$$\leq \frac{c}{t} + c \int_0^t \frac{1}{t-s} \frac{1}{1+s} ds$$

using (2). This has a logarithmic divergence but there is a more careful argument which shows that $\sup_x |u(x,t)| = O(t^{-1} \log t)$ as $t \to \infty$. See Strauss [1968] and von Wahl [1972]. For $3 < p < 5$, one gets $O(t^{-1})$. For $p \geq 5$ we do not even know how to prove boundedness. For $n > 3$ the Green's function is more singular: the analogue of (1.10) involves higher derivatives. But we cannot estimate higher derivatives, as only the first derivatives appear in the energy. So there is no analogous result known for $n > 3$.

Next consider *the case* $m > 0$, $n = 3$, $p = 3$. We still have the integral equation but we do not have (1) because of the lack of conformal invariance. (There is a term $-2m^2u^2$ in (2.17) with the wrong sign.) Instead, we use two weaker estimates: the energy estimate on a light cone (1.7) and the radial derivative estimate (2.18). Thus we have

$$\int_K u^2 d^3S < \infty \quad \text{and} \quad \iint u^4 \frac{d^3x}{r} dt < \infty.$$

The last estimate comes from integrating (2.18) over all time and bounding the first term by the energy. We may assume $u(x,t)$ vanishes outside a light cone $|x| > t+k$ and thus

$$\int^\infty f(t)dt/t < \infty \quad \text{where} \quad f(t) = \int u^4 dx.$$

This is an extremely weak statement of decay. Because t^{-1} is not integrable, $f(t)$ could not be a constant and in fact $\int_I f(t)dt$ is arbitrarily small on arbitrarily long time intervals I. The proof of decay continues from this level like a jacking-up process. The succeeding steps are that: $u(x,t)$ is arbitrarily small on arbitrarily long time intervals; $u(x,t) \to 0$ uniformly as $t \to \infty$; $\sup_x |u(x,t)|^2$ is integrable; and finally $|u(x,t)| = O(t^{-3/2})$.

The most interesting step is the uniform convergence to zero. Let ε be a positive number. Let $T = T(\varepsilon)$ be sufficiently large. By the preceding step, $|u(x,t)| < \varepsilon$ on some time interval $[t^*-T, t^*]$. Let

$$t^{**} = \sup\{s \mid |u| < \varepsilon \quad \text{in} \quad [t^*-T, s]\}.$$

If $t^{**} = \infty$, there is nothing to prove. Suppose $t^{**} < \infty$. Take a time t slightly later than t^{**}; namely $t^{**} \leq t \leq t^{**} + \delta$. Break up the right side of (1.9) into four parts. Since $t \geq t^{**} \geq T$ is large enough, $|u_0| < \varepsilon/4$. The integral over $[t^{**},t]$, the tip of the cone, is less than $\varepsilon/4$ if δ is chosen small enough. In the interval $[t-T,t^{**}]$, we have $|u(x,t)| < \varepsilon$. Since u appears in (1.9) to the third power and ε is small, we can arrange the integral over $[t-T,t^{**}]$ to be less than $\varepsilon/4$, no matter how large T is. The fourth part is over the large base of the cone $[0,t-T]$, where we do not know that u is small. However $t-s > T$ in that interval and so $R(x-y,t-s)$ is small in some sense. The kernel is actually constant on the hyperboloids μ = constant, but they bunch together very closely and contribute little to the integral. Altogether, we obtain $|u(x,t)| < 4(\varepsilon/4) = \varepsilon$, which contradicts the definition of t^{**}. This proves the uniform decay. For the details of this long proof, see Morawetz and Strauss [1972].

It immediately follows from Lemma 4.4 and $\sup_x |u| = O(|t|^{-3/2})$ that there exist asymptotic states u_+ and u_-. This is the first statement of Theorem 4.1.

Next we define the space on which S acts. Let $\mathcal{F}$ be defined as the limits of the free solutions with smooth data of compact support in the norm

$$||v||_F^2 = \sup_t ||v(t)||_e^2 + \sup_t \sup_x |v(x,t)|^2 + \int_{-\infty}^{\infty} \sup_x |v(x,t)|^2 dt.$$

This space contains the free solutions with data in Σ.

We claim that S takes $\mathcal{F}$ into $\mathcal{F}$. The proof is based in part on a variation of Theorem 5.3 with Σ replaced by $\mathcal{F}$. Thus, if $u_-(x,t)$ is a free solution at $-\infty$, we construct $u(x,t)$ for $t \leq T$. Thinking of $t = T$ as the initial time, we use some estimates related to the preceding theorem to obtain the behavior as $t \to +\infty$. See loc. cit. or Reed [1976].

Properties of S.

(a) S maps $\mathcal{F}$ one-one onto $\mathcal{F}$.

(b) S is a diffeomorphism on $\mathcal{F}$.

(c) S is Lorentz-invariant.

(d) S commutes with the free group.

(e) S is odd.

(f) $||Sf||_e = ||f||_e$.

(g) The skew form (4.10) is invariant under S.

(h) S is not a linear operator.

Proof. (a) To show S is one-one, suppose $u_+ = v_+$. Subtracting equations (1.9) for u and for v,

$$u(t) - v(t) = \int_t^\infty R(t-s)*[u^3(s) - v^3(s)]ds,$$

$$\sup_{t>T} ||u(t)-v(t)||_e \le \sup_{s>T} ||u(s)-v(s)||_2 \int_T^\infty \sup_x (|u|+|v|)^2 ds.$$

If T is taken sufficiently large, the left side of this inequality must vanish. Thus $u(t) = v(t)$ for all $t \ge T$. Hence for all t. As usual this implies $u_- = v_-$. This means S is one-one. Since time is reversible, S is onto.

(b) S is once-differentiable means that

$$S'(u_-)v_- = \lim_{\varepsilon \to 0} \frac{1}{\varepsilon} [S(u_- + \varepsilon v_-) - S(u_-)]$$

exists in $\mathscr{F}$ and is a continuous linear function of v_-, where u_- and $v_- \in \mathscr{F}$. Let $u = W_-(u_-)$ be the usual perturbed solution. Let w be the solution of the linearized equation

$$w_{tt} - \Delta w + m^2 w + 3u^2 w = 0$$

with the asymptotic state v_- at $t = -\infty$. Then $S'(u_-)v_-$ is the asymptotic state of w at $t = +\infty$. A similar description holds for all the higher derivatives of S.

For (c) and (g), see Morawetz and Strauss [1973]. (d) is the intertwining relation (4.2). (e) is a consequence of the oddness of $u \to u^3$. (f) follows directly from the conservation of energy and the fact that $\int u^4 dx \to 0$ as $|t| \to \infty$. (h) is a consequence of (5.2); if S were linear the limit could not even exist.

Most of these results are valid for $F(u) = \lambda |u|^{p-1}u$, $\lambda > 0$, $8/3 < p < 5$. (For the case of $8/3 < p < 3$, see Pecher [1974].) More generally they are valid for any uniformly bounded solution and any nonlinear term F(u) satisfying conditions $uF(u) \ge (2+\delta)G(u)$ for some $\delta > 0$, $G(u) = O(u^4)$ as $u \to 0$, $|F'(u)| = O(G(u)^\sigma)$ for u bounded, $\sigma > 0$. As we mentioned in Chapter I, numerical computations indicate the same results without a growth condition; that is, for arbitrarily large p.

All of the above is for space dimension $n = 3$. Are we too stupid to get the other dimensions, or are we living in the best of all possible worlds? The difficulty with higher dimensions $n \geq 4$ is that the Green's function is too singular to make analogous estimates, even though the decay rate is faster. The difficulty with $n = 1$ or 2 is the lack of the estimate from (2.18) as well as the slowness of the decay rate. For $n = 1$, Glassey [unpublished] has proved the estimate

$$\int_{-\infty}^{\infty}\int [uF(u) - 2G(u)](1 + |x|)^{-2}\, dxdt < \infty$$

assuming $uF(u) \geq 2G(u)$, which is a beginning but does not seem to be strong enough to prove scattering.

It is natural to ask whether a more covariant scattering theory is possible. For the linear Klein-Gordon equation the Lorentz-invariant norm is

$$||f||_L = \int\{|(m^2-\Delta)^{1/4}f_1|^2 + |(m^2-\Delta)^{-1/4}f_2|^2\}dx.$$

This defines a Hilbert space $\mathcal{H}_L$. Segal has asked whether the scattering operator S might be defined on the whole of $\mathcal{H}_L$. In this connection Segal [1976] and Strichartz [1977] have recently proved the following interesting estimate valid for free solutions:

$$\int_{-\infty}^{\infty}\int |u(x,t)|^{2+4/n}dxdt \leq c||u(0)||_L^{2+4/n}.$$

VII. OTHER INVARIANT WAVE EQUATIONS

A. The classical coupled Maxwell-Dirac equations. They are

$$(-i\gamma^\mu\partial_\mu + M)\psi = gA_\mu\gamma^\mu\psi$$
$$-\Box A_\mu = g\psi^\dagger\gamma^0\gamma_\mu\psi, \qquad \Box = \Delta-\partial_t^2$$
$$\partial^\mu A_\mu = 0$$

where $x \in \mathbb{R}^n$, n = space dimension, ψ is the Dirac spinor field, the A_μ are the electromagnetic potentials, $\gamma^\mu\gamma^\nu + \gamma^\nu\gamma^\mu = 2g^{\mu\nu}$, $\gamma^{0*} = \gamma^0$, $\gamma^{k*} = -\gamma^k$ for $k \neq 0$, $\gamma_\mu = g^{\mu\nu}\gamma^\nu$. As in Chapter II there are several conservation laws.

1. Conservation of charge:

$$\partial_t(\psi^\dagger\psi) = \sum_1^n \partial_k(-\psi^\dagger\gamma^0\gamma^k\psi)$$

whence the L^2-norm of ψ is constant in time.

2. Conservation of energy:

$$\frac{1}{2}\int \sum_0^n ((\partial_t A_k)^2 + |\nabla A_k|^2)dx + i\int \psi^\dagger\partial_t\psi dx = \text{const.}$$

3. Conservation of momentum:

$$\int(\sum_0^n (\partial_t A_k)(\partial_j A_k) + \text{Im}(\partial_j\psi^\dagger)\psi)dx = \text{const.}$$

4. The mass-zero equations are conformally invariant (Gross [1964]). In particular, for $M = 0$ the dilational conservation law is

$$\int\{\sum_0^3 [\tfrac{t}{2}((\partial_t A_k)^2 + |\nabla A_k|^2) + r(\partial_t A_k)(\partial_r A_k) + A_k(\partial_t A_k)]$$
$$- \text{Im}[\psi^\dagger(t\partial_t\psi + r\partial_r\psi)]\}\, dx = \text{const.}$$

These and the other conformal conservation laws will appear in Glassey and Strauss [1978].

5. Integrating no. 1 over a light cone gives an estimate analogous to (1.7) which may be useful:

$$\int_K \psi^\dagger(I + \sum_{k=1}^3 \frac{x_k}{r}\gamma^0\gamma^k)\psi\, d^3S \leq \int_{t=0} |\psi|^2 dx.$$

where K is the surface of any light cone. The matrix within parentheses has the eigenvalues 0, 0, 2, 2 and hence we have bounds of half of the components of the spinor field on light cones.

6. From no. 1 and no. 2 it follows that

$$\int \sum_{k=0}^{4} ((\partial_t A_k)^2 + |\nabla A_k|^2)dx \le c + c\,(\int|\nabla\psi|^2 dx)^{1/2}.$$

Therefore the existence problem is solved if we can find a bound on $\int|\nabla\psi|^2dx$. Gross [1966] (also see Chadam [1972]) solved the local existence problem for $n = 3$: there exists a solution in a time-interval $0 \le t \le T$ provided the initial data (or the coupling constant g) are sufficiently small in a natural norm. This is the only known existence theorem. We do not know whether any substantial class of solutions exists for all time if $n = 3$.

7. For $n = 2$, Chadam and Glassey [1976] proved the global existence (of smooth solutions) for solutions with zero magnetic field (curl $\vec{A} = \vec{0}$). For $n = 1$, Chadam [1973a] and Glassey and Chadam [1974] proved the global existence with no restrictions on the initial data. They also proved (for $n = 1$) a result like Theorem 4.3, which may be a consequence of the low dimension. The existence proof for $n = 1$ goes as follows. Let $\alpha = -\gamma^0\gamma^1$ and $\beta = -i\gamma^0$. From the spinor equation we have

$$\partial_t(\psi^\dagger\alpha\psi) = \partial_x(\psi^\dagger\psi) + 2M\psi^\dagger\alpha\beta\psi.$$

Thus A is eliminated. Multiply by $\psi^\dagger\alpha\psi$ and use the conservation of charge to get

$$\frac{d}{dt}\int [(\psi^\dagger\alpha\psi)^2 + |\psi|^4]dx = 8M\int (\psi^\dagger\alpha\psi)(\psi^\dagger\alpha\beta\psi)dx.$$

The right side is bounded by $\int|\psi|^4dx$. Hence it can have at most exponential growth. The boundedness of $\int|\partial_x\psi|^2dx$ now follows easily from the equations.

B. Other covariant equations. The reason why the methods of Chapter V do not apply to the Maxwell-Dirac system is that the interaction terms in the equations are only quadratic. In the notation of Chapter V, $p = 2$ and $q = 0$. If the degree p of interaction were 4 or more, then most of the results of Chapter V would be applicable. Chadam [1973b] carried this out for Maxwell-Dirac for $n = 3$ with the interaction terms:

$$gA_\mu\gamma^\mu\psi \quad \text{replaced by} \quad 2gA_\mu(\psi^\dagger\gamma^0\gamma^\mu\psi)\psi,$$
$$g\psi^\dagger\gamma^0\gamma_\mu\psi \quad \text{replaced by} \quad g(\psi^\dagger\gamma^0\gamma_\mu\psi)^2.$$

The verification of the hypotheses of Theorems 5.1, 5.2 and 5.3 follows the outline of the NLKG equation and we leave it to the interested reader.

For a system of coupled Klein-Gordon and Dirac equations, see Chadam and Glassey [1974]. For coupled pair of Dirac equations, see Chadam [1973b] and Glassey [1977b]. Glassey [unpublished] has calculated all the conformal identities for the nonlinear Dirac equation with zero mass

$$\partial_t\psi = \sum \alpha_k\partial_k\psi + ig|\psi|^{p-1}\psi.$$

They are all exact conservation laws if $p = 1 + 4/(n-1)$.

For the NLKG with $n = 3$ and $p = 2$, which is the simplest analog of the Maxwell-Dirac equations, one can prove that if the coupling constant is small, then there is global existence and uniform boundedness of solutions.

The existence of solitary waves as in Theorem 3.1 for some other models of physical interest is discussed in Strauss and Vazquez [1978].

C. The nonlinear Schrödinger equation.

(1) $$i\frac{\partial u}{\partial t} - \Delta u + F(u) = 0$$

for $x \in \mathbb{R}^n$. We assume $F(u) = g'(|u|)u/|u|$ where $g(0) = g'(0) = 0$ and we let $G(u) = g(|u|)$. Besides being a non-relativistic model of a classical field, this equation is important in the theory of lasers and other classical phenomena. It has the Lagrangian

$$\iint[-\mathrm{Im}(u_t\bar{u}) + \tfrac{1}{2}|\nabla u|^2 + G(u)]dxdt,$$

which is invariant under the transformation $u(x,t) \to \lambda^{n/2}u(\lambda x,\lambda^2 t)$ if $G = 0$. This leads to the multiplier $r\bar{u}_r + 2t\bar{u}_t + (n/2)\bar{u}$ and the *dilational identity:*

(2) $$\frac{d}{dt}\int [\tfrac{1}{2}\mathrm{Im}(ru_r\bar{u}) + t|\nabla u|^2 + 2tG(u)]dx + \frac{1}{2}\int [nF(u)\bar{u} - 2(n+2)G(u)]dx = 0.$$

Ginibre and Velo [1977] discovered the following analog of the inversional identity (2.17), which they call the *pseudo-conformal conservation law.* For the free evolution operator,

$$xU_0(-t)f = U_0(-t)(x + 2it\nabla)f.$$

The identity is

$$(3) \qquad \frac{d}{dt}\int [\tfrac{1}{2}|xu - 2it\nabla u|^2 + 4t^2 G(u)]dx$$

$$+ 2t\int [n\bar{u}F(u) - 2(n+2)G(u)]dx = 0.$$

Let $F(u) = \lambda|u|^{p-1}u$. If $p = 1 + 4/n$, both (2) and (3) are exact conservation laws. If $\lambda > 0$, (3) implies that $\int G(u)dx = O(t^{-2})$. If $\lambda > 0$ and $1 + 4/n < p < 1 + 4/(n-2)$, Ginibre and Velo use this to prove asymptotic completeness in a certain norm. Lin and Strauss [1977] also prove asymptotic completeness in the H^1 norm in the case $n = 3$ and $8/3 < p < 5$. This is the exact analog of Theorem 4.1.

It is easy to apply all the theorems of Chapters IV and V to (1). To apply Theorems 4.3 and 5.4 we use the bilinear form

$$B(f,g) = \operatorname{Im}\int f\bar{g}\, dx.$$

See Strauss [1974] for some of the details.

There are existence theorems for (1) also similar to or stronger than those for NLKG discussed in Chapter I. There are uniqueness and regularity if $\lambda > 0$ and $p < 1 + 4/(n-2)$ (p arbitrary if $n = 1$ or 2) and also if $\lambda < 0$ and $p < 1 + 4/n$. On the other hand, if $\lambda < 0$ and $p \geq 1 + 4/n$, any initial datum with negative energy gives rise to a solution which blows up in a finite time: this follows from (2).

The interesting case $\lambda < 0$, $n = 1$, $p = 3$ is the analog of the sine-Gordon equation. There are solitons, an inverse scattering method, an infinite number of conservation laws, and so on.

For these results, see the references mentioned above, as well as Glassey [1977a], Scott et al. [1973], Baillon et al. [1977], and Strauss [1977b].

For some results on equations of Schrödinger-Hartree type, see Chadam and Glassey [1975] and Glassey [1977b].

Major Open Problems

1. Uniqueness of the Cauchy problem for NLKG for large powers p ($p \geq 5$ for space dimension 3).
2. Existence of solutions for all time for the Maxwell-Dirac equations in 3 space dimensions, even for small initial data. Does it work for a special class of data, like those of positive energy?
3. Existence of the wave operators for NLKG for $p = 2$ and space dimension = 3. This is a model for Maxwell-Dirac.
4. Existence of the scattering operator for large inputs for NLKG, space dim = 1 and large p.
5. Uniform decay of the continuous part of solutions of the sine-Gordon equation.
6. Stability and interaction behavior of the solitary waves of Chapter IIIA.

REFERENCES

M. J. Ablowitz, D. J. Kaup, A. C. Newell and H. Segur.
[1974] Studies in Appl. Math. 53, 249-315.

D. L. T. Anderson. [1971] J. Math. Phys. 12, 945-952.

J.-B. Baillon, T. Cazenave and M. Figueira.
[1977] C. R. Acad. Sci. 284, 869-872.

T. Bałaban and R. Rączka. [1975] J. Math. Phys. 16, 1475-1481.

T. Bałaban, K. Jezuita and R. Rączka.
[1976] Comm. Math. Phys. 48, 291-311.

J. M. Chadam. [1972] J. Math. Phys. 13, 597-604.
[1973a] J. Funct. Anal. 13, 173-184.
[1973b] J. Applic. Anal. 3, 377-402.

J. M. Chadam and R. T. Glassey.
[1974] Arch. Rat. Mech. Anal. 54, 223-237.
[1975] J. Math. Phys. 16, 1122-1130.
[1976] J. Math. Anal. Appl. 53, 495-507.

G. H. Derrick. [1964] J. Math. Phys. 5, 1252.

A. Friedman. [1969] Partial Differential Equations (Holt, Rinehart and Winston, New York).

J. Ginibre and G. Velo. [1977] preprints.

R. T. Glassey. [1973a] Trans. A.M.S. 182, 187-200.
[1973b] Math. Zeit. 132, 182-203.
[1977a] J. Math. Phys.
[1977b] preprints.

R. T. Glassey and J. M. Chadam. [1974] Proc. A.M.S. 43, 373-378.

R. T. Glassey and W. A. Strauss. [1978] to appear.

L. Gross. [1964] J. Math. Phys. 5, 687-695.
[1966] Comm. Pure Appl. Math. 19, 1-15.

K. Jörgens. [1961] Math. Zeit. 77, 295-308.

J. B. Keller. [1957] Comm. Pure Appl. Math. 10, 523-530.

H. A. Levine. [1974] Trans. A.M.S. 192, 1-21.

J.-E. Lin and W. A. Strauss. [1977] J. Funct. Anal.

J. L. Lions. [1964] Rev. Roumaine Math. Pure Appl. 9, 11-18.

V. A. Marchenko. [1955] Doklady Akad. Nauk 104, 695.

A. Matsumura. [1976] Publ. Res. Inst. Math. Sci. Kyoto 12, 169-189.

C. S. Morawetz. [1968] Proc. Roy. Soc. A306, 291-296.
[1975] Notes on Time Decay and Scattering for some Hyperbolic Problems (Soc. Ind. Appl. Math., Philadelphia).

C. S. Morawetz and W. A. Strauss.
[1972] Comm. Pure Appl. Math. 25, 1-31.
[1973] Comm. Pure Appl. Math. 26, 47-54.

E. Noether. [1918] Nach. Ges. Göttingen Math. Phys. Kl., 235-257.

H. Pecher. [1974] Doctoral dissertation, Göttingen.

J. K. Perring and T. H. R. Skyrme. [1962] Nucl. Phys. 31, 550-555.

R. Rajaraman. [1975] Phys. Reports (Phys. Lett. C) 21, 227-313.

R. Rączka and W. A. Strauss. [1977] preprint.

M. Reed. [1976] Abstract Non-Linear Wave Equations, Lect. Notes in Math. No. 507 (Springer-Verlag, Berlin).

M. Reed and B. Simon. [1977] Methods of Modern Mathematical Physics, Vol. III (Academic Press, New York).

A. C. Scott, F. Y. F. Chu and D. W. McLaughlin.
[1973] Proc. IEEE 61, 1443-1483.

J. Scott-Russell. [1844] Proc. Roy. Soc. Edinburgh, 319-320.

I. E. Segal. [1963a] Ann. Math. 78, 339-364.
[1963b] Bull. Soc. Math. France 91, 129-135.
[1966] Proc. Conf. on Math. Th. Elem. Particles, M.I.T. Press, 79-108.
[1968] Ann. Sci. Ecole Norm. Sup. (4)1, 459-497.
[1976] Adv. Math. 22, 305-311.

W. A. Strauss. [1968] J. Funct. Anal. 2, 409-457.
[1970] Anais. Acad. Brasil. Ciências 42, 645-651.
[1974] Scattering Theory in Math. Physics (D. Reidel, edited by LaVita and Marchand), 53-78.
[1977a] Comm. Math. Phys.
[1977b] Part. Diff. Eqns. and Cont. Mech. (North-Holland Publ.).

W. A. Strauss and L. Vazquez. [1977] J. Comp. Phys.
[1978] to appear.

R. S. Strichartz. [1977] preprint.

L. Tartar. [1976] Univ. of Wisconsin, Math. Res. Ctr. Report No. 1584.

W. von Wahl. [1971] Math. Zeit. 120, 93-106.
[1972] J. Funct. Anal. 9, 490-495.

N. J. Zabusky and M. D. Kruskal. [1965] Phys. Rev. Lett. 15, 240-243.

STRUCTURE PROPERTIES OF SOLUTIONS OF CLASSICAL NON-LINEAR RELATIVISTIC FIELD EQUATIONS

C. Parenti
University of Ferrara, Ferrara, Italy

F. Strocchi
Scuola Normale Superiore, Pisa, Italy

G. Velo
Istituto di Fisica dell'Università and INFN, Bologna, Italy

These lectures contain a description of the motivations, of the basic ideas, and of physically relevant results concerning properties of the solutions of the evolution problem for some non-linear relativistic system of partial differential equations. For convenience, we have collected sketches of proofs of some results in Appendices A and B.

1. GENERAL FRAMEWORK

The need for non-perturbative methods in quantum field theory (QFT), the functional integral approach, the possibility of developing approximation methods based on the knowledge of classical solutions strongly motivate the study of classical non-linear equations as a way to get insight in the corresponding QFT problem. Moreover, by Hepp's result[1)], one learns that the classical solutions can be recovered as the $\hbar \to 0$ limit of expectation values of the quantum fields on suitable coherent states; as a consequence, one expects that the structure properties of the solutions of the QFT (stability, symmetry breaking, etc.) are shared by the solutions of the corresponding limiting classical theory. This motivates the analysis of global properties and the classification of solutions of the non-linear equation

$$\Box\phi + U'(\phi) = 0 \tag{1}$$

which will be the subject of these lectures. As we will see, most of the characteristic features of the quantum field theory are already present at the level of non-linear classical solutions.

The first preliminary question is to specify the class of solutions which are of physical interest. Intrinsic to the perturbative approach is the splitting of the energy into a free or kinetic part

$$E_{kin} = \frac{1}{2}\int_{\mathbb{R}^s} \left(|\nabla\phi|^2 + \phi^2 + \phi^2 \right) d^s x \tag{2}$$

(s = space dimensions) and a potential part E_{pot}, and to look for solutions for which both E_{kin} and E_{pot} make sense, i.e. they are finite. This is the attitude taken in the pioneering work on existence and uniqueness theorems for Eq. (1) [Jörgens[2], Segal[3]] and by all subsequent followers (see Ref. 4 for a comprehensive review). However, this framework does not include the very interesting cases of external field problem, the symmetry-breaking solutions, the soliton-like solutions, and, in general, the non-dissipative solutions, all of which do not decrease sufficiently fast at infinity to make E_{kin} finite. To cure this we replace the requirement that $E_{kin} < \infty$ by the condition

$$\int_{\Omega} \left(|\nabla\phi|^2 + \phi^2 + \dot{\phi}^2 \right) d^s x \; < \infty \tag{3}$$

for any bounded region Ω *(locally finite kinetic energy)*. The main physical motivation for condition (3) is the local character of any realizable measurement, as emphasized by Haag and Kastler[5]. Moreover, one cannot expect that all the interesting solutions of non-linear classical equations have globally finite kinetic energy, just as the non-trivial solutions of QFT cannot be expected to be globally Fock.

In the first order formalism the Cauchy problem for Eq. (1) can be rewritten more conveniently as the integral equation

$$\begin{pmatrix} \phi(t) \\ \psi(t) \end{pmatrix} = W(t) \begin{pmatrix} \phi_0 \\ \psi_0 \end{pmatrix} + \int_0^t W(t-s) \begin{pmatrix} 0 \\ -U'(\phi(s)) \end{pmatrix} ds \tag{4}$$

where W(t) is the one parameter group generated by $\begin{pmatrix} 0 & 1 \\ \Delta & 0 \end{pmatrix}$ and ϕ_0, ψ_0 are the initial data. Condition (3) implies that we are interested in initial data $\phi_0 \in H^1_{loc}(\mathbb{R}^s)$, $\psi_0 \in L^2_{loc}(\mathbb{R}^s)$, so that it is natural to look for solutions

$$u(t) \equiv \begin{pmatrix} \phi(t) \\ \psi(t) \end{pmatrix} \in X \equiv H^1_{loc}(\mathbb{R}^s) \oplus L^2_{loc}(\mathbb{R}^s)$$

continuous in time in the X topology.

Having clarified the class of solutions which is physically interesting to investigate, we have to specify the class of potentials which will be considered for the following analysis. To simplify the discussion and to be more concrete from now on we will assume that the potential U(z) satisfies:

A) ("Lower bound"):

$$U(z) \geqslant \alpha + \beta |z|^2$$

for suitable α, β (not necessarily positive);

B) For s = 1 U in an entire function;

s = 2 $U = \sum_{k=0}^{\infty} c_k z^k$ with $\sum_{k=0}^{\infty} |c_k|\, k^{k/2}\, |z|^k < \infty$

s = 3 U is a twice continuously differentiable real function such that

$$\sup_z \; (1 + |z|^2)^{-1}\, |U''(z)| < \infty$$

The following analysis remains valid for a much larger class of potentials, for which we refer to Refs. 6 and 7. There one can also find the easy extension to the case of a multicomponent field $\phi = \begin{pmatrix} \phi_1 \\ \vdots \\ \phi_n \end{pmatrix}$. Within this framework the Cauchy problem is well posed and one has

Theorem 1 [6)*)] *For any initial data* $\begin{pmatrix} \phi_0 \\ \psi_0 \end{pmatrix} \in X$ *the integral Eq. (4) has a unique solution* $\begin{pmatrix} \phi(t) \\ \psi(t) \end{pmatrix} \in C^{(0)}(\mathbb{R};X)$.

In what follows it will always be understood that when we talk of solutions of Eq. (4) we refer to solutions belonging to $C^{(0)}(\mathbb{R};X)$.

2. CLASSIFICATION AND STRUCTURE PROPERTIES OF THE SOLUTIONS

In order to classify the solutions of the non-linear problem we are interested in, it is convenient to introduce a notion of "small perturbation" of a given solution. The guiding physical idea is that initial data for which the energy difference of the corresponding solutions is not finite cannot be realized in the same physical world. It is therefore natural to partitionate the set of solutions of Eq. (4) into classes according to the following equivalence relation: $\begin{pmatrix} \phi_1(t) \\ \psi_1(t) \end{pmatrix}$ is a *small perturbation* of $\begin{pmatrix} \phi_0(t) \\ \psi_0(t) \end{pmatrix}$ if

$$\begin{pmatrix} \phi_1(t) - \phi_0(t) \\ \psi_1(t) - \psi_0(t) \end{pmatrix} \in C^{(0)}(\mathbb{R};Y) \tag{5}$$

where $Y \equiv H^1(\mathbb{R}^s) \oplus L^2(\mathbb{R}^s)$. Particularly interesting, from a physical point of view, are those classes of solutions that are invariant under time translations: $\begin{pmatrix} \phi_0(t) \\ \psi_0(t) \end{pmatrix}$ is said to belong to such a class if

$$\begin{pmatrix} \phi_0(t) - \phi_0(0) \\ \psi_0(t) - \psi_0(0) \end{pmatrix} \in C^{(0)}(\mathbb{R};Y) \tag{6}$$

Property (6) is a kind of stability of the class under time evolution.

*) It has been pointed out by L. Gärding and W. Strauss that the theorem can be generalized and its proof simplified.

In this way one gets a rigorous treatment of the Goldstone's picture[8] according to which the symmetry-breaking occurs because a physical theory is described by small perturbations around one of the minima of the potential (this property being preserved by time evolution).

The above partition into classes of the solutions of Eq. (4) can be naturally transferred into a partition of the space X considered as the space of the initial data.

Definition 1 *Two initial data* $\begin{pmatrix}\phi_0\\ \psi_0\end{pmatrix}$, $\begin{pmatrix}\phi_1\\ \psi_1\end{pmatrix}$ *of Eq. (4) are said to belong to the same* *Hilbert space sector* *(HSS) if the corresponding solutions satisfy (5) and (6).*

It is clear that a HSS is uniquely determined by any of its elements $\begin{pmatrix}\phi_0\\ \psi_0\end{pmatrix}$ and will be denoted by $\mathcal{H}_{\begin{pmatrix}\phi_0\\ \psi_0\end{pmatrix}}$. The natural question of existence of non-trivial HSS is answered by the following theorem.

Theorem 2 [7] *Let* $\begin{pmatrix}\phi_0\\ \psi_0\end{pmatrix} \in X$ *be such that* $\phi_0 \in L^\infty(\mathbb{R}^s)$; *then* $\begin{pmatrix}\phi_0\\ \psi_0\end{pmatrix}$ *defines a HSS* $\mathcal{H}_{\begin{pmatrix}\phi_0\\ \psi_0\end{pmatrix}}$ *iff:*

a) $\psi_0 \in L^2(\mathbb{R}^s)$,

b) $\Delta\phi_0 - U'(\phi_0) \equiv h \in H^{-1}(\mathbb{R}^s)$ *).

Moreover, if conditions (a) and (b) are satisfied, it follows that

$$\mathcal{H}_{\begin{pmatrix}\phi_0\\ \psi_0\end{pmatrix}} = \begin{pmatrix}\phi_0\\ \psi_0\end{pmatrix} + Y$$

A brief proof of Theorem 2 can be found in Appendix A.

It is clear that if $\begin{pmatrix}\phi_0\\ \psi_0\end{pmatrix}$ satisfies (a) and (b), so does $\begin{pmatrix}\phi_0\\ 0\end{pmatrix}$; they belong to the same HSS which can be denoted simply by $\mathcal{H}_{\phi_0}$. In Theorem 2 the condition $\phi_0 \in L^\infty(\mathbb{R}^s)$ can be replaced by a more general one. However, to discuss the concrete examples arising from physics, the simpler formulation of the theorem chosen here will be sufficient. Interesting examples of functions ϕ_0 which define HSS are the constants $\phi_0 = c$ such that $U'(c) = 0$. More generally a static solution $\begin{pmatrix}\phi_0\\ 0\end{pmatrix} \in X$ of Eq. (1) satisfies the equation

$$\Delta\phi_0 - U'(\phi_0) = 0 \tag{7}$$

and therefore defines an HSS if $\phi_0 \in L^\infty(\mathbb{R}^s)$. The soliton solutions of the Sine-Gordon equation and the solitary waves of the ϕ^4 theory are bounded solutions of Eq. (7), belonging to $H^1_{loc}(\mathbb{R}^s)$. It is important to stress that condition (b) of Theorem 2 is more general than Eq. (7), and therefore it includes non-dissipative time-dependent solutions as those found by T.D. Lee in three-dimensional scalar field theory[9].

*) That is, $\int |\hat{h}(k)|^2(1 + |k|^2)^{-1}\, d^sk < \infty$, $\hat{h}$ being the Fourier transform of h.

Every HSS $\mathcal{H}_{\phi_0}$ defined by a ϕ_0 satisfying the conditions of Theorem 2 carries a natural Hilbert space structure induced by the identification i:

$$\mathcal{H}_{\phi_0} \to Y, \qquad i\begin{pmatrix}\phi' \\ \psi'\end{pmatrix} = \begin{pmatrix}\phi' - \phi_0 \\ \psi'\end{pmatrix}$$

This justifies the name of HSS.

Similarly to the QFT case, different HSS require inequivalent definitions of observables. This can be easily seen for the energy functional which we will now discuss. The conventional expression of the energy density for the theory described by Eq. (1) is

$$\frac{1}{2}\left(|\nabla\phi|^2 + \dot{\phi}^2\right) + U(\phi) \tag{8}$$

However, if we add to (8) any function of the space variables, the equation of motion (1) is obviously unchanged and the new expression for the total energy remains (formally) conserved. This ambiguity is related to the fact that only energy differences have a physical meaning and this will be used to "renormalize" the energy in each HSS.

Let us consider a HSS $\mathcal{H}_{\phi_0} = \begin{pmatrix}\phi_0 \\ 0\end{pmatrix} + Y$ with $\phi_0 \in L^\infty(\mathbb{R}^s) \cap H^1_{loc}(\mathbb{R}^s)$, $\Delta\phi_0 - U'(\phi_0) \in H^{-1}(\mathbb{R}^s)$. Then the following result holds.

Theorem 3 [7] *Consider the energy functional defined as follows*

$$E_{\phi_0}(\phi,\psi) = \int_{\mathbb{R}^s} \left[\frac{1}{2}\left(|\nabla\phi|^2 - |\nabla\phi_0|^2\right) + \frac{1}{2}\psi^2 + U(\phi) - U(\phi_0)\right] d^s x \tag{9}$$

Then

i) If $\begin{pmatrix}\phi \\ \psi\end{pmatrix} \in \mathcal{H}_{\phi_0}$ *and* $\phi - \phi_0 \in H^1_{comp}(\mathbb{R}^s)$, $E_{\phi_0}(\phi,\psi)$ *is well defined;*

ii) The functional E_{ϕ_0} *has a unique extension to the whole* $\mathcal{H}_{\phi_0}$ *(still denoted by* E_{ϕ_0}*) which is continuous in the Hilbert space topology of* $\mathcal{H}_{\phi_0}$*. Moreover, the functional is constant in time, i.e.*

$$E_{\phi_0}(\phi(t),\psi(t)) = E_{\phi_0}(\phi(0),\psi(0)), \quad \forall t \in \mathbb{R}, \tag{10}$$

if $\begin{pmatrix}\phi(t) \\ \psi(t)\end{pmatrix}$ *is the solution corresponding to the initial data* $\begin{pmatrix}\phi(0) \\ \psi(0)\end{pmatrix} \in \mathcal{H}_{\phi_0}$.

iii) If $|\nabla\phi_0| \in L^2(\mathbb{R}^s)$*, then for any* $x_0 \in \mathbb{R}^s$ *and for any* $\omega \in C_0^\infty(\mathbb{R}^s)$ *with* $\omega \equiv 1$ *in a neighbourhood of the origin, we have*

$$E_{\phi_0}(\phi,\psi) = \lim_{\rho\uparrow+\infty} \int_{\mathbb{R}^s} \left[\frac{1}{2}\left(|\nabla\phi|^2 - |\nabla\phi_0|^2\right) + \frac{1}{2}\psi^2 + U(\phi) - U(\phi_0)\right] \omega\left(\frac{x - x_0}{\rho}\right) d^s x \tag{11}$$

Moreover, for any $\binom{\alpha}{\beta} \in \mathcal{H}_{\phi_0}$ *it follows that*

$$E_{\phi_0}(\phi,\psi) = E_\alpha(\phi,\psi) + E_{\phi_0}(\alpha,0), \quad \forall \binom{\phi}{\psi} \in \mathcal{H}_{\phi_0} . \tag{12}$$

Prôperty (11) allows one to recover the energy functional as the limit of the difference of the energy densities integrated over a finite volume as this volume invades the whole space. In a similar and even easier way one may discuss the definition of other observables, like, for example, the linear momentum. A sketch of the proof of Theorem 3 is presented in Appendix B.

Hilbert space sectors can be regarded as a sort of basic constitutents, stable under time evolution, which describe one solution and its small perturbations. For example, in the case s = 1 and

$$U(\vec{\phi}) = \lambda(\vec{\phi}\cdot\vec{\phi})^2 - m_1^2\phi_1^2 - m^2\sum_{j\neq 1}\phi_j^2 \tag{13}$$

where $m_1 < m$, $\lambda > 0$, and $\vec{\phi} = (\phi_1, \ldots, \phi_n)$ is an n-component field, the constant solutions

$$\begin{aligned} &\vec{\phi} = (\phi_1 = \pm m_1/\sqrt{2\lambda},\ \phi_2 = 0, \ldots \phi_n = 0) \equiv \Phi_{1\pm} \\ &\vec{\phi} = (\phi_1 = 0,\ \phi_j = \text{const},\ \text{with } \vec{\phi}\cdot\vec{\phi} = m^2/2\lambda) \equiv \Phi_\alpha \end{aligned} \tag{14}$$

(α labels the rotations in SO_{n-1})

and the soliton like solutions

$$\Phi_{s_i} = (0,\ldots 0, \phi_{s_i}, 0,\ldots,0), \quad \phi_{s_i} = \begin{cases} \frac{m_1}{\sqrt{2\lambda}}\ \mathrm{tgh}\ m_1 x, & i = 1 \\ \frac{m}{\sqrt{2\lambda}}\ \mathrm{tgh}\ m x, & i > 1 \end{cases} \tag{15}$$

define *different* Hilbert sectors. Because of the SO_{n-1} symmetry, two different HSS may have finite energy with respect to each other, in the sense that the difference in energy of any two elements picked up from these sectors is finite (this happens, for example, for all $\mathcal{H}_{\Phi_{s_i}}$, $\mathcal{H}_{\Phi_{s_j}}$ with $i \neq j$, i, $j \neq 1$). It is then clear from the above examples that the partition into HSS is, in general, too fine for physical purposes. In fact, even if the time evolution makes each HSS a closed world, small "external" field perturbations may cause transition between sectors which have relatively finite energy. It is therefore convenient, in order to get an interesting and reasonable physical theory, to group together those HSS satisfying the following

Definition 2 *Two HSS $\mathcal{H}_{\phi_1}$, $\mathcal{H}_{\phi_2}$, defined by functions satisfying the conditions of Theorem 2 and for which $|\nabla\phi_1|$, $|\nabla\phi_2| \in L^2(\mathbb{R}^s)$, are said to belong to the same energy sector*) $\mathcal{E}$ if*

$$U(\phi_1) - U(\phi_2) \in L^1(\mathbb{R}^s). \tag{16}$$

In each energy sector $\mathcal{E}$ one may define an energy density $\frac{1}{2}(|\nabla\phi|^2 + \psi^2) + U(\phi) - U(\phi_0)$, where ϕ_0 is a given reference element, and the integral of such a density [in the sense of Eq. (11)] provides a definition of an energy functional which is finite in $\mathcal{E}$.

In analogy to the quantum version of the theory it is convenient to consider those subcollections of HSS, belonging to the same energy sectors, which contain one and only one HSS corresponding to a constant. They correspond to pure phases and are the classical analogues of quantum field theories with a unique vacuum (translationally invariant state). For the physical motivations and the physical implications connected with the uniqueness of the vacuum, see Streater and Wightman[11] and Glimm et al.[12].

Definition 3 *A (maximal) subcollection P of HSS belonging to the same energy sector, such that it contains one and only one HSS corresponding to a constant $\mathcal{H}_{\phi=const}$ (uniqueness of the vacuum sector) is called a pure phase.*

The concept of pure phase is particularly suitable for discussing spontaneous symmetry breaking as it will be clear in the following section.

3. LOCAL INTERNAL SYMMETRIES AND THEIR SPONTANEOUS BREAKING. DYNAMICAL CHARGES

The framework described so far permits a precise discussion of the mechanism by which a symmetry of the equations of motion fails to define a symmetry of the physical theory (spontaneous breaking of the symmetry). We will be mainly concerned with internal symmetries, i.e. mappings of solutions into solutions which commute with space translations and rotations.

Definition 4 *Let $g:\mathbb{R}^n \to \mathbb{R}^n$ be a diffeomorphism of class $C^{(2)}$ such that the map*

$$T_g : \begin{pmatrix}\phi(x)\\ \psi(x)\end{pmatrix} \longmapsto \begin{pmatrix} g(\phi(x)) \\ J_g(\phi(x))\,\psi(x)\end{pmatrix}$$

where J_g is the Jacobian matrix of g, is continuous from X to X. T_g is said to be a local internal symmetry if for every solution $\begin{pmatrix}\phi(t)\\ \psi(t)\end{pmatrix} \in C^{(0)}(\mathbb{R};X)$ of Eq. (4), $T_g\begin{pmatrix}\phi(t)\\ \psi(t)\end{pmatrix}$ is again a solution of Eq. (4) [with initial data $\begin{pmatrix}\phi(0)\\ \psi(0)\end{pmatrix}$ and $T_g\begin{pmatrix}\phi(0)\\ \psi(0)\end{pmatrix}$, respectively].

*) Here we call energy sectors the strong energy sectors of Ref. 10.

In what follows we will suppose that the potential U, not identically zero, is normalized in such a way that U(0) = 0 and is so regular that the solutions of Eq. (4) are smooth [say $C^{(2)}$ both in space and time] when the initial data are sufficiently smooth [say $C^{(\infty)}$]. We have the following

Theorem 3 [7] *Let T_g be a local internal symmetry. Then g is an affine transformation*

$$g(\phi) = A\phi + a \tag{17}$$

with $a \in R^n$, $A \in GL(n;R)$ such that $A^T A = \lambda \mathbb{1}$, and

$$U(A\phi + a) = \lambda U(\phi) + U(a) \tag{18}$$

A basic property of local internal symmetries is that they map HSS into HSS, i.e.

$$T_g \quad \mathcal{H}_\phi \longrightarrow \mathcal{H}_{A\phi+a} \tag{19}$$

Since the potential has been chosen independent of the space and time variables, space translations and space rotations map solutions of Eq. (4) into solutions of the same equation, and considerations similar to the above ones can be applied.

It is now easy to understand why a physical theory (i.e. a pure phase P) may fail to be invariant under a local internal symmetry of the equations of motion.

Definition 5 *A local internal symmetry T_g is said to be spontaneously broken in a pure phase P if there is at least one element $\begin{pmatrix}\phi\\ \psi\end{pmatrix} \in P$ such that*

$$T_g \begin{pmatrix}\phi\\ \psi\end{pmatrix} \notin P \tag{20}$$

Otherwise T_g is said to define a global (internal) symmetry of the pure phase P, and the theory described by P is said to be invariant under T_g

A similar definition can be given for an energy sector $\mathcal{E}$ or for a HSS $\mathcal{H}_\phi$. In analogy with Coleman's theorem that the invariance of the vacuum is the invariance of the world, a sufficient condition for Tg to be unbroken in $\mathcal{H}_\phi$ is that there is one element $\begin{pmatrix}\phi_1\\ \psi_1\end{pmatrix} \in \mathcal{H}_\phi$ such that $T_g \begin{pmatrix}\phi_1\\ \psi_1\end{pmatrix} \in \mathcal{H}_\phi$.

Local internal symmetries define conserved charges in those pure phases P in which they are not broken. They have essentially a kinematical character since they are based on invariance properties of the equations of motion. These conserved charges are somewhat introduced in the theory from outside and do not have a dynamical origin. However, the existence of HSS allows one to introduce conserved quantities whose origin is strictly related to the dynamics of the theory *(dynamical charges)*. For a given HSS they are identified with the "behaviour at infinity" of any of its elements.

<u>*Theorem 4*</u> [7] *Let* $\binom{\phi}{\psi}$ *belong to the HSS* $\mathcal{H}_{\binom{\phi}{\psi}}$, *let* $\lim_{r\to\infty} \phi(r\omega) = a(\omega)$ *exist in almost all directions* $\omega \in S^{s-1}$. *Then, for all* $\binom{\phi'}{\psi'} \in \mathcal{H}\binom{\phi}{\psi}$, $\lim_{r\to\infty} \phi'(r\omega) = a(\omega)$ *in almost all drections* $\omega \in S^{s-1}$.

Therefore a(ω), when it exists, is the same for all the elements of a HSS; in particular, it is conserved in time since a HSS is invariant under time evolution. For s = 3, if $|\nabla\phi| \in L^2(\mathbb{R}^s)$, the limit $\lim_{r\to\infty} \phi(r\omega)$ exists and therefore dynamical charges can be introduced in each energy sector and in each pure phase. If one solves Eq. (4) in $H^1(\mathbb{R}^s) \oplus L^2(\mathbb{R}^s)$ one obtains a HSS with zero dynamical charge.

If a local internal symmetry T_g is not broken in a pure phase P, one may identify dynamical charges related by T_g and thus define dynamical charges which commute with T_g. The same identification can be performed with respect to the rotation group if P is invariant under rotations. When the charges a(ω), belonging to a given pure phase P, are continuous functions of ω, one may adopt the point of view of identifying charges which can be obtained one from the other by continuous deformations, i.e. when they are homotopic. In this way one obtains the so-called topological charges.

APPENDIX A

We first prove that conditions (a) and (b) are satisfied if $\begin{pmatrix}\phi(t)\\ \psi(t)\end{pmatrix} - \begin{pmatrix}\phi_0\\ \psi_0\end{pmatrix} \equiv \begin{pmatrix}\chi(t)\\ \zeta(t)\end{pmatrix} \in C^{(0)}(\mathbb{R};Y)$, where $\begin{pmatrix}\phi(t)\\ \psi(t)\end{pmatrix}$ is the solution of Eq. (1) [in $C^{(0)}(\mathbb{R};X)$] with initial data $\begin{pmatrix}\phi_0\\ \psi_0\end{pmatrix}$. Use of the explicit expression for W(t) yields the following integral equation for $\begin{pmatrix}\chi(t)\\ \zeta(t)\end{pmatrix}$:

$$\begin{pmatrix}\chi(t)\\ \zeta(t)\end{pmatrix} = L(t) + \int_0^t W(t-s)\begin{pmatrix}0\\ -U'(\phi_0+\chi(s)) + U'(\phi_0)\end{pmatrix} ds \tag{A.1}$$

where

$$L(t) = \begin{pmatrix}\dfrac{1-\cos\sqrt{-\Delta}\,t}{-\Delta} & \dfrac{\sin\sqrt{-\Delta}\,t}{\sqrt{-\Delta}}\\[2ex] \dfrac{\sin\sqrt{-\Delta}\,t}{\sqrt{-\Delta}} & \cos\sqrt{-\Delta}\,t - 1\end{pmatrix}\begin{pmatrix}h\\ \psi_0\end{pmatrix}. \tag{A.2}$$

Now, the assumptions on U and the boundedness of ϕ_0 imply that the function $t \mapsto -U'(\phi_0 + \chi(t)) + U'(\phi_0) \in C^{(0)}(\mathbb{R};L^2)$ and therefore $L(t) \equiv \begin{pmatrix}A(t)\\ B(t)\end{pmatrix} \in C^{(0)}(\mathbb{R};Y)$. An immediate computation yields

$$A(t) - \int_0^t B(\tau)\,d\tau = t\,\psi_0$$

which implies that $\psi_0 \in L^2(\mathbb{R}^s)$ and therefore

$$\frac{\sin\sqrt{-\Delta}\,t}{\sqrt{-\Delta}}\,\psi_0 \in C^{(0)}(\mathbb{R};H^1).$$

As a consequence we obtain

$$\frac{1-\cos\sqrt{-\Delta}\,t}{-\Delta}\,h \in C^{(0)}(\mathbb{R};H^1). \tag{A.3}$$

From (A.3) it follows immediately that $\displaystyle\int_{|k|\le 1} |\hat{h}(k)|^2\, d^s k < \infty$.

To take care of the high k behavour of $\hat{h}(k)$, we integrate (A.3) in the t-variable and obtain

$$\int_0^1 \frac{1-\cos\sqrt{-\Delta}\,t}{\sqrt{-\Delta}}\,h\,dt = \left(\frac{1}{\sqrt{-\Delta}} - \frac{\sin\sqrt{-\Delta}}{-\Delta}\right)h \in L^2(\mathbb{R}^s)$$

This relation trivially implies that $\displaystyle\int_{|k|\ge 1} |\hat{h}(k)|^2\,|k|^{-2}\, d^s k < \infty$.

This completes the proof that $\psi_0 \in L^2(\mathbb{R}^s)$ and $h = \Delta\phi_0 - U'(\phi_0) \in H^{-1}(\mathbb{R}^s)$. Conversely, let conditions (a) and (b) be satisfied. We consider the integral equation

$$\begin{pmatrix} \chi(t) \\ \zeta(t) \end{pmatrix} = W(t)\begin{pmatrix} \chi_0 \\ \zeta_0 \end{pmatrix} + L(t) + \int_0^t W(t-s)\begin{pmatrix} 0 \\ -U'(\phi_0+\chi(s)) + U'(\phi_0) \end{pmatrix} ds \tag{A.4}$$

where L(t) is given by Eq. (A.2). Since $\psi_0 \in L^2(\mathbb{R}^s)$ and $h \in H^{-1}(\mathbb{R}^s)$, it is trivial to verify that $L(t) \in C^{(0)}(\mathbb{R};Y)$. Now, as a consequence of the assumptions on U and of the boundedness of ϕ_0, the function

$$\chi \mapsto G_{\phi_0}(\chi) \equiv U(\phi_0+\chi) - U(\phi_0) - U'(\phi_0)\,\chi \tag{A.5}$$

has the following properties:

i) $G'_{\phi_0}(\chi) \in L^2(\mathbb{R}^s)$ if $\chi \in H^1(\mathbb{R}^s)$.

ii) For all $\rho > 0$ there exists a positive constant $C(\rho)$ such that

$$\| G'_{\phi_0}(\chi_1) - G'_{\phi_0}(\chi_2) ; L^2(\mathbb{R}^s) \| \leq C(\rho) \| \chi_1 - \chi_2 ; H^1(\mathbb{R}^s) \|$$

if $||\chi_k;H^1(\mathbb{R}^s)|| \leq \rho$, $k = 1, 2$.

iii) There exists a non-negative constant γ such that $G_{\phi_0}(z) \geq -\gamma |z|^2$

The above properties of G_{ϕ_0} allow one to apply a slight modification of a theorem of I. Segal[3]) to Eq. (A.4) and conclude that for any initial data $\begin{pmatrix} \chi_0 \\ \zeta_0 \end{pmatrix} \in Y$ there is a unique solution $\begin{pmatrix} \chi(t) \\ \zeta(t) \end{pmatrix} \in C^{(0)}(\mathbb{R};Y)$. A trivial uniqueness argument implies that $\begin{pmatrix} \phi(t) \\ \psi(t) \end{pmatrix} - \begin{pmatrix} \phi_0 \\ \phi_0 \end{pmatrix} = \begin{pmatrix} \chi(t) \\ \zeta(t) \end{pmatrix}$, if $\chi(0) = \zeta(0) = 0$, which shows that $\begin{pmatrix} \phi_0 \\ \psi_0 \end{pmatrix}$ defines a HSS. An analogous argument can be used to show that $\mathcal{H}_{\begin{pmatrix} \phi_0 \\ \psi_0 \end{pmatrix}} = \begin{pmatrix} \phi_0 \\ \psi_0 \end{pmatrix} + Y$.

APPENDIX B

i) Upon defining $\chi = \phi - \phi_0$, the integrand of the right-hand side of Eq. (9) can be rewritten as

$$V_{\phi_0}(\chi,\psi) + U'(\phi_0)\chi + \nabla\chi\cdot\nabla\phi_0 \tag{B.1}$$

where

$$V_{\phi_0}(\chi,\psi) = \frac{1}{2}\left(|\nabla\chi|^2 + \psi^2\right) + G_{\phi_0}(\chi) \tag{B.2}$$

with G_{ϕ_0} defined by Eq. (A.5). Now the properties of G_{ϕ_0} listed in Appendix A imply that $V_{\phi_0}(\chi,\psi) \in L^1(\mathbb{R}^s)$ for any $\chi \in H^1(\mathbb{R}^s)$ and $\psi \in L^2(\mathbb{R}^s)$. Furthermore, if $\chi \in H^1_{comp}(\mathbb{R}^s)$, then $U'(\phi_0)\chi$ and $\nabla\chi\cdot\nabla\phi_0$ both belong to $L^1(\mathbb{R}^s)$. This proves that $E_{\phi_0}(\phi,\psi)$ is well defined if $\phi - \phi_0 \in H^1_{comp}(\mathbb{R}^s)$.

ii) Still by the mentioned properties of G_{ϕ_0} the map $Y \ni \binom{\chi}{\psi} \to V_{\phi_0}(\chi,\psi) \in L^1(\mathbb{R}^s)$ is is continuous. Therefore it will be enough to show that the linear functional

$$H^1_{comp}(\mathbb{R}^s) \ni \chi \mapsto \int_{\mathbb{R}^s} U'(\phi_0)\chi\, d^s x + \int_{\mathbb{R}^s} \nabla\chi\cdot\nabla\phi_0\, d^s x \tag{B.3}$$

has a continuous extension to the whole $H^1(\mathbb{R}^s)$. This is an immediate consequence of the identity

$$\int_{\mathbb{R}^s} \left(U'(\phi_0)\chi + \nabla\chi\cdot\nabla\phi_0\right) d^s x = -\left\langle \Delta\phi_0 - U'(\phi_0), \chi\right\rangle_{H^{-1}_{loc},\, H^1_{comp}}$$

and of the assumption $\Delta\phi_0 - U'(\phi_0) \in H^{-1}(\mathbb{R}^s)$. Equation (10) is essentially the energy conservation.

iii) Equation (11) follows immediately from the identity

$$\int_{\mathbb{R}^s} \left[\frac{1}{2}\left(|\nabla\phi|^2 - |\nabla\phi_0|^2\right) + \frac{1}{2}\psi^2 + U'(\phi) - U(\phi_0)\right]\omega\left(\frac{x-x_0}{\rho}\right) d^s x =$$

$$= \int_{\mathbb{R}^s} V_{\phi_0}(\chi,\psi)\,\omega\left(\frac{x-x_0}{\rho}\right) d^s x - \left\langle \Delta\phi_0 - U'(\phi_0),\, \omega\left(\frac{x-x_0}{\rho}\right)\right\rangle_{H^{-1},\,H^1}$$

$$- \left\langle \nabla\phi_0,\, \frac{1}{\rho}(\nabla\omega)\left(\frac{x-x_0}{\rho}\right)\chi \right\rangle_{L^2,\,L^2} .$$

Equation (12) is trivial to verify.

REFERENCES

1) Hepp, K.: Commun. Math. Phys. 35, 265 (1974).

2) Jörgens, J.: Math. Z. 77, 295 (1961).

3) Segal, I.: Ann. Math 78, 339 (1963).

4) Strauss, W.A.: Non-linear scattering theory, *in* Scattering theory in mathematical physics, J.A. Lavita and J.P. Marchand, Eds., D. Reidel Publishing Company, Dordrecht-Holland (1974).
Reed, M.: Abstract Non-linear Wave Equations, Springer-Verlag, Heidelberg, 1976.

5) Haag, R., Kastler, D.: J. Math. Phys. 5, 848 (1964).

6) Parenti, C., Strocchi, F., Velo, G.: Ann. Sc. Norm. Sup. Pisa 3, 443 (1976).

7) Parenti, C., Strocchi, F., Velo, G.: Commun. Math. Phys. 53, 65 (1977).

8) Goldstone, J.: Nuovo Cimento 19, 154 (1961).

9) Lee, T.D.: *in* Extended Systems in Field Theory, J.L. Gervais, A. Neveu, Eds., Phys. Reports 23C, 1976.

10) Parenti, C., Strocchi, F., Velo, G.: Nuovo Cimento 39B, 147 (1977).

11) Streater, R.F., Wightman, A.S.: PCT, Spin and Statistics, and all That, W.A. Benjamin, New York, 1964.

12) Glimm, J., Jaffe, A., Spencer, T.: Commun. Math. Phys. 45, 203 (1975).

RELEVANCE OF CLASSICAL SOLUTIONS TO QUANTUM THEORIES

J.L. Gervais

Laboratoire de Physique Théorique de l'Ecole Normale Supérieure
24, rue Lhomond 75231 PARIS Cedex 05 FRANCE

INTRODUCTION

Those lectures are concerned with recent interesting developments of the last few years which took place as it was gradually realized that the existence of non-trivial classical solutions of field theories has rather striking implications at the quantum level. The common exciting feature of these implications is that they are beyond reach of the usual perturbative approach based on Feynman rules and thus represent a major breakthrough in our understanding of quantum field theory.

The importance of classical solutions was revealed by a few pioneering works. Nielsen and Olesen(1) pointed out that in the Higgs model there exist classical solutions of the vortex type which behave very much like relativistic strings and thus resemble the hadron picture which emerges from dual models. Dashen, Hasslacher and Neveu(2) extended WKB methods to field theory and showed that localized classical solutions in Minkowski space correspond to the existence of new quantum states which could not be described by standard perturbation. Thus, field theories may have a richer structure than expected and in view of ref. (1), this raises the hope that the hadrons, with their complicated non perturbative structure, can be fitted into the scheme of local quantum field theory. These new objects generally called solitons will be the subject of the first part of my lectures.

More recently, a different direction was advocated by Polyakov(3,4) who pointed out the importance of classical solutions of finite action in the Euclidean space obtained after continuation to purely imaginary time. Such solutions, called instantons, only exist in theories with degenerate vacua where standard perturbation exhibits spontaneous symmetry breaking. They are the signal of tunneling between these different vacua so that symmetry is restored and long-range correlation may be destroyed.

As a result, Goldstone bosons may be avoided in such a way

that the so-called η problem can be solved. The long-range correlation may disappear if the instantons are able to screen the long-range forces. In this case quarks become confined in the sense that the criterion of Wilson is satisfied. Here, again the earlier picture based on standard perturbation turn out to be misleading and the new picture which emerges is much closer to the physical reality. This will be the subject of the second part of my lectures.

The common feature of both approaches is to be based on semi-classical approximation, i.e. small $\hbar$. If γ is a typical coupling constant and ϕ is a typical field, one always finds that $\hbar \propto \gamma$ and the the classical solution ϕ_{cl} is of order $\gamma^{-1/2}$. This is why the results one gets are non-perturbative in γ.

Standard perturbation is in general established by letting $\phi = \phi_v + \bar{\phi}$ where ϕ_v is a constant which is a minimum of the potential ; ϕ_v is zero or order $\gamma^{-1/2}$ and $\bar{\phi}$ is considered $O(1)$ in the way Feynman perturbation is built up. The effect of non-trivial classical solution is studied by letting $\phi = \phi_{cl} + \tilde{\phi}$. If $\tilde{\phi}$ is considered $O(1)$ this leads again to an expansion in γ where the dominant term is the contribution of ϕ_{cl}. The non-perturbative effect with respect to the usual perturbation expansion $\phi = \phi_v + \bar{\phi}$ is under control as it is entirely contained in ϕ_{cl}. Because $\hbar \propto \gamma$ the perturbation expansion in γ one obtains is simply an expansion in $\hbar$ i.e. an estimation of quantum corrections to classical results. The fact that ϕ_{cl} is a classical solution, that is a minimum of the action, ensures that no term linear in $\tilde{\phi}$ will appear. This is necessary since otherwise those terms would give tadpole terms of order $\gamma^{-1/2}$ into the perturbation expansion of $\tilde{\phi}$ which would lead to corrections of the same order as the classical term.

Those semi-classical methods are most easily developed using Feynman path integrals. In this case we will be simply making change of variables in the path integral and the general idea is to consider the contribution of the fluctuation around non-trivial minima of the action.

For solitons we want the soliton energy to be finite compared to vacuum energy. This then must also be true at the classical level. Thus the classical soliton solution should, at large distance, tend to one of the possible vacua of the theory that is to a minimum of the potential. The one-soliton state in its rest frame corresponds to a static solution and will thus be a minimum of the energy. Unless one imposes some additional condition, a minimum of the energy can only be a classical vacuum, which is a constant field. To obtain non-trivial solutions one can either look for a minimum with a non-zero value of some conserved charge (non-topological soliton[5]) or in the case of degenerate minima impose boundary conditions which cannot be continuously deformed into the vacuum state boundary condition. If this is possible there always exists a quantity called topological charge which is conserved irrespective of the equations of motion because it can only take discrete values classically and one is looking at a minimum of the energy with non-zero value of this topological charge. The instanton being a solution in Euclidean field theory is a minimum of the action with all kinetic terms positive. Hence, an instanton solution in d dimensional space time is also a time-independent soliton solution in $d + 1$ dimensions and the classification of possible solutions can be done simultaneously for both cases. It is based on homotopy theory and I will not repeat it here[6].

The last general point is that these classical solutions always involve arbitrary parameters so that the change of field $\phi = \phi_{u} + \tilde{\phi}$ is not well defined. The solution to this problem is to consider these parameters as dynamical variables which therefore should be determined from ϕ itself[7,8]. Those dynamical variables are called collective coordinates and have natural physical interpretations as soliton position and momenta or Euclidean space time regions where tunneling occurs. I will mainly concentrate on this aspect of the semi-classical method mostly reviewing works performed over the last two years in collaboration with A. Jevicki and B. Sakita. Complementary reviews have been written by R. Rajaraman[9], S. Coleman[6] and R. Jackiw[10]. For additional references see ref.(11).

I. SOLITONS

I-A. One soliton in two dimensions

We consider the Lagrangian

$$\mathcal{L} = -\frac{1}{2}\partial_\mu\phi\,\partial^\mu\phi - V(\phi) \tag{1.1}$$

with the potential of the form

$$V(\phi) = \frac{1}{g^2}\,U(g\phi) \tag{1.2}$$

which has a classical solitary wave solution

$$\phi_c(x,t) = \phi_0\left(\frac{x - vt - x_0}{\sqrt{1-v^2}}\right) \qquad -\frac{\partial^2}{\partial x^2}\phi_0(x) + \frac{\delta V}{\delta\phi_0(x)} = 0$$

with finite energy. At the classical level, this solution can be interpreted as a particle with the mass $M_0 = \int dx\,(\partial\phi_0/\partial x)^2$, since the energy and momentum operators (' means $\partial/\partial x$)

$$H = \int dx\left[\frac{1}{2}\dot\phi^2 + \frac{1}{2}\phi'^2 + V(\phi)\right]$$

$$P = \int dx\,\dot\phi\,\phi'$$

give

$$H(\phi_c) = (p^2 + M_0^2)^{1/2} \qquad p = -P(\phi_c)$$

I will present a method for quantization of such classical solutions. In the case of weak coupling, when the soliton mass is large, we developed a systematic perturbation expansion for the one-soliton sector(7,12). With the corresponding Feynman rules, one can make perturbative calculations of transition matrix elements for the initial and final states containing one soliton and an arbitrary number of mesons associated to the field ϕ .

The transition amplitude between initial and final states described by the wave functionals $\Psi_i[\phi]$ and $\Psi_f[\phi]$ is given by the following path integral :

$$S_{fi} = \int \mathcal{D}\Pi \mathcal{D}\phi \, \Psi_f^*[\phi(x,t_f)] \Psi_i[\phi(x,t_i)] \quad \exp\left\{ i \int dx\, dt \left[\Pi \dot{\phi} - \mathcal{H}(\Pi,\phi) \right] \right\}$$

$$\mathcal{H} = \frac{1}{2}(\Pi^2 + \phi'^2) + V(\phi) \tag{1.3}$$

If, in order to develop a perturbation expansion for the one-soliton sector, one simply expands around the classical solution ϕ_0 , as in the case of spontaneous symmetry breaking, one finds divergences connected with the translation invariance of our theory. Namely, the propagator of this perturbation expansion would be the inverse of the following differential operator :

$$\frac{\partial^2}{\partial t^2} + \Omega^2 \equiv \frac{\partial^2}{\partial t^2} - \frac{\partial^2}{\partial x^2} + V''(\phi_0) \tag{1.4}$$

where

$$V''(\phi_0) = \left. \frac{\delta^2 V}{\delta \phi^2} \right|_{\phi = \phi_0}$$

Taking the space derivative of the field equation satisfied by ϕ_0 one immediately sees that ϕ_0' is eigenstate of Ω^2 with eigenvalue zero. Thus the propagator is ill defined since the differential operator (1;4) has a zero eigenvalue.

To solve this difficulty and develop a consistent perturbation expansion for the one-soliton sector, we will first separate the center-of-mass motion, extracting the total momentum and the center-of-mass coordinate. We insert the following identities into the path integral expression for the S-matrix element :

$$\int \prod_t \left\{ \mathcal{D}p(t)\, \delta(-p(t) + P) \right\} = 1 \qquad P \equiv \int dx\, \Pi \phi'$$

$$\int \prod_t \left\{ \mathcal{D}X(t)\, \delta\left(Q[\phi(x+X), \Pi(x+X)] \right) \frac{\partial Q}{\partial X} \right\} = 1 \tag{1.5}$$

The first identity which we call the constraint, serves to identify the variable $p(t)$ with the total momentum of the system while the second one is the gauge condition associated with the constraint.

$\mathcal{Q}$ can be arbitrary. We notice that $\partial\mathcal{Q}/\partial X$ is given by Poisson bracket :

$$\frac{\partial \mathcal{Q}}{\partial X} = \{\mathcal{Q}, P\} \equiv \int dx \left[\frac{\delta \mathcal{Q}}{\delta \phi}\frac{\delta P}{\delta \pi} - \frac{\delta \mathcal{Q}}{\delta \pi}\frac{\delta P}{\delta \phi}\right] \tag{1.6}$$

Next, we make a change of variables

$$\phi(x,t) = \tilde{\phi}(x - X(t), t) \equiv \tilde{\phi}(\rho, t)$$

$$\pi(x,t) = \tilde{\pi}(x - X(t), t) \equiv \tilde{\pi}(\rho, t)$$

$$\rho = x - X(t) \tag{1.7}$$

so that, using $\dot{\phi} = \dot{\tilde{\phi}} - \dot{X}\tilde{\phi}'$ and the constraint, we get

$$\int dx \left[\pi\dot{\phi} - \mathcal{H}(\pi,\phi)\right] = p(t)\,\dot{X}(t) + \int d\rho \left[\tilde{\pi}\dot{\tilde{\phi}} - \mathcal{H}(\tilde{\pi},\tilde{\phi})\right]$$

$$\Psi_{i,f}[\phi] = \exp(i\, p_{i,f} X(t_{i,f}))\, \Psi_{i,f}[\tilde{\phi}] \tag{1.8}$$

From the first expression, one sees that X is the variable conjugate to p, i.e. the C.M. position and (1.7) is a change to the moving frame attached to this center of mass. Thus, we have explicitly exhibited the total momentum and center of mass position associated with a given field configuration. If it corresponds to quantum fluctuation in the one soliton sector, X and p will be automatically the position and momentum of this soliton.

As X appears only in the term $p\dot{X}$, we can immediately integrate over X and p which leads to

$$S_{fi} = \delta(p_f - p_i) \int \mathcal{D}\tilde{\pi}\, \mathcal{D}\tilde{\phi}\, \Psi_f^*[\tilde{\phi}]\, \Psi_i[\tilde{\phi}] \prod_t [\delta(p + P)\, \delta(\mathcal{Q})]$$
$$\prod_t \{\mathcal{Q}, P\} \quad \exp\left\{ i \int d\rho\, dt \left(\tilde{\pi}\dot{\tilde{\phi}} - \mathcal{H}(\tilde{\pi},\tilde{\phi})\right)\right\} \tag{1.9}$$

where $p = p_i = p_f$. The stationary point of the action with constraints is given by the following variational equation

$$\delta\left\{\int dt\left[\int d\rho\left(\tilde{\pi}\dot{\tilde{\phi}} - \mathcal{H}\right) + \alpha(t)(p+P)\right]\right\} = 0$$

where α is a Lagrange multiplier. One obtains, for the lowest energy stationary point $(\dot{\phi}_c = 0)$, exactly the soliton solution

$$\phi_c = \phi_0\left((\rho - a)\sqrt{1+\frac{p^2}{M_0^2}}\right) \qquad \pi_c = -\frac{p}{\sqrt{p^2+M_0^2}}\,\phi_c' \tag{1.10}$$

where ϕ_0 is solution of

$$-\phi_0'' + \frac{\delta V}{\delta \phi_0} = 0 \tag{1.11}$$

and the constant a is fixed by the $\mathcal{G}$ condition. The corresponding classical energy is found to be

$$E_c = \left(p^2 + M_0^2\right)^{1/2} \tag{1.12}$$

At this point, we observe that, due to the property (1.2) of our potential, ϕ_c is of the order of $1/g$; accordingly, M_0 is of the order of $1/g^2$. We can develop the perturbation expansion in g around the classical solution

$$\tilde{\phi} = \phi_c(\rho) + \chi(\rho, t) \qquad \tilde{\pi} = \pi_c(\rho) + \varpi(\rho, t) \tag{1.13}$$

Here, χ and ϖ are considered order zero and represent small quantum fluctuations around the classical solution. In the case when the initial and final states contain only one soliton, the shift (1.13) gives, in the first approximation, the relativistic form for the soliton energy and we may develop a perturbation theory for the soliton energy.

On the other hand, if the momentum is considered $O(1)$, one can alternatively shift by the zero momentum classical solution $\phi_0(\rho)$ because our method only requires that the function used in the shift be a classical solution to leading order so that in the final expression, one does not get zeroth order terms linear in X. Since this last case is much simpler, I shall only discuss the corresponding perturbation expansion briefly in what follows. (See ref.(12) for details).

Next, one has to choose the gauge condition. Although an arbitrary choice leads to the consistent perturbation expansion free of the infrared divergences, we prefer to choose a linear gauge condition

$$\mathcal{G}[\phi(x+X,t),\Pi(x+X,t)] \equiv \int dx\, f(x)\,\phi(x+X,t)$$

$$\frac{\partial \mathcal{G}}{\partial X} = \int dx\, f(x)\,\phi'(x+X,t) \tag{1.14}$$

in order to eliminate the zero-energy mode in the simplest possible way. Here, f is still an arbitrary function, and identifying it later with the zero-frequency eigenfunction, we will completely eliminate the zero-energy mode from our functional integral. Now, before making the shift $\tilde{\Phi} = \phi_0 + X$, we linearize the constraint which is quadratic in fields by making the following change of variables

$$\tilde{\Pi}(\rho,t) = -f(\rho)\,\frac{p + \int \bar{\Pi}(\rho,t)[\tilde{\Phi}' - fc]\,d\rho}{\int f\tilde{\Phi}'\,d\rho} + \bar{\Pi}(\rho,t) \tag{1.15}$$

Then, the constraint becomes :

$$\delta\left(p + \int \tilde{\Pi}\tilde{\Phi}'\,d\rho\right) = \delta\left(c\int f(\rho)\,\bar{\Pi}(\rho,t)\,d\rho\right)$$

Computing the Jacobian of this transformation, we get

$$\det\left[\frac{\delta\tilde{\Pi}}{\delta\bar{\Pi}}\right] = \prod_t \left(\int d\rho\, f\tilde{\Phi}'\right)^{-1}$$

so, it exactly cancels out the $\partial\mathcal{G}/\partial X$ which is given by (1.14). Now, the Hamiltonian becomes more complicated :

$$H \equiv \int d\rho\, \mathcal{H} = \frac{\left(p + \int d\rho (\tilde{\phi}' - fc)\bar{\pi}\right)^2}{2\left(\int d\rho\, f\phi'\right)^2} + \int d\rho \left[\frac{\bar{\pi}^2}{2} + \frac{\tilde{\phi}'^2}{2} + V(\tilde{\phi})\right] \tag{1.16}$$

We used the normalization $\int d\rho\, f(\rho)^2 = 1$. The transition amplitude is now of the form

$$S_{fi} = \delta(p_f - p_i) \int \mathcal{D}\bar{\pi}\, \mathcal{D}\tilde{\phi}\; \Psi_f^*[\tilde{\phi}]\Psi_i[\tilde{\phi}]$$

$$\delta\left(\int f\tilde{\phi}\, d\rho\right) \delta\left(\int f\bar{\pi}\, d\rho\right) \exp\left\{i \int dt\, d\rho \left[\bar{\pi}\dot{\tilde{\phi}} - \mathcal{H}\right]\right\} \tag{1.17}$$

and since both the gauge condition and the constraint are linear in fields one can easily develop a perturbation expansion.

We now continue our discussion by making the shift

$$\tilde{\phi}(\rho,t) = \phi_0(\rho) + \chi(\rho,t) \qquad \bar{\pi}(\rho,t) = \varpi(\rho,t) \tag{1.18}$$

with the choice of f, c

$$f = \frac{\phi_0'}{\sqrt{M_0}} \equiv \psi_0 \qquad c = \sqrt{M_0} \tag{1.19}$$

The final result reads

$$H = M_0 + \frac{\left(p + \int \varpi \chi' d\rho\right)^2}{2M_0\left(1 + \xi/M_0\right)^2} + \int d\rho \left[\frac{\varpi^2}{2} + \frac{\chi'^2}{2} + V - V(\phi_0) - \frac{\delta V}{\delta\phi}\bigg|_{\phi_0}\chi\right] + \Delta V$$

with

$$\xi = \int d\rho \, \Phi_0'(\rho) \, \chi'(\rho, t)$$

(1.21)

$$\Delta V = \frac{1}{8}\left[\frac{-3(\psi_0', \psi_0')}{(\tilde{\Phi}_0', \psi_0)^2} + 2\frac{(\psi_0', \tilde{\Phi}'')}{(\psi_0, \tilde{\Phi}')^3} + \frac{(\psi_0', \tilde{\Phi}')^2}{(\psi_0, \tilde{\Phi}')^4} + \sum_{n,m\neq 0} \frac{|(\psi_m, \psi_m')|^2}{(\psi_0, \tilde{\Phi}')} \right]$$

$$\Omega^2 \psi_m = \omega_m^2 \psi_m \qquad \text{(see below)} \qquad (1.22)$$

where we noted in general for two artibrary functions h_1, h_2

$$(h_1, h_2) \equiv \int d\rho \, h_1(\rho) \, h_2(\rho)$$

$\Delta V[X]$ is an additional potential which starts contributing at the two-loop level. It is not obtained if one performs the changes of variables (1.7), (1.15), (1.18) into the action as we have done here.

We have presented the procedure in these three steps because we want to illustrate the general method on this particular case. Here, on the other hand, one can directly obtain the result by making the change of field[12]

$$\Phi = \Phi_0(x-X) + \chi(x-X, t)$$

$$\Pi = \Pi_0 + \varpi(x-X, t)$$

$$\Pi_0 \equiv -\frac{\Phi_0'(x-X)}{M_0 + \xi}\left(p + \int \varpi \chi' d\rho \right) \qquad (1.23)$$

with the condition

$$\int d\rho \, \varpi \Phi_0' = \int d\rho \, \chi \Phi_0' = 0 \qquad (1.24)$$

The need of an additional term was firts pointed out by Tomboulis[13] who performed the canonical transformation (1.23) in the operator formalism. In the functional method ΔV also arises if the change of variables (1.23) is done with enough care[14]. The essential point is that in a functional integral, the derivatives which appear in the exponential of the action are not derivatives in the usual sense.

Consider for instance a transition probability in quantum mechanics

$$\int \mathcal{D}q(t) \exp\left\{ i \int dt \left(\frac{1}{2} \dot{q}^2 - V(q) \right) \right\}$$

the real meaning of this integral is that we should take time intervals ε and write $(q_\ell = q(\ell\varepsilon))$

$$\lim_{\varepsilon \to 0} \int \prod_\ell dq_\ell \exp\left\{ i\varepsilon \sum_\ell \frac{1}{2} \left(\frac{q_{\ell+1} - q_\ell}{\varepsilon} \right)^2 - V(q_\ell) \right\}$$

The q_ℓ 's are idependent variables. However, the first term of the exponent has a finite limit only for $q_{\ell+1} - q_\ell = O(\sqrt{\varepsilon})$. Hence, if we perform a canonical transformation $Q = F(q)$, we must expand F up to second order derivative when we compute $Q_{\ell+1} - Q_\ell$ in terms of $q_{\ell+1} - q_\ell$. The additional term $\frac{1}{2} \frac{d^2F}{dq_\ell^2} [q_{\ell+1} - q_\ell]^2$ being of order ε gives a correction to the potential. This is in contrast with the naive computation where one would just write

$$\dot{Q} = \frac{dF}{dq} \dot{q}$$

The Feynman rules of perturbation expansion can now be obtained easily[12]. The propagator is determined from the quadratic part of the Hamiltonian by expanding in terms of eigenfunctions of the following differential equations

$$\Omega^2 \psi_n \equiv \left(- \frac{d^2}{d\rho^2} + V''(\phi_0) \right) \psi_n = \omega_n^2 \psi_n \tag{1.25}$$

The zero-energy eigenfunction is given by ψ_0
We chose f to be precisely given by ψ_0 in such a way that $\omega_0 = 0$ mode disappears from the eigenfunction expansion of χ and ϖ because of the δ condition in (1.17).

Since we use first-order formalism, this perturbation expansion involves three different propagators $\langle 0|T(\chi\chi)|0\rangle$, $\langle 0|T(\varpi, \varpi)|0\rangle$, $\langle 0|T(\chi, \varpi)|0\rangle$. The Hamiltonian (1.20) contains products of χ

and ϖ at the same point and therefore there are ordering problems if we want to write H as an operator as was done by Tomboulis[13]. In the functional formalism this ordering problem appears also in practice because the perturbation expansion will contain the mixed propagator $\langle 0|T(\chi,\varpi)|0\rangle$ with zero time separation which is ambiguous. This ambiguity is seen if one looks more carefully into the meaning of (1.17) again because derivative terms in the action are to be handled with care. Indeed, if we take discrete time intervals ε, it is not clear how we should interpret the term $\int \varpi\dot{\chi}\,dt$ of (1.17). Different choices lead to different expressions for ΔV. The expression (1.22) correspond to the so-called mid-point definition namely we choose field variables $\chi(x,t_{2\ell})$, $\varpi(x,t_{2\ell+1})$; $t_{\ell+1}-t_\ell = \varepsilon$, and write

$$\int dx \int \varpi\dot{\chi}\,dt \equiv \sum_\ell \int dx\; \varpi(x,t_{2\ell+1})\left[\chi(x,t_{2\ell+2}) - \chi(x,t_{2\ell})\right]$$

In operator formalism ΔV is the term associated with Weyl's ordering for the expression (1.20) of H. In perturbation theory this means that the mixed propagator $\langle 0|T(\chi,\varpi)|0\rangle$ for zero-time separation is taken to be zero, i.e; all closed loops of mixed propagators are to be dropped. This choice is different from the one of ref. (13), however, it is more suitable for deriving Feynman perturbation theory explicitly. See ref. (14) for details.

In this perturbation expansion Lorentz invariance is not manifest, but one can show that higher order corrections in coupling constant sum up to restore Lorentz invariance at least at the tree level[12]. Since $M_0 = O(g^{-2})$ while p is considered $O(1)$ we get the non relativistic expansion for the energy. At the lowest order we obtain, for instance, $E = M_0 + \frac{p^2}{2M_0} \simeq (p^2 + M_0^2)^{1/2}$.

The renormalization of the one-soliton sector can be carried out in a straightforward manner, by adding the mass renormalization counter term

$$H_{\delta m} = -\frac{1}{2}\delta m^2 \int d\rho\,(\phi_0 + \chi)^2$$

where δm^2 is the mass counter term computed in the zero-soliton sector.

In the example of ϕ^4 theory, the first two one-loop contributions to the soliton energy have been computed[12]. After renormalization one obtains the finite answer

$$E \simeq M_0 + \Delta M + \frac{p^2}{2M_0} - \frac{p^2}{2M_0^2}\Delta M \simeq \left(p^2 + (M_0 + \Delta M)^2\right)^{1/2}$$

where ΔM is the first quantum correction to the soliton mass calculated originally by Dashen, Hasslacher and Neveu[2]. For the diagrammatic representation of the Feynman rules, and also for the specific computations the reader is referred to ref. (12).

With this systematic perturbation expansion, one can make perturbative calculations of other quantities in the one-soliton sector besides the soliton energy and mass corrections. As an example, we consider the ϕ field matrix elements investigated by Goldstone and Jackiw[15]. Their assumption about the leading order of $\langle p', \{k'\} | \phi_{op} | p, \{k\} \rangle$ can be easily seen to be true[16]. Let us compute the leading term in the simplest matrix element $\langle p' | \phi_{op} | p \rangle$ From the path integral expression for this matrix element, we obtain the corresponding operator form

$$\langle p' | \phi_{op}(x,0) | p \rangle = \langle p' | \phi_0(x - X_{op}) | p \rangle + \langle p' | \chi_{op} | p \rangle \tag{1.26}$$

Inserting identity, we obtain the leading term to be

$$\langle p' | \phi_{op}(x,0) | p \rangle \cong \int dy \; e^{iy(p-p')} \phi_0(x-y) \tag{1.27}$$

which is the result of ref. (15). Higher order corrections can now be systematically computed, and similarly, one has a complete perturbation expansion for the arbitrary ϕ_{op} -field matrix element in the one-soliton sector[12].

The two-loop correction to soliton mass has been computed[17] in Sine-Gordon theory and found to agree with the conjecture of Dashen Hasslacher and Neveu[18] that WKB is exact for the ratio between soliton mass and fundamental field mass.

Note that expression (1.17) a priori looks like a highly non-renomalizable Hamiltonian since it involves vertices with an arbitrary number of legs. It is remarkable that finite results are in fact obtained to any order by just using the same counter term as in the usual sector. At the two-loop level this already involves remarkable cancellation among highly divergent integrals.

Let us now turn to different possible choices for $\mathcal{G}$. First of all, we can choose $\mathcal{G}$ in such a way that the zeroth mode disappears if we perform the shift (1.13) in order to avoid tadpole terms completely. We have to do that if the momentum is of order $1/g$, i.e. if the velocity of the soliton is finite for small g . The function f and constant C are now, instead of (1.19)

$$f = \frac{\phi_c'}{(p^2 + M_0^2)^{1/2}} \qquad C = (p^2 + M_0^2)^{1/2} \tag{1.28}$$

where ϕ_c is given by (1.10).

One obtains

$$H = (p^2 + M_0^2)^{1/2} + \frac{1}{2\sqrt{p^2+M_0^2}} \frac{\left(p + \int \varpi \chi' d\rho\right)^2}{\left(1 + \xi/\sqrt{p^2+M_0^2}\right)^2} + \int d\rho \left[\frac{\varpi^2}{2} + \frac{\chi'^2}{2} + V - V(\phi) - \frac{\delta V}{\delta \phi}\chi\right] + \tilde{\Delta V}$$

$$\xi \equiv \int d\rho\, \phi_c'(\rho)\, \chi'(\rho, t)$$

Now, to zeroth order we get $E = (p^2 + M_0^2)^{1/2}$ directly. This perturbation theory is however more difficult to handle since the quadratic term reads

$$H^{(2)} = v \int d\rho\, \varpi \chi' + \frac{1}{(p^2+M_0^2)^{1/2}} \frac{3}{2} v^2 (\phi_c' \chi')$$

$$+ \int d\rho \left[\frac{\varpi^2}{2} + \frac{\chi'^2}{2} + V''[\phi]\frac{\chi^2}{2}\right]$$

$$p = \frac{M_0 v}{\sqrt{1-v^2}} \tag{1.29}$$

This quadratic form can be diagonalized by introducing Lagrange multipliers for the constraints and working in first order formalism. The basic tool is the introduction of boosted solutions $\psi_m\left(x\sqrt{1+\frac{p^2}{M_0^2}}\right)$ instead of (1.28). The real problem is the determination of $\widetilde{\Delta V}$ because (1.23) is now replaced by

$$\Pi = \Pi_0 + \varpi$$

$$\phi = \phi_0\left((x-X)\sqrt{1+\frac{p^2}{M_0^2}}\right) + \chi \qquad (1.30)$$

Hence it is no more a point canonical transformation since involves both X and p. At the quantum level the meaning of $\phi_0\left((x-X_{op})\sqrt{1+\frac{P_{op}}{M_0^2}}\right)$ is rather unclear. However, since everything worked out fine with the choice (1.18), (1.23), one should be able to handle (1.28), (1.30) exactly as well. Note that this problem only arises at the two-loop level where $\widetilde{\Delta V}$ starts to contribute.

More generally, one can choose f to be different from the zeroth mode one wants to eliminate. This mode then appears in the propagator but because the result does not depend on g there will be Ward-like identities which will tell us that the zeroth mode contribution vanishes. Indeed Faddeev and Korepin[19] have pointed out that the singularity of the propagator due to the zeroth mode is rather mild. They remarked that if we look at the resolvent

$$R(\varepsilon) = (\hat{H}^{(2)} + i\varepsilon)^{-1}$$

$$\hat{H}^{(2)} \equiv \frac{\partial^2}{\partial t^2} - \frac{\partial^2}{\partial x^2} + V''(\phi_0) \qquad (1.31)$$

one gets

$$R(x,t;x',t') = \frac{i}{2\sqrt{i\varepsilon}}\, e^{-i\sqrt{i\varepsilon}|t'-t|}\, \psi_0(x)\psi_0(x') + \tilde{R} \qquad (1.32)$$

where $\tilde{R}$ is well behaved for $\varepsilon \to 0$ and corresponds to the contribution of the other modes besides the zeroth mode. Thus, the singularity is not a pole in ε , it is rather of the type $1/\sqrt{\varepsilon}$ and in ref. (19) it was proposed to approximate

$$\frac{e^{-i|t-t'|\sqrt{i\varepsilon}}}{2\sqrt{i\varepsilon}} \simeq \frac{1}{2\sqrt{i\varepsilon}} - \frac{i}{2}|t-t'| + o(\sqrt{i\varepsilon}) \tag{1.33}$$

and to use as a propagator

$$\frac{1}{2\sqrt{i\varepsilon}}\,\psi_0(x)\,\psi_0(x') - \frac{i}{2}|t-t'|\,\psi_0(x)\,\psi_0(x') + \lim_{\varepsilon\to 0}\tilde{R}$$

Up to two-loop level for a soliton mass, it has been checked that the $1/\sqrt{\varepsilon}$ term cancels. This reflects the existence of the Ward identities that we mentioned earlier as it shows that the zeroth mode indeed drops out.

More recently, Jevicki[20] pointed out that this procedure is ambiguous as higher order term in (1.33) can give finite contribution when multiplied by several terms of order $1/\sqrt{\varepsilon}$. He proposed a different method which does not introduce $X(t)$ as a bona fide quantum variable but only extracts a constant parameter a from ϕ by letting

$$1 = \int da\; \delta\left(\int dx\,dt\; \phi(x,t)\;\phi_0'(x-a)\right) J \tag{1.34}$$

instead of (1.5).

Up to two-loop level for soliton mass the zeroth mode again cancels and the result agrees with what one gets from (1.20). This method however can only be used if the soliton momentum is conserved. It does not apply, for instance, to matrix elements of the ϕ_{op} field as in (2.16).

I-B. Several solitons - General collective coordinate method :

For the case of several solitons we obviously need to extract more collective coordinates. We thus discuss a method for doing this which, in fact, is very general[21] and can be applied to any problem in which collective coordinates are relevant.

Let $\phi(x)$ and $\pi(x)$ be a canonical field and its conjugate momenta respectively. Let $H[\pi,\phi]$ be the Hamiltonian of a system under consideration. We consider a group $\mathcal{G}$ of transformations generated by a set of n generators $P_a[\pi,\phi]$ through Poisson brackets. We assume that the Lie algebra closes namely

$$\{P_a, P_b\} = C^d_{ab} P_d \tag{1.35}$$

For arbitrary group element specified by parameters X_α, the transform of $A[\pi,\phi]$ is given by

$$\tau_{[X]}(A) \equiv A_{[X]} = \sum_{m=0}^{\infty} \frac{1}{m!} \{\{\cdots\{\{A, G_{\{X\}}\}, G_{\{X\}}\}\cdots\}, G_{\{X\}}\} \tag{1.36}$$

where

$$G_{\{X\}} \equiv \sum_\alpha X_\alpha P_\alpha[\pi,\phi] \tag{1.37}$$

Since $\mathcal{G}$ is a Lie group, we have

$$\tau_{[Y]} \cdot \tau_{[X]} = \tau_{[Z]}$$

where

$$Z^a = f^a(Y^1,\cdots,Y^n; X^1,\cdots,X^n) \tag{1.38}$$

Let us define V^a_b and its inverse U^a_b by

$$V^a_b(X) = \frac{\partial f^a}{\partial y^b}(y,X)\Big|_{y=0}$$

$$V^a_b\, U^b_c = \delta^a_c$$

The structure constant C^d_{ab} is related to U and V by

$$C^d_{ab} = -V^e_a V^f_b \left(\frac{\partial U^d_f}{\partial X^e} - \frac{\partial U^d_e}{\partial X^f} \right)$$

Considering an infinitesimal transformation, one obtains

$$\{ \phi_{[X]}, P_a \} = \frac{\partial \phi_{[X]}}{\partial X^b} V^b_a(X)$$

(1.39)

Now that we have enough machinery, we introduce X as a dynamical variable through making the change of variable

$$\tilde{\phi}(x,t) = \phi_{[X(t)]}(x,t) \qquad \tilde{\Pi}(x,t) = \Pi_{[X(t)]}(x,t)$$

(1.40)

This is the generalization of (1.7).

The choice of P_α is a priori arbitrary. We assume that it is such that the X_α are the relevant collective coordinates as we shall discuss below. Hence, we denote by X_α the collective coordinates while one may say that $\tilde{\phi}$ and $\tilde{\Pi}$ are fields in the body fixed coordinate system (or moving coordinate system).

Next, we insert (1.40) into the Hamiltonian to obtain a new Hamiltonian as a function of $\tilde{\phi}$ $\tilde{\Pi}$ and X 's

$$\tilde{H}[\tilde{\Pi}, \tilde{\phi}, X] \equiv H[\tilde{\Pi}_{[-X]}, \tilde{\phi}_{[-X]}]$$

(1.41)

where $[-X]$ is the inverse transformation of (1.40)

Let us now consider a new system with Hamiltonian $\tilde{H}$, canonical variables $X, \tilde{\Phi}$, and canonical momenta p and $\tilde{\Pi}$. We now show that this new system is equivalent to the old one if we impose the constraints[22]

$$p_\alpha + P_\beta[\tilde{\Pi},\tilde{\Phi}]\, U^\beta_\alpha \equiv F_\alpha[\tilde{\Pi},\tilde{\Phi},p,X] = 0$$

It is obvious that we have to impose n constraints since the new system has n more dynamical variables than the old one.

In the sense of Dirac, those constraints are first class since one can check that (these Poisson brackets also involve X_α and p_β derivatives)

$$\{F_\alpha, F_\beta\} = 0 \qquad \{\tilde{H}, F_\alpha\} = 0$$

The existence of first class constraints, precisely, reflects the gauge invariance of the new system under the canonical transformations generated by F_α, which is simply

$$\tilde{\Phi} \to \tilde{\Phi}_{[-Y]} \quad , \quad \tilde{\Pi} \to \tilde{\Pi}_{[-Y]} \quad , \quad [X] \to [X+Y]$$

where the sign $[X+Y]$ means group multiplication of the elements and where Y is an arbitrary function of t. That this transformation leaves $\tilde{H}$ invariant is obvious since it does not change the fields Φ and Π we started from.

The effective Hamiltonian has the form

$$H_{eff} = H + \sum_\alpha \lambda_\alpha F_\alpha \tag{1.42}$$

where λ_α are Lagrange multipliers. λ_α is determined from the equation $\dot{X}_\alpha = \delta H_{eff}/\delta p_\alpha$ which gives $\lambda_\alpha = \dot{X}_\alpha$. Choosing X_α determines λ_α through this equation and thus fixes the gauge.

For $X_\alpha \equiv 0$, we find back the old system since then $\tilde{\Phi} = \Phi$, $\tilde{\Pi} = \Pi$, $\lambda_\alpha = 0$. Therefore, since the physical contents of the theory is gauge independent, the new description is equivalent to the old one.

The quantization of the new system can be done following Faddeev's method[23]. One adds n additional gauge fixing conditions

$Q_\alpha = 0$, and write the transition matrix element as

$$\int \mathcal{D}\tilde{\phi}\, \mathcal{D}\tilde{\pi}\, \mathcal{D}X\, \mathcal{D}p \prod_\alpha \delta(F_\alpha) \prod_\beta \delta(Q_\beta)\, \det\{F_\alpha, Q_\beta\} \times$$

$$\times \exp\left\{ i \int dt \left(p_\alpha \dot{X}_\alpha + \int dx\, \tilde{\pi} \dot{\tilde{\phi}} - H \right) \right\} \qquad (1.43)$$

assuming that the Q 's are such that $\det\{F_\alpha, Q_\beta\} \neq 0$

However, in order to be self-contained and to show the generalization of our one-soliton method, we rederive the quantization procedure starting from the transition matrix element

$$\int \mathcal{D}\phi\, \mathcal{D}\pi\, \Psi_f^*[\phi]\, \Psi_i[\phi]\, \exp\left\{ i \int dt \left(\int dx\, \pi \dot{\phi} - H \right) \right\}$$

and introducing into the functional integral

$$\int \prod_{\alpha\beta} \mathcal{D}X_\alpha\, \mathcal{D}p_\beta\, \delta(p_\alpha + P_\alpha[\pi_{[X]}, \phi_{[X]}])\, \delta(Q_\beta[\pi_{[X]}, \phi_{[X]}])\, J = 1$$

$$J = \prod_t \det\left(\frac{\delta Q_\beta}{\delta X_\gamma} \right) = \prod_t \det\{P_\alpha, Q_\beta\}$$

$$(1.44)$$

In this proof, we consider only, to simplify, the case of the abelian canonical group. We immediately obtain

$$\int \mathcal{D}\phi\, \mathcal{D}\pi \prod_{\alpha\beta} \mathcal{D}p_\alpha\, \mathcal{D}X_\beta\, \delta(p_\alpha + P_\alpha)\, \delta(Q_\beta)\, \Psi_f^* \Psi_i\, \det\{P, Q\}$$

$$\exp\left\{ i \int dt \left(\int dx\, \pi \dot{\phi} - H \right) \right\} \qquad (1.45)$$

If we first integrate over ϕ and π , we can make the change of variable (1.40) for fixed X and one sees that the Jacobian is one since (1.40) is a canonical transformation[24]. Next, consider the term

$$\int_{t_i}^{t_f} dx\, dt\, \pi \dot{\phi} = \int_{t_i}^{t_f} dx\, dt\, \tilde{\pi}_{[-X]} \frac{\partial}{\partial t}\left(\tilde{\phi}_{[-X]} \right) \equiv K_{[-X]}$$

By an infinitesimal change δX, one can verify that

$$\delta K = -\int_{t_i}^{t_f} dt\, P_\alpha \delta \dot{X}_\alpha + \int dx \left(\text{surface terms} \right) \Big|_{t_i}^{t_f}$$

dropping the surface term, we obtain $K = \int dt\, P_\alpha \dot{X}_\alpha + K_0$ where $K_0 = K_{[X=0]} = \int dx\, dt\, \tilde{\pi} \dot{\tilde{\phi}}$ so that finally

$$\int dx\, dt\, \pi \dot{\phi} = \int dt\, p_\alpha \dot{X}_\alpha + \int dx\, dt\, \tilde{\pi} \dot{\tilde{\phi}}$$

p_α and X_α are just conjugate variables. We finally obtain

$$\int \mathcal{D}\tilde{\pi}\, \mathcal{D}\tilde{\phi}\, \mathcal{D}X_\alpha\, \mathcal{D}p_\beta\, \delta(p_\beta + P_\beta[\tilde{\pi},\tilde{\phi}])\, \delta(Q_\beta[\tilde{\pi},\tilde{\phi}])\, \det\{P,Q\}$$

$$\tilde{\Psi}_f^*[\tilde{\phi}]\, \tilde{\Psi}_i[\tilde{\phi}]\, \exp\left\{ i \int dt \left(p_\alpha \dot{X}_\alpha + \int dx\, \tilde{\pi} \dot{\tilde{\phi}} - \tilde{H}[\tilde{\phi},\tilde{\pi},X] \right) \right\}$$

(1.46)

which agrees with (1.43) and is the generalization of (1.9).

In this expression, we have replaced Ψ_i, Ψ_f by $\tilde{\Psi}_i$, $\tilde{\Psi}_f$, in order to take into account the surface terms which we dropped in the computation together with the change of argument of $\Psi_{i,f}$. It is likely [23], though no general proof exists that simply transformed from $\tilde{\Psi}_{i,f}$ by $\Psi_{i,f}$

$$\tilde{\Psi}_{i,f}[\phi] = \left(\exp\left\{ i \sum_\alpha X_\alpha(t_{i,f})\, P_{\alpha\,\text{op.}} \right\} \Psi_{i,f} \right)[\phi]$$

(1.47)

namely, is, as one expects, obtained from $\Psi_{i,f}$, by the unitary transformation associated with the canonical transformation introduced by (1.40).

The procedure we have followed is exactly the generalization of our discussion for one soliton. There, for small g, the soliton has a large mass compared to the mass associated with the quanta of χ. Hence, for small g the soliton position moves much more slowly than the other degrees of freedom. This is the standard criterion for introducing collective coordinates; it is the so-called adiabatic approximation. In general, if we assume that the X_α in (1.46) vary much more slowly than the other degrees of freedom, we can determine an effective potential by first solving the dynamics of the other degrees of freedom with fixed X_α, p_β. In functional formalism this is formally done by assuming that the $\Psi_{i,f}$ are eigenstates of $P_{\alpha\,op}$ with eigenvalues $p_{\alpha,i}$, $p_{\alpha,f}$, and computing for fixed X_α, p_β

$$\exp\left\{-i\int dt\, H_{eff}[X,p]\right\} \equiv \int \mathcal{D}\tilde{\Pi}\,\mathcal{D}\tilde{\Phi}\;\delta(p_\alpha + P_\alpha)\,\delta(Q_\beta)\,\det\{P,Q\}$$

$$\Psi_f^*[\tilde{\Phi}]\,\Psi_i[\tilde{\Phi}]\,\exp\left\{i\int dt\left(\int dx\,\tilde{\Pi}\dot{\tilde{\Phi}} - \tilde{H}[\tilde{\Pi},\tilde{\Phi},X]\right)\right\}$$

(1.48)

In the adiabatic approximation, the transition probability is given by

$$\int \mathcal{D}X_\alpha\,\mathcal{D}p_\beta\; e^{-i p_{\alpha,f} X_\alpha(t_f)}\; e^{i p_{\beta,i} X_\beta(t_i)}$$

$$\exp\left\{i\int dt\left(p_\alpha \dot{X}_\alpha - H_{eff}\right)\right\}$$

If the P_α are not constants of motion (i.e. $\{P_\alpha, H\} \neq 0$), $\tilde{H}$ depends explicitly on X_α and the dynamics of the collective coordinates is complicated. If on the other hand $\{P_\alpha, H\} = 0$, we have $\tilde{H} = H[\tilde{\Pi},\tilde{\Phi}]$, independent of X and the dynamics of X_α, p_β is trivial. The eigenstates are plane waves. In this case, it is better to do the other way round as we did for the one soliton case, namely, choosing again $\Psi_{i,f}$ to be eigenstates of $P_{\alpha\,op.}$, we first integrate over X_α and p_β, immediately obtaining

$$\prod_\alpha \delta(p_{\alpha,i} - p_{\alpha,f}) \int \mathcal{D}\tilde{\Pi}\,\mathcal{D}\tilde{\Phi}\;\delta(p_{\alpha,i} + P_\alpha)\,\delta(Q_\beta)\,\det\{P,Q\}$$

$$\Psi_f^*\,\Psi_i\;\exp\left\{i\int dt\left(\int dx\,\tilde{\Pi}\dot{\tilde{\Phi}} - H\right)\right\}$$

(1.49)

In order to apply the semi-classical method, we look for the minimum of the action, but now we have constraints. At the classical level, if P_α are constants of motion, they can be given arbitrary values by a suitable choice of boundary conditions. In addition, starting from a classical solution with given p_α, we can generate an infinite set of classical solutions with the same p_α by applying an arbitrary transformation of $\mathcal{G}$ with parameters γ_α.

Hence, we have general classical solutions of the form $\phi_{cl}(x,t;p_\alpha,\gamma_\beta)$ and we can satisfy the constraints by choosing $p_\alpha = p_{\alpha,i}$ and by fixing γ_β such that all the Q's vanish. In order to obtain a consistent perturbation we will let

$$\tilde{\phi} = \phi_{cl}^{m}(x,t;p_{\alpha,i},\gamma_\beta) + \chi \tag{1.50}$$

where ϕ_{cl}^{m}, which is the classical ground state of the sector considered, is the lowest energy classical solution satisfying the constraints. It will, in general, involve no additional parameters besides γ and $p_{\alpha,i}$, so that, because of the constraints, (1.49) is a well defined change of fields and no zeroth mode problem will be encountered in the perturbation theory for χ.

In the more general case when P_α are not constants of motion, semi-classical methods can be applied to (1.48) and again, one will be looking at a minimum of the action with constraints. Now since the constraints do not commute with the Hamiltonian, this problem is totally different from looking at minima of the action alone. In particular, one may find solutions which exist only when the constraints are imposed on the system.

For solitons, only the case of the constant of motion has been used to discuss the scattering of solitons in two dimensions. The collective coordinate approach to soliton scattering has been developed[25] on the example of binary collisions in Sine-Gordon theory[26]. The general case has been investigated following the same method for Sine-Gordon theory[27] and non-linear Schrödinger equation[28].

These works are based on the existence of an infinite set of constants of motion[26] at the classical level. Those are associated with general canonical transformations which mix "positions" and "momenta". Therefore, it is not clear whether they can really be performed at the quantum level and we have no way to determine the additional term ΔV which would arise from a more careful treatment in the one-soliton case. So far, however, only the one-loop approximation has been studied in which we do not expect ΔV to contribute.

Finally, there is a special feature of the soliton quantization in four-dimensional space-time which introduces a new type of collective coordinates. These solutions exist only for Higgs models with a non-trivial unbroken gauge group. They involve massless vector bosons as a consequence. If we look at a soliton monopole solution which has an electric charge[29], it follows from Gauss theorem that the electric field at infinity decreases like r^{-2} . As a result the usual field equations do not follow from the variations of the standard action since one cannot drop the surface term one obtains by partial integration of the kinetic term. This is a very general problem which arises whenever one studies field configurations with non-zero total electric charge. The solution[30] is to introduce new degrees of freedom at infinity such that with a modified action the field equations can really be derived. For the dyon solution one of these degrees of freedom is the collective coordinate conjugate to the charge and charge quantization conditions come out naturally, see ref.(30) for details.

II - INSTANTONS

II-A. The example of quantum mechanics

We first illustrate the general ideas on the example of a symmetric potential with two minima as drawn below

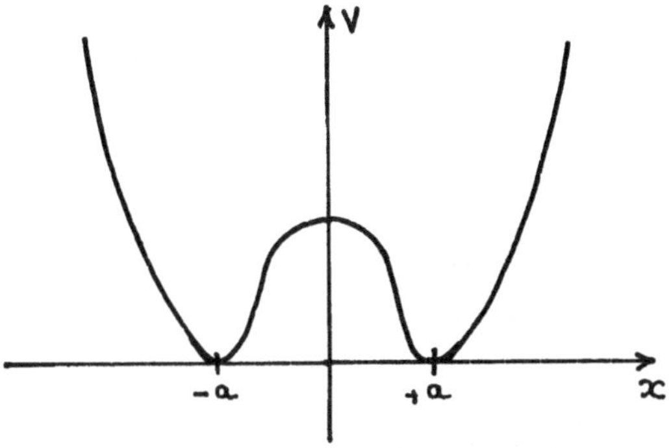

Naively, one would quantize the theory by considering quantum fluctuations around one of the two minima of V. In this way one would conclude that there are two possible ground states for the theory. The usual reasoning would then say that since we must pick up one of the two in order to quantize the theory, we have a spontaneous symmetry breakdown of the invariance by $x \to -x$. Choosing for instance the right minimum, one would let $x = a + \tilde{x}$ and develop perturbation by assuming $\tilde{x} \ll a$. Of course we all know that this is wrong since there is tunneling between states localized around $x = +a$ and states localized around $x = -a$. Hence, it does not make sense to assume that $\tilde{x}$ is small since, if we wait long enough, the important values of $\tilde{x}$ will be precisely $\tilde{x} \sim -2a$. The tunneling actually restores symmetry and the true ground state is in fact a symmetric combination of states localized around $x = +a$ and $x = -a$. If tunneling is a small effect, the two lowest eigenstates are just the symmetric and anti-symmetric combinations of the two ground states we would have if there was no tunneling. The energy difference $\Delta E \sim e^{-c/\hbar}$ is a non-analytic function of $\hbar$ near $\hbar = 0$ since classically, it is impossible to go from $x = a$ to $x = -a$. It is well known that none-theless ΔE can be estimated by semi-classical methods if one uses classical trajectories with imaginary time[31]. Starting from the classical equation with energy E

$$\frac{1}{2}\dot{x}^2 + V(x) = E$$

and letting $t = i\tau$ leads to

$$\frac{1}{2}\left(\frac{dx}{d\tau}\right)^2 - V(x) = -E$$

Hence, V and E have changed signs. The classically forbidden region becomes the classically allowed region. In particular there is a solution with $E=0$, $x_0(\tau)$ such that

$$\tau - b = \int_0^{x_0(\tau)} \frac{dy}{\sqrt{2V(y)}} \tag{2.1}$$

which is the instanton solution of this problem. It contains an arbitrary parameter b which is the "time" when classically we go from $x<0$ to $x>0$.

Standard WKB applied to this problem shows that if we introduce the classical action associated with $x_0(\tau)$

$$I = \int_{-\infty}^{+\infty} d\tau \left[\frac{1}{2}\left(\frac{dx_0}{d\tau}\right)^2 + V(x_0) \right]$$

we have

$$\Delta E \sim e^{-I/\hbar} \tag{2.2}$$

This result can also be obtained by using semi-classical methods in functional integral after continuing to pure imaginary time[31].

If we denote by $|\pm\rangle$ the two ground states we would have if there was no tunneling, eq.(2.2) shows that we cannot consider them as true ground states because $\langle +|H|-\rangle \sim e^{-I/\hbar}$. Until recently it was believed that no such phenomenon could occur in field theory because there is an infinite number of degrees of freedom. The feeling one had was that the transition probability between two ground states

$|\alpha\rangle, |\beta\rangle$ would be essentially given by the product of the probabilities for each degree of freedom leading to $\langle\alpha|H|\beta\rangle \sim e^{-(\infty)/\hbar} = 0$
We now know that this is not correct. If there is a classical solution for imaginary time with finite action S_{cl} which goes from $|\alpha\rangle$ to $|\beta\rangle$ we have instead

$$\langle\alpha|H|\beta\rangle \sim e^{-S_{cl}/\hbar} \tag{2.3}$$

and we are in the same situation as in potential theory.

II-B. Tunneling in field theory

Vacuum to vacuum amplitudes are easily studied in functional methods by looking at the functional integral of the exponential of the action for very large time since this procedure automatically projects out all but the vacuum state matrix elements. This functional integral is continued to imaginary time by using standard arguments and the resulting integral is evaluated by expanding around the minima of the action which are now Euclidean classical solutions(3).

We discuss the example of pure Yang-Mills field theory with internal symmetry group SU(2).

$$\mathcal{L} = -\frac{1}{4}\left(G^a_{\mu\nu}\right)^2$$
$$G^a_{\mu\nu} = \partial_\mu A^a_\nu - \partial_\nu A^a_\mu + g\,\varepsilon_{abc} A^b_\mu A^c_\nu \tag{2.4}$$

Since in Section III we will deal simultaneously with Euclidean and Minskowski quantities we will distinguish Euclidean field theory symbols by putting a hat on them, for instance V_μ means $V_0, \vec{V}$ while $\hat{V}$ means $\vec{V}, V_4$

Because the outcome will be of the form (2.3) we are only interested in classical solutions with finite Euclidean action. Hence for $\hat{x}^2 \equiv x_4^2 + \vec{x}^2 \to \infty$ all solutions should become pure gauges. Any field configuration of this type is classified by the topological charge (Pontryagin index)

$$\hat{Q} = \frac{g^2}{64\pi^2}\,\hat{\varepsilon}_{\mu\nu\rho\sigma} \int \hat{d}_4 x\, \hat{G}^a_{\mu\nu}\, \hat{G}_{\rho\sigma} \tag{2.5}$$

As argued in ref. (4), since obviously

$$\left(\hat{G}^a_{\mu\nu} \pm \frac{1}{2}\, \hat{\varepsilon}_{\mu\nu\rho\sigma}\, \hat{G}_{\rho\sigma} \right)^2 \geqslant 0$$

one has

$$\int \hat{d}_4 x\, \mathcal{L} \geqslant \frac{8\pi^2}{g^2}\nu$$

where ν is the value of $\hat{Q}$. The equality sign is only reached when

$$G^a_{\mu\nu} = \pm \frac{1}{2}\, \hat{\varepsilon}_{\mu\nu\rho\sigma}\, \hat{G}_{\rho\sigma} \tag{2.6}$$

Since it corresponds to the minimum of the action, any solution of (2.6) is also a solution of the Euclidean classical field equation. So far, all instanton solutions which have been exhibited[32] satisfy (2.6).

A particular solution is the one instanton solution of ref. (4)

$$\hat{A}^P_\mu(\hat{x}, \lambda) \equiv \hat{A}^{Pa}_\mu \frac{\tau^a}{2} = \frac{i}{g}\, \omega^{-1} \partial_\mu \omega\, \frac{\hat{x}^2}{\hat{x}^2 + \lambda^2}$$

$$\omega = \frac{x_4 - i\vec{x}\cdot\vec{\tau}}{\sqrt{\hat{x}^2}} \tag{2.7}$$

λ is a scale parameter. It has $\nu = 1$

The tunneling process is most easily seen in the gauge $A^a_4 = 0$ where $\hat{Q}$ - can be written as

$$\hat{Q} = \hat{\rho}(x_4 = +\infty) - \hat{\rho}(x_4 = -\infty)$$

$$\hat{\rho}(x_4) = \frac{g^2}{16\pi^2} \int d_3x \left[A_i^a \partial_j A_k^a + \frac{g}{3} \varepsilon_{abc} A_i^a A_j^b A_k^c \right] \varepsilon_{ijk}$$

(2.8)

Using time independent gauge transformations one may choose $\hat{\rho}(-\infty)$ to be an integer. Then, $\hat{\rho}(+\infty)$ will also be an integer. Since for $x_4 = \pm\infty$ we must be at a classical ground state, we conclude that one has to consider a discrete set of vacua labeled by an integer $|n\rangle$. In functional integral, we must integrate over field configurations with finite action which are also classified by the values of ν. Selecting a particular value ν of the Pontryagin index, we conclude that

$$\langle n+\nu| e^{-HT} |n\rangle \underset{T\to\infty}{\sim} \int (\mathcal{D}\hat{A})_\nu \, e^{-\int \hat{d}_4x \hat{\mathcal{L}}} \quad \text{gauge fixing}$$

(2.9)

This is the argument developed in ref. (33) where the functional integral on the right-hand side was estimated by expanding around instanton classical solutions. The result is typically of the form (2.3) and shows that tunneling indeed occurs between the different ground states.

In this calculation, collective coordinates must be introduced for instanton sizes and positions. For instance the most general one-instanton solution is $\hat{A}^{p}(\hat{x}-\hat{X},\lambda)$ where $\hat{X},\lambda$ are the instanton position and size and we must avoid the zeroth mode associated with small variations of $\hat{X}$ and λ. For the ground states however, this simply corresponds to a small gauge transformation so that the momenta conjugate to $\hat{X}$ and λ are always zero. Thus, $\hat{X}$ and λ are not treated as true quantum mechanical operators defined for each time. They are rather numbers extracted from the field integrated over all space-time as in formula (1.34).

In general, one is led to the study of the equivalent of a partition function of statistical mechanics where g is replaced by the temperature[3,34]. We illustrate the role of instantons on one-dimensional Ising model which corresponds to tunneling in quantum mechanics.

At $T=0$ there are two ground states

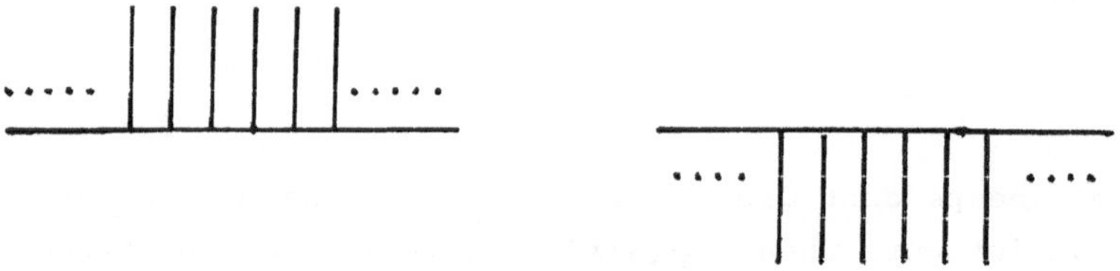

Let us pick up the one on the left. For small temperature we have "instanton" configurations

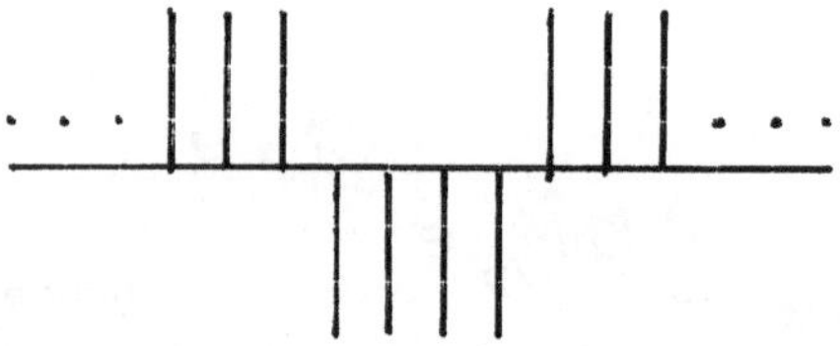

they have a higher energy than the ground state but there are many more such configurations since the flipping of spin can occur anywhere, any even number of times. In the partition function

$$Z = \sum_{\text{configurations}} e^{-E/T}$$

the instanton configurations in fact dominate the vacuum contribution. Indeed, it is well known that the dominant contributions to Z are those with lowest free energy $F = E - TS$, $S \sim \ln$ (nb of states).

Thus instantons destroy the long range order of the vacuum completely in this case (it is well known that there is no phase transition in one dimension). The ground state becomes symmetric between up and down in agreement with the well-known fact that in quantum mechanics the ground state is always symmetric as we recalled it in sect. (II.A).

In general, Goldstone bosons associated with spontaneous breakdown of a continuous symmetry reflect the existence of fluctuations around ground states with very long wavelength. By destroying long-range order in the vacuum state, instantons may therefore prevent Goldstone bosons from appearing. Indeed, evidences have been given[37] that, in this way, the axial $U(1)$ symmetry of quark model could be spontaneously broken without predicting a ninth axial boson with a small mass thus solving the so-called η problem.

Since, they can spoil long-range order, instantons may lead to quark confinement in the sense of Wilson, namely, the expectation value $\langle 0| exp(-\oint dl_\mu A^\mu)|0\rangle$ for a large closed loop mass decreases like the exponential of the area enclosed by the loop[3,34].

Finally, by tunneling from the vacuum, pairs of fermions can be created leading to various possibilities of violation for baryon conservation laws[37].

The general problem of this method, when applied to pure Yang-Mills field theory, is that it is a small coupling approximation in a theory where the effective coupling constant depends on the scale considered. The final results involve an integration over instanton size which is out of control for large instantons where this effective coupling constant is not small[33,35] so that the approximation presumably breaks down.

III - WKB WAVE FUNCTION

So far, we mostly discussed methods based on Feynman path integral quantization. This type of approach was originally developed in potential theory(36) in order to obtain WKB results without introducing wave function explicitly. The motivation was that WKB wave functions for more than one degree of freedom were thought to be too complicated to be handled explicitly. Moreover, the path integral formalism leads more easily to Feynman rules as we developed for one soliton case.

In general, however, for reason of complexity one will not be able to really go beyond the first quantum correction and lowest order WKB will be a sufficient approximation.

Moreover, one cannot always avoid the use of wave functions. For instance, in the semi-classical treatment of soliton scattering by path integrals(25), eigenstates of Hamiltonian had to be built so as to establish scattering formalism, although the method used there is only formal because the eigenstates so obtained involve the momentum of the field. The use of wave functions seems also unavoidable if one wants to discuss matching at turning points. This problem arises especially in connection with the semi-classical approach to vacuum tunneling in field theory which we discussed in Sect. II.

In two remarkable papers(37), Banks, Bender and Wu have studied this last question in particular systems with two degrees of freedom. Their basic idea is that barrier penetration occurs mostly in small tubes in configuration space around certain classical solutions so that WKB is essentially one dimensional and they could determine the ground state wave function. These classical paths correspond to classical solutions with pure imaginary time and are the instanton solutions we discussed above.

Up to now, vacuum tunneling in field theory has been studied mostly by path integrals in Euclidean space-time. In recent preprints(38) we proposed an interpretation of Euclidean classical solutions in Minkowski space-time which, as we later realized, is the generalization to field theory of the ideas of Banks, Bender and Wu. In Euclidean field theory the problem of matching is avoided through the gas approximation(33) recalled above. As I shall argue later on, this is not satisfactory for theories with no mass scale and one seems to be forced to really handle the matching at turning points in field theory.

In this section, I will show that, contrary to the common belief, general WKB eigenfunctions, to first two orders in $\hbar$ and for a given classical trajectory, are rather simple objects which can be systematically written down once one has solved the classical problem of small fluctuations around the classical trajectory considered[38]. We should remark here that if one applies this WKB method to the soliton problems, the basic formalism is similar to that of Christ and Lee[10].

In order to simplify the writing, I will use the language of quantum mechanics though I also have in mind field theory implications which will be briefly discussed in sect. III-B.

III-A. Determination of WKB Wave Function

Let us consider a system with N degrees of freedom. We denote the generalized coordinates by $\vec{R}$ and potential by $V(\vec{R})$. The Lagrangian of the system is assumed to be $\mathcal{L} = \frac{1}{2}\dot{\vec{R}}^2 - V(\vec{R})$. Field theory can be regarded formally as a system of infinite degrees of freedom $N \to \infty$. The simplest way to deal with $\hbar$ expansion is to introduce a parameter g such that V can be written as

$$V(\vec{R}) = \frac{1}{g^2}\mathcal{V}(g\vec{R}) \tag{3.1}$$

where $\mathcal{V}$ does not depend on g. This means that in V, the nth power of $\vec{R}$ has a coefficient proportional to g^{n-2}. From (3.1) one sees that any classical solution and classical action is, respectively, proportional to g^{-1} and g^{-2}. Letting $\hbar = 1$, we thus see that g^2 plays the role of $\hbar$ and semi-classical approximation will mean expansion in g.

As it is well known in the leading order in WKB, Schrödinger equation reduces to Hamiltonian-Jacobi equation. Namely, if we let

$$H\psi = E\psi \qquad\qquad \psi = e^{i\epsilon W}$$

we get, to leading order,

$$\frac{\epsilon^2}{2}\left(\frac{\partial W}{\partial R_i}\right)^2 + V(\vec{R}) = E \tag{3.2}$$

Obviously, we can choose $\varepsilon^2 = 1$ if $E > V$ (classically allowed region) and $\varepsilon^2 = -1$ if $E < V$ (classically forbidden region). Equation (3.2) is the Hamiltonian-Jacobi equation with potential $\varepsilon^2 V$ and energy $\varepsilon^2 E$. By the standard method[39] solutions of (3.2) are obtained as

$$W(\vec{R}) = \int^{s_1} ds \sqrt{2\left| E - V(\vec{x}(s)) \right|}$$

$$\left(\frac{d\vec{x}}{ds}\right)^2 = 1 \qquad \vec{x}(s_1) = \vec{R} \tag{3.3}$$

where the curve $\vec{x}$ is such that the integral is stationary. The classical meaning of $\vec{x}$ is best shown by introducing another parametrization denoted by τ such that $(\vec{x}_\tau \equiv d\vec{x}/d\tau)$

$$\frac{1}{2\varepsilon^2} \vec{x}_\tau^{\,2} + V(\vec{x}) = E \tag{3.4}$$

Then $\vec{x}$ should satisfy

$$\frac{1}{\varepsilon^2} \vec{x}_{\tau\tau} = - \vec{\nabla} V(\vec{x}) \tag{3.5}$$

Hence, $\varepsilon\tau$ plays the role of time for a classical trajectory with energy E and potential energy V. In the forbidden region $\varepsilon\tau$ is purely imaginary. Note that ψ is a stationary state and it is thus clear that τ has nothing to do with the true time.

In order to obtain the functional form of $W(\vec{R})$ by (3.3) we must know the general solution of (3.4) and (3.5), i.e. the trajectory passing through a point $\vec{R}_0$ and an arbitrary point $\vec{R}$ in configuration space. In practice, however, especially in field theories, one knows explicitly only a limited class of classical trajectories. So we assume that only a classical solution $\vec{x}(\tau)$ with energy E_0 is given, and consider the wave function in the vicinity of the classical trajectory in configuration space.

Then (3.3.) is simply the WKB exponent for one degree of

freedom which is the position along the trajectory. Hence the dominant effect due to the existence of a classical trajectory $\vec{\pi}(\tau)$ is contained into the quantum mechanics of this degree of freedom. Following our general method recalled above we introduce it as a collective coordinate, that is, we extract it out of $\vec{R}$ through the change of variable.

$$\vec{R} = \vec{\pi}(f(q)) + \sum_{a=2}^{N} \vec{n}_a(f(q))\eta^a \tag{3.6}$$

f is an arbitrary given function which fixes the parametrization of the curve, q is the new coordinate which indicates the position on the curve, $\vec{n}_a(\tau)$ together with $\vec{\pi}_\tau(\tau)$ form a moving local reference frame at the point $\vec{\pi}(\tau)$. We choose it such that

$$\vec{n}_a(\tau).\vec{n}_b(\tau) = \delta_{ab} \qquad \vec{n}_a(\tau)\vec{\pi}_\tau(\tau) = 0$$

At this point it is useful to note that indeed (3.6) is analogous to (1.23)-(1.24) if we replace the soliton position X by the collective coordinate q.

Equation (3.6) can actually only represent a small neighbourhood of the classical curve in configuration space because the vectors $\vec{\pi}_\tau, \vec{n}_a$ form only a local reference frame. Consistency will be achieved at the end when we will obtain the wave function which decreases away from the classical trajectory with an exponential decrease of order g^0. Indeed, if this is verified, the relevent values of η^a are such that $|\eta^a| \ll |\vec{\pi}|$ because $\vec{\pi}(\tau)$, being a classical solution, is of order g^{-1}. We shall come back to this point later on. For the reader who is more familiar with $\hbar$ expansion we note that $\vec{\pi}$ and η^a are of order $\hbar^0$ and $\hbar^{1/2}$, respectively, so that the same picture also emerges.

A straightforward computation shows that

$$\vec{P} \equiv -i\vec{\nabla} = \frac{\vec{\pi}_\tau}{f'(q)[\vec{\pi}_\tau^2 - \vec{\pi}_{\tau\tau}\cdot\vec{\eta}]}\left(p - f'(q)\sum_{a,b}\eta^a T_a^b \zeta^b\right) + \sum_a \vec{n}_a \zeta^a \tag{3.8}$$

$$p \equiv -i\frac{\partial}{\partial q} \qquad \zeta^a = -i\frac{\partial}{\partial \eta^a} \qquad \Gamma^b_{\ a} = \vec{n}_b \cdot \vec{n}_{a\tau}$$

$$\vec{\eta} = \sum_a \eta^a \vec{n}_a \qquad f'(q) = \frac{df}{dq}$$

(3.9)

In the above expression $\vec{\pi}$ and $\vec{n}_a$ are to be considered for $\tau = f(q)$, τ indices mean taking derivatives with respect to τ before replacing τ by $f(q)$. We use the same conventions hereafter.

We insert (3.6) and (3.8) into the Hamiltonian and expand in power of g. To the order we are working, i.e. g^0 order, the ordering of operators is irrelevent. One gets

$$H = \frac{p^2}{2f'^2\vec{\pi}_\tau^2} + V^{(0)}(q) + \frac{p^2}{f'^2(\vec{\pi}_\tau^2)^2}\vec{\pi}_{\tau\tau}\cdot\vec{\eta} + V^{(1)}_a\eta^a + \frac{1}{2}(\zeta^a)^2$$
$$+ \frac{1}{2}V^{(2)}_{ab}\eta^a\eta^b - \frac{p\,\Gamma^b_{\ a}\eta^a\zeta^b}{f'\vec{\pi}_\tau^2} + \frac{3}{2}\frac{p^2}{f'^2}\frac{(\vec{\pi}_{\tau\tau}\cdot\vec{\eta})^2}{(\vec{\pi}_\tau^2)^3}$$

(3.10)

We have expanded the potential.

$$V(\vec{R}) = V^{(0)}(q) + \eta^a V^{(1)}_a(q) + \frac{1}{2}\eta^a\eta^b V^{(2)}_{ab}(q) + \cdots$$

(3.11)

It is easy to see that $V^{(m)}(q) = O(g^{m-2})$

Let us now solve the Schrödinger equation to first two leading orders by letting

$$H\Psi = (E_0 + E_1)\Psi \qquad E_0 = O(g^{-2}) \qquad E_1 = O(g^0) \tag{3.12}$$

$$\Psi(q,\vec{\eta}) = e^{i\varepsilon S_0}\,\tilde{\chi}(q,\vec{\eta}) \qquad S_0 = O(g^{-2}) \qquad \tilde{\chi} = O(g^0) \tag{3.13}$$

The Schrödinger equation to orders g^{-n}, $n = 4, 2, 1, 0$, lead, respectively, to the equations

$$\frac{\partial S_0}{\partial \eta^a} = 0 \tag{3.14a}$$

$$\left(\frac{\partial S_0}{\partial q}\right)^2 = 2 f'^2 \vec{\pi}_\tau^{\,2}\, |E_0 - V^{(0)}| \tag{3.14b}$$

$$(\varepsilon \vec{\pi}_{\tau\tau} \cdot \vec{m}_a + V_a^{(1)})\eta^a = 0 \tag{3.14c}$$

$$\left\{-\frac{1}{2}\frac{\partial^2}{\partial \eta^{a2}} - i\varepsilon\left(\frac{1}{f'}\frac{\partial}{\partial q} - \Gamma^b_a \eta^a \frac{\partial}{\partial \eta^b}\right) + \frac{1}{2} W_{ab}\eta^a\eta^b - i\frac{\varepsilon}{2f'}\left(\frac{\partial}{\partial q}\ln(f'\vec{\pi}_\tau^{\,2})\right) - E_1\right\}\tilde{\chi} = 0 \tag{3.14d}$$

$$W_{ab} = V^{(2)}_{ab} + \frac{3\varepsilon^2}{\vec{\pi}_\tau^{\,2}}\pi^a_{\tau\tau}\pi^b_{\tau\tau} \qquad \pi^a_{\tau\tau} \equiv \vec{\pi}_{\tau\tau}\cdot\vec{m}_a \tag{3.15}$$

Equation (3.14b) is as expected the leading WKB equation for q degree of freedom and we get

$$S_0(q) = \int^q dq' \sqrt{(2f'^2\vec{\pi}_\tau^{\,2})\,|E_0 - V^0|} \tag{3.16}$$

It is readily checked to be of order g^{-2} if f is of order g^0

Next projecting equation (3.5) onto the vectors $\vec{m}_a(\tau)$ $a=2,\dots N.$, one sees that (3.14c) is indeed satisfied since $\vec{\pi}(\tau)$ is a classical solution.

Our task is now to solve equation (3.14d). For this we first remark that as one could have expected, we only have the combination $\frac{1}{f'}\frac{\partial}{\partial q}$ so that it is simpler to reexpress $\tilde{\chi}$ as a function of τ redefined by $\tau = f(q)$ in any region where f is single valued. Note that f should be chosen such that f' is always non-vanishing and we assume f' to be positive. The next to last term in (3.14d) corresponds to the standard WKB factor of order zero in g for q quantum mechanics. It goes away if we redefine $\tilde{\chi}$ as

$$\tilde{\chi} = \frac{\chi}{\sqrt{f' \vec{\pi}_\tau^2}} = \frac{\chi}{\sqrt{S_0'}} \tag{3.17}$$

and we have to solve the equation

$$\mathcal{H}\chi = E_1 \chi \tag{3.18}$$

$$\mathcal{H} \equiv -i\epsilon\left(\frac{\partial}{\partial \tau} - \Gamma_a^b \eta^a \frac{\partial}{\partial \eta^b}\right) - \frac{1}{2}\frac{\partial^2}{\partial \eta^{a2}} + \frac{1}{2} W_{ab}\eta^a \eta^b \tag{3.19}$$

This is a non - trivial problem since both Γ and W are functions of τ . The crucial point of our method is that (3.18) can be solved if one knows a complete set of solutions for the equation of small fluctuations around $\vec{\pi}(\tau)$. Denote such a solution by $\vec{v}$ From (3.5) it satisfies

$$\frac{1}{\epsilon^2} v^i_{\tau\tau} = \left.\frac{\partial^2 V}{\partial R^i \partial R^j}\right|_{\vec{R}=\vec{\pi}(\tau)} v^j \tag{3.20}$$

We shall assume that the matrix $\partial^2 V/\partial R^i \partial R^j |_{\vec{\pi}}$ is positive definite. Hence, (3.20) will have solutions with real exponential (oscillating exponential) behaviour for $\varepsilon^2 = -1$ $(\varepsilon^2 = +1)$ Expand $\vec{v}$ in the moving frame by

$$\vec{v} = (\vec{v}\cdot\vec{\pi}_\tau)\frac{\vec{\pi}_\tau}{\vec{\pi}_\tau^{\,2}} + \sum_a u_a \vec{n}_a \tag{3.21}$$

Taking the derivative of (3.5) with respect to τ, one sees that $\vec{\pi}_\tau$ is also a solution of (3.20). From Wronskien argument one gets

$$\frac{d}{d\tau}\left(\vec{\pi}_\tau \vec{v}_\tau - \vec{\pi}_{\tau\tau}\vec{v}\right) = 0$$

So we can choose $\vec{v}$ such that

$$\vec{\pi}_\tau \vec{v}_\tau = \vec{\pi}_{\tau\tau}\cdot\vec{v} \tag{3.22}$$

From this one can check that (3.20) implies for u_a the equation

$$D_{ab} D_{bc} u_c + \varepsilon^2 W_{ab} u_b = 0 \tag{3.23}$$

$$D_{ab} \equiv \delta_{ab}\frac{\partial}{\partial\tau} + T^b_a \tag{3.24}$$

The method of solving (3.18) is based on the remark that if u satisfies (3.23), the operator

$$A = e^{-i\nu\tau/\varepsilon}\left[u_a \frac{\partial}{\partial\eta^a} - i\varepsilon\,(Du)_a \eta^a\right] \tag{3.25}$$

is such that

$$[\mathcal{H}, A] = -\nu A \tag{3.26}$$

Hence, if ν is positive (negative) A acts as a destruction (creation) operator on the eigenfunctions of $\mathcal{H}$. If ν is zero, A is not interpretable in terms of creation-annihilation operator as it commutes with $\mathcal{H}$. This will be related to the zeroth mode phenomenon we discussed in sect. II, which is linked to symmetry properties of V .

The set of ν and operators A which can appear will be specified by the boundary conditions of the region of configuration space considered. We shall illustrate this point with two specific examples : periodic orbit in allowed region, and penetration problem for quantum fluctuations around a local minimum of V . The later example is obviously relevant for vacuum tunneling. We shall not discuss a true WKB matching at a turning point since we have not yet studied this problem in detail. In order to simplify the discussion, we shall further assume that none of the ν's encountered vanishes. Some comments on the general case are given at the end of this section.

In the first case we will have

$$\vec{\pi}(\tau) = \vec{\pi}(\tau + T) \tag{3.27}$$

and A should be periodic of period T so that both functions (here $\varepsilon = \pm 1$)

$$u\, e^{-i\nu\tau/\varepsilon} \equiv g \qquad\qquad (Du)\, e^{-i\nu\tau/\varepsilon} \equiv f \tag{3.28}$$

must be periodic with period T . At this point, the discussion proceeds along lines similar to ref. (25). From (3.23) and (3.28), we see that ν is such that

$$\mathcal{B}\begin{pmatrix} g \\ f \end{pmatrix} = \nu \begin{pmatrix} g \\ f \end{pmatrix}$$

$$\mathcal{B} \equiv i \begin{pmatrix} \varepsilon D_{ab} & -\delta_{ab} \\ W_{ab} & \varepsilon D_{ab} \end{pmatrix} \tag{3.29}$$

The periodicity condition makes $\mathcal{B}$ hermitian with inner product

$$(2|1) \equiv \int_0^T d\tau \, (g_2^*, f_2^*) \, \sigma_2 \begin{pmatrix} g_1 \\ f_1 \end{pmatrix}$$

(3.30)

and the ν's are the set of eigenvalues of the operator $\mathcal{B}$. They are necessarily real since $\mathcal{B}$ is hermitian. Since $\mathcal{B}$ is purely imaginary we see that if $\binom{g}{f}$ is eigenvector of $\mathcal{B}$ with eigenvalue ν $\binom{g^*}{f^*}$ is also eigenvector but with eigenvalue $-\nu$. Let ν_m be the set of all positive ν's and u^m be the set of corresponding small fluctuations. We define

$$A_m \equiv e^{-i\nu_m \tau/\varepsilon} \left[u^m_a \frac{\partial}{\partial \eta^a} - i\varepsilon (Du^m)_a \eta^a \right]$$

(3.31)

From the hermiticity of $\mathcal{B}$ it is straightforward to check that if we normalize u^m by

$$i\varepsilon\left((Du^{m*})_a u^m_a - u^{m*}_a (Du^m)_a \right) = 1$$

(3.32)

we have

$$[A_m, A_n] = [A_m^+, A_n^+] = 0 \qquad [A_m, A_n^+] = \delta_{mn}$$

(3.33)

Equation (3.23) has $2N-2$ independent solutions so we get $N-1$ creation-annihilation operators.

Next, we discuss penetration problem $(\varepsilon = \pm i)$ for quantum fluctuations around a local minimum of V. We choose E_0 to be equal to the value of V at the minimum. Since $\varepsilon^2 = -1$ the classical trajectory corresponds to a maximum of potential energy. If the potential is harmonic near its minimum, it takes an infinite τ interval to reach the stability point. For definiteness we choose the corresponding limit to be $\tau \to -\infty$. S_0 is an integral with a fixed lower bound so it tends to $-\infty$ in the limit. Near the minimum the term involving e^{-S_0} is a decreasing function of the distance to the stability point. It must be of order 1 as it will be matched to the oscillator wave function of small oscillation near the minimum which has the same behaviour. On the contrary, the term involving e^{S_0} is an increasing function of the distance to the stability point. It will be matched to an exponentially small component of the wave function of small oscillations which has similar behaviour and which appears in solving the Schrödinger equation near the minimum because, due to tunneling, the energies differ from exact harmonic oscillator energies. We shall only discuss the matching of e^{-S_0} term here.

Introduce

$$\Lambda_{ab} = \lim_{\tau \to -\infty} V_{ab}^{(2)}$$

We can then obtain a set of solutions of (3.23) such that

$$u_a^{m\,(\pm)} \to f_a^m \, e^{\pm \omega_m \tau}$$

$$\Lambda_{ab} f_b^m = \omega_m^2 f_a^m \qquad\qquad f_a^m f_a^n = \frac{\delta_{mn}}{2\omega_n} \tag{3.34}$$

The creation-annihilation operators will be defined by

$$A_m = e^{-\omega_m \tau} \left[u_a^{m(+)} \frac{\partial}{\partial \eta^a} + (Du^{m(+)})_a \eta^a \right]$$

$$\tilde{A}_m = - e^{-\omega_m \tau} \left[u_a^{m(-)} \frac{\partial}{\partial \eta^a} + (Du^{m(-)})_a \eta^a \right] \tag{3.35}$$

Indeed, using the Wronskien together with (3.34) one can show that

$$[A_m, A_n] = [\tilde{A}_m, \tilde{A}_n] = 0 \qquad [A_n, \tilde{A}_m] = \delta_{nm} \tag{3.36}$$

From (3.34) it follows that

$$A_m \xrightarrow[\tau \to -\infty]{} f_a^m \left[\frac{\partial}{\partial \eta^a} + \omega_m \eta^a \right]$$

$$\tilde{A}_m \xrightarrow[\tau \to -\infty]{} f_a^m \left[-\frac{\partial}{\partial \eta^a} + \omega_m \eta^a \right] \tag{3.37}$$

Hence, $A_m, \tilde{A}_m$ tend to the creation-annihilation operators of quantum fluctuations around the local minimum of V. The WKB eigenstates will thus match to the eigenstates of the allowed region.

Finally, going back to the general discussion $(\varepsilon^2 = \pm 1)$ we determine the ground state wave function. For both signs of ε^2 we have $N-1$ annihilation operators of the form

$$A_m = e^{-i\nu_m \tau/\varepsilon} \left[U_a^m \frac{\partial}{\partial \eta^a} - i\varepsilon (DU^m)_a \eta^a \right] \tag{3.38}$$

The ground state χ_0 is the solution of

$$A_m \chi_0 = 0 \qquad m = 2, \ldots, N \tag{3.39}$$

which gives[41]

$$\chi_0 = d(\tau)\, e^{-\frac{1}{2}\Omega_{ab}(\tau)\eta^a \eta^b}$$

$$\Omega_{ab}(\tau) \equiv -i\varepsilon \sum_m U^{-1}{}_a^m (DU^m)_b \tag{3.40}$$

where U^{-1} is such that

$$\sum_m U^{-1}{}_a^m U_b^m = \delta_{ab} \tag{3.41}$$

The matrix exists because the U_a^m, $m = 2, \ldots N$, are linearly independent vectors in the $N-1$ dimensional space orthogonal to $\vec{\pi}_\tau$. It is furthermore easy to check that Ω is a symmetric matrix.

Equation (3.39) determines χ_0 up to an arbitrary function $d(\tau)$ which we compute by inserting (3.40) into the equation $\mathcal{H}\chi_0 = E_1^0 \chi_0$. Combining everything we finally obtain the ground state wave function

$$\psi_0 = \frac{e^{i\epsilon S_0}}{\sqrt{f' \vec{\pi}_\tau^2}} \frac{e^{iE_1^0 \tau/\epsilon}}{\sqrt{\det U}} e^{-\frac{1}{2}\Omega_{ab}\eta^a\eta^b} \tag{3.42}$$

This is the generalization of the formula obtained by Banks, Bender and Wu[37]. The excited state wave functions are obtained by applying the creation operators to ψ_0. They will involve, in addition the standard Hermite polynomials.

As we explained earlier, our method makes sense if ψ vanishes rapidly away from the classical path. This will be the case if $\mathrm{Re}\,\Omega$ is a positive definite matrix. In the allowed region this condition is readily checked to hold using the orthogonality relations due to the hermiticity of $\mathcal{B}$ in our previous example. The normalization condition (3.32) gives

$$2\,\mathrm{Re}\,\Omega_{ab} = \sum_m U^{-1}{}_a^{m*}\, U^{-1}{}_b^m \tag{3.43}$$

which is indeed positive definite. In the case of forbidden region, we have no such proof and indeed Ω is not always positive definite. However, if V varies very slowly and if the classical path is very close to a straight line, one has

$$\Omega_{ab} \simeq \sum_m U^{-1}{}_a^m\, \nu^m U_a^m \tag{3.44}$$

which is positive definite since $\nu^m > 0$. The condition of positive definiteness will hold whenever Ω does not differ drastically from (3.44).

Finally, we comment on the case where some of the ν's are zero. This will always be the case if the potential has a symmetry so that in configuration space we have a continuous set of classical solutions with the same classical action. Then the Hamiltonian commutes with the corresponding infinitesimal generators, and this was reflected in the fact that the operators with $\nu = 0$ commute with $\mathcal{H}$. This is the zeroth mode problem of semi-classical methods which can be solved by introducing collective coordinates as discussed above. In the present context, it is simply equivalent to performing the standard separation of variable in Schrödinger equation for symmetric potential before applying WKB method and we shall not elaborate upon it here.

III-B. Applications

This result obviously has many applications in various potential and field theory problems. We briefly indicate two of them which are related to our two examples of Sec. III.

1) Quantization condition for periodic solutions :

As we discussed above, a state is characterized by the occupations numbers n_i $i=2,\dots,N$. (Again we assume that none of the ν's vanishes). The wave function must be periodic with period T The creation-annihilation operators were chosen to be periodic in such a way that the quantization condition does not depend on n_i but only on the particular classical trajectory considered. As we have seen before, $u^m e^{-i\nu_m \tau/\varepsilon}$ is periodic of period T. From this we conclude

$$\det\left[\mathcal{U}(\tau+T)\right] = \det\left[\mathcal{U}(\tau)\right] e^{i\sum_n \nu^n T/\varepsilon} \tag{3.45}$$

Formula (3.42) leads to

$$\Psi_0(\tau+T) = \Psi_0(\tau) \exp\left\{ i\left(\varepsilon W(E_0) + \frac{E_1^0 T}{\varepsilon} - \sum_n \nu^n \frac{T}{2\varepsilon}\right)\right\}$$

where $W(E)$ is the action integral over a period

$$W(E) \equiv \oint ds \sqrt{2(E-V)} \tag{3.46}$$

The quantization condition reads (here $\varepsilon = \pm 1$)

$$W(E_o) + T\left(E_1^\circ - \sum_m \frac{\nu^m}{2}\right) = 2m\pi \qquad m \text{ integer}$$

This can be put into the same form as in ref.(2) since for any level

$$E_1 = E_1^\circ + \sum_i n_i \nu_i$$

$$W(E) = W(E_o) + E_1 T$$

and we get

$$W(E) = 2m\pi + \sum_i \left(n_i + \frac{1}{2}\right)\nu_i T \tag{3.47}$$

which agrees with the result of ref.(2).

2) Vacuum tunneling in field theories :

It will be associated with solutions with $\varepsilon^2 = -1$ (forbidden region) such that E_o is the energy of the classical vacuum. In Lorentz invariant field theories we are thus led to consider classical solutions of Euclidean field equations according to eq. (3.5), i.e. instanton solutions. We shall briefly discuss two typical examples.

Example 1 : Two-dimensional Higgs model

The Lagrangian is given by

$$\mathcal{L} = -\frac{1}{4}\left(\partial_\mu A_\nu - \partial_\nu A_\mu\right)^2 - \left|(\partial_\mu - ieA_\mu)\phi\right|^2 + \frac{1}{2g^2}\left(1 - g^2|\phi|^2\right)^2 \tag{3.48}$$

where ϕ is a complex scalar field. The Euclidean solutions are the vortex solutions $\hat{A}_{cl}, \hat{\Phi}_{cl}$. (From now on we use again a hat symbol on any quantity which is relative to Euclidean space time, so as to distinguish from Minkowski quantities since we shall handle both at the same time.) The vortex solutions are classified by the topological index (magnetic flux number)

$$\hat{Q} = \frac{e}{2\pi} \int \hat{d}_2 x \left[\partial_4 \hat{A}_1 - \partial_1 \hat{A}_4 \right] \tag{3.49}$$

Our other example (example II) will be the SU_2 Yang-Mills model described in sec. II-B.
Throughout the discussion we choose the $A_0 = 0$ gauge so the classical solution will be considered in the $\hat{A}_4 = 0$ gauge.

As an example we shall look only at the quantum meaning of one instanton (one vortex) solutions. In example II, one can check in $A_4 = 0$ gauge, that this solution of ref. (4) takes the form

$$\hat{A}_{cl\,j}(\vec{x}, x_4) \equiv \hat{A}^a_{cl\,j} \frac{\tau^a}{2} = O \hat{A}^P_j O^{-1} - \frac{i}{g} (\partial_j O) O^{-1} \qquad j = 1,2,3.$$

$$O = \exp\left\{ \frac{i \vec{x}.\vec{\tau}}{\sqrt{\vec{x}^2 + \lambda^2}} \left[\tan^{-1}\left(\frac{x_4}{\sqrt{\vec{x}^2 + \lambda^2}} \right) - \frac{\pi}{2} \right] \right\} \tag{3.50}$$

where $\hat{A}^P_\mu$ is given by (2.7).
The gauge condition $\hat{A}_4 = 0$ does not break gauge invariance by a time independent gauge transformation. As a result, in (3.50), we could have replaced the term $-\pi/2$ by an arbitrary function of $\vec{x}$ and there is an arbitrariness in the definition of $\tan^{-1}$. We choose it such that $\tan^{-1}(-\infty) = (n + \frac{1}{2})\pi$ and call $\hat{A}^n_{cl}$ the function so obtained. One can see that

$$\hat{A}^n_{cl\,j} \underset{\tau \to -\infty}{\simeq} \frac{i}{g} K^n \nabla_j (K^{-n}) \equiv \hat{A}^n_{v\,j}$$

$$\hat{A}^n_{cl\,j} \underset{\tau \to +\infty}{\simeq} \hat{A}^{n+1}_{v\,j} \qquad K = \exp\left(\frac{i\pi \vec{x}.\vec{\tau}}{\sqrt{\vec{x}^2 + \lambda^2}} \right) \tag{3.51}$$

As expected, for $\tau \to \pm\infty$ we go to the allowed region and $\hat{A}_{\mu}$ tend to a pure gauge term that is to a classical ground state of the theory. This is the situation of our second example of Sec. III-A. These ground states are related by the gauge transformation K successively. The same discussion can be carried out in example I.

From our general discussion, we know that the field theory eigenstate of Hamiltonian will be nonzero only in a neighbourhood of the classical path in configuration space parametrized by the new quantum variable q through $x_4 = f(q)$

In the present examples one gets a better insight by choosing f such that q coincides with the variable $\hat{\varrho}$ introduced in (2.8) for example II. In $\hat{A}_4 = 0$ gauge, one has

$$\hat{Q} = \int_{-\infty}^{+\infty} dx_4 \frac{\partial}{\partial x_4} \hat{\rho} = \hat{\rho}(+\infty) - \hat{\rho}(-\infty)$$

(3.52)

$$\hat{\rho} = \frac{e}{2\pi} \int dx_1 \hat{A}_1 \qquad \text{(example I)}$$

$$\hat{\rho} = \frac{g^2}{16\pi^2} \int d_3 x \left[\hat{A}_i^a \partial_j \hat{A}_k^a + \frac{g}{3} \varepsilon_{abc} \hat{A}_i^a A_j^b A_k^c \right] \varepsilon_{ijk}$$

(example II)

(3.53)

We can choose f from the equation

$$q = \hat{\rho} \left[\hat{A}_{\mu}^{m} (\vec{x}, f(q)) \right]$$

(3.54)

in the interval of q where one has a solution. Differentiating (3.12) we get

$$1 = \frac{df}{dq} h(q)$$

$$h(q) = \frac{e}{2\pi} \int_{-\infty}^{+\infty} dx_1 \partial_4 \hat{A}_{\mu 1}^{n} \qquad \text{(example I)}$$

$$h(q) = \frac{g^2}{16\pi^2} \int d_3 \vec{x} \, (\partial_4 \hat{A}_{\mu i}^{m}) \hat{G}_{\mu jk}^{a\, n}$$

(example II) (3.55)

one can check that h does not depend on n as it is invariant by time independent gauge transformations. Equation (3.55) can be rewritten as

$$\frac{1}{2}\left(\frac{df}{dq}\right)^2 - \frac{1}{2h^2} = 0 \qquad f'h > 0$$

(3.56)

We have a mechanical analogue to a "point" with "position" f "time" q, zero "energy" and "potential" $u = -1/2h^2$. Since $u<0$ its "velocity" never vanishes and it always moves toward the right or left depending on the sign of h. For an instanton $h>0$ so $f'>0$, we look at this case as a specific example. From equation (3.10), one finds

$$\int_{-\infty}^{+\infty} \frac{df}{\sqrt{-2u}} = \hat{Q}$$

(3.57)

The Pontryagin index corresponds to the "time" Δq required by the "point" to move from $f=-\infty$ to $f=+\infty$. Thus $\Delta q = 1$ is the interval where (3.54) can be used for given n. Patching together the results obtained from (3.54) for all values of n one defines f for all values of q. It is found to be periodic of period 1 due to K gauge transformations and to be such that for $\sigma \to 0$, $f(n-\sigma) \to -\infty$, $f(n+\sigma) \to +\infty$, $\sigma > 0$ n integer. The classical path is finally given by $A_{cl}(\vec{x}, f(q))$ which is defined for arbitrary q by

$$\hat{A}_{cl}(\vec{x}, f(q)) = \hat{A}^n_{cl}(\vec{x}, f(q)) \qquad n \leq q \leq n+1$$

so that $q \to q+1$ is equivalent to a K gauge transformation on $\hat{A}_{cl}$.

Example I is similar and we end up with trajectories in configuration space which are periodic up to a gauge transformation K. Because our theory must be gauge invariant the state described by Ψ must satisfy this property. Since a fixed phase factor in a wave function is unobservable we can have in general

$$\Psi(q+1) = e^{i\theta}\Psi(q)$$

where θ is an arbitrary angle. In this way one finds very naturally the degeneracy of the vacuum[33, 42]. Moreover, since we have the excited state function we can study the spectrum of excitations of the theory which is the physically relevant problem.

Because $q \to q+1$ is equivalent to a gauge transformation, the q quantum mechanics is equivalent to that of a periodic potential. Hence, θ arises as in Bloch waves of one dimensional crystal.

The matching problem and the determination of wave function are possible to handle for example II, since, due to $O(5)$ invariance of classical solution[43], the equation for small fluctuations is entirely solvable.

Finally, we note a crucial difference between example I (mass scale) and II (no mass scale). In example I, f cannot be computed explicitly but, since $h(q)$ is the integral of magnetic field, one has

$$h(f) \underset{f \to \infty}{\simeq} C e^{-\mu f}$$

where C is a constant and μ is the mass of vector field. This leads to the following behaviour for the inverse function

$$q \underset{f \to \pm\infty}{\simeq} n + \frac{C}{\mu} e^{-\mu |f|} \tag{3.58}$$

In example II, $q(f)$ can be computed

$$q(f) = \frac{1}{4} \left[\frac{3\lambda^2 f + 2f^3}{(\lambda^2 + f^2)^{3/2}} \right] + \text{cste} \tag{3.59}$$

which leads to

$$q \underset{f \to \pm\infty}{\simeq} n + \frac{15}{16} \left| \frac{f}{\lambda} \right|^{-4} \tag{3.60}$$

Thus in example I, we have an exponential behaviour while in example II we have a power behaviour.

$|f| \to \infty$ corresponds to approaching the minima of the potential for q quantum mechanics. The two different behaviours (3.58), (3.59) show that these potentials behave in a very different way in these two cases. In fact, it is harmonic near the minimum in example I while the potential is much flatter in example II. As a result WKB matching will lead to rather different results. In example I, one would obtain a result equivalent to the dilute gas approximation[33] of Euclidean field theory[44]. In example II, a different result may come out. This question is important for the problem of quark confinement, since in dilute gas approximation the Yang-Mills theory does not seem [45] to confine quarks, contrary to the initial hopes of Polyakov[14]. According to Sec. II, one performs the canonical transformation[46]

$$A_i(\vec{x}) = \hat{A}_{ci}(\vec{x}, f(q)) + \tilde{A}_i(\vec{x})$$

$$\int d\vec{x}\ \tilde{A}_i(\vec{x})\ \partial_q \hat{A}_{ci}(\vec{x}, f(q)) = 0$$

Note that we are in Schrödinger representation. Hence, A and $\tilde{A}$ do not involve the true time of the problem.

REFERENCES

(1) H.B. Nielsen and P. Olesen, Nucl. Phys. B61 (1973) 45
(2) R. Dashen, B. Hasslacher and A. Neveu, Phys. Rev. D10 (1974) 4114 ; 4130 ; 4138
(3) A.M. Polyakov, Phys. Letters 59B (1975) 82
(4) A.A. Belavin, A.M. Polyakov, A.S. Schwartz, Y.S. Tyupkin, Phys. Letters 59B (1975) 85
(5) See e.g. T.D. Lee in ref. (11)
(6) For a general discussion, see Coleman's 1975 Erice Lectures
(7) J.L. Gervais, B. Sakita, Phys. Rev. D11 (1975) 2943
(8) L.D. Faddeev, P.P. Kulish, V.E. Korepin, Pizma JETP 21 (1975) 302
(9) R. Rajaraman, Phys. Report 21C (1975) 227
(10) R. Jackiw, Rev. of Mod. Phys. 49 (1977) 681
(11) Extended Systems in Field Theory, edited by J.L. Gervais and A. Neveu, Phys. Report 23C (1976)
(12) J.L. Gervais, A. Jevicki, B. Sakita, Phys. Rev. D12 (1975) 1038
(13) E. Tomboulis, Phys. Rev. D12 (1975) 1678
(14) J.L. Gervais, A. Jevicki, Nucl. Phys. B110 (1976) 93
(15) J. Goldstone, R. Jackiw, Phys. Rev. D11 (1975) 1486
(16) p, p' ; k, k' are soliton and meson momenta
(17) A. De Vega, Nucl. Phys. B115 (1976) 428
(18) R. Dashen, B. Hasslacher and A. Neveu, Phys. Rev. D11 (1975) 3424
(19) L.D. Faddeev, V.E. Korepin, Phys. Lett. 63B (1976) 435
(20) A. Jevicki, Nucl. Phys. B117 (1976) 365
(21) J.L. Gervais, A. Jevicki, B. Sakita, in ref. (11)
(22) A. Hosoya, and K. Kikkawa, Nucl. Phys. B101 (1975) 271
(23) L.D. Faddeev, Theor. and Math. Phys. 1 (1970) 1
(24) See e.g. A. Katz, Classical Mechanics, Quantum Mechanics Field Theory (Academic Press, 1965)
(25) J.L. Gervais, A. Jevicki, Nucl. Phys. B110 (1975) 113
(26) A.S. Scott, F.Y.F. Chu, D.W. McLaughlin, Proc. I.E.E.E.61 (1973) 1443
(27) M.T. Jaekel, Nucl. Phys. B118 (1977) 506
(28) J. Honerkamp, M.Schlindwein, A. Wiesler, Nucl. Phys. B121 (1977) 531
(29) B. Julia and A. Zee, Phys. Rev. D11 (1975) 2227
(30) J.L. Gervais, B. Sakita, S. Wadia, Phys. Letters 63B (1976) 55
(31) D. McLaughlin, J. Math. Phys. 13 (1972) 1099
(32) E. Witten, Phys. Rev. Lett. 38 (1977) 121
't Hooft, unpublished
R. Jackiw, C. Nohl, C. Rebbi, Phys. Rev. D15 (1977) 1642
For very interesting recent developments on this problem, see Stora's lecture note in this volume
(33) C. Callan, R. Dashen, D. Gross, Phys. Letters 63B (1976) 334
(34) A.A. Belavin, A.M. Polyakov, Nucl. Phys. B123 (1977) 429
(35) G. 't Hooft, Phys. Rev. Letters 37 (1976) 8, and Phys. Rev. D 14 (1976) 3432
(36) J. Keller, Ann. Phys. 4 (1958) 180
M. Gutzwiller, J. Math. Phys. 2 (1970) 21 ; 11 (1970) 1791 ; 10 (1969) 1004 ; 8 (1967) 1979
M. Maslov, Theor. Math. Phys. 2 (1970) 21. Theory of disturbancies and asymptotic methods (Moscow Univ. Press, 1965) ; Théorie de perturbations et méthodes asymptotiques (Dunod Paris, 1972)
(37) T. Banks, C.M. Bender, T.T. Wu, Phys. Rev. D8 (1973) 3346 ;
T. Banks , C.M. Bender, Phys. Rev. D8, (1973) 3366
(38) J.L. Gervais, B. Sakita, CCNY preprints HEP 76/11 (1976) ; HEP 77/8 (1977)
(39) N. Christ, T.D. Lee, Phys. Rev. D12 (1975) 1606
(40) See e.g. L. Landau, E. Lifchitz, Classical Mechanics
(41) Hereafter, all sums over u^m's or v^m's only run over the ones

with $\nu^m > 0$
(42) R. Jackiw, C. Rebbi, Phys. Rev. Lett. 37 (1976) 172
(43) R. Jackiw, C. Rebbi, Phys. Rev. D14, (1976) 517
(44) This has been checked for one degree of freedom by C. Callan (unpublished)
(45) C. Callan, R. Dashen, D. Gross, Phys. Lett. 66B (1977) 375

YANG MILLS INSTANTONS, GEOMETRICAL ASPECTS

R. STORA
Centre de Physique Théorique, CNRS, Marseille

Lectures given at the International School of Mathematical Physics, Erice, 27 June - 9 July, 1977.

- FOREWORD -

These notes are based on seminar notes prepared during the year 1976-1977 at the Centre de Physique Théorique du CNRS, Marseille, by : W. Franklin, C.P. Korthals-Altes, J. Madore, J.L. Richard, R. Stora, and private lectures by I.M. Singer to the author, to whom, however all incorrections should be attributed.

R. Stora

- ACKNOWLEDGEMENT -

The author is indebted to I.M. Singer for illuminating comments, to L. Gärding for fruitful afternoon discussions, and to the participants to the Marseille seminar who have provided me with most of my yet imperfect knowledge of the subject.

I - INTRODUCTION

The word instanton [1] has been coined by analogy with the word soliton. They both refer to solutions of elliptic non linear field equations with boundary conditions at infinity (of euclidean space time in the first case, euclidean space in the second case)lying on the set of classical vacua in such a way that stable topological properties emerge, susceptible to survive quantum effects, if those are small. Under this assumption, instantons are believed to be relevant to the description of tunnelling effects between classical vacua [2] and signal some characteristics of the vacuum at the quantum level, whereas solitons should be associated with particles, i.e. discrete points in the mass spectrum . In one case the euclidean action is finite, in the other case, the energy is finite. From the mathematical point of view, the geometrical phenomena associated with the existence of solitons have forced physicists to learn rudiments of algebraic topology [3] . The study of euclidean classical Yang Mills fields involves naturally mathematical items falling under the headings :

- differential geometry (fibre bundles, connections)
- differential topology (characteristic classes, index theory)

and, more recently

- algebraic geometry.

Most of the machinery is old enough so that it can be learnt from mathematical books or sets of lecture notes where complete bibliographies can be found. It is out of question to give here a complete review of the mathematical apparatus. We shall rather pick out some of the results and show how they apply to the specific case at hand.

These notes are divided as follows :

Section II is devoted to a description of the physicist's views
Section III is devoted to the mathematician's views.

These notes are sketchy in the sense that very few technical details are fully described. Displaying them all would have required reproducing large portions of mathematical books. Emphasis has been put on some details of the 19th century geometry which is not easily accessible anymore, and not currently known to physicists. The more accessible mathematical items are referred to as accurately as possible, including chapters, paragraphs, page numbers. It is thus hoped that these notes can be used as a guide through the recent literature.

II - THE PHYSICIST'S VIEWS

The problem to be solved is the following : find euclidean Yang Mills fields $A_\mu^\alpha(x)$ which minimize locally the euclidean action

$$S = \frac{1}{4g^2} \sum_{\substack{\alpha,\beta \\ \mu,\nu}} \int d^4x \; F_{\mu\nu}^\alpha \, g_{\alpha\beta} \, g^{\mu\rho} g^{\nu\sigma} \, F_{\rho\sigma}^\beta \tag{1}$$

The notations are as follows

$x \in E^4$ four dimensional euclidean space.

α : labels an orthonormal basis of the Lie algebra $\mathcal{G}$ of a simple compact Lie group G; unless otherwise specified $G = SU2$ for which there is the largest available information.

$g_{\alpha\beta}$: Killing form of

$g^{\mu\nu}$: flat riemannian metric in E^4

d^4x : volume element in E^4 corresponding to $g^{\mu\nu}$

$$F_{\mu\nu}^\alpha = \partial_\mu A_\nu^\alpha - \partial_\nu A_\mu^\alpha + f_{\beta\gamma}^\alpha A_\mu^\beta A_\nu^\gamma \tag{2}$$

$f_{\beta\gamma}^\alpha$: structure constants of $\mathcal{G}$.

The first class of instantons found by Belavin et al [4] is by now well known. It has the following characteristics :

$$S < \infty \qquad F_{\mu\nu}(x) \xrightarrow[|x|\to\infty]{} 0$$

$$A_\mu(x) \underset{|x|\to\infty}{\sim} g^{-1}(\hat{x}) \, \partial_\mu \, g(\hat{x})$$

where the homotopy class of $S_\infty^3 \ni x \xrightarrow{g} G$ corresponds to the integer $n = \pm 1$. Both cases, $n = \pm 1$ are treated together, by considering a Yang Mills field $A_\mu^{\alpha\beta}$ with value in the Lie algebra of $SO4$ which is the direct sum of two copies of the $SU2$ Lie algebra. The topological number n is related to a Chern number (the integral of a Chern characteristic class) :

$$2n = \frac{1}{4\pi^2} \int F_{\mu\nu}^\alpha \, \epsilon^{\mu\nu\rho\sigma} \, g_{\alpha\beta} \, F_{\rho\sigma}^\beta \; d^4x \tag{3}$$

For given n , absolute minima of S are reached for

$$F_{\mu\nu}^\alpha = \pm \frac{1}{2} \epsilon_{\mu\nu}{}^{\rho\sigma} \; F_{\rho\sigma}^\alpha = \pm \left(*F\right)_{\rho\sigma}^\alpha \tag{4}$$

which in particular imply the usual field equations

$$\nabla^{\mu} F^{\alpha}_{\mu\nu} = 0 \tag{5a}$$

$$\nabla^{\mu} (*F)^{\alpha}_{\mu\nu} = 0 \tag{5b}$$

However, all solutions which have been so far constructed saturate the absolute bound

$$S = \frac{2\pi^2}{g^2} |n| \tag{6}$$

deduced from the identity

$$\int \left(F^{\alpha}_{\mu\nu} \pm *F^{\alpha}_{\mu\nu} \right) g_{\alpha\beta}\, g^{\mu\rho} g^{\nu\sigma} \left(F^{\beta}_{\rho\sigma} \pm *F^{\beta}_{\rho\sigma} \right) d^4x \geq 0 \tag{7}$$

The $n=1$ solutions assume several equivalent forms [4],[5],[1]

$$\begin{aligned} A^{\alpha}_{\mu} &= \frac{x^2}{x^2+\lambda^2} \left[g^{-1}(x)\, \partial_{\mu} g(x) \right]^{\alpha} \\ &= - \eta^{\alpha}_{\mu\nu} \frac{2x^{\nu}}{x^2+\lambda^2} \end{aligned} \tag{8}$$

where

$$\begin{aligned} g(x) &= \frac{x_4 + i\vec{x}.\vec{\sigma}}{\sqrt{x^2}} = \frac{\tilde{x}}{\sqrt{x^2}} \\ \eta^{\alpha}_{\mu\nu} &= \frac{1}{2} \operatorname{tr} \tilde{\sigma}_{\mu} \underline{\sigma}_{\nu} \sigma^{\alpha} \end{aligned} \tag{9}$$

Through a conformal transformation which leaves both the euclidean action and the topological invariant unchanged, or a gauge transformation one gets the following equivalent form [1] :

$$A^{\alpha}_{\mu} = \eta^{\alpha}_{\mu\nu}\, \partial^{\nu} \log \left(1 + \frac{\lambda^2}{x^2} \right) \tag{10}$$

later generalized by 't Hooft [6] for higher n-values :

$$\begin{aligned} A^{\alpha}_{\mu} &= \eta^{\alpha}_{\mu\nu}\, \partial^{\nu} \log \rho \\ \frac{\Box \rho}{\rho} = 0 \qquad \rho &= 1 + \sum_{1}^{n} \frac{\lambda_i^2}{(x-x_i)^2} \end{aligned} \tag{11}$$

The SO_4 version which puts together solutions pertaining to opposite n's reads :

$$A_\mu^{\alpha\beta} = \Sigma_{\mu\nu}^{\alpha\beta}\, \partial^\nu \log \rho \tag{12}$$

where the $\Sigma_{\mu\nu}^{\alpha\beta}$'s are the matrix elements of the SO_4 Lie algebra :

$$\Sigma_{\mu\nu}^{\alpha\beta} = \delta_\mu^\alpha\, \delta_\nu^\beta - \delta_\mu^\beta\, \delta_\nu^\alpha \tag{13}$$

This collection of solutions has been enlarged by Jackiw, Nohl and Rebbi [7] into a 5n+4 parameter family with

$$\rho = \frac{1}{(x-x_0)^2} + \sum_1^n \frac{\lambda_i^2}{(x-x_i)^2} \tag{14}$$

It was also argued by these authors that there ought to be solutions depending on 5n + 3(n-1) = 8n - 3 parameters corresponding to n-1 relative orientations of isospin axis, for instanton number n and this was checked in the neighbourhood of the known solutions, in the linear approximation [8]. This situation has been further analyzed by Brown, Carlitz, Lee [9] who relate the dimensionality of instanton fluctuations to that of minimally coupled massless fermions belonging to the adjoint representation. The latter is connected to the Adler anomaly, through an argument of S. Coleman [10], and hence to the instanton number.

Although the fermion problem is interesting in itself [1] and can be handled for an arbitrary compactification of E^4 ,[11] , it is only directly related to the instanton problem in the case where the metric is flat. The argument can then be summarized as follows :

Let

$$A_\mu^\alpha = \mathring{A}_\mu^\alpha + a_\mu^\alpha \tag{15}$$

and let us impose the Landau gauge [11] condition in the background field $\mathring{A}$ which we assume to correspond to a self dual solution :

$$\mathring{\nabla}^\mu a_\mu = 0 \tag{16}$$

The linearized system then reads

$$\tilde{\nabla}\, \underset{\sim}{a} = 0 \tag{17}$$

where

$$\begin{aligned} \tilde{\nabla} &= \tilde{\sigma}_\mu \nabla^\mu \\ \underset{\sim}{a} &= \underset{\sim}{\sigma}_\mu\, a^\mu \end{aligned} \tag{18}$$

Since every quaternion $\mathfrak{a}$ is determined by its first column $\underline{a}$, one has

$$\tilde{\nabla}\, \underline{a} = 0 \tag{19}$$

Conversely, for each solution $\underline{a}$ of this spinor equation, there corresponds a two dimensional real manifold of solutions of the initial equation, corresponding to the one dimensional complex manifold of solutions $\lambda \underline{a}$, λ complex. This in turn is equivalent to the massless Dirac equation

$$\not{\nabla}\, \psi = 0 \tag{20}$$

together with the chirality condition

$$\psi = \gamma_5 \psi \tag{21}$$

(in the Weyl representation).

The rest of the argument which fits very well within the methods to be described in the next section involves several steps :

i) for a given self or anti-self duality property of the gauge field, the Dirac equation possesses only chiral or anti-chiral solutions

ii) the difference between the number of chiral and anti-chiral solutions can be evaluated in terms of the Adler anomaly, i.e. the instanton number.

This developping subject owes much to physicists who have first made a number of remarkable guesses. It seems however that mathematicians have taken over with powerful - and rigorous - techniques. It is to be noticed that one of the first contributors, A.S. Schwarz [4],[11] left a name in the theory of characteristic classes and was the first to have used the powerful index theory as early as April 1976 [12]. Later, M.F. Atiyah and I.M. Singer, the main contributors in this ten year old theory, and collaborators [13] , have both reproduced A.S. Schwarz's work and gone beyond with the help of the hitherto unused techniques of algebraic geometry [14].

Some mathematical aspects dug out by physicists have not been exploited so far, namely, those related to the general conformal invariance of the problem : the function ρ involved in the 't Hooft ansatz can be identified with the conformity factor [15] occurring in the line element of a non compact manifold conformal to E^4 (flat for self or anti-self dual $F^{\alpha}_{\mu\nu}$, with constant curvature in the case of general solutions of the field equations). These remarks have not been fully exploited yet, because much of the mathematics used so far relies on the compactness of the manifolds that are used.

III - THE MATHEMATICIAN'S VIEWS

It is a matter of philosophy whether in principle a Yang Mills fields ought to be associated with a connection on a principal fibre bundle [16]. It is a fact that Yang Mills fields considered in the previous section are of this type and that the corresponding mathematical apparatus can be used either to streamline previously obtained results or to obtain new results.

We shall now review the various items enumerated in the previous section from a more mathematical point of view.

1. The n=1 instantons, a geometrical description [5],[15],[18].

Let us first map $E^4 \cup \infty$ into $S^4 \subset E^5$ through a stereographic projection. Call a , F , the differential forms

$$a = A^{\alpha}_{\mu}\, dx^{\mu}\, e_{\alpha} \tag{22}$$

$$F = \frac{1}{2}\, F^{\alpha}_{\mu\nu}\, dx^{\mu} \wedge dx^{\nu}\, e_{\alpha} \tag{23}$$

where e_{α} is a basis of $\mathcal{G}$. We shall not distinguish the forms on $E^4 \cup \infty$ and their inverse images on S^4 . Since the stereographic projection is conformal, it preserves

$$S = \frac{1}{4} \int (F, *F) \tag{24}$$

$$2n = \frac{1}{4\pi^2} \int (F, F) \tag{25}$$

where $*$ denotes the dual for whatever Riemannian metric is involved, and (,) is the Killing form of $\mathcal{G}$.

There are concrete examples of fibre bundle with structure group either $SO4$ or $SU2 \times SU2$ pertinent to the instanton antiinstanton doubling or $SU2$, pertinent to single instanton or antiinstanton description :
the $SO4$ principal bundles with basis S^4 are known to depend on two integers [17]. The simplest non trivial one is $SO5$ $(SO5 \backslash SO4 = S^4)$. The Maurer Cartan form on $SO5$, $\omega = g^{-1}dg$ with value in the Lie algebra of $SO5$ can be restricted to the Lie algebra of $SO4$ [18] and one can check by choosing coordinates that it is the n = 1 instanton in its initial version, which is $SO5$ invariant. The conformal transforms of this solution are obtained by restricting the Maurer Cartan form on $SO(5,1)$ to a right coset modulo $SO5$, and then to the Lie algebra of $SO4$. We thus obtain a five parameter family of solutions indexed by a point of $SO(5,1) \backslash SO5$, each solution being invariant under left translation by a subgroup of $SO(5,1)$ conjugate

to $SO5$. In this version one has to go from S^4 to E^4 by a stereographic projection, and it is actually much more direct to work with the covering groups $USp2$ of $SO5$ - the 2 x 2 unitary group with quaternion elements - and $SL(2,H)$ - the 2 x 2 unimodular group with quaternion elements - of $SO5,1$. The quotient $USp2\backslash SU2\times SU2$ is the projective quaternionic line $P_1(H)$, i.e. the set of pairs of quaternions (x,y) under the equivalence relation $(x,y)\sim(qx,qy)$ where q is an arbitrary non vanishing quaternion. $P_1(H)$ can be used naturally as a model of compactified E^4 and the formulae given by Jackiw and Rebbi [5] are directly recovered by the constructions indicated above. In particular, it is easy to verify that the corresponding curvature fulfills the self-duality condition [4], by using its expression in terms of the Maurer Cartan form on the one hand, and local coordinates on the other hand [18].

One can similarly deal with the SU2 version by considering $S^7 = USp2\backslash SU2$ and appropriately restricting the Maurer Cartan form [18].

2. The 't Hooft instantons as connections on principal bundles [6],[19].

On S^4 (resp. $P_1(H)$), a has n singularities at x_i, $i=1\dots n$ and, in the neighbourhood of such a singularity

$$a \underset{x\to x_i}{\sim} g_i^{-1}\, dg_i \tag{26}$$

where g_i is the translated by x_i of g given by formula (9). Cover S^4 (resp. $P_1(H)$) by $n+1$ open sets : n ball-neighbourhoods Ω_i of x_i ($\Omega_i\cap\Omega_j=\emptyset$), $\Omega_0 = \complement \bigcup_i \Omega_i/2$

Define

$$\begin{aligned} a_0 &= a_{|\Omega_0} \\ a_i &= \operatorname{ad} g_i^{-1}\, a + g_i\, dg_i^{-1} \end{aligned} \tag{27}$$

Then, by the conventional construction of principal bundles, there is a bundle with transition functions

$$g_{i0} = g_i^{-1} \quad \text{in } \Omega_i\cap\Omega_0 \tag{28}$$

and a connection defined by $a_0, a_i, i=1\dots n,$ on it (cf. Kobayashi Nomizu [18], p. 66).

Although there are canonical examples of $SO4$ fibre bundles over S^4 for arbitrary allowed topologies [20] - indexed by two integers - they have not suggested so far any geometrical characterization of the connections which minimize the euclidean action.

3. The manifold of connections minimizing the Euclidean action, local aspects.

There are two essentially equivalent versions of [11] [13] of the study of the manifold of solutions of the self or antiself duality condition Eq. 4 .

One is based on the linearized system

$$\tilde{\overset{\circ}{\nabla}} \tilde{a} = 0 \tag{29}$$

in the neighbourhood of a solution $\overset{\circ}{a}$, a connection on some principal bundle $\mathcal{B}$ over S^4 . $\tilde{a}$ is a section of the $SO4\times G$ bundle $T^{*1}(S^4) \underset{\mathcal{B}}{\times} \mathcal{G}$ with basis S^4 associated with $\mathcal{B}$ and the riemannian structure on S^4. $\tilde{\overset{\circ}{\nabla}}$ is a first order elliptic operator i.e. its first order symbol $\tilde{\xi}$, (obtained by replacing $\frac{\partial}{\partial x}$ by $i\xi$ in the higher degree terms) is invertible for $\xi \neq 0$. It maps sections of $T^{*1}(S^4) \underset{\mathcal{B}}{\times} \mathcal{G}$ into sections of $(T^{*0} \oplus T_+^{*2})(S^4) \underset{\mathcal{B}}{\times} \mathcal{G}$ where $T_+^{*2}(S^4)$ denotes the space of self dual 2-forms. Since $\tilde{\overset{\circ}{\nabla}}$ is elliptic, the dimensionality of the space of solutions ker $\tilde{\overset{\circ}{\nabla}}$ is finite [23] since S^4 is compact. The index theorem can be applied : [22] [23] [24] [25]

$$\operatorname{Ind} \tilde{\overset{\circ}{\nabla}} = \dim \ker \tilde{\overset{\circ}{\nabla}} - \dim \ker \tilde{\overset{\circ}{\nabla}}^T \tag{30}$$

can be computed in terms of topological data ($\tilde{\overset{\circ}{\nabla}}^T$ now maps $(T^{*0} \oplus T_+^{*2})(S^4) \underset{\mathcal{B}}{\times} \mathcal{G}$ into $T^{*1}(S^4) \underset{\mathcal{B}}{\times} \mathcal{G}$, and is the usual adjoint) because of its "universality". The calculation proceeds through a formal algebra of characteristic classes whose terms factor out into factors involving the $\mathcal{G}$ fibration, expressible by means of the character of the adjoint representation, and factors involving the basis, expressible in terms of the character of the $SO4$ representation in $T^{*1}(S^4)$ and $(T^{*0} \oplus T_+^{*2})(S^4)$ respectively. For $G = SUN$ the formula reads

$$\operatorname{Ind} \tilde{\overset{\circ}{\nabla}} = 4\,N\,n(\mathcal{B}) - \Big(\underbrace{\frac{\chi(S^4)}{2}}_{=1} + \underbrace{\frac{p^1(T(S^4))}{6}}_{=0} \Big)(N^2-1) \tag{31}$$

χ = Euler Poincaré characteristic ; $\chi(S^4) = 2$; $p^1(T(S^4))$ = Pontrjagin number of S^4 ; $p^1(T(S^4)) = 0$; $n(\mathcal{B})$ = Chern number of $\mathcal{B}$.

Next one shows that

$$\dim \ker \tilde{\overset{\circ}{\nabla}}^T = 0 \tag{32}$$

The calculation is a bit lengthy and repeatedly makes use of two arguments which are schematized below : a positivity argument classical in the Hodge theory of harmonic forms [23] ,[29] , and an irreducibility argument concerning $\overset{\circ}{a}$.

From

$$\overset{\circ}{\nabla}{}^{T} H = 0 \qquad \begin{array}{l} H = (h, f) \\ h \in T^{*0} \underset{\mathcal{B}}{\times} \mathcal{G} \\ f \in T_{+}^{*2} \underset{\mathcal{B}}{\times} \mathcal{G} \end{array} \tag{33}$$

we deduce

$$\int_{S^4} \left(\overset{\circ}{\nabla}{}^{T} H, \overset{\circ}{\nabla}{}^{T} H \right) = 0 \tag{34}$$

hence, from positivity

$$\overset{\circ}{\nabla} h = 0 \tag{35}$$

which in turn implies

$$\overset{\circ}{\nabla}{}^{2} h = [\overset{\circ}{F}, h] = 0 \tag{36}$$

Actually Eq. 36 is equivalent to Eq. 35 :

$$0 = \int_{S^4} (h, \overset{\circ}{\nabla}{}^{2} h) = \int (\overset{\circ}{\nabla} h, \overset{\circ}{\nabla} h) \tag{37}$$

hence, Eq. 35 follows from positivity.

If $\overset{\circ}{F}$ spans the $SU2$-Lie algebra everywhere, (the irreducibility property, which is true here, it follows that h vanishes. This argument incidentally shows why the background Landau gauge does not leave any gauge freedom (Eq. 36 has no non-vanishing solution).

The method used by Atiyah, Hitchin and Singer [13] is essentially equivalent : they consider the elliptic complex [23][24][25]

$$0 \rightarrow T^{*0}(S^4) \underset{\mathcal{B}}{\times} \mathcal{G} \xrightarrow{\overset{\circ}{\nabla}} T^{*1}(S^4) \underset{\mathcal{B}}{\times} \mathcal{G} \xrightarrow{\overset{\circ}{\nabla}_{+}} T_{+}^{*2}(S^4) \underset{\mathcal{B}}{\times} \mathcal{G} \rightarrow 0 \tag{38}$$

($\overset{\circ}{\nabla}_{+} \nabla_{+} = 0$ by the self-duality of $\overset{\circ}{F}$).
Under the irreducibility hypothesis for $\overset{\circ}{F}$, the topological index of this complex is identical with the one previously computed and so is its analytical index (sum of dimensions of various kernels). The formulation is however slightly different : instead of removing the gauge freedom by fixing the background Landau gauge condition, the gauge freedom is eliminated by subtracting $\dim \operatorname{Im} \overset{\circ}{\nabla}$, which is

correct in all cases.

Once the linearized system has been analyzed, there remains to prove that each solution of the linearized system gives rise to a true solution, in a neighbourhood of $\mathring{a}$. This requires the use of the infinite dimensional implicit function theorem on top of

$$\dim \ker \tilde{\nabla}^T = 0$$

in the first version,

$$\dim \ker \mathring{\nabla}_+^T = 0$$

in the second version [25] where the question is to study the deformation of the complex (38).

The last question [11] connected with this has to do with the solutions of the massless Dirac equation Eqs. (20, 21). $\mathring{\not\nabla}$ is an elliptic operator from positive chirality fields to negative chirality fields, and coincides with its adjoint, the two spaces being interchanged. The index of $\mathring{\not\nabla}$, $n_+ - n_-$, ($n_\pm = \#$ positive/negative chirality of fermion zero modes) can be computed again using topological data [22], [11], [13], [9] . On the other hand, it is related to the Adler Bardeen anomaly, as remarked by S. Coleman [10] : it can be computed by investigating the corresponding Laplacian and diffusion operator [25] [27] [30] :

$$\mathrm{Ind}\, \mathring{\not\nabla} = O(t^\circ)\left[\mathrm{tr}\, e^{-t\frac{1+\gamma_5}{2}\mathring{\not\nabla}^2} - \mathrm{tr}\, e^{-t\frac{1-\gamma_5}{2}\mathring{\not\nabla}^2}\right] \tag{39a}$$

$$= O(t^\circ)\left[\mathrm{tr}\, \gamma_5\, e^{t\mathring{\not\nabla}^2}\right. \tag{39b}$$

where the symbol $O(t^\circ)$ means : selecting the zeroth order term in the asymptotic expansion [28] of the indicated quantity for $t \to 0+$, from which the negative powers of t involved in each term of (39a) drop out. This method of calculation turns out to be quite close to the physicist's version [9] based on the evaluation of Feynman graphs, to which it provides a firm foundation. The result can also be obtained by the purely algebraic methods involving the relevant representations of $SU2 \times SU2 \times \mathcal{G}$. The coefficient of n comes out a half of what it was in the Yang Mills case, but the coefficient of dim G changes significantly, in particular the term proportional to $\chi(S^4)$ is missing : in non flat space the relationship between the Dirac and Yang Mills problems is unclear, since here again positivity and the irreducibility of the connection take care of the absence of solutions of the adjoint system.

4. The manifold of connections minimizing the Euclidean action, global aspects.

Progress has recently been made [31] [32] towards a global study of self or antiself duality equations (Eq. 4)

$$F^{\alpha}_{\mu\nu} = \pm (*F)^{\alpha}_{\mu\nu}$$

So far, only $G = SU2$ has been dealt with. Since this is a quasilinear elliptic system, one expects solutions to possess analyticity properties. These analyticity properties have been found [31] and restrict the differential geometry framework to the algebraic geometry framework. Most of the geometry involved is related to general views put forward by R. Penrose [33], which ought to apply to a general class of conformal invariant euclidean field theories.

It may be of interest, for the purpose of orientation to review a simpler problem which bears some resemblance to the Yang Mills problem, namely the non linear σ model in two dimensions [34].

One looks for minima of

$$S = \int \partial_\mu \vec{\varphi}(x) . \partial_\mu \vec{\varphi}(x) \, d^2x \tag{40}$$

with the constraint

$$\vec{\varphi}^2(x) = \sum_1^3 \varphi_i^2(x) = 1 \tag{41}$$

and the boundary condition

$$\vec{\varphi}(x) \xrightarrow[|x| \to \infty]{} \vec{\varphi}_0 \qquad \vec{\varphi}_0^{\,2} = 1 \tag{42}$$

Thus

$$R^2 \cup \infty \ni x \longrightarrow \vec{\varphi}(x) \in S^2$$

defines a mapping from $R^2 \cup \infty$ to S^2 whose degree is given by

$$n = \int (\vec{\varphi}, \partial_\mu \vec{\varphi}, \tilde{\partial}_\mu \vec{\varphi}) \, d^2x \tag{43}$$

$$\tilde{\partial}_\mu = \epsilon_{\mu\nu} \partial^\nu$$

Given n, S reaches an absolute minimum for

$$\partial_\mu \vec{\varphi} = \pm \vec{\varphi} \times \tilde{\partial}_\mu \vec{\varphi} \tag{44}$$

$$n = \pm |n|$$

as stems from the positivity condition

$$\int \sum_{\mu} \left(\partial_\mu \vec{\varphi} \pm \vec{\varphi} \times \tilde{\partial}_\mu \vec{\varphi} \right)^2 d^2x \geq 0 \tag{45}$$

$$S \geq \mp n$$

It is convenient to use the variables

$$\begin{aligned} z &= x_1 + i x_2 \\ \zeta &= \xi_1 + i \xi_2 \end{aligned} \tag{46}$$

where ξ_1 , ξ_2 are the coordinates of the stereographic projection of $\vec{\varphi}$ on R^2 . In terms of these variables

$$\begin{aligned} S &= \int \left(\left|\frac{\partial \zeta}{\partial z}\right|^2 + \left|\frac{\partial \zeta}{\partial \bar{z}}\right|^2 \right) \frac{dz\, d\bar{z}}{(1+|\zeta|^2)^2} \\ n &= \int \left(\left|\frac{\partial \zeta}{\partial z}\right|^2 - \left|\frac{\partial \zeta}{\partial \bar{z}}\right|^2 \right) \frac{dz\, d\bar{z}}{(1+|\zeta|^2)^2} \end{aligned} \tag{47}$$

and (44) reads

$$\frac{\partial \zeta}{\partial z} = 0 \qquad \text{or} \qquad \frac{\partial \zeta}{\partial \bar{z}} = 0 \tag{48}$$

according to the sign of n .

The general solution reads

$$\zeta(z) = \sum_1^n \frac{\lambda_p}{z - z_p} + \zeta_0 \tag{49}$$

in the holomorphic case. z should be replaced by $\bar{z}$ in the antiholomorphic case. λ_p , z_p , are arbitrary complex numbers, ζ_0 , the stereographic projection of $\vec{\varphi}_0$.

Now, it has been often argued that there are similarities between the non linear σ model in two dimensions and the Yang Mills model in four dimensions. There is an obvious analogy here in the derivation of Eq. 48, to be compared with the self-duality condition Eq. 4, from the positivity condition (45) analogous to Eq. 7.

On the other hand, there is a substantial difference between S^2 and S^4 : there is a unique complex structure [35] on S^2 (up to a sign), which makes it an analytic manifold, and is $SO2$ invariant. It is associated with the system of isotropic lines on the sphere. On S^4 , there is no global complex structure. Locally, the analogue of the isotropic generators, of S^2 is provided by any isotropic 2-plane. Such two planes are parametrized by a point on $P^1(\mathbb{C})$, i.e. S^2, and so are, locally, the complex structures of S^4 : working with $P^1(H)$, for convenience, an isotropic two plane is parametrized by

$$T(P^1(H)) \ni \tilde{X} = u \otimes \lambda \quad (\text{resp } \lambda \otimes u \tag{50}$$

with u fixed spinor up to scaling, λ variable spinor :

$$(x,x) = \det \tilde{X} = 0 \tag{51}$$

Thus the direction of this two plane is parametrized by $P^1(\mathbb{C})$. Similarly, a complex structure compatible with $SO4$

$$\tilde{X} \to \widetilde{JX} \qquad J^2 = -1 \qquad JJ^T = 1 \tag{52}$$

can be parametrized by

$$\begin{aligned} &\tilde{X} \to u(J)\,\tilde{X}\,v(J) \\ &u(J),\ v(J) \in SU2 \end{aligned} \tag{53}$$

Now, all possible J's are of the form [35]

$$J = S J_0 S^{-1} \qquad S \in SO4$$

where J_0 is a special solution.
So, choosing J_0 :

$$\tilde{X} \to \widetilde{J_0 X} = \tilde{X}\, i\sigma_2$$

we can parametrize J by

$$\tilde{X} \to \tilde{X}\, v(J) \qquad , \quad v(J) \in SU2\ , \quad v^2(J) = -1$$

$v(J)$ is uniquely determined by its eigensubspace pertaining to the eigenvalue $+i$, so that the $SO4$ invariant complex structures are locally labelled by $P^1(\mathbb{C})$.

Now, $P^1(H)$ can be fibred by its complex structures into $P^3(\mathbb{C})$ which is a complex analytic manifold :

$$P^3(\mathbb{C}) \xrightarrow{P^1(\mathbb{C})} P^1(H) \tag{54}$$

The fibration can be described as follows. Let

$$\mathbb{C}^4 \ni Z = \begin{pmatrix} u \\ v \end{pmatrix} \qquad u, v \in \mathbb{C}^2 \tag{55}$$

$$P^3(\mathbb{C}) = \mathbb{C}^4 / \{ \tilde{Z} \sim \lambda Z \,,\ \mathbb{C} \ni \lambda \neq 0 \}$$

The involution σ on $\mathbb{C}^4$

$$\begin{pmatrix} u \\ v \end{pmatrix} \rightarrow \begin{pmatrix} i\sigma_2 u^* \\ i\sigma_2 v^* \end{pmatrix} \tag{56}$$

with square -1 induces an involution $[\sigma]$ with square +1 on $P^3(\mathbb{C})$. This involution has no fixed point, but fixed lines images of

$$\lambda \begin{pmatrix} u \\ v \end{pmatrix} + \mu \begin{pmatrix} i\sigma_2 u^* \\ i\sigma_2 v^* \end{pmatrix} \qquad \lambda, \mu \in \mathbb{C} \tag{57}$$

$$= \begin{pmatrix} q(u)\omega \\ q(v)\omega \end{pmatrix} \quad ; \quad \omega = \begin{pmatrix} \lambda \\ \mu \end{pmatrix} \in P^1(\mathbb{C}) \qquad \begin{array}{l} q(u) = (u, i\sigma_2 u^*) \\ q(v) = (v, i\sigma_2 v^*) \end{array}$$

These lines are labelled by $P^1(H)$ since the change

$$\begin{array}{l} q(u) \rightarrow q(u)\, q \\ q(v) \rightarrow q(v)\, q \end{array} \qquad 0 \neq q \in H \tag{58}$$

only corresponds to a different parametrization.

The subgroup of $SL(4,\mathbb{C})$, which acts on $P^3(\mathbb{C})$, and commutes with $[\sigma]$ is just $SL(2,H)$ which therefore acts on the real lines images of

$$Z_x = \begin{pmatrix} x v \\ v \end{pmatrix} \qquad v \in P^1(\mathbb{C}) \tag{59}$$

according to

$$\left\{ SL(2,H) \ni \begin{pmatrix} A & B \\ C & D \end{pmatrix} ,\ x \in H \right\} \longrightarrow Ax + B \frac{1}{Cx+D} \in H \tag{60}$$

as expected.

Another useful description of these real lines is through the introduction of their Plücker coordinates. Let

$$F = Z_x \wedge \sigma Z_x \tag{61}$$

Then it is easy to check that, up to an overall scale which depends on v alone, F is determined by x and that

$$\mathrm{tr}\ F * F = 0 \tag{62}$$

The components of F are called the Plücker coordinates of X [36], [37] and Eq. 62 is the equation of the Plücker quadric Q_4 in P^5, which is left invariant by $SO(5,1)$, .

This concludes a brief summary of the euclidean version of the geometrical framework introduced by R. Penrose.

The relevance of isotropic two planes is the following [32], [38], [39] : if one assumes that the solutions to our problem can be analytically continued in some neighbourhood of X, the self duality condition Eq. 4 means that the curvature vanishes on all isotropic two-planes. One thus infers that the connection, restricted to such a two plane is a gauge. If u is the projective spinor defining the direction of an isotropic two-plane, and $\tilde{a}$ the quaternion associated to a,

$$u\,\tilde{a} = H^{-1}(x,u)\, u \frac{\partial}{\partial \tilde{x}}\, H(x,u) \tag{63}$$

The fundamental theorem of the theory, due to Atiyah and Ward [31] actually shows how the gauge function $H(x,u)$ can be found, in principle :

To each principal $SU2$ bundle, and connection with (anti) self dual curvature, there corresponds in a unique way a holomorphic (algebraic) $\mathbb{C}^2$ bundle over $P^3(\mathbb{C})$. The complex structure is defined uniquely by the connection.

The construction is summarized by the following diagram [40]

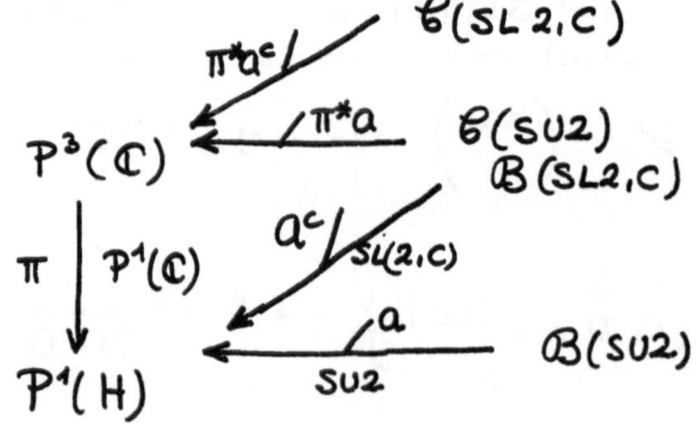

(64)

where :

- $\mathcal{B}(SU2)$ is the principal $SU2$ bundle over $P^1(H)$, a the connection defining horizontal subspaces.
- $\mathcal{B}(SL2,C)$ is the unique SL2,C extension of $\mathcal{B}(SU2)$, a^c the extension of

- $\mathcal{C}(SU2)$, $\mathcal{C}(SL2,C)$ are the inverse images of $\mathcal{B}(SU2)$, $\mathcal{B}(SL2,C)$ by the projection π : $P^3(\mathbb{C}) \xrightarrow{\pi} P^1(H)$ previously described ; $\pi^* a$, $\pi^* a^c$, the inverse images of a , a^c respectively.

An almost complex structure J is defined on $\mathcal{C}(SL2,C)$ by the complex structure of $SL(2,C)$, along the fibres, and by lifting the complex structure of $P^3(\mathbb{C})$ to the horizontal subspaces defined by $\pi^* a^c$.
The integrability condition [35]

$$J\left([X,Y]-[JX,JY]\right)=[JX,Y]+[X,JY] \tag{65}$$

where X , Y are arbitrary vector fields, which insures that the Lie bracket of two holomorphic vector fields $X+iJX$, $Y+iJY$ is again holomorphic, i.e. of the form $Z+iJZ$ is known to be sufficient for J to define a complex structure. It can be checked here by choosing for X , and Y either horizontal or vertical fields. The self-duality condition is part of the integrability condition (63), for X , Y horizontal, $[X,Y]$ vertical, the other parts being trivially checked. This shows that $\mathcal{C}(SL2,C)$ is holomorphic. $\mathcal{C}(SL2,C)$ has two further properties : a reality property related to its construction from $\mathcal{B}(SU2)$ of which it is a kind of complexification, and a triviality property : the restriction of $\mathcal{C}(SL2,C)$ to real lines of $P^1(H)$ is trivial. This leads to the construction of the gauge function $H(x,u)$ in Eq. 63 as indicated by Ward [32] and also used by other authors [38],[39], as follows. Cover $P^1(\mathbb{C})$ by the two open sets Ω_0 , Ω_∞:

$$\Omega_{\substack{0\\ \infty}} = \left\{ u=\begin{pmatrix} z \\ 1 \end{pmatrix} , \quad z \neq \substack{\infty \\ 0} \right\} \tag{66}$$

The restriction of $\mathcal{C}(SL2C)$ to a real line x is defined by a transition function $F(x,u)$ holomorphic in $\Omega_0 \cap \Omega_\infty$, with value in $SL(2,\mathbb{C})$.
On the other hand $F(x,u)$ is not arbitrary :

$$F(x,u) = G(v,u)\Big|_{v=xu} \tag{67}$$

where G is the transition function of $\mathcal{C}(SL2C)$ in $\Omega_0^\varepsilon \cap \Omega_\infty^\varepsilon$, where $\Omega^\varepsilon_{\substack{0\\ \infty}}$ denotes a neighbourhood of $\Omega_{\substack{0\\ \infty}}$ in $P^3(\mathbb{C})$. Thus, F is invariant under the translation $x \to x+\lambda \otimes u\sigma_2$ and thus fulfills the differential equation

$$D_u F = 0 \tag{68}$$

where

$$D_u = u \frac{\partial}{\partial \tilde{x}} \tag{69}$$

is the operator appearing in Eq. 63. Of course G is homogeneous of degree 0 in u, v, and so F is homogeneous of degree 0 in u. Triviality means that one may split F according to

$$F = F_0 \, F_\infty^{-1} \tag{70}$$

$$F_{\infty}^{0} \text{ in } SL(2,C) \text{, holomorphic in } \Omega_{\infty}^{0}$$

It follows that

$$F_0^{-1} D_u F = F_\infty^{-1} D_v F \tag{71}$$

both sides continue each other in $\Omega_0 \cup \Omega_\infty$ and because they are homogeneous of degree 1 in u, they are linear in u, so that one may write

$$u\tilde{a} = F_0^{-1} D_u F = F_\infty^{-1} D_u F_\infty \tag{72}$$

where $\tilde{a}$ depends only on x and is identified with the connection; furthermore the identity

$$D_u (F_0^{-1} D_v F) - D_v (F_0^{-1} D_u F) - [F_0^{-1} D_u F, F_0^{-1} D_v F] = 0 \tag{73}$$

yields the self duality condition on the curvature. This essentially summarizes Ward's argument [32].

Now the structure of $\mathcal{E}(SL2,C)$ contains lots of information into which the author cannot go by mere lack of competence. For instance, the lines x on which triviality breaks down are represented by intersecting the Plücker quadric with an algebraic surface of degree n

$$P_n(F) = 0 \qquad P_n \text{ homogeneous polynomial of degree } n \tag{74}$$

Such lines are called jumping lines, and will appear as singularities of a, since (72) will break down.
In particular, the n=1 instanton is associated with a hyperplane section of Q_4. The set of n=1 instantons is parametrized by such hyperplanes (5 parameters) and transformed into one another by $SO5{,}1\backslash SO5$, $SO5$ being the stabilizer of one real hyperplane. The corresponding model for the associated $\mathbb{C}^2$ bundle, the so-called null correlation bundle [36],[41] is explicit enough so that one may reconstruct the known solution [18], for which now uniqueness is proved.

Whereas the information on the manifold of singularities of a is far from complete, nothing is known about their nature. On the other hand, the problem can be linearized in principle through a sequence of nested Ansatz [31],[42] which generalize the 't Hooft Ansatz through the introduction of massless free fields with higher spins. However the gauge freedom

$$F_0 \to G_0 F_0 \qquad \text{i.e.} \quad F \to G_0 F G_\infty^{-1} \tag{75}$$
$$F_\infty \to G_\infty F_\infty$$

where $G_{\substack{0\\ \infty}}$ belongs to $SL(2,C)$, is holomorphic in $\Omega_{\substack{0\\ \infty}}$ and fulfills

$$D_u G_{\substack{0\\ \infty}} = 0 \tag{76}$$

does not seem to have lead so far to very tractable formulae. The occurrence of massless fields is a consequence of Eq. 68 together with detailed properties of $\mathcal{E}(SL2,C)$ according to which F can be gauged into a triangular form characterized by its non diagonal element.

This concludes a rather imperfect description of the subject. Many omissions, in particular in this last part, are due to the author's poor understanding of the already published results for which written proofs are yet to come but it is hoped that the bibliography will help the puzzled reader to obtain information at the right source.

- REFERENCES -

[1] G. 't HOOFT
Phys. Rev. Lett. 37, 8 (1976)
Phys. Rev. D 14 , 3432 (1976)

[2] R. JACKIW, C. REBBI
Phys. Rev. Lett. 37, 172 (1976)

G.G. CALLAN, R. DASHEN, D.J. GROSS
Phys. Lett. B 63 , 334 (1976)

[3] S. COLEMAN
"Classical Lumps and their Quantum Descendants", International School of Subnuclear Physics, Ettore Majorana, Acad.Press New York, 1975 (Vol. 11)

[4] A. BELAVIN, A. POLYAKOV, A.S. SCHWARZ, Y. TYUPKIN
Phys. Lett. B 59, 85 (1975)

[5] R. JACKIW, C. REBBI
Phys. Rev. D 14, 517 (1976)

[6] G. 't HOOFT
unpublished

[7] R. JACKIW, C. NOHL, C. REBBI
Phys. Rev. D 15, 1642 (1977)

[8] R. JACKIW, C. REBBI
Phys. Rev. Lett., B 67, 189 (1977)

[9] L.S. BROWN, R.D. CARLITZ, C. LEE
University of Washington preprint, RLO 1388, 724, April 1977

R. JACKIW, C. REBBI
unpublished

N.K. NIELSEN, B. SCHROER
CERN preprint TH 2317, May 1977

[10] S. COLEMAN
unpublished

C.P. KORTHALS ALTES
private communication

[11] A.S. SCHWARZ
Phys. Lett. B 67, 172 (1977)

[12] A.S. SCHWARZ
private communication and JINR preprint D2-9788, Proceedings of the Alushta Symposium, April 1976.

[13] M.F. ATIYAH, N.J. HITCHIN, I.M. SINGER
Oxford preprint, 1977
Proc. Nat. Ac. Sci. USA,

[14] M.F. ATIYAH, R.S. WARD
Comm.Math.Phys. 55, 117 (1977)

R.S. WARD
Phys.Rev.Lett. A 61, 81 (1977)

[15] L.P. EISENHARDT
Riemannian Geometry

T. EGUCHI, P.G.O. FREUND
Phys.Rev.Lett. 37, 1251 (1976)

A. BELAVIN, D.E. BURLANKOV
Phys.Rev.Lett. A 58, 7 (1976)

C.P. KORTHALS ALTES, J. MADORE, J.L. RICHARD
unpublished

[16] For a brief summary, see R. STORA in Cargèse Lectures, 1976

[17] N. STEENROD
The topology of fibre bundles, Princeton University Press, Princeton, 1974

[18] S. KOBAYASHI, K. NOMIZU
Foundations of Differential Geometry, Interscience, 1963, vol. I, p. 103.

R. DURIEUX
Thèse de 3e cycle, Marseille, in preparation

[19] T. YONEYA
City College preprint (1977)

[20] W. GREUB, S. HALPERIN, R. VANSTONE
Connections, Curvaturen and Cohomology, Acad. Press 1976, vol. II, p. 476

[21] J.W. MILNOR, J.D. STASCHEFF
Characteristic Classes, Annals of Mathematics Studies (P.U.P.) n° 76, p. 245

[22] R.S. PALAIS
Seminar on the Atiyah Singer Index Theorem, Ann.Math.Stud., n° 57 :
Eq(2), p. 2 , Eq.(6), p. 31, § 5.2, p. 42, § 3, p. 32-39

[23] R.O. WELLS, Jr.
Differential Analysis on Complex Manifolds, Prentice Hall 1973

I. VAISMAN
Cohomology and Differential Forms, M. Dekker 1973, pp. 256, 269 (for the Bochner Lichnerowicz theorem).

[24] Cartan Schwarz Seminar, Secrétariat Mathématique, 1963-1964, Théorème d'Atiyah Singer sur l'indice d'un opérateur différentiel elliptique, Exposés 18, 19, T.2.

[25] P.B. GILKEY
The Index Theorem and the Heat Equation, Mathematical Lecture Series, vol. 4 , publish or perish, 1974.

[26] Ph. A. GRIFFITHS in Proceedings of the Conference on Complex Analysis, Minneapolis 1964, Springer 1965, pp. 113 and ff. Especially Th.2.1, Th.2.2.

A. KUMPERA, D. SPENCER
Lie Equations I, General Theory, Ann.Math.Studies, n° 73 (1972)

[27] M. ATIYAH, R. BOTT, V.K. PATODI
Invent. Math. 19, 279, (1973)

[28] M. BERGER, P. GAUDUCHON, E. MAZET
Le spectre d'une variété Riemannienne, Springer Lecture Notes in Mathematics, vol. 194.

[29] S.I. GOLDBERG
Curvature and Homology, Ac.Press, New York (1962)

[30] J. MADORE
unpublished

[31] M.F. ATIYAH and R.S. WARD
Instantons and Algebraic Geometry, Comm.Math.Phys., 55, 117 (1977)

[32] R.S. WARD
On Self Dual Gauge Fields, Phys.Lett. A 61, 81 (1977)

[33] R. PENROSE, M.A.H. Mac CALLUM
"Twistor Theory...", Physics Reports 6, 241 (1972)

[34] A. BELAVIN, A. POLYAKOV
JETP Letters 22, 303 (1975)

[35] S. KOBAYASHI, K. NOMIZU
Foundations of Differential Geometry, Vol. II, Ch. IX.

[36] H. SCHWERDTFEGER
Introduction to Linear Algebra and the Theory of Matrices, Noordhoff 1961, Appendix A.

[37] R. PENROSE
General Relativity and Gravitation 7, 31 (1976)

[38] C.N. YANG
Condition of Self-Duality for SU2 Gauge Fields..., Phys.Rev.Lett. 38, 1377 (1977)

[39] A.A. BELAVIN, V.E. ZAKHAROV
Phys.Lett.

[40] I.M. SINGER
Lecture given at the International CNRS Colloquium on Operator Algebras and their Mathematical Applications, Marseille June 20-25, 1977, and private communication

[41] W. BARTH
Math. Ann. 226, 125 (1977)

[42] I.M. SINGER
private communication

A. LICHNEROWICZ
private communication

QUANTUM THEORY OF NON-LINEAR INVARIANT WAVE (FIELD) EQUATIONS

OR: SUPER SELECTION SECTORS IN CONSTRUCTIVE QUANTUM FIELD THEORY*†

Jürg Fröhlich§
Department of Mathematics
Princeton University
Princeton, N. J. 08540

TABLE OF CONTENTS

*Research partially supported by the NSF under grant MPS 75-11864
†Lectures delivered at the International School of Mathematical Physics "Ettore Majorana", Erice, Sicily, Summer 1977

§A. Sloan Foundation Fellow

QUANTUM THEORY OF NON-LINEAR INVARIANT WAVE (FIELD) EQUATIONS
OR: SUPER SELECTION SECTORS IN CONSTRUCTIVE QUANTUM FIELD THEORY

General Introduction to Parts 1 and 2

In these lecture notes we first give a brief summary of the general framework presently underlying the constructions of non-trivial relativistic quantum fields (r.q.f.'s). Completely absent from previous reviews of the general framework of constructive quantum field theory (c.q.f.t.) is a discussion of a pragmatic, rigorous theory of super selection sectors useful for applications to specific r.q.f. models: that is to say, previous reviews have exclusively concentrated on describing the general (Euclidean) framework underlying the construction of r.q.f.'s in the vacuum ("Wightman") representation. The phenomena of quantum solitons and non-trivial topological charges in standard r.q.f. models in two space-time dimensions and higher dimensional gauge theories have taught us that this is not enough.

We hope to close the gap described here in these notes, at least to some extent.

Another topic that may not have received the attention it deserves, in some of the recent reviews, is new developments in scattering theory useful for applications to r.q.f. models. We make a few remarks on this subject (mainly concerning multisoliton scattering theory), but we can unfortunately not cover the most interesting recent developments in the subject.

The rigorous connections between the classical and the quantum theory of non-linear, invariant wave equations is only treated, in these notes, in the form of references. This is unfair, because the problems coming up here are most interesting. But a valuable review of this topic would go beyond the limits of these notes.

However, heuristic connections between the classical and the quantum theory will be mentioned at several places, in particular in the discussion of instantons and solitons. We shall also meet some limitations to these connections. We now briefly indicate what sort of heuristic connections we have in mind. (The reader is advised to also consult Gervais' contribution to these proceedings):

Most parts of this review are motivated by the study of relativistic Hamiltonian systems. The classical Hamiltonian functionals of these systems provide Hamilton equations of motion which are classical, non-linear, invariant wave equations; (see Strauss' contribution to these proceedings).

In order to get some qualitative insight into the quantum theory of such relativistic Hamiltonian systems we shall determine the critical points (absolute and local minimas) of their classical Hamilton functionals: Their structure will teach us some qualitative properties of the quantum theory, in particular properties of the physical vacuum state and of the super selection rules.

But it turns out (especially in the study of gauge theories) that it is not always enough to investigate the critical points of the classical Hamilton functional in order to predict all qualitative features of the quantum theory. This is because of the possibility of quantum mechanical tunnelling between different absolute minima of the classical Hamilton functional. If the (classical) Euclidean action has local minima at non-constant field configurations (instantons) interpolating different, constant field configurations corresponding to absolute minima then there is tunnelling, and the structure of the quantum mechanical ground state (vacuum) is not correctly predicted by the minima of the classical Hamilton functional.

A well known example is the one dimensional anharmonic oscillator with Hamilton function

$$p^2 - \frac{1}{4} x^2 + g x^4 + \frac{1}{64g} \tag{1}$$

where g is strictly positive.

One sees immediately that this Hamilton function is non-negative. The potential

$$-\frac{1}{4} x^2 + g x^4 + \frac{1}{64g} \tag{2}$$

has two degenerate minima at $X = \pm (8g)^{-\frac{1}{2}}$ (where it vanishes) separated by a (for $0 < g << 1$) large barrier. The curvature of the potential at $x = \pm (8g)^{-\frac{1}{2}}$ is $= 1$.

If we did not know better these properties of the classical Hamilton function might mislead us to make the following wrong conjecture: For $0 < g << 1$, the groundstate and the entire spectrum of the corresponding quantum mechanical Hamilton operator are two fold degenerate, (and the spectrum resembles the one of two uncoupled, harmonic oscillators).

That this is wrong can be seen by considering the Euclidean action of the anharmonic, double hump oscillator, defined by

$$S = \int_{-\infty}^{+\infty} \{\dot{x}(t)^2 - \frac{1}{4} x(t)^2 + g\, x(t)^4 + \frac{1}{64g}\}\, dt \tag{3}$$

It has two absolute minima at

$$x(t) \equiv \pm \frac{1}{\sqrt{8g}} \tag{4}$$

and local minima at

$$x(t) = \pm \frac{1}{\sqrt{8g}} \tanh\left(\frac{t}{\sqrt{8g}}\right) \tag{5}$$

representing homotopy classes of ("field") configurations x(t) different from the ones represented by (4). The solutions (5) (of the Euler-Lagrange equations obtained by varying S) interpolate the solutions (4), $x(t) \equiv \pm (8g)^{-\frac{1}{2}}$, and signalize quantum mechanical tunnelling between the two absolute minima (classical groundstates) (4) of the classical Hamilton function(1). It is well known that, indeed, there is tunnelling (barrier penetration). The groundstate of the quantum mechanical Hamilton operator is unique, the spectrum is simple. However, it does resemble - for small $g > 0$ - the one of two uncoupled, harmonic oscillators: Eigenvalues come in pairs separated by a small, but positive gap $\propto \exp(-\text{const.}\ g^{-1})$ and perturbation theory in $g^{\frac{1}{2}}$ about either of the two degenerate minima $x = \pm (8g)^{-\frac{1}{2}}$ is asymptotic to both eigenvalues in one such pair; see [R1] . Moreover, A. Sokal has shown [S1] that these eigenvalues as functions of $g^{\frac{1}{2}}$ have singularities along a horn-shaped region

$$\{z: \mathrm{Re}z > 0, |\arg z| < \varepsilon(\mathrm{Re}z) \cdot \mathrm{Re}z\}, \tag{6}$$

with $\varepsilon(\mathrm{Re}z) \searrow 0$, as $\mathrm{Re}z \searrow 0$, accumulating at $g = 0$.

These analyticity properties are incompatible with Borel summability [S1].

The situation described here for the anharmonic, double hump oscillator appears to be typical for gauge theories with instantons [H1 , C1] - such as the two dimensional abelian Higgs model which we discuss in [F1] ; see also [C2] . We summarize this somewhat vague discussion: A study of the critical points (absolute and local minima) of the Euclidean action of a Hamiltonian system (and of the classical Euler-Lagrange equations obtained by varying the action) appears to be a useful guide for guessing the qualitative properties of the groundstate and the low-lying spectrum of the corresponding quantum system. Unfortunately we cannot discuss the recent developments of this topic ("instanton physics," see [P1, B1, P2, H2, C3, J1] in these notes for reasons of "space-time limitations";

Next we want to briefly indicate in what sense in a Hamiltonian system the critical points of the classical Hamilton functional yield information about the structure of the super-selection sectors of the corresponding quantum system. We do that by considering a specific r.q.f. model in two space-time dimensions. Here some notation: Space-time points in M^2 are denoted $x = (\underline{x}, t)$ (with t the time coordinate). Partial derivatives with respect to $\underline{x}$, resp. t are denoted $\partial_{\underline{x}}$, resp. ∂_t. Furthermore $\phi(x)$ is a real, scalar c-number field on M^2, and $\pi(x)$ denotes the momentum (field) canonically conjugated to ϕ. We now define the classical Hamilton density $\mathcal{H}$ of the well known $g\,\phi_2^4$ - theory.

$$\mathcal{H}(\pi, \phi) = \mathcal{H}_o(\pi,\phi) + \mathcal{H}_I(\phi), \tag{7}$$

with

$$\mathcal{H}_o(\pi(x), \phi(x)) = \frac{1}{2}\{\pi(x)^2 + (\partial_{\underline{x}}\phi(x))^2\}, \tag{8}$$

and

$$\mathcal{H}_I(\phi(x)) = g\ \phi(x)^4 - \frac{1}{4}\phi(x)^2 + \frac{1}{64g} \tag{9}$$

The constant term in the "potential" $\mathcal{H}_I$ is so chosen that $\mathcal{H} \geqq 0$. The Hamilton equations of motion derived from (7) - (9) are

$$\left.\begin{aligned} \pi(x) &= \partial_t \phi(x), \\ \Box\phi(x) &= -4\,g\,\phi(x)^3 + \frac{1}{2}\phi(x) \end{aligned}\right\} \tag{10}$$

For a complete existence theory for solutions of (10) see the contributions of Gårding and Strauss to these proceedings, and refs. given there.

A finite-energy solution of (10) is a solution ϕ_0 for which the Hamilton functional

$$H(\pi,\phi) = \int_{-\infty}^{+\infty} d\underline{x}\, \mathcal{H}(\pi(\underline{x},t), \phi(\underline{x}, t)) \geqq 0 \tag{11}$$

takes a finite value

$$E_0(\phi_0) = \int_{-\infty}^{+\infty} d\underline{x}\, \mathcal{H}(\partial_t\phi_0(\underline{x},0), \phi_0(\underline{x}, 0)) < \infty$$

These finite-energy solutions fall into four homotopy classes represented by the stationary solutions

$$\phi_+(x) \equiv (8g)^{-\frac{1}{2}}, \; \phi_-(x) \equiv -(8g)^{-\frac{1}{2}} \tag{12}$$

$$\phi_s(x) = (8g)^{-\frac{1}{2}} \tanh((8g)^{-\frac{1}{2}} \underline{x}), \quad \phi_{\bar{s}}(x) = -(8g)^{-\frac{1}{2}} \tanh((8g)^{-\frac{1}{2}} \underline{x}) \tag{13}$$

The solutions ϕ_s and $\phi_{\bar{s}}$ are called soliton - , resp. anti-soliton solutions. They interpolate between the constant solutions ϕ_+ and ϕ_-. (Note that ϕ_s and $\phi_{\bar{s}}$ are equal to the "instanton solutions" (5) of the anharmonic, double hump oscillator. At this point we meet the following rather general principle: A theory in d space-time dimensions with solitons corresponds to an analogous theory in d-1 space-time dimensions which has instantons).

From heuristic and rigorous work concerning quantum solitons that has been done over the past few years emerged the following picture: The homotopy classes of finite energy solutions to some classical Hamiltonian equations of motion (non-linear, invariant field equations) are, for small values of g ($\propto$ Planck's constant), in a one - one correspondence with inequivalent super-selection sectors of the corresponding quantum field theory, provided the homotopy classes represented by the constant solutions (in $g\,\phi_2^4$, ϕ_+ and ϕ_-) are in one - one correspondence with the critical points of the Euclidean action (in $g\,\phi_2^4$ given by $\phi(x) = \pm (8g)^{-\frac{1}{2}}$), i.e., provided there is no tunnelling leading to a more complicated structure of the physical vacuua.

In Part II we shall show that this picture is correct for the $g\,\phi_2^4$ - theory in two space-time dimensions: In this r.q.f. model the homotopy classes represented by ϕ_+ and ϕ_- correspond to two inequivalent vacuum sectors, whereas the ones represented by ϕ_s and $\phi_{\bar{s}}$ correspond to two (inequivalent) soliton sectors with non-vanishing, opposite "topological charge"

$$Q = \int_{-\infty}^{+\infty} d\underline{x}\, (\partial_{\underline{x}}\phi)(x).$$

By considering the two space-time dimensional, abelian Higgs model one can also show that this picture may be false in the strong coupling regime (where there may exist superselection sectors not predicted by the classical field equations)[Fl]. This is perhaps bad news for many heuristic approaches to the problem of quantizing solitary waves which are usually based on some sort of expansion in g ($\propto \hbar$ = Planck's constant), only valid for small g.

So far we have pretended that the Euclidean action (more generally: the Euclidean description of r.q.f.t.) provides information concerning the properties of the quantum mechanical ground state, in r.q.f.t. the vacuum sectors of a r.q.f. model, but that we must investigate the Hamilton functional (more generally the Hamiltonian description of r.q.f.t.) in order to obtain insight into the structure of superselection sectors inequivalent to the vacuum sectors. While this principle is of considerable, heuristic value, it is conceptually not satisfactory.

Our main contention is that with the help of somewhat more sophisticated mathematical analysis it is possible to extract all information about a r.q.f. theory from its Euclidean description. This contention is detailed in Part I.

In Part I we summarize the general framework underlying the construction of non-trivial r.q.f.'s and their super-selection sectors. In Part II we shall discuss an explicit model (the $g\phi_2^4$ theory) exemplifying thereby the general framework of Part I and emphasizing the role played by classical soliton solutions in obtaining insight into the general features of the quantum theory.

These notes are supposed to be light reading and do not contain any lengthy proofs or difficult estimates.

Part 1

The general framework of constructive quantum field theory (including a pragmatic theory of super-selection sectors)

1.1. Introduction

One of the things a general framework of c.q.f.t. ought to teach us is how to attempt to solve a non-linear, invariant field equation of the form

$$\Box\phi(x) = N\left[F(\phi(x))\right], \tag{1.1}$$

where $\Box$ is the d'Alembertian, $F = -P'$ and P' is the derivative of a polynomial P bounded below (e.g. $P(\phi) = g\,\phi^4 - \frac{1}{4}\phi^2 + \frac{1}{64g}$), $x \varepsilon M^\nu$ with $\nu = 1, 2, 3, (4)$ (the number of space-time dimensions), and N denotes "normal ordering," a prescription of how to define higher powers of $\phi(x)$; finally the solution $\phi(x)$ of (1.1) is supposed to be a real, scalar field on M^ν with values in the "operator-valued distributions" on a "suitable Hilbert space".

The following are some of the central problems arising in attempting to solve (1.1) under these requirements:

--What is a "suitable Hilbert space"?

--What is a "suitable" topological vector space of operator - valued distributions within which solutions to equation (1.1) for some (hopefully interesting) class of F's can be constructed? What class of Cauchy data (at time $t = 0$, say) can be imposed?

--Do physics or mathematics impose additional requirements on the class of F's and on the solutions $\phi(x)$?

Clearly, the c-number solutions to (1.1) constructed in the lectures of Gårding and Strauss can be viewed as operator-valued distributions on a one-dimensional Hilbert space and hence represent a solution to the problem posed above. But they do not have anything to do with quantum mechanics, resp. the quantum theory of non-linear, invariant field equations.

In quantum mechanics observables do generally not commute with each other, and there is no reason why, quantum mechanically,

$$\left[\phi(x), \phi(y)\right] \equiv \phi(x)\ \phi(y) - \phi(y)\ \phi(x) = 0 \tag{1.2}$$

(in the sense of distributions).

If we impose (1.2) as a condition on the solutions of (1.1) we fall back into the classical theory. Quantum mechanics tells us that, in general, the r.h.s. of (1.2) ought to be $\neq 0$ and proportional to $\hbar$ (Planck's constant).

Quantum physics does however impose specific restrictions on $[\phi(x), \phi(y)]$ that should be obeyed by the solutions to (1.1). One of them is (Einstein) causality:

$$[\phi(x), \phi(y)] = 0, \quad \text{for} \quad (x-y)^2 < 0. \tag{1.3}$$

Here $(x-y)^2 = (x^0-y^0)^2 - (\underline{x} - \underline{y})^2$; ($x = (x^0,\underline{x})$ with x^0 the time-component of x, $\underline{x} \in R^{\nu-1}$).

However, for $(x-y)^2 > 0$, $[\phi(x), \phi(y)] \neq 0$. (1.4)

It is well known that for many years the combination of (1.1) with (1.3) - (1.4) looked like an ill-defined problem, or at least a very difficult one, except for $\nu = 1$, (the case of the anharmonic oscillator) where (1.3) is void.

This lead to frustration, and frustation lead to (among other things) axiomatic field theory which attempted to formulate general principles (axioms) any r.q.f.t., and in particular any quantum theory of (1.1), ought to satisfy.

One of the main purposes of formulating axioms was to show that the basic principles of r.q.f.t. are compatible with each other (and with the requirement of non-trivial scattering) or, more concretely, to pose the problem of developing a quantum theory of field equations such as (1.1) in a mathematically precise way.

Most suitable for our purposes are the Wightman axioms [S2, J2] which we briefly recall here, for the convenience of the reader. (At this point we note, however, that the developments of the past two or three years, in particular the discovery of quantum solitons, call for a slight modification of the framework laid down by the Wightman axioms. A possible such modification or extension, inspired by the Haag-Kastler axioms [H3] - and the author's work on quantum solitons in r.q.f. models in two dimensional space-time [F2, F3] - is described later in Part 1 and exemplified in Part 2).

First we give the <u>Hilbert space formulation</u> of the <u>Wightman axioms</u>, postulates which - it is proposed - ought to be obeyed by all admissible solutions of a relativistic field equation such as (1.1):

(Wo) The states of a physical system are the unit rays of a separable Hilbert space $\mathcal{H}$ (more generally, of a family of such Hilbert spaces: the super-selection sectors).

(Wi) To each test function f in the Schwartz space $\mathcal{S}(R^d)$ there corresponds an unbounded operator $\phi(f)$ on $\mathcal{H}$ with $\phi(f)^* \supseteq \phi(\bar{f})$. There is a domain D dense in $\mathcal{H}$ which is in the domain of all the operators $\{\phi(f);\ f \in \mathcal{S}(R^d)\}$ and left invariant by them.

(Wii) There is a continuous, unitary representation $U: (a,\Lambda) \in \mathcal{P}_+^\uparrow \longmapsto U(a,\Lambda)$ of the Poincaré group $\mathcal{P}_+^\uparrow$ on $\mathcal{H}$ such that

$$U(a,\Lambda)\ \phi(f)\ U(a,\Lambda)^* = \phi(f_{(a,\Lambda)})$$

$$f_{(a,\Lambda)}(x) = f(\Lambda^{-1}(x-a))$$

(Wiii) The spectrum of the infinitesimal generator (H,P) (the energy-momentum operator) of $\{U(a,1): a \in R^{\nu}\}$ is contained in the forward light cone $\overline{V}_+$, and (0,0) is an eigenvalue of (H,P).

(Wiv) On the domain D of (Wi),

$$[\phi(f), \phi(g)] = 0,$$

if the supports of f and g are space-like separated; (see (1.3)).

(Wv) Let $\mathcal{P}(\phi)$ be the *algebra of polynomials in $\{\phi(f): f \in \mathcal{S}(R^d)\}$. Then there is a vector Ω in the eigenspace of (H,P) corresponding to the eigenvalue (0,0) such that $\Omega \in D$ and $\{\mathcal{P}(\phi)\Omega\}$ is dense in $\mathcal{H}$.

From now on we choose $D = \{\mathcal{P}(\phi)\Omega\}$. In this formulation the field equation is equivalent to the requirement that

$$\{\Box\phi(x) - N[F(\phi(x))]\}\Omega = 0 \tag{1.5}$$

in the sense of $\mathcal{H}$-valued distributions. This is a consequence of the Reeh-Schlieder theorem [S2, J2].

Among (Wo) - (Wv) we feel (Wv) is the least plausible axiom; and it is the one that will require modifications in models with quantum solitons.

Wightman's reconstruction theorem says that a r.q.f.t. satisfying (Wo - Wv) is completely determined by the vacuum expectation values (v.e.v.'s)

$$W_n(x_1,\ldots,x_n) = \langle\Omega, \phi(x_1)\ldots\phi(x_n)\Omega\rangle, \tag{1.6}$$

n = 0, 1, 2,..., and equation (1.6) is to be understood as an equation between tempered distributions, thanks to (Wi) and (Wv). The W_n's are called Wightman distributions. They satisfy - see [S2, J2] for explanations and details:

(W0) $W_0 = 1$ ($\Leftrightarrow$ $\langle\Omega,\Omega\rangle = 1$); for all $n \geq 1$, W_n is a tempered distribution, ($\Leftrightarrow$ (Wi), (Wv) + nuclear theorem).

(W1) Positivity ($\Leftrightarrow$ the scalar product $\langle -, - \rangle$ of $\mathcal{H}$ is positive definite).

(W2) Invariance: The distributions W_n are Poincaré-invariant ($\Leftrightarrow$ W(ii), W(v)).

(W3) Spectrum condition: Support properties of the Fourier transform $\hat{W}_n$ of W_n, n = 1, 2, ... ($\Leftrightarrow$ spec $(H,P) \subseteq \overline{V}_+$).

(W4) Locality: For $(x_i - x_{i+1})^2 < 0$ $W_n(\ldots x_i, x_{i+1},\ldots) = W_n(\ldots x_{i+1}, x_i\ldots)$, for all n ($\Leftrightarrow$ (Wiv)).

Wightman has shown the equivalence of (Wo) - (Wv) and (W0) - (W4).

One of the main problems of constructive quantum field theory is to construct models of r.q.f.'s with non-trivial scattering satisfying (W0) - (W4), if possible in a space-time of dimension 4. Conventionally one looks for such models which are formally parametrized by a non-linear, invariant field equation, e.g. of the form (1.1), (1.5);

(nowadays other parametrizations, formally equivalent to a field equation, are however usually preferred).

In dimension $\nu = 4$ this main problem is still unsolved, but, for $\nu = 1,2,3$, plenty of models with non-trivial scattering satisfying (W0) - (W4) and (1.5) exist; (for these models unitarity of the scattering operator at high energies is still an open problem. Recently Zamolodchikov [Z1] and a group of people in Berlin [T1] explicitly constructed some non-trivial, unitary scattering operators for r.q.f. models in two space-time dimensions, a truly admirable result. However it is not yet proven that these scattering operators come from a r.q.f. theory satisfying (W0) - (W4), and, moreover, they are trivial in so far as they do not describe inelastic processes, and all momenta are conserved. This makes these constructions special to two space-time dimensions, in contrast to some constructive q.f.t. results [S3, O1, E1]).

In pursuit of the main problem of c.q.f.t. it has turned out to be extremely difficult to attempt a direct construction of the Wightman distributions $\{W_n\}_{n=0}^{\infty}$ of some formal r.q.f.t.model, e.g. one formally defined in terms of a field equation such as (1.1), (1.5); see [G1, H4].

This lead to the following discovery [S4, N1] : It is always possible to convert the hyperbolic problem of solving a non-linear, invariant field equation such as (1.1), (1.5) subject to the requirement that the solutions satisfy (W0) - (W4) into an elliptic problem + analytic continuation.

Part of the discovery (in a sense its real depth) is the following recipe: Try to solve the elliptic problem by means of probabilistic potential theory, (the theory of generalized, stochastic processes).

This discovery has a prehistory in quantum mechanics where it is often easier to construct the unitary time evolution group e^{itH} of a quantum mechanical system by first constructing the semigroup e^{-tH}, $t \geq 0$, using path space techniques and the Feynman-Kac formula and then doing an analytic continuation in the time t. (This is by now a well established technique for proving selfadjointness of quantum mechanical Hamiltonians H).

Before illustrating the discovery described here by considering some examples we explain on what mathematical facts it is based:

From axioms (W2) and (W3) it follows, according to the Bargmann-Hall-Wightman theorem [S2, J2], that the Wightman distributions $W_n(z_1,\dots,z_n)$ are holomorphic in a large domain of $\mathbb{C}^{\nu n}$ (called the extended tube) which contains the domains

$$\mathcal{T}_n = \{z_1,\dots,z_n : \operatorname{Im}(z_j - z_{j-1}) \in V_+\}$$

(and, by (W4), also the domains obtained by permuting the arguments) and the so called Euclidean points

$$\mathcal{E}_n = \{z_1,\dots,z_n : z_j = (\underline{x}_j, it_j),\ z_i \neq z_j,\ \text{for } i \neq j\}\ , \tag{1.7}$$

where $\underline{x}_j \in R^{\nu-1}$ denotes the space - and it_j, $t_j \in R$, the time-component of z_j.

On $\mathfrak{E}_n$ one defines

$$S_n (x_1,\dots,x_n) = W_n (\underline{x}_1, it_1,\dots, \underline{x}_n, it_n), \quad x_j = (\underline{x}_j, t_j). \tag{1.8}$$

These are called the Euclidean Green's or Schwinger functions. For $t_1 < t_2 <\dots<t_n$, $S_n (x_1,\dots,x_n)$ is formally given by

$$S_n (x_1,\dots,x_n) = < \Omega, \prod_{j=1}^{n-1} (\phi (\underline{x}_j, 0) \; e^{-(t_{j+1} - t_j) H}) \; \phi (\underline{x}_n, 0) \; \Omega>. \tag{1.9}$$

Most of the properties of the functions S_n can be guessed by studying the formal equation (1.9). Among these properties - which follow from (W0) - (W4) - are:

The Schwinger functions S_n are Euclidean invariant, (to be compared with (W2));

they have a positivity property, called Osterwalder - Schrader positivity [O2], (a rigorous expression for the facts that, according to (1.9) they can be written as scalar products and that e^{-tH}, $t \geqq 0$, is a selfadjoint contraction semigroup);

they are symmetric under permutations of their arguments (a consequence of locality (W4)).

The discovery [S4, N1] described above can now be rephrased as follows: A construction of generalized functions $\{S_n\}_{n=0}^{\infty}$ satisfying the properties stated above (and some additional technical conditions [N1, O2, F4]; this will turn out to be an "elliptic problem") automatically yields Wightman distributions $\{W_n\}_{n=0}^{\infty}$, by analytic continuation in the time-variables, such that equation (1.8) holds.

The past five years of c.q.f.t. and theoretical physics ("instanton-physics" [P2, H2, C1, C2]) have shown that it is advantageous to first try to construct the Schwinger functions $\{S_n\}_{n=0}^{\infty}$ of some r.q.f.t., using the theory of functional integrals, (renormalized Fredholm determinants) and stochastic processes.

This raises the following natural question: What precise properties (including the ones already mentioned) of a sequence $\{S_n\}_{n=0}^{\infty}$ of generalized functions are sufficient for the S_n's to be the Schwinger functions of W_n's satisfying (W0) - (W4), in the sense of equation (1.8) (see also (1.9))?

This question has been answered by Osterwalder and Schrader in a very satisfactory way [O2]. Here we briefly recall a somewhat weaker version of their main theorem ("Osterwalder-Schrader reconstruction").

We define

$$R^{\nu}_{+,\varepsilon} = \{ (\underline{x},t): t \geqq \varepsilon, \; \varepsilon \geqq 0\} \tag{1.10}$$

$$\mathcal{S}(R^{\nu}_{+,\varepsilon}) = \{ f : f \in \mathcal{S}(R^{\nu}), \; \mathrm{supp}\, f \subseteq R^{\nu}_{+,\varepsilon} \},$$
$$\text{and } \mathcal{S}_{+} = \mathcal{S}(R^{\nu}_{+,\varepsilon=0}) \tag{1.11}$$

For $f \in \mathcal{S}(R^\nu)$, define

$$\text{Exp}\, f = \{f_n\}_{n=0}^{\infty}\text{, with } f_o = 1,\ f_1 = f, \ldots$$

$$f_n(x_1,\ldots,x_n) = (n!)^{-1} \prod_{j=1}^{n} f(x_j) \equiv (n!)^{-1} f^{\otimes n}(x_1,\ldots,x_n) \qquad (1.12)$$

Following Osterwalder and Schrader [O2] we postulate - as in [D1] - the following properties on $\{S_n\}_{n=0}^{\infty}$: (E0) $S_o = 1$, $S_n \in \mathcal{S}'(R^{\nu n})$; for all $f \in \mathcal{S}(R^\nu)$ for which $|f| < 1$, where $|\cdot|$ is some norm continuous on $\mathcal{S}(R^\nu)$,

$$\sum_{n=o}^{\infty} (n!)^{-1} S_n(f^{\otimes n}) \equiv S(\text{Exp}\, f) \qquad (1.13)$$

converges absolutely.

(This is much more than has to be assumed [O2]; but it is convenient for the purposes of this review).

(E1) The S_n's are Euclidean invariant, for all n = 1, 2, 3, ..., or

$$S(\text{Exp}\, f) = S(\text{Exp}\, f_\beta), \qquad (1.14)$$

for all proper Euclidean motions β of R^ν, where $f_\beta(x) \equiv f(\beta^{-1}x)$, and $|f| < 1$.

(E2) Let $f_\theta(\underline{x},t) \equiv f(\underline{x},-t)$.

For $f^j \in \mathcal{S}_+$, j = 1,...,N, arbitrary N = 1, 2, 3, ...

$$M_{ij} \equiv S(\text{Exp}\, \overline{f^i}_\theta \times \text{Exp}\, f^j) \equiv \sum_{m,n=o}^{\infty} (m!)^{-1}(n!)^{-1} S_{m+n}((\overline{f^i}_\theta)^{\otimes m} \otimes (f^j)^{\otimes n})$$

are the matrix elements of a positive semidefinite matrix M.

Remark: This is a stronger version of the usual Osterwalder-Schrader positivity condition, [O2].

(E3) For all n = 1, 2, 3, ..., $S_n(x_1,\ldots,x_n)$ is symmetric under permutations of its arguments.

Remark: Postulates (E0) - (E3), as formulated here, are suitable for the Euclidean description of relativistic, neutral, scalar fields. For fields with charge and (or) spin and (or) internal degrees of freedom they have to be modified in a standard way. For gauge fields the modification is not standard, and this is still an important research topic!

Theorem 1.1, [O2]

For generalized functions $S_n(x_1,\ldots,x_n)$, n = 0, 1, 2, ..., to be the Schwinger functions, in the sense of equation (1.8) - (1.9), of a r.q.f.t. satisfying (W0) - (W4) it is sufficient that $\{S_n\}_{n=0}^{\infty}$ satisfy (E0) - (E3).

Remark: Osterwalder and Schrader [O2] (and Glaser [G2]) prove a more general theorem of this type, with considerably weaker hypotheses, but identical conclusions.

A few remarks on the proof of Theorem 1.1 - which consists in essence of continuing the Schwinger functions S_n - and estimates on S_n - analytically in the time-variables back to the (real time) Minkowski region and verifying (W0) - (W4) for the boundary values so obtained - can be found in the next section. This result tells us that a r.q.f.t. satisfying (W0) - (W4) can be described in terms of its Schwinger functions, i.e. there exists a purely Euclidean description of r.q.f.t. (There are other versions of (E0) - (E3) equivalent to (W0) - (W4), [O2]).

Next, we explain how this Euclidean description of r.q.f.t. may be related to functional integrals and stochastic processes:

Postulates (E0) - (E3) (in particular (E3)) are obviously compatible with the existence of a probability measure μ on (the σ - algebra Σ generated by the Borel cylinder sets of) $\mathcal{S}' \equiv \mathcal{S}'_{real}(R^\nu)$ of real-valued, tempered distributions such that

$$\begin{aligned} S_n(f_1,\dots,f_n) &\equiv \int S_n(x_1,\dots,x_n) \prod_{i=1}^{n} f_i(x_i)\, d^\nu x_i \\ &= \int \prod_{i=1}^{n} \phi(f_i)\, d\mu(\phi), \end{aligned} \tag{1.15}$$

for arbitrary $f_1,\dots,f_n$ in $\mathcal{S}(R^\nu)$.

Here $\{\phi(f) : f \in \mathcal{S}(R^\nu)\}$ are the "coordinate functions" defined by $\phi(f)[T] = T(f)$, for all $T \in \mathcal{S}'$.

Given (E0) and (E3), the existence of a probability measure μ on $\mathcal{S}'$ is equivalent to the following inequality, usually referred to as Nelson-Symanzik positivity:

(NS) For arbitrary $f^1,\dots,f^N$ in $\mathcal{S}(R^\nu)$ with $|f^j| < 1/2$; $N = 1, 2, 3, \dots,$

$$\tilde{M}_{ij} \equiv S(\overline{\mathrm{Exp}\, f^i} \times \mathrm{Exp}\, f^j) \tag{1.16}$$

are the matrix elements of a positive (semi-) definite matrix $\tilde{M}$.

Note: Condition (NS) does not follow from the Wightman axioms (W0) - (W4). In rare cases (NS) follows from the combination of (W0) - (W4) with a field equation, such as (1.1), (1.5).

Nevertheless, condition (NS) holds in most of the r.q.f. models thus far constructed, and it has been a powerful tool for constructing them.

By Minlos' theorem [M1] , (1.16), (E0) and (E3) imply that $S(\mathrm{Exp}\, f)$, $|f| < 1$, is the Laplace transform of a probability measure μ on $\mathcal{S}'$:

$$S(\mathrm{Exp}\, f) = \int_{\mathcal{S}'} e^{\phi(f)}\, d\mu(\phi), \quad |f| < 1. \tag{1.17}$$

In terms of the measure μ, postulates (E1) and (E2) can be reformulated as follows:

Let F be a $\sum$ - measurable function on $\mathcal{S}'$.
Let β be some Euclidean motion of R^ν. We define $\beta F(\phi) = F(\phi_\beta)$, where

$$\phi_\beta(f) = \phi(f_\beta),\ f \in \mathcal{S}(R^\nu).$$

From (E1) we then get

$$\int_{\mathcal{S}'} \beta F(\phi)\, d\mu(\phi) = \int_{\mathcal{S}'} F(\phi)\, d\mu(\phi), \tag{1.18}$$

For all β; i.e. μ is Euclidean invariant.

Let $\sum_\pm$ be the smallest σ - algebras on $\mathcal{S}'$ such that all the coordinate functions

$$\{\phi(f) : f \in \mathcal{S}(R^\nu),\ \mathrm{supp}\, f \subseteq \{t \gtrless 0\}\}$$

are $\sum_\pm$ - measurable.

We define the Euclidean time-reflection θ by

$$\theta F(\phi) = F(\phi_\theta),\ \phi_\theta(f) = \phi(f_\theta),\ f \in \mathcal{S}(R^\nu),$$

for arbitrary $\sum$ - measurable functions F on $\mathcal{S}'$. Clearly θ defines an isomorphism between $\sum_+$ and $\sum_-$.

<u>Lemma 1.2</u>, [F4] :

Suppose the Schwinger functions $\{S_n\}_{n=0}^{\infty}$ satisfy (E0) - (E3) and (NS). Then there exists a probability measure μ on $\mathcal{S}'$ such that

(1) Equations (1.17) and (1.15) hold, i.e. the moments of μ are the Schwinger functions S_n.

(2) μ is Euclidean invariant, in the sense of equation (1.18).

(3) For arbitrary $\sum_+$ - measurable functions F on $\mathcal{S}'$

$$\int \overline{\theta F(\phi)}\ F(\phi)\, d\mu(\phi) \geqq 0 \tag{1.19}$$

(This is the probabilistic version of Osterwalder-Schrader positivity).

<u>Remark</u>: Every $\sum_+$ - measurable function $F \in L^2(\mathcal{S}', \sum, d\mu)$ can be approximated in L^2 by functions of the form

$$\sum_{i=1}^{m} c_i\ e^{\phi(f_i)},\ c_i \in \mathbb{C},\ f_i \in \mathcal{S}(R^\nu),$$

$|f_i| < 1/2$; see [F4]. So Lemma 1.2 (3) follows from (E2), (NS) and (1.17).

The class of probability measures described in Lemma 1.2 is denoted $\mathcal{M}_{q.m.}$. These measures are called <u>quantum measures</u> [F5 ,F6].

Now we can rephrase the main problem of the quantum theory of non-linear, invariant field equations as follows:

Consider the equation (1.1),

$$\Box\phi\,(x) = N\left[F(\phi(x))\right],$$

where $F = -P'$, and P is e.g. a polynomial bounded from below, and degree $(P) = 4$, for $\nu = 3$. (For $\nu = 1,2$, much larger classes of functions P can be studied).

Let $\phi(x)$ be the stochastic process given by some measure $\mu \in \mathcal{M}_{q.m.}$.

Let $N_\mu\left[F(\phi(x))\right]$ be some normal (Ito) ordered version of F (i.e. N_μ is a prescription, depending on the measure μ, of how to define powers $\phi(x)^n$ of the process $\phi(x)$ with distribution μ). In the cases understood at this time (e.g. the $\lambda\phi^4$ - models in 2 or 3 space-time dimensions) the operator N_μ depends on μ in a manner that can be determined a priori, and, as a formal operation on powers, it coincides with N. Let Δ be the ν-dimensional Laplacean, and define

$$N_\mu\left[P(\phi + f) - P(\phi)\right] \equiv \int d^\nu x\, N_\mu\left[P((\phi(x) + f(x)) - P(\phi(x))\right]$$

Theorem 1.3:

In $\nu = 1,2,3$ space-time dimensions, and for the class of functions F specified above, any solution $\mu \in \mathcal{M}_{q.m.}$ (a quantum measure) satisfying the "Radon-Nikodym" equations

$$(\mathrm{RN})\qquad \frac{d\mu(\phi + f)}{d\mu(\phi)} = e^{\phi(\Delta f) + \frac{1}{2}(f,\,\Delta f)} \times e^{-N_\mu\left[P(\phi + f) - P(\phi)\right]}, \quad F = -P',$$

for arbitrary $f \in \mathcal{S} \equiv \mathcal{S}_{real}(R^\nu)$, provides a solution $\phi(x)$ to the non-linear, invariant field equation (1.1) satisfying the Wightman axioms (Wo) - (Wv) (resp. (W0) - (W4))and the equation (1.5).

Remarks: By Osterwalder-Schrader reconstruction, Theorem 1.1, and Lemma 1.2, the moments of any measure $\mu \in \mathcal{M}_{q.m.}$ are the Schwinger functions, in the sense of equations (1.8), (1.9), of a unique r.q.f.t. satisfying the Wightman axioms (W0) - (W4). The proof of Theorem 1.3 is therefore reduced to showing that equation (RN) implies the field equation (1.5). Here is the outline of a formal proof of (1.5):

Let $f_j \in \mathcal{S}(R^\nu_{+,\varepsilon})$, for some $\varepsilon > 0$. Using equation (RN) in infinitesimal form ("integration by parts on function space", see [D2, F7]) one finds

$$\Delta_x \int_{\mathcal{S}'} \phi(x) \prod_{j=1}^{n} \phi(f_j)\, d\mu(\phi) = \int_{\mathcal{S}'} N_\mu\left[F(\phi(x))\right] \prod_{j=1}^{n} \phi(f_j)\, d\mu(\phi), \qquad (1.20)$$

for arbitrary n = 0,1,2,..., in the sense of distributions on $\mathcal{S}(R^\nu_-) \equiv \{f: f_\theta \in \mathcal{S}(R^\nu_+)\}$.

Equation (1.5) follows from equation (1.20) by Osterwalder-Schrader reconstruction (analytic continuation in the time variable, [O2]. See also [F7]).

Theorem 1.3 tells us that the main problem of the quantum theory of equation (1.1) can be solved by constructing solutions $\mu \in \mathcal{M}_{q.m.}$ to equation (RN). In one dimension the problem of solving equation (1.1), resp. (1.5), in the class of fields ϕ satisfying (W0) - (W4) is equivalent to solving equation (RN) in the class $\mathcal{M}_{q.m.}$ of quantum measures on $\mathcal{S}' = \mathcal{S}'_{real}$ (R). This is the quantum mechanical anharmonic oscillator, and the equivalence follows from the Feynman-Kac formula [N2, S5]. In the case $\nu = 2,3$, $F = -P'$, $P(\phi) = g\phi^4 + \frac{m^2}{2}\phi^2$, $0 \leqq g << 1$, $m^2 > 0$, this equivalence follows from [G3 , E2 , M2] (for g > 0: under one additional, most natural assumption), but it is strongly believed to be true for the class of functions F specified above, without restrictions. (In $\nu = 4$ dimensions the situation is very unclear),

A very formal "solution" (generally meaningless, but of great heuristic power) of equation (RN) is given by the so called Euclidean Gell'Mann- Low formula:

$$d\mu(\phi) = Z^{-1} e^{-\frac{1}{2}\int \{(\nabla\phi)^2 (x) + N_\mu[2P(\phi(x))]\} d^\nu x} \prod_{x \in R^\nu} \mathcal{D}\phi(x), \tag{1.21}$$

where Z is an infinite normalization factor.

If $P(\phi) = \frac{m^2}{2}\phi^2$ then equation (1.21) can be given a rigorous mathematical meaning: In this case μ is simply the Gaussian measure on $\mathcal{S}'$ with mean 0 and covariance $(-\Delta + m^2)^{-1}$; ϕ is then called the "free, Euclidean field"; see [N3] . The moments of μ are the Schwinger functions of the relativistic, neutral, scalar field ϕ of mass m. (It is easy to show that μ is indeed a quantum measure, etc.; see e.g. [F6] for a discussion of the free Euclidean field (and some interacting fields in $\nu = 1$ and 2 dimensions) from the point of view adopted in these notes).

Further solutions $\mu \in \mathcal{M}_{q.m.}$ of equation (RN) have been constructed for the following choices of P, setting F = -P'; (we omit the discussion of the case $\nu = 1$ which is standard quantum mechanics).

1) $\nu = 2$, P an arbitrary polynomial bounded from below. See [G3 ,G4 ,S5] and refs. given there.

2) $\nu = 2$, $P(\phi) = \lambda \cos(\varepsilon\phi + \theta) + \frac{m^2}{2}\phi^2$, $m^2 \geqq 0$, λ real, $\varepsilon^2 < 4\pi$, $\theta \in [0,2\pi)$; see [F8 , F9] and refs. given there.

$\nu = 2$, $P(\phi) = \lambda \operatorname{Cosh}(\varepsilon\phi) + \frac{m^2}{2}\phi^2$, or $P(\phi) = \lambda \exp(\varepsilon\phi) + \frac{m^2}{2}\phi^2$, $m^2 \geqq 0$, $\lambda > 0$, $\varepsilon^2 < 4\pi$; see [A1 , F10] and refs.

3) $\nu = 3$, $P(\phi) = \lambda\phi^4 + \frac{m^2}{2}\phi^2$, $\lambda > 0$, m^2 real; see [G5, M3 , F11] and refs. given there.

In some of these cases (e.g. 3) with $m^2 = -1$, $\lambda > 0$ sufficiently small) it is known that there are at least two solutions μ_+ and μ_- to equations (RN) which are mutually singular. On the other hand, in case 1) uniqueness theorems are known for

$P(\phi) = \lambda Q(\phi) + \frac{m^2}{2}\phi^2$, $m^2 > 0$ (fixed), Q a polynomial bounded from below, and $0<\lambda \ll 1$. Part 2 of these notes is devoted to a discussion of examples, where the solutions of equation (RN) are non-unique, and of the consequences of such non-uniqueness.

The most substantial results concerning the construction of measures $\mu \in \mathcal{M}_{q.m.}$ satisfying (RN) (resp., heuristically, (1.21)) for a large class of functions F are due to Glimm, Jaffe and Spencer (see [G3 , G5 , S6 , G6 ,G7]) to Guerra, Rosen and Simon (see [G4 , G8]) and to Nelson (see [N4]).

In these references, however, only the construction of solutions $\mu \in \mathcal{M}_{q.m.}$ to equation (RN) and, as a consequence, of a quantum field $\phi(x)$ satisfying (W0) - (W4) and equation (1.5) is studied. In other words, in these references only the "vacuum representation" of a quantum field $\phi(x)$ satisfying (1.1) is constructed. In some cases (two dimensional r.q.f. models in the multiple phase region and gauge theories in two, three or four dimensions) this does not appear to yield a complete quantum theory of the non-linear, invariant field equation in question, because of the existence of non-trivial Poincaré -covariant super-selection sectors orthogonal to the vacuum sector [C4,F2,F3] (see the heuristic explanations given in the introduction to Parts 1 and 2). A general theory of such super-selection sectors is presented in Section 1.2.4.

1.2. Outline of a general theory of super selection sectors in relativistic quantum field theory

1.2.1. Remarks on Osterwalder-Schrader reconstruction

For the purposes of developing a general theory of super selection sectors starting from the Euclidean description of r.q.f.t. which is concrete enough to be useful in the study of special models (e.g. the $\lambda\phi^4$ - model in two dimensions) it is convenient to depart from a yet somewhat stronger version of the Osterwalder-Schrader axioms[02] than the one formulated in postulates (E0) - (E3). Thus we first present a reformulation of these postulates.

Let $\{h_\alpha\}_{\alpha\varepsilon Z^\nu}$ be a C^∞ partition of the identity on R^ν with the following properties: For all $\alpha\varepsilon Z^\nu$, $0 \leqq h_\alpha \leqq 1$, and h_α is C^∞ with supp h_α contained in a cube centered at α with faces parallel to the coordinate hyperplanes and sides of length 3/2.

Let $||\cdot||$ be a norm continuous on $\mathcal{S}(R^\nu)$ with the following properties:

1) $||\cdot||$ is translation invariant.
2) If χ is the characteristic function of a compact rectangle in R^ν, $||\chi|| < \infty$.
3) For $f(\underline{x}) \varepsilon \mathcal{S}(R^{\nu-1})$ and $\chi_{[0,T]}$ the characteristic function of the strip

$$\{x : x = (\underline{x},t),\ 0 \leqq t \leqq T\},$$

$$|| f \otimes \chi_{[0,T]} ||^2 \leqq o(T)\, |f|^2, \tag{2.1}$$

where $|\cdot|$ is some norm continuous on $\mathcal{S}(R^{\nu-1})$.

Examples:

$$\text{(a)}\ ||f|| = ||f||_p \equiv \left(\int |f(x)|^p\, d^\nu x\right)^{1/p},\ p < 2. \tag{2.2}$$

$$\text{(b)}\ ||f||^2 = \int_0^\infty d\rho\,(m^2)\,(|f|, (-\Delta + m^2)^{-1}\,|f|),$$

where $(\cdot,\cdot)$ is the L^2 scalar product, Δ is the ν dimensional Laplacean and ρ is a positive measure on $[0,\infty]$ with

$$\int_0^\infty \frac{d\rho(m^2)}{m^2} < \infty\ ; \tag{2.3}$$

see [D1].

We now state a stronger version of the regularity condition (E0).

(E0') Exponential bound

("Stability under linear perturbations of the dynamics")

Let $f\varepsilon\,\mathcal{S}(R^\nu)$; set $f_\alpha \equiv f\cdot h_\alpha$, $\alpha \varepsilon Z^\nu$.

Suppose that $|| f_\alpha|| < 1$, for all $\alpha \varepsilon Z^\nu$.

Then

$$S(\text{Exp } f) \equiv \sum_{n=o}^{\infty} (n!)^{-1}\, S_n\,(f^{\otimes n})$$

converges absolutely, and

$$|S\,(\mathrm{Exp}\ f)\,| \leqq K_1 \exp\Big[K_2 \sum_{\alpha\varepsilon Z^\nu} \|f_\alpha\|\Big]\ , \tag{2.4}$$

for some finite constants K_1 and K_2.

The remaining postulates (E1) - (E3) are unchanged, i.e.

(E1) Euclidean invariance

$$S(\mathrm{Exp}\ f) = S(\mathrm{Exp}\ f_\beta) \tag{2.5}$$

(E2) Osterwalder-Schrader positivity

For $f_j \varepsilon \mathcal{S}_+$, $j=1,\ldots,N$, $N=1,2,3,\ldots$

$S\,(\mathrm{Exp}\ \overline{f_\theta^i} \times \mathrm{Exp}\, f^j)$ are the matrix elements of a positive semi-definite matrix.

(E3) Symmetry

$S_n\,(x_1,\ldots,x_n)$ is symmetric under permutations of its arguments.

Remark: By (E3)

$$S\,(\mathrm{Exp}\ f \times \mathrm{Exp}\ g) = S\,(\mathrm{Exp}\ (f + g)), \tag{2.6}$$

where, by (E0'), both sides are well defined for f and g in $\mathcal{S}(R^\nu)$ with $\|f_\alpha\| < 1/2$, $\|g_\alpha\| < 1/2$, for all $\alpha \varepsilon Z^\nu$.

Let $\underline{\mathcal{S}}$ be the linear space of finite sequences F of test functions $f_n \varepsilon \mathcal{S}(R^{\nu n})$ with $f_o \varepsilon \mathbb{C}$ and $f_n \equiv 0$, for all $n > n_o$ (F), for some finite n_o (F).

Let $\underline{\mathcal{S}}_+$ be the subspace of sequences $F \varepsilon \underline{\mathcal{S}}$ with the property that

$$\mathrm{supp}\ f_n \subseteq R^{\nu n}_+\ ,\ \text{for all } n = 1,2,\ldots,n_o(F),$$

$$\text{where } R^{\nu n}_+ = \{x_1,\ldots,x_n\ ;\ x_j = (\underline{x}_j,t_j),\ t_j \geqq 0,\ j=1,\ldots,n\}. \tag{2.7}$$

$$\text{We set } S(F) \equiv \sum_{n=o}^{\infty} S_n\,(f_n), \tag{2.8}$$

$$S\,(F \times \mathrm{Exp}\ g \times H) \equiv \sum_{m,j,n} (j!)^{-1}\, S_{m+j+n}\,(f_m \otimes g^{\otimes j} \otimes h_n) \tag{2.9}$$

Standard arguments [F7 , D1] show that (E0') implies that the r.h.s. of (2.9) converges absolutely for F and H in $\underline{\mathcal{S}}$ and $g \varepsilon \mathcal{S}\,(R^\nu)$ with $\|g_\alpha\| < 1$, for all $\alpha \varepsilon Z^\nu$.

For arbitrary Euclidean motions β of R^ν we define

$$F_\beta = \{f_{n,\beta}\}_{n=o}^{n_o(F)} \quad \text{with}$$

$$f_{0,\beta} = f_0, \tag{2.10}$$

$$f_{n,\beta}(x_1,\ldots,x_n) = f_n(\beta^{-1} x_1,\ldots,\beta^{-1} x_n);$$

$\beta = \theta$ denotes reflection at $t = 0$, and $\beta = t$ translation by $(\underline{0},t)$ (time-translations).

We set $\overline{F} = \{\overline{f}_n\}_{n=o}^{n_o(F)}$ (2.11)

From (E1) we then obtain

$$S(F_\beta) = S(F), \tag{2.12}$$

and from (E2)

$$S(\overline{F}_\theta \times F) \geqq 0, \tag{2.13}$$

for all $F \in \underline{\mathcal{S}}_+$.

By (2.9) and (2.13), $S(\overline{F}_\theta \times G)$ defines a positive semi-definite inner product on $\underline{\mathcal{S}}_+$. Let N be its kernel and consider

$$D = \underline{\mathcal{S}}_+ / N \tag{2.14}$$

Then S defines a scalar product on D . Given $F \in \underline{\mathcal{S}}_+$, we denote by W(F) the equivalence class of F modulo N. We define

$$\langle W(F), W(G)\rangle \equiv S(\overline{F}_\theta \times G), \tag{2.15}$$

for F and G in $\underline{\mathcal{S}}_+$.

Completing D in the norm $||\cdot||$ given by the scalar product $\langle\cdot,\cdot\rangle$ yields a separable Hilbert space $\mathcal{H}_W$. By construction D is dense in $\mathcal{H}_W$; $\mathcal{H}_W$ turns out to be the physical (Wightman) Hilbert space of some r.q.f.t. satisfying (W0) - (W4) with Schwinger functions given by S_n, n=0,1,2,... We set

$$\left.\begin{array}{l} \Omega = W(1), \text{ where } 1 \equiv \{f_n\} \quad , \\ f_o = 1,\ f_m = 0, \text{ for all } m \geqq 1; \end{array}\right\} \tag{2.16}$$

Ω turns out to be the physical vacuum.

Using (E0') and (2.9) it is easy to show - see [F4 , D1] - that the map W can be extended to sequences of test functions of the form

$$\begin{array}{l} \operatorname{Exp} f \times F \equiv \{g_n\}_{n=o}^{\infty} \quad , \\ g_n \equiv \sum_{k=o}^{n} (k!)^{-1} f^{\otimes k} \otimes f_{n-k} \, , \end{array} \tag{2.17}$$

where $F = \{f_n\}_{n=0}^{n_0(F)} \in \underline{\mathcal{S}}_+$ and $f \in \mathcal{S}_+$ with $\|f_\alpha\| < 1/2$, for all $\alpha \in Z^\nu$, in such a way that

$$W(\operatorname{Exp} f \times F) \in \mathcal{H}_W. \tag{2.18}$$

This defines a dense subspace of $\mathcal{H}_W$ containing D.

If $F \in \underline{\mathcal{S}}_+$ then $F_t \in \underline{\mathcal{S}}_+$, for all $t \geqq 0$; hence

$$T_t : F \dashrightarrow F_t, \quad t \geqq 0,$$

defines a semigroup on $\underline{\mathcal{S}}_+$.

For all F and G in $\underline{\mathcal{S}}_+$

$$S(\overline{F}_\theta \times G_t) \overset{(E1)}{=} S((\overline{F}_\theta)_{-t} \times G)$$

$$= S((\overline{F_t})_\theta \times G). \tag{2.19}$$

Thus, for $G \in N$,

$$|S(\overline{F}_\theta \times G_t)| = |S((\overline{F_t})_\theta \times G)|$$

$$\leqq S((\overline{F_t})_\theta \times F_t)^{1/2}\, S(\overline{G}_\theta \times G)^{1/2}$$

$$= 0, \text{ i.e. } G_t \in N,$$

so that N is invariant under T_t, and T_t can be lifted to $\underline{\mathcal{S}}_+/N$.

This permits us to define a semigroup P_t, $t \geqq 0$, on D by

$$P_t\, W(F) = W(T_t F) \equiv W(F_t),\ t \geqq 0\ . \tag{2.20}$$

<u>Lemma 2.1</u>:

(1) P_t, $t \geqq 0$, is a densely defined, <u>symmetric</u> semigroup on $\mathcal{H}_W$.

(2) For all $\psi \in D$, $\operatorname{s-lim}_{t \downarrow 0} P_t\, \psi = \psi$.

(3) $\|P_t\, \psi\| \leqq \|\psi\|$, for all $\psi \in D$.

<u>Proof</u>:

(1) Since D is dense in $\mathcal{H}_W$, (2.20) shows that P_t, $t \geqq 0$, is densely defined. Furthermore

$$\begin{aligned} \langle W(F), P_t\, W(G)\rangle &= \langle W(F), W(G_t)\rangle \\ &= S(\overline{F}_\theta \times G_t) \\ &= S((\overline{F_t})_\theta \times G) \end{aligned}$$

$= \langle W(F_t), W(G)\rangle$
$= \langle P_t W(F), W(G)\rangle.$

(2) Clearly $f_{n,t} \to f_n$, as $t \searrow 0$, in $\mathcal{S}(R^{\nu n})$, for all $n = 0,1,2,\ldots$.
Since $S_n(x_1,\ldots,x_n) \in \mathcal{S}'(R^{\nu n})$, for all n, and $\underline{\mathcal{S}}_+$ consists of finite sequences of test functions, we conclude that

$$S(\overline{G}_\theta \times F_t) \to S(G_\theta \times F), \text{ as } t \searrow 0,$$

for all F and G in $\underline{\mathcal{S}}_+$. From this and the definitions of W and $\langle\cdot,\cdot\rangle$ (2) follows.
(3) Let $\psi = W(F) \in D$.
Then

$$\begin{aligned}
||P_t \psi||^2 &= \langle P_t W(F), P_t W(F)\rangle \\
&= \langle W(F), P_{2t} W(F)\rangle, \text{by (1)} \\
&= \langle W(F), W(F_{2t})\rangle, \text{ by (2.20)} \\
&\leqq \langle W(F), W(F)\rangle^{\frac{1}{2}} \langle W(F_{2t}), W(F_{2t})\rangle^{\frac{1}{2}}, \text{ by the Schwarz inequality} \\
&= ||\psi||\,||P_{2t}\psi||, \text{ by (2.20)} \\
&= ||\psi||\,\langle\psi, P_{4t}\psi\rangle^{\frac{1}{2}} \\
&\leqq \cdots\cdots \\
&\leqq ||\psi||^{2\sum_{n=1}^{N} 2^{-n}}\; ||P_{2^N t}\,\psi||^{2\cdot 2^{-N}} \,.
\end{aligned} \quad (2.21)$$

By (EO') (in particular the translation invariance of the norm $||\cdot||$)

$$||P_{2^N t}\,\psi||^{2\cdot 2^{-N}} = \langle W(F_{2^N t}), W(F_{2^N t})\rangle^{2^{-N}}$$

converges to 1, as $N \to \infty$, for all $F \in \underline{\mathcal{S}}_+$.
Thus, the r.h.s. of (2.21) converges to $||\psi||^2$, as $N \to \infty$; see [O2]. Q.E.D.

Corollary 2.2:

P_t, $t \geqq 0$, is a strongly continuous contraction semigroup on $\mathcal{H}_W$. Its generator H is a densely defined, positive selfadjoint operator on $\mathcal{H}_W$.

Remarks:

1. H turns out to be the relativistic Hamilton operator.
2. Clearly P_t, $t \geqq 0$, is the restriction of a holomorphic contraction semigroup $P_z = e^{-zH}$, $\mathrm{Re}\, z > 0$, to the positive, real axis. (2.22)

Let $g \equiv \{g_n\} \in \underline{\mathcal{S}}_+$, $g_n = 0$, for all $n \neq 1$, $g_1 = g \in \mathcal{S}_+$. Let $F \equiv \{f_n\} \in \underline{\mathcal{S}}_+$, and set

$$g \times F = \{g \otimes f_n\} \in \underline{\mathcal{S}}_+. \quad (2.23)$$

Suppose that supp $g \subseteq \{x : x = (\underline{x},t),\ 0 \leqq t \leqq \varepsilon\}$.

We define an operator $\phi_\varepsilon(g)$ on $\mathcal{H}_W$:

$$\phi_\varepsilon(g)\, W(F) = W(g \times T_\varepsilon F)$$

$$= W(g \times F_\varepsilon), \text{ for all } F \in \underline{\mathcal{S}}_+.$$

Then

$$\langle W(G), \phi_\varepsilon(g)\, W(F)\rangle = S(\overline{G}_\theta \times g \times F_\varepsilon)$$

$$\overset{(E1)}{=} S(\,(\overline{G}_\theta)_{-\varepsilon} \times g_{-\varepsilon} \times F)$$

$$\overset{(E3)}{=} S(\,\overline{(\overline{g}_{-\varepsilon,\theta} \times G_\varepsilon)_\theta} \times F)$$

$$= S(\,\overline{(\overline{g}_{\theta,\varepsilon} \times G_\varepsilon)_\theta} \times F)$$

$$= \langle \phi_\varepsilon(\overline{g}_{\theta,\varepsilon})\, W(G), W(F)\rangle ,$$

i.e. $\phi_\varepsilon(g)^* \supseteq \phi_\varepsilon(\overline{g}_{\theta,\varepsilon})$. (2.24)

We now consider the special case, where $\varepsilon = 1/n$,

$$g = h \otimes \delta_n,\ h \in \mathcal{S}(R^{\nu-1}),\ 0 \leqq \delta_n \in \mathcal{S}_+$$

with supp $\delta_n \subseteq \{x : 0 \leqq t \leqq 1/n\}$,

$$\overline{(\delta_n)_{\theta,1/n}} = \delta_n, \text{ and } \delta_n(t) \to \delta(t) \qquad (2.25)$$

(the Dirac measure), as $n \to \infty$, in $\mathcal{S}'(R)$.

Then

$$\langle P_{t_1} W(G), \phi_{1/n}(h \otimes \delta_n)\, P_{t_2} W(F)\rangle = \langle \phi_{1/n}(\overline{h} \otimes \delta_n)\, P_{t_1} W(G), P_{t_2} W(F)\rangle \qquad (2.26)$$

by (2.24) and (2.25).

This equation shows that either side is continuous in $t_1 \geqq 0$ and $t_2 \geqq 0$. The l.h.s. of (2.26) is analytic in t_1 on $\{z : \text{Re}z > 0\}$; see (2.22). On this domain it is bounded by

$$||W(G)||\ ||\phi_{1/n}(h \otimes \delta_n)\, P_{t_2} W(F)|| ,$$

since e^{-zH} is a contraction, for $\text{Re}z > 0$.

Similarly, the r.h.s. is analytic in t_2 on $\{z : \text{Re}z > 0\}$ and bounded there by

$$||\phi_{1/n}(\bar{h} \otimes \delta_n) P_{t_1} W(G)||\ ||W(F)||.$$

By the generalized tube theorem (see [O2] for such applications) we obtain joint analyticity in t_1 and t_2, for $|\arg t_1| + |\arg t_2| < \frac{\pi}{2}$.

From these analyticity properties, the definition of $\phi_{1/n}(h \otimes \delta_n)$ (in particular linearity in $h \otimes \delta_n$) and time-translation invariance of S (a special case of (E1)) we conclude that

$$\lim_{n\to\infty} \langle P_{t_1} W(G), \phi_{1/n}(h \otimes \delta_n) P_{t_2} W(F)\rangle$$
$$\equiv \langle P_{t_1} W(G), \phi_o(h) P_{t_2} W(F)\rangle \qquad (2.27)$$

exists, for all F and G in $\underline{\mathcal{S}}_+$, $t_1 > 0$, $t_2 > 0$. The limit defines a sesquilinear form $\phi_o(h)$, linear in h, on $D_+ \times D_+$, where

$$D_+ = \bigcup_{\varepsilon>o} e^{-\varepsilon H} D$$

is a dense domain in $\mathcal{H}_W$. By (2.24)

$$\langle\psi_1, \phi_o(h)\psi_2\rangle = \langle\phi_o(\bar{h})\psi_1, \psi_2\rangle, \qquad (2.28)$$

for all ψ_1, ψ_2 in D_+.

The form $\phi_o(h)$ turns out to be the time 0 - quantum field; see [F4,D1] . Next we consider perturbations of the semigroup $P_t = e^{-tH}$, $t \geqq 0$.

Let $h \in C_o^\infty(R^{\nu-1})$, (i.e. $h \in \mathcal{S}(R^{\nu-1})$, supp h compact), and suppose that, for all $\alpha \in Z^\nu$, $||(h \otimes 1)_\alpha|| < 1/2$. Let $F \in \underline{\mathcal{S}}_+$. We define

$$P_t^h W(F) = W(\text{Exp}(h \otimes \chi_{[o,t]}) \times T_t F). \qquad (2.29)$$

This definition makes sense for all $t \geqq 0$, as a consequence of postulate (E0') and our assumption that $||(h \otimes 1)_\alpha|| < 1/2$, for all α; see (2.18). Furthermore,

$$P_t^h P_s^h W(F) = W(\text{Exp}(h \otimes \chi_{[o,t]} \times T_t(\text{Exp}(h \otimes \chi_{[o,s]}) \times T_s F))$$
$$= W(\text{Exp}(h \otimes \chi_{[o,t]}) \times \text{Exp}(h \otimes \chi_{[t,t+s]}) \times T_{t+s} F)$$
$$\overset{(2.6)}{=} W(\text{Exp}(h \otimes \chi_{[o,t+s]}) \times T_{t+s} F)$$
$$= P_{t+s}^h W(F). \qquad (2.30)$$

Similar calculations show that

$$<W(G),\ P_t^h\ W(F)> = <P_t^{\overline{h}}\ W(G),\ W(F)> \ . \tag{2.31}$$

Theorem 2.3, [F4 , D1]:

For $h \in C_0^\infty (R^{\nu-1})$ with $||(h \otimes 1)_\alpha|| < 1/2$, for all $\alpha \in Z^\nu$, P_t^h is an exponentially bounded semigroup - selfadjoint for real h - on $\mathcal{H}_W$. There exists a norm $|\cdot|$ continuous on $C_0^\infty (R^{\nu-1})$ such that

$$||P_t^h|| \leqq e^{|h|\cdot t} \ . \tag{2.32}$$

The infinitesimal generator A^h of P_t^h is a sectorial operator on $\mathcal{H}_W$, and

$$\mathrm{Re}A^h \geqq - |h|\cdot 1 \ . \tag{2.33}$$

In the sense of sesquilinear forms on $D_+ \times D_+$

$$A^h = H - \phi_0 (h). \tag{2.34}$$

For real $h \in C_0^\infty (R^{\nu-1})$ with $||(h \otimes 1)_\alpha|| < 1/2$, $A^{\pm h}$ is selfadjoint, and

$$\pm\ \phi_0 (h) \leqq H + |h|, \text{ on } Q(H) \tag{2.35}$$

(the quadratic form domain of H).

Remarks:

The proof [D1] of Theorem 2.3 is an elaboration of a result of [F4]. The basic ingredients of this proof are:

1. Generalized Feynman-Kac formula [D1]:
For $\psi = W(F) \in D$,

$$<\psi,\ P_t^h\ \psi> = S(\overline{F}_\theta \times \mathrm{Exp}\ (h \otimes \chi_{[0,t]}) \times F_t) \tag{2.36}$$

2. the inequality:

$$|<\Omega,\ P_t^h\ \Omega>| = |S\ (\mathrm{Exp}\ (h \otimes \chi_{[0,t]}))\ | \leqq K'\ e^{\frac{1}{2}|h|t} \ . \tag{2.37}$$

where $|h| \equiv K'' \sum_{\alpha \in Z^\nu} ||(h \otimes \chi_{[0,1]})_\alpha||$, and K', K'' are finite constants.

Clearly, inequality (2.37) follows from the support properties of $h \otimes \chi_{[0,t]}$ and postulate (E0') (by the translation invariance of the norm $||\cdot||$). It turns out [D1] that (2.36) - (2.37) combined with a general Reeh-Schlieder argument, first used in

[S7 , M4] in a somewhat different context, and successive applications of the Schwarz inequality as in (2.21) prove that

$$||P_t^h|| \leqq e^{|h|t}, \quad t \geqq 0$$

which is (2.32). From this (2.33) follows by general arguments. The infinitesimal generator A^h of P_t^h, $t \geqq 0$, is identified with $H - \phi_o(h)$ on $D_+ \times D_+$ by a direct calculation, using the generalized Feynman-Kac formula (2.36), (EO') and inequality (2.1) (to bound error terms); see [F4 , D1]. The rest of the proof is standard.

So far we have (explicitly) only used the time-translation invariance of $S = \{S_n\}_{n=o}^{\infty}$. Let $\underline{x}$ denote space-like translation by $(\underline{x},o)$, and assume, as in (E1), that S is space-translation invariant. Then

$$U(\underline{x},o)\, W(F) \equiv W(F_{\underline{x}}), \quad F \in \mathcal{S}_+ \tag{2.38}$$

defines a strongly continuous group on D converging to the identity when $\underline{x} \to 0$. Moreover

$$\begin{aligned} ||U(\underline{x},o)\, W(F)||^2 &= S((\overline{F}_{\underline{x}})_\theta \times F_{\underline{x}}) \\ &= S((\overline{F}_\theta)_{\underline{x}} \times F_{\underline{x}}) \\ &= S(\overline{F}_\theta \times F), \text{ by (E1)} \\ &= ||W(F)||^2 \quad . \end{aligned}$$

Hence $U(\underline{x},o)$ is unitary. Since $(F_{\underline{x}})_t = (F_t)_{\underline{x}}$,

$$\left[e^{-tH}, U(\underline{x},o) \right] = 0, \tag{2.39}$$

for all $\underline{x}$, all $t \geqq 0$. We denote the infinitesimal generator of $U(\underline{x},o)$ by $\underline{P}$; $(H,\underline{P})$ turns out to be the energy-momentum operator of a r.q.f.t. with Schwinger functions $\{S_n\}$. By (2.39) H and $\underline{P}$ commute, so that

$$U(\underline{x},t) \equiv e^{itH}\, U(\underline{x},o) = e^{i(tH - \underline{x}\cdot\underline{P})} \quad .$$

Moreover, the quadratic form domain Q(H) of H is invariant under $U(\underline{x},o)$. Thus, the linearity of $\phi_o(h)$ in h and inequality (2.35) imply that there is some norm $|||\cdot|||$ on $\mathcal{S}(R^{\nu-1})$ such that, for real $h \in \mathcal{S}(R^{\nu-1})$,

$$\pm\, \phi_o(h) \leqq |||h|||\,(H + 1), \text{ on } Q(H), \tag{2.40}$$

see [D1].

The relativistic quantum field $\phi(f)$, $f \in \mathcal{S}(R^\nu)$, can now be defined as a quadratic form on Q(H) by means of the following weak integral:

$$\phi(f) = \int_{-\infty}^{+\infty} dt\, e^{itH}\, \phi_o\, (f(\cdot,t))\, e^{-itH} \quad . \tag{2.41}$$

From this and (2.40) we obtain

$$\pm\, \phi\, (f) \leqq |f|_{\mathcal{S}}\, (H + 1), \text{ on } Q(H), \tag{2.41}$$

for some norm $|\cdot|_{\mathcal{S}}$ continuous on $\mathcal{S}(R^{\nu})$ and arbitrary, real $f \varepsilon \mathcal{S}(R^{\nu})$.

Glimm and Jaffe [G3] have shown that inequality (2.41) implies that $\phi(f)$ is essentially selfadjoint on any core for H and that all vacuum expectation values

$$W_n\, (f_1,\ldots,f_n) \equiv \langle\Omega, \prod_{j=1}^{n} \phi(f_j)\, \Omega\rangle \tag{2.42}$$

exist.

At this point we can refer to Nelson's basic paper [N1] (see also [O2, F4]) for a proof that the analytic continuations of the distributions $\{W_n\}$ to imaginary times - see equation (1.8) - are precisely the generalized functions $\{S_n\}_{n=o}^{\infty}$ from which we started and that, if the latter satisfy (E1) - (E3), the former satisfy the Wightman axioms (W0) - (W4). We have given enough background, here, to enable the interested reader to do this last step himself.

Remark:

The version of Osterwalder-Schrader reconstruction reviewed here is taken from [F4, D1]. But many basic steps in our arguments come from the original papers [O2 , N1]. At the price of postulating the much more restrictive version (E0'), (E1) - (E3) of the Osterwalder-Schrader axioms our proofs are mathematically more elementary than the original ones of [O2]. Furthermore postulates (E0'), (E1) - (E3) automatically imply (contrary to the original Osterwalder-Schrader axioms) that the r.q.f.t. reconstructed in this section satisfies all axioms of Haag and Kastler [H3], i.e. it can be formulated as a relativistic quantum theory of local observables. This turns out to be an important fact in our discussion of super selection rules.

We emphasize, at this point, that all models constructed so far satisfy (E0'), (E1) - (E3) and that even the $\lambda\phi_4^4$ - theory satisfies these postulates provided it exists as a limit of "lattice theories"; see [F4]. This makes (E0'), (E1) - (E3) a reasonable starting point for a general investigation.

Next we show how to construct algebras of local observables.

1.2.2 Reconstruction of algebras of local observables

Theorem 2.4, [D1]:

Suppose ϕ is some r.q.f. on some Hilbert space $\mathcal{H}$ satisfying the Wightman axioms (Wo) - (Wv). Let H denote the Hamilton operator of this r.q.f.t. Assume that

$$\pm \phi (f) \leqq |f|_{\mathcal{S}} (H + 1) \tag{2.43}$$

for some norm $|\cdot|_{\mathcal{S}}$ continuous on $\mathcal{S}(R^{\nu})$ and all $f \in \mathcal{S} \equiv \mathcal{S}_{real}(R^{\nu})$.

Then $D(\phi(f)) \supseteq D(H)$, and $\phi(f)$ is essentially selfadjoint on any core for H, [G9 , N5].

Let f and g be two test functions in $\mathcal{S}$ with space-like separated supports, so that Wightman's form of locality, (Wiv), implies

$$[\phi(f), \phi(g)] = 0,$$

on the dense domain $D = \{\mathcal{P}(\phi)\Omega\} \subset \mathcal{H}$.

Then all bounded functions of $\phi(f)$ and $\phi(g)$ commute with each other, in particular

$$e^{i\phi(f)} e^{i\phi(g)} = e^{i\phi(g)} e^{i\phi(f)} .$$

Remarks:

In [D1] this theorem has been derived as the consequence of a general theorem concerning the commutativity of unbounded operators. Generalizations of this result with applications to canonical field theories were discussed in [F12]. Under the same hypotheses as in Theorem 2.4, all time-ordered products of field operators exist [N5]. This is useful for scattering theory, [E1]. By (2.41), (2.42), the hypotheses of Theorem 2.4 are a consequence of (E0'), (E1) - (E3).

Definition 1:

A double cone in Minkowski space M^{ν} is here defined to be the intersection of some forward light cone V_+ and some backward light cone V_-. Let $\mathcal{O}$ be a union (bounded or unbounded) of double cones. Let $\mathfrak{A}(\mathcal{O})$ be the von Neumann algebra ("local algebra") generated by

$$\{e^{i\phi(f)} : f \in \mathcal{S}, \text{ supp } f \subset \mathcal{O}\} ,$$

on $\mathcal{H}$. By construction of $\dot{\mathfrak{A}}(\mathcal{O})$ and Theorem 2.4

$$\dot{\mathfrak{A}}(\mathcal{O})' \supseteq \dot{\mathfrak{A}}(\sim\mathcal{O}). \tag{2.44}$$

Here $\dot{\mathfrak{A}}(\mathcal{O})'$ is the commutant (commuting algebra) of $\dot{\mathfrak{A}}(\mathcal{O})$, and

$$\sim\mathcal{O} = \{y \in M^{\nu}: (x-y)^2 < 0, \text{ for all } x \in \mathcal{O}\} \tag{2.45}$$

is the causal complement of $\mathcal{O}$.

Let $\mathring{\mathfrak{A}}$ denote the norm closure of $\bigcup_{\{\mathcal{O}\}} \mathring{\mathfrak{A}}(\mathcal{O})$, where $\{\mathcal{O}\}$ denotes the net of all bounded, open double cones. This is a C* algebra, called the algebra of all local observables.

Let U: $\xi \in \mathcal{P}_+^{\uparrow} \dashrightarrow U(\xi)$ be the unitary representation of the Poincaré group $\mathcal{P}_+^{\uparrow}$ on $\mathcal{H}_0$ guaranteed to exist by (Wo) - (Wv), resp. by (E0'), (E1) - (E3). For $\xi \in \mathcal{P}_+^{\uparrow}$ and $A \in \mathring{\mathfrak{A}}$, define

$$\tau_{\xi}(A) = U(\xi)\, A\, U(\xi)^{*}. \tag{2.46}$$

This defines a representation of $\mathcal{P}_+^{\uparrow}$ as a group of *automophisms of $\mathring{\mathfrak{A}}$. Let

$$\mathcal{O}(\xi) = \{x \in M^{\nu}: \xi^{-1} x \in \mathcal{O}\}.$$

Then

$$\tau_{\xi}(\mathring{\mathfrak{A}}(\mathcal{O})) = \mathring{\mathfrak{A}}(\mathcal{O}(\xi)), \tag{2.47}$$

by construction of $\{\mathring{\mathfrak{A}}(\mathcal{O})\}$.

It is shown in [D1] that $<\{\mathring{\mathfrak{A}}(\mathcal{O})\}, \mathring{\mathfrak{A}}, \tau_{\xi}>$ satisfies all Haag-Kastler axioms and that (E0'), (E1) - (E3), resp. (Wo) - (Wv) and (2.43) imply that the hypotheses of a basic theorem due to Bisognano and Wichmann [B2] hold. This theorem allows to introduce another local net $\{\mathfrak{A}(\mathcal{O})\}$ and a C* algebra $\mathfrak{A} = \overline{\bigcup_{\{\mathcal{O}\}}\mathfrak{A}(\mathcal{O})}$, where the local albegras $\mathfrak{A}(\mathcal{O})$ are obtained as intersections of "wedge algebras" (see [B2]) such that $<\{\mathfrak{A}(\mathcal{O})\}, \mathfrak{A}, \tau_{\xi}>$ satisfies all Haag-Kastler axioms and, in addition, duality

$$\mathfrak{A}(\mathcal{O})' = \mathfrak{A}(\sim\mathcal{O}). \tag{2.48}$$

Equation (2.48) is the main result of the deep analysis [B2] and is important in the theory of super selection sectors [D3].

We now summarize these results in

Corollary 2.5:

Let $\{S_n\}_{n=0}^{\infty}$ be a sequence of generalized functions satisfying (E0'), (E1) - (E3).

Then the S_n's are the Schwinger functions, in the sense of equation (1.8), of a unique r.q.f.t. satisfying the Wightman axioms which, by (2.43) - (2.47), determines a quantum theory of local observables satisfying all Haag-Kastler axioms and duality (equation (2.48)).

1.2.3 Vacuum Super Selection Sectors

In the following a C* algebra $\mathfrak{A}$ is called an algebra of local observables iff it is the norm closure of the union of local von Neumann algebras satisfying the Haag-Kastler axioms, (2.44) - (2.47).

Let Ω (= W(1)) be the physical vacuum of some r.q.f.t. We define a state ω on $\mathfrak{A}$ by

$$\omega(A) = \langle\Omega, A\Omega\rangle, \ A \in \mathfrak{A} \ . \tag{2.49}$$

Definition 2:

A state ω on $\mathfrak{A}$ satisfying (2.49), for some Ω which is the physical vacuum of an r.q.f.t. is hence forth called a Wightman state on $\mathfrak{A}$, i.e. ω is a Wightman state if $(\omega, \mathfrak{A})$ are determined, via Definition 1 and (2.49), by some r.q.f.t. obeying axioms (W0)- (W4). For given $(\mathfrak{A}, \tau_\xi)$ the class of all Wightman states on $\mathfrak{A}$ is denoted W. The structure of W will determine what are here called the vacuum super selection rules.

Note that under the hypotheses of Theorem 2.4, in particular (2.43), the Hilbert space $\mathcal{H}$ obtained from $\mathfrak{A}$ and $\omega \in W$ by means of the Gel'fand-Naimark-Segal construction (see e.g. [L1]) coincides with the physical (Wightman) Hilbert space $\mathcal{H}_W$ obtained by Wightman-, resp. Osterwalder-Schwader reconstruction, [G9].

Let $\bar{\mathfrak{A}}$ be the von Neumann algebra generated by $\mathfrak{A}$ on $\mathcal{H}$, (i.e. $\bar{\mathfrak{A}}$ = weak closure of $\mathfrak{A}$ = $(\mathfrak{A}')'$). Araki [A2] has shown that $\bar{\mathfrak{A}}$ is a type I_∞ von Neumann algebra.

Definition 3:

Let $\mathfrak{Z}$ denote the center of $\mathfrak{A}$ which is the subalgebra of operators $Z \in \bar{\mathfrak{A}}$ with the property that

$$[Z, A] = 0, \quad \text{for all } A \in \bar{\mathfrak{A}} \ .$$

Clearly $\mathfrak{Z}$ is an abelian algebra, closed in norm. By the Gel'fand isomorphism (see e.g. [L1]) it can be realized as the algebra of continuous functions C(X) on some compact Hausdorff space X (which is henceforth called the space of "pure phases", in analogy with terminology used in statistical mechanics). The positive, linear functionals ρ on C(X), normalized such that $\rho(1) = 1$, are given by the regular Borel probability measures on X, [H5].

Definition 4:

Two Wightman states ω_1 and ω_2 are said to be equivalent ($\omega_1 \simeq \omega_2$) if they determine the same Wightman distributions.

The following theorem is due to Araki [A2] and Borchers [B3]; see also [B4, F4].

Theorem 2.6:

Let $\mathfrak{A}$ be an algebra of local observables such as constructed in Section 2.2. Given $\omega \in W$, there exists a probability measure ρ on X and a measurable function $\omega : X \to W$, $x \longmapsto \omega_x \in W$, such that, for all $A \in \mathfrak{A}$,

$$\omega(A) = \int_X d\,\rho(x)\,\omega_x\,(A). \tag{2.50}$$

For ρ-almost all $x \in X$, ω_x is a pure Wightman state, i.e. the representation determined by ω_x is irreducible, and ω_x satisfies the cluster decomposition properties:

$$\omega_x\,(A\,\tau_a\,(B)) \dashrightarrow \omega_x\,(A)\,\omega_x\,(B), \tag{2.51}$$

as $|a| \to \infty$ in a space-like direction.

Let Δ be a measurable subset of X, and define $\bar{\Delta} \equiv \{x \in X : \exists\, x' \in \Delta$ such that $\omega_x \simeq \omega_{x'}\} \supset \Delta$. Then $\bar{\Delta}\setminus\Delta$ has ρ-measure 0.

Remark:

Mathematical precision would require a more careful statement of the hypotheses of Theorem 2.6; (e.g. it suffices to assume inequality (2.43)). See also [B3 , H6,A2,F4]. Next, suppose a Wightman state ω comes from a quantum measure $\mu \in \mathcal{M}_{q.m.}$, in the sense that the Schwinger functions determined by ω - see (1.8), (1.9) - are the moments of μ. (If the Schwinger functions satisfy (E0) - (E3) and Nelson-Symanzik positivity (1.16) then they are the moments of some $\mu \in \mathcal{M}_{q.m.}$).

By definition of $\mathcal{M}_{q.m.}$, μ is Euclidean invariant. The Euclidean group acts as a σ-algebra isomorphism of the σ-algebra Σ on $\mathcal{S}'$, so that, given a measurable subset Γ of $\mathcal{S}'$ and a Euclidean motion β, Γ_β is the motion of Γ under β and is measurable, as well. See (1.17) - (1.18). Let $\{\beta \equiv t \equiv (1, (\underline{0}, t))\}$ be the subgroup of the proper Euclidean group consisting of translations in the Euclidean time direction.

Definition 5:

We define a σ-algebra of "time translation invariant" subsets of $\mathcal{S}'$ by

$$\Sigma_{inv} \equiv \{\Gamma \in \Sigma : \mu\,(\Gamma_t \,\Delta\, \Gamma) = 0, \text{ for all } t\},$$

i.e. each set Γ in $\Sigma_{inv.}$ has the property that $\Gamma_t \Delta \Gamma$, the symmetric difference of Γ_t and Γ, has μ-measure 0; Σ_{inv} is trivial if, for all $\Gamma \in \Sigma_{inv.}$, $\mu(\Gamma) = 0$ or 1.

Theorem 2.7, [F4]:

Suppose that $\Sigma_{inv.}$ is non-trivial. Then there exists a non-trivial probability measure ρ on $\Sigma_{inv.}$,

$$\rho = \mu \restriction \Sigma_{inv.} ,$$

and a $\sum_{inv.}$ - measurable mapping m from $\mathcal{S}'$ to probability measures on $(\mathcal{S}', \sum)$, m: $\chi \in \mathcal{S}' \dashrightarrow \mu_\chi$, such that, for all $\Gamma \in \sum$,

$$\mu(\Gamma) = \int_{\mathcal{S}'} d\rho(\chi)\, \mu_\chi(\Gamma) \quad , \tag{2.52}$$

and, for ρ-almost all χ,

(1) μ_χ is ergodic under the action of Euclidean time-translations.

(2) Given $\Delta \in \sum_{inv.}$, define

$$\Delta' = \{\chi : \mu_\chi = \mu_{\chi'}, \text{ for some } \chi' \in \Delta\} \supset \Delta \quad .$$

Then $\rho(\Delta' \setminus \Delta) = 0$; (this means that, typically, μ_χ and $\mu_{\chi'}$ are mutually singular, for $\chi \neq \chi'$).

(3) The moments of the measure μ_χ are the Schwinger functions, in the sense of equations (1.8) - (1.9), of a unique r.q.f.t. satisfying all Wightman axioms (Wo) - (Wv), including uniqueness of the physical vacuum ($\leftrightarrow$ (0,0) is a simple eigenvalue of $(H,\underline{P})$ $\longleftrightarrow$ cluster properties; see (2.51)).

(4) There is a 1-1 correspondence between the decomposition (2.52) of μ into time-translation ergodic components and the decomposition (2.50) given in Theorem 2.6 for the Wightman state ω determined by μ (by Osterwalder-Schrader reconstruction).

Remark:

The general form of postulates (EO'), (E1) - (E3) and estimates like

$$\pm\, \phi(f) \leqq |f|_{\mathcal{S}}\, (H + 1), f \in \mathcal{S} \quad ,$$

(see (2.43)) are stable under the decompositions (2.50), resp. (2.52). See [B4 ,F7].

Definition 6:

The equivalence classes of extremal (= pure = clustering) Wightman states in W are called the vacuum super selection rules.

Theorems 2.6 and 2.7 express that every $\omega \in W$ can be obtained as a convex combination of **extremal** Wightman states.

Examples of vacuum super selection rules will be mentioned in Part 2.

Definition 7:

Let G be some compact, topological group.

(R) G is an internal symmetry group of an r.q.f.t. described by a C* algebra $\mathfrak{A}$ of local observables and a family W of Wightman states ω on $\mathfrak{A}$ if G is represented as a group $\{\tau_g : g \in G\}$ of *automorphisms of $\mathfrak{A}$ commuting with the representation $\{\tau_\xi : \xi \in \mathcal{P}_+^\uparrow\}$ of the Poincaré group, i.e. if for all $A \in \mathfrak{A}$

$$\tau_\xi (\tau_g (A)) = \tau_g (\tau_\xi (A)), \forall g \in G, \forall \xi \in \mathcal{P}_+^\uparrow .$$

(E) In the Euclidean description of an r.q.f.t. whose Schwinger functions satisfy Nelson-Symanzik positivity (NS) - see (1.16) - and are therefore the moments of a quantum measure $\mu \in \mathcal{M}_{q.m.}$ internal symmetry groups can be defined in terms of certain properties of $\mathcal{M}_{q.m.}$:

G is an internal symmetry group iff G has a representation as a group of σ-algebra isomorphisms of $\Sigma : \Gamma \in \Sigma \dashrightarrow \Gamma_g \in \Sigma$, for all $g \in G$, such that when $\mu \in \mathcal{M}_{q.m.}$ the measure μ_g defined by

$$\mu_g (\Gamma) \equiv \mu (\Gamma_g), \Gamma \in \Sigma \quad , \tag{2.53}$$

is also a quantum measure. In addition one requires that μ and μ_g satisfy the same equation. E.g.

$$\frac{d\mu_g (\phi + f)}{d\mu_g \quad (\phi)} = \frac{d\mu (\phi + f)}{d\mu \quad (\phi)} \quad , \tag{2.54}$$

for all real test functions f on R^ν of compact support, so that all measures $\{\mu_g : g \in G\}$ solve the same "Radon-Nikodym" equation (RN) of Theorem 1.3 that μ satisfies.

Remarks:

1. There are various relations between the relativistic and the Euclidean definition of internal symmetries, (R) and (E); see e.g. [F4].

2. The determination of the manifold of solutions $\mu \in \mathcal{M}_{q.m.}$ of a given Radon-Nikodym equation, i.e. of the vacuum super selection rules of a given model, is one of the exciting problems of constructive quantum field theory (uniqueness versus non-uniqueness of solutions, phase transitions [G6,G7,F5,F13]). Insight into this problem is often obtained by studying the internal symmetry group (in the sense of Definition 7, (E)) of the model.

In the following we usually adopt Definition 7, (R), even though Definition 7, (E) is more useful for the study of specific models of Bose quantum fields.

As mentioned, there is often an intimate relationship between internal symmetries and the structure of vacuum super selection rules. This is because of spontaneous symmetry breaking.

Definition 8:

Let G be an internal symmetry group of some r.q.f.t. described by $(\mathfrak{A}, \tau_\xi, W)$. Let ω be a pure Wightman state in W. The internal symmetry group G is said to be spontaneously broken in ω iff the isotropy group of ω

$$H_\omega \equiv \{g : \omega \circ \tau_g = \omega\}$$

is a proper subgroup of G.

As a consequence of the fact that

$$\tau_g(\tau_\xi(A)) = \tau_\xi(\tau_g(A)),$$

$A \in \mathfrak{A}$, $\xi \in \mathcal{P}_+^\uparrow$, $g \in G$, one has that, for $\omega \in W$, $\omega \circ \tau_g \in W$, i.e. W is a "G-space". Let $\mathfrak{E}(W)$ denote the pure states (extreme points) of W. Since, for each $g \in G$, τ_g is a *automorphism of $\mathfrak{A}$, $\mathfrak{E}(W)$ is also a G-space.

Particularly simple and interesting for field theory is the situation where the isotropy groups of all states $\omega \in \mathfrak{E}(W)$ are conjugated to the same subgroup $H \subset G$. (If all isotropy groups are equal to H then H is a normal subgroup of G, but this situation is not typical).

When the action of G on $\mathfrak{E}(W)$ is transitive then all isotropy groups are conjugated to the same subgroup H, and $\mathfrak{E}(W) \cong G/H$. (The orbit of each $\omega \in \mathfrak{E}(W)$ is then all of $\mathfrak{E}(W)$). Fix some $\omega_0 \in \mathfrak{E}(W)$. Let ω be an arbitrary state in W. By Theorem 2.6,

$$\omega = \int_X d\rho(x)\, \omega_x \quad .$$

In the present situation X can be chosen to be G/H, $d\rho$ is a measure on G/H, and

$$\omega_x = \omega_0 \circ \tau_g \quad ,$$

where g is in the equivalence class mod. H_ω labelled by x. The only G-invariant state $\omega \in W$ is then given by $\omega = \int_{G/H} d\rho(x)^o \omega_x$, where ρ is the measure on G/H induced by the Haar measure on G.

The situation described here arises in some models of classical statistical mechanics (ferromagnetic Ising models [S8, L2]) closely related to relativistic quantum field models of the sort considered in Theorem 1.3, [G4 , S9]. It is believed to arise in the models studied in Part 2. However, there exist models in two space-time dimensions for which $\mathfrak{E}(W)$ contains at least two states but is not related to any internal symmetries, [F3].

If the internal symmetry group G is a connected Lie group to which there corresponds local, locally conserved currents (typically Noether currents) and if this symmetry is spontaneously broken in some Wightman state ω then the r.q.f.t. reconstructed from $(\mathfrak{A}, \tau_\xi, \omega)$ describes at least one scalar 0-mass particle, the Goldstone boson. (In general, the number of Goldstone bosons is equal to the number of "broken generators"). This situation can only arise in three or more space-time dimensions. This is the content of the Goldstone theorem [K1 , E3 , S10, F14].

Examples for this situation are known in three space-time dimensions [F13].

If G is a discrete group which is spontaneously broken in a pure state $\omega \in \mathcal{E}(W)$ and the dimension of space-time is two then there are, in general, further (charged) super-selection sectors, (which in [F4 , F2 ,F3] we have called soliton sectors, because of their formal relations to solitary waves in classical, non-linear field equations). This is described in the next section.

There are very convenient sufficient conditions for some Wightman state ω not to be a pure state and for the spontaneous breaking of a symmetry group G in some pure Wightman states $\{\omega_x : x \in \Gamma \subset X, \rho(\Gamma) > 0\}$, in the case where $\omega = \int_X d\rho(x)\, \omega_x$ is G-invariant.

Such sufficient conditions play a significant role in the analysis of phase transitions in specific r.q.f. models. They are discussed e.g. in refs. [F4 , G6 , D4 , F3], to which we refer the reader.

Peculiar vacuum super selection rules, apparently not connected to an internal symmetry group, arise in gauge quantum field theory. They were discovered in [H2 , C3 , J1] as an effect of the instanton and are discussed on a more rigorous basis in [F1].

This is a very recent and fascinating subject which we can unfortunately not discuss in these notes. But we emphasize that the phenomenon alluded to, here, fits well into the general formalism developed in this section.

2.2.4 Charged Super Selection Sectors*

In this section it is investigated - in an abstract form - in what sense a relativistic system given in terms of an algebra $\mathfrak{A}$ of local observables generated by all bounded functions of selfadjoint, local fields (Theorem 2.4), a representation τ_ξ of $\mathcal{P}_+^\uparrow$ as a group of *automorphisms of $\mathfrak{A}$ and a family W of Wightman states on $\mathfrak{A}$ (the vacuum super selection rules; Section 2.3) determines all its super selection rules, i.e. its own "charged" super selection sectors "orthogonal" to its vacuum sectors.

In the following we must assume a vague familiarity of the reader with the general ideas of the Haag-Kastler [H3] and the Doplicher-Haag-Roberts [D3] treatment of super selection sectors and with §6 of our paper on quantum solitons [F2]. Otherwise the motivation at the basis of our analysis is obscure. Some knowledge of [R2] and of [F3] might be helpful.

Since most of the concepts and results presented in the following are due to Doplicher, Haag and Roberts, any result which is stated without reference is theirs.

Our main contention is that the structure of charged super selection rules of ($\mathfrak{A}$, τ_ξ , W) is determined completely by a particular family of strictly local (resp. bi-local) observables which we now define.

Let $\mathcal{O}$ be a bounded, open double cone in Minkowski space M^ν. We recall that

$$\mathcal{O}(\xi) \equiv \{x \in M^\nu : \xi^{-1} x \in \mathcal{O}\}$$

is the image of $\mathcal{O}$ under the Poincaré transformation ξ. We let $\mathcal{O}_\xi$ denote the smallest connected , convex union of bounded, open double cones containing $\mathcal{O} \cup \mathcal{O}(\xi)$.

Definition 9:

A local Poincaré cocycle Γ is a mapping from the Poincaré group $\mathcal{P}_+^\uparrow$ into the algebra $\mathfrak{A}$ of all local observables with the properties:

(1) There exists some bounded, open double cone $\mathcal{O}$ such that, for all $\xi \varepsilon \mathcal{P}_+^\uparrow$, $\Gamma : \xi \dashrightarrow \Gamma(\xi)$ is a unitary operator in the local von Neumann algebra $\mathfrak{A}(\mathcal{O}_\xi)$, and, in every normal representation of the local von Neumann algebras, $\Gamma(\xi)$ is strongly (or weakly) continuous in ξ. Henceforth $\mathcal{O}$ is called the localization cone of Γ;

(2)

$$\Gamma(\xi_1 \cdot \xi_2) = \Gamma(\xi_1)\, \tau_{\xi_1}(\Gamma(\xi_2))$$

(cocycle identity)

*This section is dedicated to A.S. Wightman who encouraged me to think about super selection sectors and quantum solitons at different instances. Most of the concepts reviewed here are either explicitly contained in or have grown out of [D3 , R2 , F2]; some of them were also pioneered in [P3 , S11]. The first applications to interacting r.q.f.t's in two space-time dimensions were made in [F2, F3].

We will shortly introduce an equivalence relation on the family of all local Poincaré cocycles. It will then turn out that the charged super selection sectors are in 1-1 correspondence with equivalence classes of ("non-trivial") local Poincaré cocycles.

A rather simple consequence of Definition 9 is

Theorem 2.8:

In $\nu \geq 3$ space-time dimensions a local Poincaré cocycle Γ with localization cone $\mathcal{O}$ has the property that

$$\Gamma(\xi) \in \mathfrak{A}(\mathcal{O} \cup \mathcal{O}(\xi)), \text{ for all } \xi \in \mathcal{P}_+^{\uparrow}.$$

Remarks:

1. The proof of Theorem 2.8 proceeds by first considering the space translation cocycles

$$\Gamma(\underline{a}) \equiv \Gamma(\xi = (1, (\underline{a}, o))), \ \underline{a} \in R^{\nu-1} \quad ,$$

for which Theorem 2.8 is an immediate consequence of Definition 9, (1) and the cocycle identity (2). This has already been noticed in [R2]. The general case can be reduced to the case of space translations by a somewhat lengthy chain of arguments; see [F15]. (Although Theorem 2.8 is not difficult we are not aware of any proof of the complete result in the literature).

2. In two space-time dimensions, Theorem 2.8 is in general false (because $\sim \mathcal{O}$ is not path-connected), as we shall see in Part 2; see also [F2, F3].

3. In gauge theories the concept of local (Poincaré) cocycles has to be modified. One must consider path-dependent versions of Γ which satisfy, for closed paths, a generalized cocycle identity expressing the intrinsic geometry of gauge fields, [R2 , F15]. Such path-dependent, local operators have a rather long history in heuristic quantum field theory. The case of quantum electrodynamics is special! [F15]

In the following we speak of $\mathcal{O}$-local cocycles when we mean local Poincaré cocycles with localization cone $\mathcal{O}$. Space-time translation cocycles are denoted

$$\Gamma(\underline{a}, t) \equiv \Gamma(\xi = (1, (\underline{a}, t))).$$

Definition:

A *morphism σ of $\mathfrak{A}$ is a linear mapping from $\mathfrak{A}$ into $\mathfrak{A}$ with the properties

$$\sigma(A \cdot B) = \sigma(A) \cdot \sigma(B), \quad \sigma(A^*) = \sigma(A)^* \quad ; \tag{2.55}$$

it is said to be irreducible iff $\overline{\sigma(\mathfrak{A})} = \overline{\mathfrak{A}}$; σ is called a *automorphism of $\mathfrak{A}$ iff it is a *morphism with inverse σ^*, and σ^* satisfies (2.55).

Theorem 2.9:

Let Γ be an $\mathcal{O}$-local cocycle. Then, for all t and all $A \in \mathfrak{A}$,

$$\sigma(A) \equiv \lim_{\underline{a}\to\infty} \Gamma(\underline{a},t)\, A\, \Gamma(\underline{a},t)^*$$

exists and defines a *morphism of $\mathfrak{A}$ independent of t. In $\nu \geqq 3$ space-time dimensions, σ is independent of the direction in which $\underline{a}$ tends to infinity. In $\nu = 2$ dimensions, σ generally depends on whether $\underline{a} \to \infty$ in the positive (right) or in the negative (left) direction. (Geometrically, this reflects the fact that, for $\nu = 2$, $\sim \mathcal{V}$ is not connected).

Remarks:

This result is an immediate consequence of the localization properties of the $\mathcal{O}$-local cocycle Γ and the cocycle identity, Definition 9, (2). A proof can be found in [R2, F15], but we propose as a simple exercise to the reader to construct his own proof.

Definition 10:

Let ρ be some state on $\mathfrak{A}$, (i.e. ρ is a positive, linear functional on $\mathfrak{A}$ with $\rho(1) = 1$). (We assume that, for all bounded, open double cones $\mathcal{O}$, ρ is a normal state on $\mathfrak{A}(\mathcal{O})$. Such states are called locally normal). The Gel'fand-Naimark-Segal construction[L1] associates with $(\mathfrak{A}, \rho)$ a *representation $\pi = \pi_\rho$ of $\mathfrak{A}$ on a Hilbert space $\mathcal{H}_\rho$ and a vector $\Omega_\rho \in \mathcal{H}_\rho$ such that

$$\rho(A) = (\Omega_\rho, \pi_\rho(A)\, \Omega_\rho), \text{ and} \tag{2.56}$$

$$\{\pi_\rho(A)\, \Omega_\rho : A \in \mathfrak{A}\} \text{ is dense in } \mathcal{H}_\rho .$$

A state ρ is said to be Poincaré-covariant iff there exists a continuous, unitary representation $U_\rho : \xi \in \mathcal{P}_+^\uparrow \dashrightarrow U_\rho(\xi)$ on $\mathcal{H}_\rho$ such that

$$\pi_\rho(\tau_\xi(A)) = U_\rho(\xi)\, \pi_\rho(A)\, U_\rho(\xi)^* = U_\rho(\xi)\, \pi_\rho(A)\, U_\rho(\xi^{-1}) .$$

A state ρ is said to satisfy the relativistic spectrum condition iff ρ is Poincaré-covariant, and the spectrum of the infinitesimal generators $(H_\rho, \underline{P}_\rho)$ (the energy-momentum operator) of the space-time translation subgroup $U_\rho((1, (\underline{a},t)))$, $(\underline{a},t) \in M^\nu$, is contained in the closed forward light cone $\overline{V}_+$.

A state ρ on $\mathfrak{A}$ is called a vacuum state iff there exists some Wightman state $\omega \in W$ and a vector $\psi \in \mathcal{H}_\omega$ such that, for all $A \in \mathfrak{A}$,

$$\rho(A) = (\psi, A\, \psi). \tag{2.57}$$

A state ρ is called charged iff ρ is a Poincaré-covariant state satisfying the relativistic spectrum condition which is not a vacuum state. (This is "equivalent" to saying that ρ is charged if there is no $\psi \in \mathcal{H}_\rho$ for which

$$U_\rho(\xi)\,\psi = \psi, \text{ for all } \xi \in \mathcal{P}_+^\uparrow).$$

Given a relativistic system $(\mathfrak{A}, \tau_\xi, W)$ we are interested in finding all charged states on $\mathfrak{A}$.

They yield, via the Gel'fand-Naimark-Segal construction (2.56), all charged super selection sectors. Formulated like that, this is a vast problem. As recognized by Doplicher, Haag and Roberts in a series of fundamental papers [D3], it becomes somewhat more tractable if we only try to find all those charged states ρ on $\mathfrak{A}$ which have the form

$$\rho(A) = \omega \circ \sigma(A) \equiv \omega(\sigma(A)),\ A \in \mathfrak{A}, \qquad (2.58)$$

where ω is some state in W - (we may assume, ω is a pure Wightman state) - and σ is a *morphism of $\mathfrak{A}$, see (2.55). (An equivalent definition of states satisfying (2.58) can be found in [D3]).

A *(auto) morphism σ is then said to be Poincaré-covariant, to satisfy the relativistic spectrum condition, to be charged iff $\omega\circ\sigma$ is Poincaré-covariant,..., charged, for all $\omega \in W$!

A *(auto) morphism σ is said to be local iff

$$\sigma(A) = A, \text{ for all } A \in \mathfrak{A}(\sim \mathcal{O}), \qquad (2.59)$$

for some bounded, open double cone $\mathcal{O}$, called the support of σ ; see [D3].

This section is devoted to an analysis of those charged super selection sectors which are given in terms of local, charged *(auto) morphisms.

It is believed [D3] that, with the exception of the charged sectors in quantum electrodynamics (and other gauge theories) and the soliton sectors in two dimensional r.q.f.t.'s [F2 , F3] , all charged super selection sectors arise in this way. The case of $\nu = 2$ dimensions is special and will require further comments; see also §6 of [F2].

Local charged morphisms σ will turn out to be, via Theorem 2.9, in a 1-1 correspondence with non-trivial local Poincaré cocycles (a term defined below).

In establishing this correspondence, duality $(\mathfrak{A}(\mathcal{O})' = \mathfrak{A}(\sim \mathcal{O})$, equation (2.48) of section 2.2) is essential. (As mentioned there, it can be assumed without loss of generality, thanks to [B2]).

The following result is a first step.

Theorem 2.10, [F2, F3, R2]:

Let σ be the *morphism determined - via Theorem 2.9 - by some $\mathcal{O}$-local cocycle Γ.

(1) For $\nu \geqq 3$,

$\sigma(A) = A$, for all $A \in \mathfrak{A}(\sim \mathcal{O})$;

and for all $A \in \mathfrak{A}(\mathcal{O}_1)$ with $\mathcal{O}_1 \cap \mathcal{O} \neq \phi$, $\sigma(A) \in \mathfrak{A}(\mathcal{O}_1 \cup \mathcal{O})$.

(2) Consider the case of $\nu = 2$ dimensions. Let $\mathcal{O}_L$ denote the component of $\sim \mathcal{O}$ to the left of $\mathcal{O}$, i.e.

$$\mathcal{O}_L = \{x \in M^2 : (x-y)^2 < 0,\ \underline{x} < \underline{y}, \text{ for all } y \in \mathcal{O}\}$$

and $\mathcal{O}_R$ the one to the right of $\mathcal{O}$. Then

$$\sigma(A) = A,$$

for all $A \in \mathfrak{A}(\mathcal{O}_L)$ or for all $A \in \mathfrak{A}(\mathcal{O}_R)$, (but in general not for $A \in \mathfrak{A}(\mathcal{O}_L)$ and $A \in \mathfrak{A}(\mathcal{O}_R)$). More precisely, if

$$\sigma(A) = \lim_{\underline{a} \to \pm\infty} \Gamma(\underline{a})\ A\ \Gamma(\underline{a})^*$$

then

$$\sigma(A) = A, \text{ for all } A \in \mathfrak{A}(\mathcal{O}_{\substack{L\\R}}).$$

In order to decide whether a state $\omega \circ \sigma$ obtained by composing $\omega \in W$ with a local *morphism σ arising from a local Poincaré cocycle Γ is a vacuum state or not we must introduce the notions of equivalence and triviality of local cocycles. We give the definitions for systems in $\nu \geqq 3$ dimensions. The case of two dimensions is special and will require further comments.

Definition 11:

(1) Two local Poincaré cocycles Γ_1 and Γ_2 are said to be equivalent ($\Gamma_1 \approx \Gamma_2$) iff there exists a bounded, open double cone $\mathcal{O}_{12}$ and a unitary operator $V \in \mathfrak{A}(\mathcal{O}_{12})$ such that, when $\mathcal{O}_{12}$ and $\mathcal{O}_{12}(\underline{a})$ are space-like separated

$$\Gamma_1(\underline{a})\ \Gamma_2(\underline{a})^* = V^*\ \tau_{\underline{a}}(V), \quad (\tau_{\underline{a}} \equiv \tau_\xi = (1, (\underline{a}, o))).$$

(2) A cocycle Γ is said to be trivial iff it is equivalent to the identity cocycle 1, i.e.

$$\Gamma(\underline{a}) = V^*\ \tau_{\underline{a}}(V) \quad V \in \mathfrak{A}(\mathcal{O}_{12}).$$

Remark:

If Γ is trivial then $\mathcal{O}_{12}$ is precisely the localization cone of Γ. (In [R2] such cocycles are called "local coboundaries", a most reasonable terminology). Clearly there is an abundance of trivial local cocycles.

Lemma 2.11:

If σ_1 and σ_2 are *morphisms arising from two equivalent local cocycles Γ_1 and Γ_2 then

$$\sigma_1 (A) = V^* \ \sigma_2 (A) V, \quad V \in \mathfrak{A}(\mathcal{O}_{12}).$$

The relation $\Gamma_1 \approx \Gamma_2$ is an equivalence relation. If Γ is a trivial $\mathcal{O}$-local cocycle then

$$\sigma(A) = \lim_{\underline{a}\to\infty} \Gamma(\underline{a}) \ A \ \Gamma(\underline{a})^* = V^* A V, \quad V \in \mathfrak{A}(\mathcal{O}).$$

Proof: (Serves as an example)

Since $\mathfrak{A}$ is the norm closure of $\bigcup_{\{\mathcal{O}\}} \mathfrak{A}(\mathcal{O})$, it suffices to prove the lemma for all $A \in \mathfrak{A}(\mathcal{O}')$, where $\mathcal{O}'$ is an arbitrary, bounded, open double cone. Let $\mathcal{O}_1$ and $\mathcal{O}_2$ be the localization cones of Γ_1 and Γ_2. By definition of σ_1 and σ_2,

$$\begin{aligned}
\sigma_1 (A) &= \lim_{\underline{a}\to\infty} \Gamma_1 (\underline{a}) \ A \ \Gamma_1 (\underline{a})^* \\
&= \lim_{\underline{a}\to\infty} (\Gamma_1 (\underline{a}) \ \Gamma_2 (\underline{a})^*) \ \Gamma_2 (\underline{a}) \ A \ \Gamma_2 (\underline{a})^* \ (\Gamma_1 (\underline{a}) \ \Gamma_2 (\underline{a})^*)^* \\
&= \lim_{\underline{a}\to\infty} V^* \tau_{\underline{a}} (V) \ \Gamma_2 (\underline{b}) \ A \ \Gamma_2 (\underline{b})^* \ (V^* \tau_{\underline{a}} (V))^* \ ,
\end{aligned}$$

provided $\underline{b}$ is large enough, by Theorem 2.9 and Definition 11,

$$\begin{aligned}
&= \lim_{\underline{a}\to\infty} V^* \tau_{\underline{a}} (V) \ \sigma_2 (A) \ \tau_{\underline{a}} (V)^* \ V \\
&= V^* \sigma_2 (A) \ V,
\end{aligned}$$

since, by Theorem 2.10, (1) and the fact that

$$\tau_{\underline{a}} (V) \in \mathfrak{A}(\mathcal{O}_{12} (\underline{a})), \ \left[\sigma_2 (A), \tau_{\underline{a}} (V)\right] = 0 \ ,$$

for $\underline{a}$ large enough. The rest of the proof is obvious.

Q.E.D.

A simple way of defining equivalent cocycles in $\nu = 2$ dimensions is as follows: Γ_1 and Γ_2 are equivalent in a representation π of the algebra $\mathfrak{A}$ iff there exists a unitary operator $V \in \pi (\mathfrak{A})''$(the weak closure of $\pi(\mathfrak{A})$) such that

$$\sigma_1 (A) = V^* \ \sigma_2 (A) \ V,$$

where σ_1 and σ_2 are the *(auto) morphisms arising from Γ_1 and Γ_2. Furthermore Γ is trivial in a representation π of $\mathfrak{A}$ iff there exists a unitary operator $V \in \pi(\mathfrak{A})''$ such that $\Gamma(\xi) = V^* \tau_\xi (V)$.

We emphasize that in $\nu = 2$ dimensions the notions of equivalence and triviality of cocycles are thus representation dependent, (in any reasonable definition. In Part 2 we shall meet some local cocycle Γ which is trivial in one and non-trivial in another representation; see also [F2 , F3]). This situation makes a general, axiomatic analysis of the two dimensional theories more cumbersome but facilitates the construction of examples of cocycles, non-trivial in the physical representation, by constructing them in a representation in which they are trivial!

Progress in a general analysis of non-trivial, local Poincaré cocycles in two dimensional r.q.f.t.'s can (and has been) made by relating them to spontaneously broken, internal symmetry groups, (Section 2.3).

Let G be an internal symmetry group of a two dimensional r.q.f.t., (i.e. $\tau_g (\tau_\xi (A)) = \tau_\xi (\tau_g (A))$, for all $g \in G$, $\xi \in \mathcal{P}_+^\uparrow$).

One may consider mappings from G into local Poincaré cocycles, $G \ni g \longmapsto \Gamma_g$, with the following properties:

Let $\mathcal{O}$ be the localization cone of Γ_g and $\underline{a}$ a space-translation. If $\underline{a}$ is large and positive then the causal complement $\sim (\mathcal{O} \cup \mathcal{O}(\underline{a}))$ of $\mathcal{O}$ and $\mathcal{O}(\underline{a})$ consists of three regions

$$\mathcal{O}_L, \quad \mathcal{O}_R \cap \mathcal{O}(\underline{a})_L \quad \text{and} \quad \mathcal{O}(\underline{a})_R,$$

(with $\mathcal{O}_L$, $\mathcal{O}_R$ defined in Theorem 2.10, (2)).

One requires that

$$\Gamma_g (\underline{a})\, A\, \Gamma_g (\underline{a})^* = A, \text{ for } A \in \mathfrak{A}(\mathcal{O}_L \cup \mathcal{O}(\underline{a})_R) \tag{2.60}$$

$$\Gamma_g (\underline{a})\, A\, \Gamma_g (\underline{a})^* = \tau_g (A), \text{ for } A \in \mathfrak{A}(\mathcal{O}_R \cap \mathcal{O}(\underline{a})_L) \tag{2.61}$$

Analogous requirements are made for $\underline{a} < 0$.

One says that such cocycles are indexed by G. If Γ_g satisfies (2.60), (2.61) then

$$\sigma_{g,\pm} (A) = \lim_{\underline{a} \to \pm\infty} \Gamma_g (\underline{a})\, A\, \Gamma_g(\underline{a})^*$$

has the properties

$$\sigma_{g,\pm} (A) = A, \text{ for } A \in \mathfrak{A}(\mathcal{O}_{\substack{L \\ R}}) \tag{2.62}$$

$$\sigma_{g,\pm} (A) = \tau_g (A), \text{ for } A \in \mathfrak{A}(\mathcal{O}_{\substack{R \\ L}}) \ . \tag{2.63}$$

The double cone $\mathcal{O}$ is called the support of $\sigma_{g,\pm}$ (in analogy to our terminology in $\nu \geqq 3$ dimensions, (2.59)).

We now suppose that G is spontaneously broken in some (pure) Wightman state $\omega \varepsilon W$; (because of the Goldstone theorem this means, in general, that G is a discrete group).

Theorem 2.12, [F2 , F15]:

Assume that g is not in the isotropy subgroup of ω (so that $\omega \circ \tau_g$ determines a vacuum sector $\mathcal{H}_{\omega \circ \tau_g}$ which is "orthogonal" to $\mathcal{H}_{\omega}$).

Then $\omega \circ \sigma_{g,\pm}$ is not a vacuum state, (i.e. $\mathcal{H}_{\omega \circ \sigma_{g,\pm}}$ does not contain any vector ψ invariant under the Poincaré group, or, $(\psi, \cdot \psi) \notin W$, for all $\psi \varepsilon \mathcal{H}_{\omega o \sigma_{g,\pm}}$).

Remarks:

1. It may happen that there is some other Wightman state $\omega_o \varepsilon W$ such that g is in the isotropy subgroup of ω_o; (this situation arises in the $(\lambda\phi^6 - \lambda^{\frac{1}{2}} \phi^4 + \sigma \phi^2)_2$-quantum field model for small $\lambda > o$ and a particular value of σ, [G10]. In this model $G = Z_2 = \{1, -1\}$, with $\tau_1 (\phi) = \phi$, $\tau_{-1} (\phi) = -\phi$. For small $\lambda > o$ and some $\sigma = \sigma(\lambda)$ there exist three pure vacuum states, ω_+, ω_- and ω_o, and G is spontaneously broken in $\omega_\pm$, but not in ω_o). One expects that in the Gel'fand-Naimark-Segal representation π_{ω_o} of $\mathfrak{A}$, determined by ω_o, Γ_g is trivial, and $\omega_o \circ \sigma_{g,\pm}$ are vector states given by unit rays in $\mathcal{H}_{\omega_o}$. (In addition one expects cocycles which are not indexed by G).

2. (In contrast to [R2]) we do not expect that non-trivial, local Poincaré cocycles in two dimensional r.q.f.t.'s are always indexed by a spontaneously broken, internal symmetry group. This is because there are r.q.f.t.'s with degenerate physical vacua (i.e. W contains at least two different states) without broken symmetries[F3]. For such theories one expects that there exist charged soliton sectors: There ought to exist non-trivial, local Poincaré cocycles and charged super selection sectors. (A somewhat artificial construction of charged soliton sectors in such models can be given. It is based on ideas developed in §5 of [F2]).

Next, we wish to investigate the properties of the states $\omega \circ \sigma$, where σ is a morphism arising from a non-trivial, local Poincaré cocycle.

Theor 2.13:

Let σ be an irreducible morphism of the algebra $\mathfrak{A}$ arising from a non-trivial, local Poincaré cocycle. Then, for all $\omega \varepsilon W$, $\omega \circ \sigma$ is not a vacuum state.

Remarks:

To prove Theorem 2.13 one may assume, without loss of generality, that ω is pure, hence clustering. Suppose now that $\omega \circ \sigma$ is a vacuum state. Using the cluster properties of ω and the localization properties of σ one easily shows that this would imply that $\mathcal{H}_{\omega \circ \sigma} = \mathcal{H}_\omega$ and that σ is an inner automorphism. But this contradicts the non-triviality of the cocycle from which σ arises; see also [F15].

Theorem 2.13 suggests that the states $\{\omega \circ \sigma : \omega \in W\}$ may precisely be the charged states of the theory. For our applications in Part 2 it suffices to consider morphisms σ which are *automorphisms. One may ask quite generally, whether morphisms arising from local Poincaré cocycles are *automorphisms, see (2.55). This is not so. Only when the statistics of a super selection sector [D3] is ordinary Bose - or Fermi statistics the sector is generated by a *automorphism. (If the theory has a "gauge group of the first kind" such a sector is labelled by a one dimensional representation of this group). See [D3].

Next we shall investigate the covariance properties of the states $\{\omega\circ\sigma : \omega\in W\}$.

Theorem 2.14 (Intertwining relations), [D3 , F2]:

Let σ be a *morphism arising from an $\mathcal{O}$-local Poincaré cocycle Γ. Then

(1) $\sigma(\tau_\xi(A)) = \tau_\xi(\Gamma(\xi^{-1})^* \ \sigma(A)\ \Gamma(\xi^{-1}))$, for all $\xi \in \mathcal{P}_+^\uparrow$ and all $A \in \mathfrak{A}$.

(2) If σ is a *automorphism then σ has an inverse σ^* with the same support as σ, and $\sigma^*(A) = \lim_{\underline{a}\to\infty} [\sigma^*(\Gamma(\underline{a})^*)]\ A\ [\sigma^*(\Gamma(\underline{a}))]$, where $\sigma^*(\Gamma(\xi)^*)$ is an $\mathcal{O}$- local Poincaré cocycle.

The proof of Theorem 2.4 is quite straightforward; it is based on repeated applications of the cocycle identity, Definition 9, (2), and the support properties of σ. See [D3,F2]. As an example for the type of arguments involved we sketch the proof of (1):

$$\begin{aligned}\sigma(\tau_\xi(A)) &= \lim_{\underline{a}\to\infty} \Gamma(\underline{a})\ \tau_\xi(A)\ \Gamma(\underline{a})^* \\ &= \lim_{\underline{a}\to\infty} \tau_\xi(\tau_\xi^{-1}(\Gamma(\underline{a}))\ A\ \tau_\xi^{-1}(\Gamma(a))^*) \\ &= \lim_{\underline{a}\to\infty} \tau_\xi(\Gamma(\xi^{-1})^*\ \Gamma(\xi^{-1}\cdot(1,(\underline{a},0)))\ A \\ &\qquad \times \Gamma(\xi^{-1}\cdot(1,(\underline{a},0)))^*\ \Gamma(\xi^{-1})),\end{aligned}$$

by the cocycle identity.

Let $\xi^{-1} = (\Lambda, b)$. As $\underline{a}\to\infty$ in a spatial direction, $\xi^{-1}(\underline{a},0) = \Lambda(\underline{a},0) + b$ tends to ∞ in a space-like direction (for arbitrary, fixed $\xi \in \mathcal{P}_+^\uparrow$). Combining this fact with another application of the cocycle identity

$$\Gamma(\xi^{-1}\cdot(1,(\underline{a},0))) = \Gamma(\Lambda\cdot(\underline{a},0))\ \tau_{\Lambda\cdot(\underline{a},0)}[\Gamma((1,-\Lambda(\underline{a},0))\ \xi^{-1}\ (1,(\underline{a},0)))],$$

and using the localization properties of Γ one shows that

$$\lim_{\underline{a}\to\infty} \Gamma(\xi^{-1}\cdot(1,(\underline{a},0)))\ A\ \Gamma(\xi^{-1}\cdot(1,(\underline{a},0)))^* = \sigma(A).$$

This completes the proof of (1).

Corollary 2.15, [D3]:

Let σ be a *morphism arising from a non-trivial, local Poincaré cocycle Γ and decomposable into irreducible morphisms. Then $\omega \circ \sigma$ is Poincaré covariant, satisfies the relativistic spectrum condition and is not a vacuum state. If σ is irreducible $\omega \circ \sigma$ is a charged state.

In two space-time dimensions the same conclusions hold for all those $\omega \in W$ for which Γ is non-trivial. In particular, if $\Gamma = \Gamma_g$ is indexed by an internal symmetry group G, and $\sigma_{g,\pm}$ is a *automorphism then $\omega \circ \sigma_{g,\pm}$ is a charged state iff g is not in the isotropy subgroup of ω, [F2, F15].

Remarks:

For the expert Corollary 2.15 is a direct consequence of Theorems 2.13 and 2.14.

Poincaré covariance of $\omega \circ \sigma$ is easy to prove. It obviously suffices to consider the case where σ is irreducible. We then define a unitary operator T_σ : $\mathcal{H}_\omega \rightarrow \mathcal{H}_{\omega \circ \sigma}$, intertwining the representations $\pi_\omega \circ \sigma$ and $\pi_{\omega \circ \sigma}$ of $\mathfrak{A}$. Let $\Omega \in \mathcal{H}_\omega$ and $\Omega_\sigma \in \mathcal{H}_{\omega \circ \sigma}$ denote the cyclic vectors for $\mathfrak{A}$ corresponding to $(\omega, \mathfrak{A})$, $(\omega \circ \sigma, \mathfrak{A})$, respectively, by the G.N.S. construction (2.56). In the following we omit reference to the representation of $\mathfrak{A}$ under consideration, writing A for both, the abstract element of $\mathfrak{A}$ and its representatives $\pi_\omega(A)$, resp. $\pi_{\omega \circ \sigma}(A)$. By the G.N.S. construction

$$\{A\, \Omega_{(\sigma)} : A \in \mathfrak{A}\} \quad \text{is dense in } \mathcal{H}_{\omega(\circ \sigma)}. \tag{2.64}$$

Furthermore, since σ is irreducible,

$$\{\sigma(A)\, \Omega : A \in \mathfrak{A}\} \quad \text{is dense in } \mathcal{H}_\omega \quad \text{, too.} \tag{2.65}$$

We now define T_σ by the equation

$$T_\sigma \sigma(A)\, \Omega = A\, \Omega_\sigma \,. \tag{2.66}$$

It follows immediately from (2.64) - (2.66) that T_σ is unitary. We define a unitary representation U_σ of $\mathcal{P}_+^\uparrow$ on $\mathcal{H}_{\omega \circ \sigma}$ by

$$U_\sigma(\xi)\, T_\sigma\, \sigma(A)\Omega = T_\sigma \Gamma(\xi)\, U(\xi)\, \sigma(A)\, \Omega \,, \tag{2.67}$$

where U is the unitary representation of $\mathcal{P}_+^\uparrow$ on $\mathcal{H}_\omega$ implementing $\{\tau_\xi : \xi \in \mathcal{P}_+^\uparrow\}$. Using the cocycle identity we find

$$\Gamma(\xi_1)\, U(\xi_1)\, \Gamma(\xi_2)\, U(\xi_2) = \Gamma(\xi_1)\, \tau_{\xi_1}(\Gamma(\xi_2))\, U(\xi_1 \cdot \xi_2) = \Gamma(\xi_1 \cdot \xi_2)\, U(\xi_1 \cdot \xi_2), \tag{2.68}$$

so that U_σ is indeed a representation of $\mathcal{P}_+^\uparrow$. Since T_σ, $\Gamma(\xi)$ and $U(\xi)$ are unitary, $U_\sigma(\xi)$ is unitary, for all ξ. Since $\Gamma(\xi)$ is weakly continuous in ξ in every locally normal representation of $\mathfrak{A}$ (see Definition 9), in particular on $\mathcal{H}_\omega$, and since $U(\xi)$ is continuous in ξ and T_σ unitary, we conclude that $U_\sigma(\xi)$ is weakly, hence strongly continuous in ξ on $\mathcal{H}_{\omega o \sigma}$. That $U_\sigma(\xi)$ implements $\{\tau_\xi : \xi \in \mathcal{P}_+^\uparrow\}$ unitarily on $\mathcal{H}_{\omega o\sigma}$ is now a direct consequence of the intertwining relation of Theorem 2.14, (1).

A simple proof of the relativistic spectrum condition can be found in [D3 , F2]. The basic ingredients in this proof are the existence of "(charge-) conjugate" morphisms [D3] and the cluster properties of pure Wightman states.

3. We have now shown that one can construct charged super selection sectors out of Wightman states $\omega \in W$ and strictly local (or bi-local) observables, namely the non-trivial, local Poincaré cocycles. All charged states so obtained are of the form $\{\omega \circ \sigma : \omega \in W\}$, where σ is a charged morphism of compact support determined by a nontrivial cocycle, as in Theorem 2.9.

A natural question is now whether each charged morphism arises from a non-trivial, local Poincaré cocycle. The answer is contained in the next result.

Theorem 2.16, [D3 , F15]:

Let σ be a Poincaré covariant morphism of bounded support, (i.e. $\sigma(A) = A$, for all $A \in \mathfrak{A}(\sim\mathcal{O})$, where $\mathcal{O}$ = supp σ is some bounded, open double cone).

Then σ arises from a local Poincaré cocycle $\Gamma = \Gamma_\sigma$,

$$\sigma(A) = \lim_{\underline{a}\to\infty} \Gamma_\sigma(\underline{a})\, A\, \Gamma_\sigma(\underline{a})^*, \quad A \in \mathfrak{A}.$$

Moreover, Γ_σ is non-trivial if and only if $\omega \circ \sigma$ is not a vacuum state (for any $\omega \in W$). If, in addition, σ is an irreducible morphism then $\omega o \sigma$ is charged and it is a pure state if and only if ω is pure.

Remarks:

1. Such a result has been proven for the $\nu \geqq 3$ dimensional case in [D3]. The two dimensional version of Theorem 2.16 is contained in [F2 , F15]. See also [R2].

2. The main idea of the proof of Theorem 2.16 is as follows: We define

$$\sigma_\xi(A) = \tau_\xi(\sigma(\tau_{\xi^{-1}}(A))), \quad \xi \in \mathcal{P}_+^\uparrow,\ A \in \mathfrak{A}. \tag{2.69}$$

Since σ is a Poincaré covariant morphism σ_ξ and σ are equivalent, i.e. there exists a unitary operator $\Gamma_\sigma(\xi)$ in $\mathfrak{A}$ such that

$$\sigma_\xi(A) = \Gamma_\sigma(\xi^{-1})^*\ \sigma(A)\ \Gamma_\sigma(\xi^{-1}) \tag{2.70}$$

One can choose Γ_σ such that it fulfills the cocycle identity [R2]. The localization properties of Γ follow from duality ($\mathfrak{A}(\mathcal{O})' = \mathfrak{A}(\sim \mathcal{O})$), and the continuity properties of $\Gamma_\sigma(\xi)$ in ξ from the Poincaré covariance of σ.

A better understanding of the relation between σ and Γ_σ can be achieved as follows: Suppose σ is irreducible. Then one may define a unitary operator T^*_σ from $\mathcal{H}_{\omega\circ\sigma}$ onto $\mathcal{H}_\omega$ by

$$T^*_\sigma \; A \, \Omega_\sigma \;=\; \sigma(A) \, \Omega; \text{ see (2.66).} \tag{2.71}$$

Since σ is Poincaré covariant, there exists a continuous, unitary representation U_σ of $\mathcal{P}^\uparrow_+$ on $\mathcal{H}_{\omega\circ\sigma}$ implementing $\{\tau_\xi\}$, for all $\omega \in W$. One sets

$$\Gamma_\sigma \;(\xi) = T^*_\sigma \quad U_\sigma\,(\xi)\; T_\sigma \quad U\,(\xi^{-1}); \tag{2.72}$$

Γ_σ turns out to be independent of the choice of ω.

The following question is of considerable interest in the discussion of specific examples of r.q.f.t.'s with charged super selection sectors: Are there convenient (necessary and) sufficient conditions for a given morphism σ of $\mathfrak{A}$ to be Poincaré covariant, <u>not</u> explicitly involving local Poincaré cocycles?

An answer to this question can be found in [F15], where some (possibly useful) sufficient conditions for σ to be Poincaré-covariant in terms of locally correct implementations of the Poincaré automorphisms by inner automorphisms are derived. These conditions have grown out of our experiences with two dimensional models [F3] and were found independently and priorly by S. Doplicher, (private communication).

Next we consider the structure of the class of <u>all</u> charged morphisms of an r.q.f.t. described in terms of ($\mathfrak{A}$, τ_ξ, W). This class is denoted $\mathcal{C}$ ($\mathfrak{A}$).

<u>Theorem 2.17</u>,[D3]:

The class $\mathcal{C}$ ($\mathfrak{A}$) forms a <u>semigroup</u>.

The class $\mathcal{C}_a(\mathfrak{A})$ of all charged *automorphisms forms a <u>group</u> (a subgroup of Aut $\mathfrak{A}$/In $\mathfrak{A}$).

In the case of r.q.f.t.'s in two space-time dimensions it generally suffices to consider charged automorphisms, and Theorem 2.17 is contained in [F2]; $\mathcal{C}_a$ ($\mathfrak{A}$) is then called "<u>soliton group</u>".

The total Hilbert space of the theory is now constructed as follows: Let $\varepsilon\mathcal{C}$ ($\mathfrak{A}$) denote the equivalence classes of $\mathcal{C}$ ($\mathfrak{A}$). In each equivalence class one representative, i.e. some localized morphism σ is chosen. Given some pure Wightman state $\omega \in W$, the total Hilbert space of the theory is given by

$$\mathcal{H}_{tot.} = \bigoplus_{\sigma \in \varepsilon\mathcal{C}(\mathfrak{A})} \mathcal{H}_{\omega\circ\sigma}$$

On this Hilbert space there exists then a continuous unitary representation of the

Poincaré group, and the relativistic spectrum condition is satisfied.

Doplicher, Haag and Roberts have carried out a deep analysis of the structure of $\mathcal{C}(\mathfrak{A})$, [D3]. In particular, they have found that each $\sigma \in \mathcal{C}(\mathfrak{A})$ can be characterized by a statistics parameter which determines e.g. the statistics of particles created by σ out of the vacuum. The covariance properties of $\omega \circ \sigma$ under Lorentz transformations can be discussed in terms of the local Poincaré cocycle from which σ arises. DHR have established a standard connection between spin and statistics in this general framework. Moreover, they have found that, to each irreducible $\sigma \in \mathcal{C}(\mathfrak{A})$ there belongs a conjugate morphism $\bar{\sigma} \in \mathcal{C}(\mathfrak{A})$ with the same statistics parameter and the same "spin" such that the representation $\pi_{\omega\circ\sigma\,\circ\bar{\sigma}}$ of $\mathfrak{A}$ contains the vacuum representation π_ω of $\mathfrak{A}$ precisely once as a subrepresentation. The conjugation $\sigma \longrightarrow \bar{\sigma}$ has the physical interpretation of charge conjugation.

Finally, they were able to develop a Haag-Ruelle scattering theory under the usual hypotheses concerning the energy-momentum spectrum on the sectors $\mathcal{H}_{\omega\circ\sigma}$, $\sigma \in \mathcal{C}(\mathfrak{A})$; [D3].

Their results are only valid when $\nu \geqq 3$. In two space-time dimensions their analysis of the statistics of charged super-selection sectors is not applicable, and arbitrary, intermediate statistics may arise, [S12]. It appears that, in the case $\nu = 2$, it generally suffices to consider charged *automorphisms. The existence of *automorphisms $\bar{\sigma}$ (charge -) conjugate to σ is then trivial : $\bar{\sigma} = \sigma^*$ is the inverse of σ. The Haag-Ruelle scattering theory requires modifications, too [F2 , F15]: If there exists an internal symmetry group G such that each $\sigma \in \mathcal{C}_a(\mathfrak{A})$ is of the form $\sigma_{g,\pm}$, $g \in G$, i.e. all charged automorphisms are indexed by G, then one can construct a standard Haag-Ruelle scattering theory in terms of the operators $\{T_{\sigma_{g,\pm}} : g \in G\}$, intertwining the representation $\pi_{\omega\circ\sigma_{g,\pm}}$ of $\mathfrak{A}$ with the representation $\pi_\omega \circ \sigma_{g,\pm}$ of $\mathfrak{A}$, and the observables in the maximal, G-invariant subalgebra $\mathfrak{A}_G$ of $\mathfrak{A}$; ($A \in \mathfrak{A}_G$ iff $\tau_g(A) = A$, for all $g \in G$).

In order to be able to construct a complete scattering theory one must then assume (in a concrete model : prove) that the physical vacuum Ω is cyclic in $\mathcal{H}_{tot.}$ for the algebra generated by $\{T_{\sigma_{g,\pm}} : g \in G\}$ and $\mathfrak{A}_G$. In this case the construction of the scattering states then becomes standard. A preliminary outline of this multi-soliton scattering theory has been given in [F2], a clearer account may be found in [F15].

This completes our review of the general framework of relativistic and constructive quantum field theory. We have shown that, under suitable assumptions (the stronger version (E0'), (E1) - (E3), Section 2.1, of the Osterwalder-Schrader axioms [O2]) on the Euclidean Green's functions of the fundamental (observable) fields of a r.q.f.t., these Green's functions determine in principle, the complete structure of the theory, including its local observable algebras (Section 2.2) and its charged super selection sectors.

Part 2

The super selection sectors of the $g\,\phi_2^4$ - theory

In this section we briefly exemplify the general formalism of r.q.f.t. and c.q.f.t. described in Part 1 in the context of the well known $g\,\phi_2^4$ relativistic quantum field model. It is impossible to give a self-contained presentation of the construction and analysis of this model on the following pages, as the amount of mathematical analysis going into this task is enormous; see e.g. [V1 , S5]. Our review is therefore descriptive, and no proofs are given. Nevertheless, we hope that the mere description of the properties of the $g\,\phi_2^4$ - theory brings the abstract discussion of Part 1 a little more down to earth.

2.1 Introduction

In our review of the $g\,\phi_2^4$ - theory we follow the general strategy presented in Part 1. Accordingly we first describe the construction of the vacuum sectors of this model, then describe its algebra of local observables and finally proceed to outline the construction of its charged states. Before going into some details we want to recall once more what the classical ϕ_2^4 - field theory suggests about the corresponding quantum theory. We adopt the notations of the introduction to Parts 1 and 2, formulas (6) - (13).

The classical Hamilton density is given by

$$\mathcal{H}(\pi,\phi) = \mathcal{H}_o\,(\pi,\phi) + \mathcal{H}_I(\phi) \quad , \tag{1.1}$$

$$\begin{aligned} \mathcal{H}_o\,(\pi(x),\,\phi(x)) &= \tfrac{1}{2}\,\{\pi(x)^2 + (\partial_{\underline{x}}\,\phi\,(x))^2\} \\ \mathcal{H}_I\,(\phi(x)) &= \phi(x)^4 + \tfrac{\sigma}{2}\,\phi(x)^2 + \varepsilon(\sigma), \end{aligned} \tag{1.2}$$

where

$$\varepsilon\,(\sigma) = \begin{cases} 0, \text{ for } \sigma \geqq 0 \\ \dfrac{\sigma^2}{16}\ , \text{ for } \sigma < 0, \end{cases} \tag{1.3}$$

so that $\mathcal{H}\,(\pi,\phi) \geqq 0$.

The Hamilton functional is

$$H\,(\pi,\phi) = \int_{-\infty}^{+\infty} d\underline{x}\ \mathcal{H}\,(\pi(\underline{x},0),\,\phi(\underline{x},0)) \geqq 0 \tag{1.4}$$

The Hamilton equations of motion derived from H yield the field equation

$$\Box\,\phi\,(x) = -\,4\,\phi(x)^3 - \sigma\,\phi(x) \tag{1.5}$$

As announced in the introduction, the homotopy classes of finite energy solutions ϕ_o of (1.5) are expected to be in 1-1 correspondence with inequivalent super selection sectors of the corresponding quantum field theory, provided Planck's constant $\hbar \equiv g > 0$ is very small.

For $\sigma \geqq 0$, $\phi_0(x) = 0 = \pi_0(x)$ (1.6)

is the only minimum of $H(\pi,\phi)$, and all finite energy solutions of (1.5) are in the homotopy class of $\phi_0(x) = 0$. Hence we expect that, for $\sigma \overset{>}{(=)} 0$ and $0 < g \ll 1$, this theory has one unique vacuum sector and no other super selection sectors, in particular no charged sectors.

For $\sigma < 0$, $H(\pi,\phi)$ has two absolute minima,

$$\pi_\pm(x) = 0,\ \phi_\pm(x) = \pm \frac{1}{2}\sqrt{|\sigma|}\quad , \tag{1.7}$$

and two local minimas in homotopy classes different from the ones represented by $(\pi_\pm, \phi_\pm)$:

$$\begin{aligned} \pi_s(x) &= \pi_{\bar{s}}(x) = 0 \\ \phi_s(x) &= \frac{1}{2}\sqrt{|\sigma|}\tanh\left(\frac{\sqrt{|\sigma|}}{2}\,\underline{x}\right) \\ \phi_{\bar{s}}(x) &= -\frac{1}{2}\sqrt{|\sigma|}\tanh\left(\frac{\sqrt{|\sigma|}}{2}\,\underline{x}\right) \end{aligned} \tag{1.8}$$

with

$$\begin{aligned} Q_s &\equiv \int_{-\infty}^{+\infty} d\underline{x}\ \partial_{\underline{x}}\phi_s(x) = \sqrt{|\sigma|}\quad , \\ Q_{\bar{s}} &\equiv \int_{-\infty}^{+\infty} d\underline{x}\ \partial_{\underline{x}}\phi_{\bar{s}}(x) = -\sqrt{|\sigma|} = -Q_s \end{aligned} \tag{1.9}$$

The solutions (1.7) signalize two inequivalent vacuum sectors and the solutions (1.8) are expected to correspond to two charged super selection sectors, a soliton - and an anti-soliton sector with topological charges $\approx Q_s$, resp. $Q_{\bar{s}}$.

Since the possibility of quantum mechanical tunnelling between different classical minima cannot be ruled out, a priori, we must also study the classical Euclidean action S of the ϕ_2^4 - model and determine its minima:

$$S(\phi) = \hat{S}_0(\phi) + \hat{S}_I(\phi)\ , \quad \text{with} \tag{1.10}$$

$$\hat{S}_0(\phi) = \frac{1}{2}\int |\nabla\phi(x)|^2\, d^2x \tag{1.11}$$

$$\hat{S}_I(\phi) = \int \{\phi(x)^4 + \frac{\sigma}{2}\phi(x)^2 + \varepsilon(\sigma)\}\, d^2x \tag{1.12}$$

The Euler-Lagrange equation obtained by variation of S is

$$-\Delta\phi(x) + 4\phi(x)^3 + \sigma\phi(x) = 0\ . \tag{1.13}$$

This equation has the following solutions:

$$\text{For } \sigma \geqq 0,\ \phi(x) = 0 \quad ; \tag{1.14}$$

$$\text{for } \sigma < 0,\ \phi(x) = \pm \frac{1}{2}\sqrt{|\sigma|} \quad , \tag{1.15}$$

and no other solutions of finite, total action. The solutions (1.14), (1.15) are in 1-1 correspondence with the absolute minima (1.6), resp. (1.7) of the Hamilton functional (1.4). It is therefore expected that there will not be quantum mechanical tunnelling between the classical groundstates: "The ϕ_2^4 - theory has no instantons". For small $\hbar \equiv g$, the structure of minima of the classical Hamilton functional should therefore correctly predict the structure of super selection sectors of the corresponding r.q.f.t..

For discussions of models with instantons see [P1 , P2 , H2 , C1 , C2] and [F1].

For the purpose of constructing the vacuum sectors of the ϕ_2^4 quantum field model we must expand the classical action about its minima. First we compute the curvature κ of S at its minima.

For $\sigma \geqq 0$,

$$\kappa = \sigma \quad . \tag{1.16}$$

For $\sigma < 0$,

$$\kappa = 2\ |\sigma| \tag{1.17}$$

We then obtain:

For $\sigma \geqq 0$,

$$S = S_0 + S_I, \text{ where}$$

$$\left.\begin{aligned} S_0 &= \frac{1}{2} \int \{|\nabla \phi(x)|^2 + \sigma\, \phi(x)^2\}\, d^2x \\ S_I &= \int \phi(x)^4\, d^2x \end{aligned}\right. \tag{1.18}$$

For $\sigma < 0$, we set $\tilde{\phi} = \phi \pm \frac{1}{2}\sqrt{|\sigma|}$ and obtain

$$\left.\begin{aligned} S(\phi) &\equiv \tilde{S}(\tilde{\phi}) = \tilde{S}_0(\tilde{\phi}) + \tilde{S}_I(\tilde{\phi}), \\ \tilde{S}_0 &= \frac{1}{2} \int \{ |\nabla \tilde{\phi}(x)|^2 + 2|\sigma|\, \tilde{\phi}(x)^2\}\, d^2x \\ \tilde{S}_I &= \int \{\tilde{\phi}(x)^4 \pm 2\sqrt{|\sigma|}\, \tilde{\phi}(x)^3\}\, d^2x \quad . \end{aligned}\right. \tag{1.19}$$

It is expected that, for small $g \equiv \hbar$, the dominant features of the $g\,\phi_2^4$ - quantum field model can be predicted by setting S_I, resp. $\tilde{S}_I = 0$ and studying the resulting free (Gaussian) field theories.

When $\sigma < o$ we also define (for later purposes)

$$S(\phi) = S_o(\phi) + S_I(\phi), \text{ where}$$

$$S_o = \frac{1}{2} \int \{|\nabla\phi(x)|^2 + 2|\sigma|\phi(x)^2\} \, d^2x \qquad (1.20)$$

$$S_I = \int \{\phi(x)^4 + \frac{3}{2}\sigma\phi(x)^2 + \varepsilon(\sigma)\} \, d^2x$$

Quantum mechanically speaking, the most difficult situation is met when $\sigma = 0$ ("scaling limit") which is still beyond complete comprehension, [Gl1].

We therefore restrict our attention to the two cases $\sigma = 1$ and $\sigma = -\frac{1}{2}$.

2.2 Construction of the vacuum sectors

In this section we follow the probabilistic methods reviewed in Section 1.1 of Part 1 and summarized in Theorem 1.3 and in the formal equation (1.21): We propose to construct a quantum measure $\mu \in \mathcal{M}_{q.m.}$ (a probability measure on $\mathcal{S}' = \mathcal{S}_{real}(R^2)'$ satisfying Lemma 1.2) which solves the equation

$$\left.\begin{aligned}\frac{d\mu(\phi + f)}{d\,\mu(\phi)} &= \exp - \frac{1}{g}\left[N_\mu \left(S(g^{\frac{1}{2}}\phi + f) - S(g^{\frac{1}{2}}\phi)\right)\right] \\ &= e^{\phi(\Delta f) + \frac{1}{2}(f,\Delta f)}\; e^{-\frac{1}{g}N_\mu\left[S_I(g^{\frac{1}{2}}\phi + f) - S_I(g^{\frac{1}{2}}\phi)\right]},\end{aligned}\right\}\tag{2.1}$$

for all $f \in \mathcal{S} \equiv \mathcal{S}_{real}(R^2)$; see equation (RN), Theorem 1.3, Part 1.

First we solve (2.1) with $S_I = 0$.

(a) The free Euclidean field

If $S_I = 0$, equation (2.1) has a unique solution $\mu = \mu_0$ which is the Gaussian measure on $\mathcal{S}'$ with mean 0 and covariance (operator) $(-\Delta + 1)^{-1}$. The measure μ_0 satisfies Lemma 1.2, Part 1. Let

$$S^0(\text{Exp } f) \equiv \int_{\mathcal{S}'} d\mu_0(\phi)\, e^{\phi(f)} = e^{\frac{1}{2}(f, (-\Delta + 1)^{-1} f)} \tag{2.2}$$

As the reader checks quite easily, S^0 satisfies postulates (E0'), (E1) - (E3) of Section 2.1, Part 1, so that from S^0 a unique r.q.f.t. satisfying all Wightman axioms (W0) - (W4) can be reconstructed. This is the theory of the free, neutral, scalar field:

From Section 2.1, Part 1, (2.13) - (2.18) it follows that S^0 (resp. μ_0) determines a mapping W from $\underline{\mathcal{S}}_+$ to a separable Hilbert space $\mathcal{H}_W$, in the following denoted $\mathfrak{F}$ and called Fock space, such that

$$\langle W(F), W(G)\rangle = S^0(\overline{F}_\theta \times G).$$

By Lemma 1.2, Section 1, Part 1, W can be defined on $L^2(\mathcal{S}', \Sigma_+, d\mu_0)$ - see (1.18) - (1.19), Section 1, Part 1 - and for F and G in $L^2(\mathcal{S}', \Sigma_+, d\mu_0)$

$$\langle W(F), W(G)\rangle = \int_{\mathcal{S}'} \overline{\theta F(\phi)}\, G(\phi)\, d\mu_0(\phi)\quad .$$

The vector $\Omega_0 \equiv W(1)$ is called the bare vacuum. (2.3)

Furthermore there exists a strongly continuous contraction semigroup P_t^0, $t \geq 0$, on $\mathfrak{F}$ such that

$$\langle W(F), P_t^0 W(G)\rangle = \int_{\mathcal{S}'} \overline{\theta F(\phi)}\,(tG)(\phi)\, d\mu_0(\phi) \tag{2.4}$$

where tG denotes the translate of G by the vector $(\underline{0},t) \varepsilon R^2$; see (2.20) and Lemma 2.1, Part 1. The infinitesimal generator of P_t^o is a selfadjoint, positive operator H_o with the property that $H_o \Omega_o = 0$; H_o is the free energy operator (Hamiltonian).

Finally, to each $h \in \mathcal{S}(R)$, there corresponds a densely defined quadratic form $\phi_o(h)$ on $\mathfrak{F} \times \mathfrak{F}$, the free time 0-quantum field. In this case it can be shown to determine a densely defined, closed operator, selfadjoint for real h, which is also denoted by $\phi_o(h)$.

For a detailed discussion of the free field, the Gaussian measure μ_o on $\mathcal{S}'$ and the functional S^o, see [N1, S5, F6].

(b) Normal ordering (N_μ) and the cutoff interacting action

Let $\eta(k) = \eta(|k|)$ be a C^∞ function on R^2 with $\eta(k) = 1$, for $0 \leq |k| \leq \frac{1}{2}$, $\eta(k) = 0$, for $|k| \geq 1$, and $0 \leq \eta \leq 1$.

Let $h_K(x) = (2\pi)^{-1} \int \eta(\frac{k}{K}) e^{ikx} d^2k$; h_K is a test function. Therefore the convolution of h_K with an arbitrary tempered distribution ϕ is well defined:

$$\phi_K(x) \equiv (h_K * \phi)(x)$$

Let $< - >_o$ denote the expectation determined by μ_o ($<F>_o \equiv \int_{\mathcal{S}'} F(\phi)\, d\mu_o(\phi)$, for arbitrary $\sum$ - measurable functions F on $\mathcal{S}'$).

We now define normal ordering N_μ:

$$N_\mu(\phi_K(x)^n) \equiv :\phi_K(x)^n:$$

$$= \sum_{m=o}^{[\frac{n}{2}]} \frac{n!}{m!\,(n-2m)!} \phi_K(x)^{n-2m} \left(-\frac{1}{2} <\phi_K(x)^2>_o\right)^m ,$$

in particular,

$$:\phi_K(x)^4: = \phi_K(x)^4 - 6 <\phi_K(x)^2>_o \phi_K(x)^2 + 3 <\phi_K(x)^2>_o^2 .$$

Let Λ be some compact rectangle in R^2 and χ_Λ its characteristic function. Since, for $K < \infty$, $\phi_K(x)$ is C^∞ in x,

$$(\phi_K \chi_\Lambda)(x) \equiv \phi_K(x) \cdot \chi_\Lambda(x)$$

is well defined.

This permits us to define the interacting Euclidean action of the $g\phi_2^4$ - model with "ultraviolet cutoff" $K < \infty$ and "space-time cutoff" Λ as follows:

$$S_I(\varepsilon,g,\Lambda,K) \equiv N_\mu \left[\frac{1}{g} S_I(g^{\frac{1}{2}}\phi_K\chi_\Lambda)\right] + \varepsilon \int_\Lambda \phi_K(x)\, d^2x$$

Theorem 2.1, (see e.q. [N1, D2 , S5]):

For all $g > 0$, real ε and arbitrary, compact rectangles Λ,

(1) $\lim_{K\to\infty} S_I(\varepsilon,g,\Lambda,K) \equiv S_I(\varepsilon,g,\Lambda)$ exists in $L^p(\mathcal{S}',\Sigma,d\mu_0)$, for all $1 \leq p < \infty$;

(2) $\exp[-S_I(\varepsilon,g,\Lambda)] \in L^p(\mathcal{S}',\Sigma,d\mu_0)$, for all $1 \leq p < \infty$; furthermore

$$1 \leq \langle \exp[-p\, S_I(\varepsilon,g,\Lambda)]\rangle_0^{1/p} \leq e^{K(\varepsilon,g,p)|\Lambda|}$$

for some finite constant $K(\varepsilon,g,p)$ independent of Λ; here $|\Lambda|$ denotes the area of Λ.

Remarks:

The proof of (1) is straight forward; it involves estimating simple Gaussian integrals, or, in a physicists language, Euclidean region Feynman diagrams. The lower bound in (2) is trivial: It follows from $\langle S_I(\varepsilon,g,\Lambda)\rangle_0 = 0$, by Jensen's inequality. The upper bound is non-trivial. The first proof of such an estimate is due to Nelson [N6] and was extended by Glimm [G12] ; see [N1 , D1 , S5].

Theorem 2.1 breaks down in $\nu \geq 3$ dimensions ("ultraviolet divergences"). In three dimensions an analogous result for the "renormalized" cutoff Euclidean action has been proven by Glimm and Jaffe [G5] by an admirable amount of hard analysis. Simplifications were proposed in [M2 , G13]. In four dimensions nothing interesting is known.

Theorem 2.1 permits us to define

$$d\mu_{\varepsilon,g,\Lambda}(\phi) \equiv Z(\varepsilon,g,\Lambda)^{-1} e^{-S_I(\varepsilon,g,\Lambda)}\, d\mu_0(\phi) ,$$

where $$Z(\varepsilon,g,\Lambda) = \langle e^{-S_I(\varepsilon,g,\Lambda)}\rangle_0 \quad ; \tag{2.6}$$

$d\mu_{\varepsilon,g,\Lambda}$ is a probability measure on $(\mathcal{S}', \Sigma)$.

(c) Existence of solutions of the Radon-Nikodym equations (2.1)

Before presenting more details concerning the vacuum super selection rules of the $g\phi_2^4$ - theory with $\sigma = -\frac{1}{2}$ we quote, without proof, a basic existence theorem.

Theorem 2.2, [G3 , G4 , F5] :

For arbitrary $g > 0$, $\sigma = 1$ or $\sigma = -\frac{1}{2}$,

(1) the limits

$$S^{\pm}(\mathrm{Exp}\, f) \equiv \lim_{\varepsilon\to 0\pm}\ \lim_{\Lambda\uparrow R^2} \int_{\mathcal{S}'} e^{\phi(f)}\, d\mu_{\varepsilon,g,\Lambda}(\phi)$$

exist and satisfy postulates (E0'), (E1) - (E3) of Section 2.1, Part 1;

(2) the functionals $S^{\pm}$ are the Laplace transforms of probability measures $\mu_{\pm,g}$ on $(\mathcal{S}', \Sigma)$,

$$S^{\pm}(\text{Exp } f) = \int_{\mathcal{S}'} e^{\phi(f)} \, d\mu_{\pm,g}(\phi) \quad ,$$

and the measures $\mu_{\pm,g}$ are quantum measures in $\mathcal{M}_{q.m.}$ satisfying the Radon-Nikodym equations (2.1).

(3) For $0 < g \ll 1$, $\sigma = 1$, $\mu_{+,g} = \mu_{-,g} \equiv \mu_g$; for $0 < g \ll 1$, $\sigma = -\frac{1}{2}$, $\mu_{+,g} \neq \mu_{-,g}$, but $d\mu_{+,g}(\phi) = d\mu_{-,g}(-\phi)$, [G6, G7].

The proof of this theorem is much too long and sophisticated to be even only sketched here.

Remarks:

1. It is believed that, for $0 < g \ll 1$ and $\sigma = 1$, μ_g is the only quantum measure satisfying (2.1). Furthermore, for $0 < g \ll 1$ and $\sigma = -\frac{1}{2}$, one expects that every measure $\mu \in \mathcal{M}_{q.m.}$ solving equation (2.1) is a convex combination of the extremal (ergodic) measures $\mu_{+,g}$ and $\mu_{-,g}$. The measures $\mu_{+,g}$ and $\mu_{-,g}$ are mutually singular.

2. For $\sigma = 1$ the moments of μ_g, i.e. the Euclidean Green's functions of the $g\phi_2^4$ - theory (see Sections 1 and 2.1, Part 1), have Taylor series expansions in g at g = 0 which are asymptotic and coincide with ordinary Feynman perturbation theory [D5]. For small enough $g > 0$ these moments are equal to the Borel sum of their Taylor series expansions, i.e. they are uniquely determined by perturbation theory [E2], even though the Taylor series in g at g = 0 diverges [J3].

3. For sufficiently small $g > 0$ and $\sigma = 1$ the theory reconstructed from μ has stable one particle states of positive mass $m(g) \simeq 1$ [G3] and no two particle bound states [S13]. The Haag-Ruelle scattering theory applies [G3] and the scattering operator is non-trivial [O1, E1]. Elastic two particle scattering is complete [S3, S14].

The matrix elements of the scattering operator have an asymptotic Taylor series expansion ing at g = 0 given by Feynman perturbation theory [E1]. It is believed that the scattering operator is unitary.

4. For small $g > 0$ and $\sigma = -\frac{1}{2}$ the theories reconstructed from $\mu_{\pm,g}$ have a positive mass gap $m(g) \simeq 1$, [G7]. The moments of $\mu_{\pm,g}$, i.e. the Euclidean Green's functions corresponding to the two pure Wightman states of the $g\phi_2^4$ - theory at $\sigma = -\frac{1}{2}$, have Taylor series expansions in $g^{\frac{1}{2}}$ at g = 0 (see (2.11) and (1.19)) which are asymptotic and are given by Feynman perturbation theory about mean field theory [G7]. Perturbation theory does not appear to be Borel summable, [B5].

5. Further results concerning expansions in g near g = 0, in particular a detailed discussion of the classical limit $g \propto \hbar \searrow 0$, may be found in [H7, E4]. These results might be very significant for approximate calculations.

From now on we concentrate on the discussion of the $g\phi_2^4$ - theory for $\sigma = -\frac{1}{2}$ and $0 < g \ll 1$. This is done in two steps:

Step 1: Construction of the vacuum sectors

Step 2: Construction of the charged super selection (soliton) sectors.

2.3 The vacuum super selection rules of the $g\phi_2^4$ - theory

Throughout the rest of these notes we describe the $g\phi_2^4$ - theory with $0 < g \ll 1$ and $\sigma = -\frac{1}{2}$. The classical theory then predicts the existence of precisely two inequivalent vacuum sectors and two inequivalent charged sectors, a "soliton -" and an "anti-soliton" sector. We shall indicate that these four sectors can be grouped into two pairs, each consisting of one vacuum - and one charged sector, which are in some sense "mirror images" of each other describing the same physics. (For this reason it is sometimes claimed that this model has only two sectors).

First we describe a somewhat more detailed version of Theorem 2.2 which we shall need in the construction of the charged super selection (soliton) sectors.

Let $\phi_\pm$ denote the minima of the classical action

$$|\nabla\phi|^2 + g\,\phi^4 - \frac{1}{4}\phi^2 ,$$

i.e.

$$\phi_\pm = \pm (8g)^{-\frac{1}{2}} \tag{3.1}$$

We propose to construct two quantum measures $\mu_{\pm,g}$ satisfying the Radon-Nikodym equations of the $g\phi_2^4$ - theory, such that

$$\int_{\mathcal{S}'} \phi(x)\, d\mu_{\pm,g}(\phi) \simeq \phi_\pm \quad , \tag{3.2}$$

and

$$d\mu_{+,g}(\phi) = d\mu_{-,g}(-\phi) \tag{3.3}$$

From them two pure Wightman states $\omega_\pm$ can be reconstructed with

$$\omega_\pm(\phi(x)) \equiv W_{1,\pm}(x) \simeq \phi_\pm$$

which break the internal symmetry of the $g\phi_2^4$ - theory taking ϕ to $-\phi$ spontaneously.

The measures $\mu_{\pm,g}$ are constructed as limits of space-time cutoff measures (see Theorems 2.1 and 2.2) with "± boundary conditions", in analogy with procedures known from the statistical mechanics of classical spin systems such as the Ising model.

Let $\Lambda \equiv L\times T = \{x = (\underline{x},t) : -\frac{L}{2} \leq \underline{x} \leq \frac{L}{2}, -\frac{T}{2} \leq t \leq \frac{T}{2}\}$

With LxT we associate a space-time cutoff Euclidean action

$$S_I(g, L\times T) = \int_{L\times T} \{g{:}\phi(x)^4{:} - \frac{3}{4} : \phi(x)^2 :\}\, d^2x \tag{3.5}$$

Let $\chi_L(\underline{x})$ be a C^∞ function on R with the properties $0 \leq \chi_L(\underline{x}) \leq 1$ and

$$\chi_L(\underline{x}) = \begin{cases} 1, \text{ on } \left[-\frac{L}{2}, \frac{L}{2}\right] \\ 0, \text{ on } \left(-\infty, \frac{L+1}{2}\right] \cup \left[\frac{L+1}{2}, \infty\right) \end{cases}$$

We define boundary terms corresponding to "$\pm$ boundary conditions" at $\underline{x} = \pm \frac{L}{2}$:

$$\delta S_{\underset{--}{++}}(g, L\times T) = \int_{-T/2}^{T/2} dt \left(\int_{\frac{-L+1}{2}}^{-\frac{L}{2}} + \int_{\frac{L}{2}}^{\frac{L+1}{2}}\right) d\underline{x} \, \{\phi_\pm \phi(x)$$

$$- 1/2\, \phi_\pm^2 \chi_L(x)\} (-\partial_{\underline{x}}^2 + 1) \chi_L(\underline{x}) \tag{3.6}$$

and (for later use)

$$\delta S_{-+}(g, L\times T) = \int_{-T/2}^{T/2} dt \int_{-\frac{L+1}{2}}^{-\frac{L}{2}} d\underline{x} \, \{\phi_- \phi(x) - 1/2\, \phi_-^2 \chi_L(\underline{x})\}$$

$$\times (-\partial_{\underline{x}}^2 + 1) \chi_L(\underline{x}) + \int_{\frac{L}{2}}^{\frac{L+1}{2}} d\underline{x} \, \{\phi_+ \phi(x)$$

$$- 1/2\, \phi_+^2 \chi_L(\underline{x})\} (-\partial_{\underline{x}}^2 + 1) \chi_L(\underline{x}) \,. \tag{3.7}$$

Furthermore

$$Z_{\underset{-}{+}+}(g, L\times T) = \langle e^{-S_I(g, L\times T) - \delta S_{\underset{-}{+}+}(g, L\times T)} \rangle_0 \,, \tag{3.8}$$

and

$$\langle F \rangle_{\underset{-}{+}, g}(L\times T) = Z_{\underset{--}{++}}(g, L\times T)^{-1} \langle F\, e^{-S_I(g, L\times T) - \delta S_{\underset{--}{++}}(g, L\times T)} \rangle_0 \tag{3.9}$$

for arbitrary Σ - measurable functions F on $\mathcal{S}'$; see also (2.6), Theorem 2.1 and [B5].

Theorem 3.1, [G7] :

For all $f \in \mathcal{S}(R^2)$ the limits

$$\langle e^{\phi(f)} \rangle_{\underset{-}{+}, g}(L) = \lim_{T\to\infty} \langle e^{\phi(f)} \rangle_{\underset{-}{+}, g}(L\times T)$$

and

$$\langle e^{\phi(f)} \rangle_{\underset{-}{+}, g} = \lim_{L\to\infty} \langle e^{\phi(f)} \rangle_{\underset{-}{+}, g}(L)$$

exist, and $\langle - \rangle_{\underset{-}{+}, g}(L)$, $\langle - \rangle_{\underset{-}{+}, g}$ are the expectations determined by probability measures $\mu_{\pm, g, L}$, $\mu_{\pm, g}$, respectively, defined on $(\mathcal{S}', \Sigma)$.

The two measures $\mu_{+, g}$ and $\mu_{-, g}$ are distinct quantum measures satisfying equations (2.1), (3.2) and (3.3). Furthermore $\mu_{\pm, g, L}$ is invariant under translations in the Euclidean time direction, (see (1.18), Lemma 1.2, Part 1) and satisfies Osterwalder-Schrader positivity

$$\int_{\mathcal{S}'} \overline{\theta F(\phi)}\, F(\phi)\, d\mu_{\pm,g,L}(\phi) \geqq 0, \tag{3.10}$$

for arbitrary Σ_+ - measurable functions F on $\mathcal{S}'$; (see (1.19), Lemma 1.2, Part 1).

Theorem 3.1 was proven in [G7]. The $\pm$ boundary conditions used here to "select pure phases" appear in [B5] , but are equivalent to the ones used in [G7] .

We now refer the reader once more to Section 2.1, Part 1, for hints to the proofs of the following standard facts:

By (3.10) and Section 2.1, Part 1 (equations (2.13) - (2.18)), the measure $\mu_{\pm,g,L}$ determines a physical Hilbert space which can be seen to be the closure of $L^2(\mathcal{S}', \Sigma_+, d\mu_{\pm,g,L})$ in the norm given by the positive definite, sesquilinear form

$$F,G \mapsto \int_{\mathcal{S}'} \overline{\theta F(\phi)}\, G(\phi)\, d\mu_{\pm,g,L}(\phi) \quad .$$

It turns out (see e.g. [S5]) that there is a natural isomorphism between this Hilbert space and the Fock space $\mathfrak{F}$ of the free, neutral, scalar field of mass 1; see Section 2.2, (a).

Furthermore, since $\mu_{\pm,g,L}$ is invariant under translations in the Euclidean time direction and because of (3.10), there exists a continuous contraction semigroup $P_t^{\pm,g,L}$, $t \geqq 0$, on $\mathfrak{F}$ such that

$$\langle W(F), P_t^{\pm,g,L} W(G)\rangle = \int_{\mathcal{S}'} \overline{\theta F(\phi)}\, (tG)(\phi)\, d\mu_{\pm,g,L}(\phi) \quad .$$

Its infinitesimal generator is denoted $H_{\underline{++}}(L)$. We set $\Omega_\pm(L) = W(1)$. By construction

$$H_{\underline{++}}(L)\, \Omega_\pm(L) = 0 \tag{3.11}$$

As an operator on Fock space $\mathfrak{F}$ $H_{\underline{++}}(L)$ can be shown [S5] to be the closure of

$$H_o + H_I(L) + \delta H_{\underline{++}}(L) - E_{\underline{++}}(L), \tag{3.12}$$

where H_o is the energy operator of the free field constructed in Section 2.2, (a), and

$$H_I(L) = \int_{-\frac{L}{2}}^{\frac{L}{2}} d\underline{x}\, \{g{:}\phi_o(\underline{x})^4{:} - \tfrac{3}{4} : \phi_o(\underline{x})^2 : + (64g)^{-1}\}$$

$$\delta H_{\underline{++}}(L) = \left(\int_{-\frac{L+1}{2}}^{-\frac{L}{2}} + \int_{\frac{L}{2}}^{\frac{L+1}{2}}\right) d\underline{x}\, \{\phi_\pm\, \phi_o(\underline{x}) - \tfrac{1}{2}\phi_\pm^2\, \chi_L(\underline{x})\} \times (-\partial_{\underline{x}}^2 + 1)\, \chi_L(\underline{x}) \quad ,$$

$$E_{\underline{++}} (L) = \inf \; \text{spec} \, (H_0 + H_I(L) + \delta \, H_{\underline{++}}(L))$$

We also define

$$\delta H_{-+}(L) = \int_{-\frac{L+1}{2}}^{-\frac{L}{2}} d\underline{x} \; \{\phi_- \, \phi_0(\underline{x}) - \tfrac{1}{2} \phi_-^2 \, \chi_L(x)\} \, (-\partial_{\underline{x}}^{\,2} + 1) \, \chi_L(\underline{x})$$

$$+ \int_{\frac{L}{2}}^{\frac{L+1}{2}} d\underline{x} \; \{\phi_+ \, \phi_0(\underline{x}) - \tfrac{1}{2} \phi_+^2 \, \chi_L(\underline{x})\} \, (-\partial_{\underline{x}}^2 + 1) \, \chi_L(\underline{x}),$$

and

$$\tilde{H}_{\underline{++}} (L) = H_{\underline{++}} (L) + E_{\underline{++}} (L);$$

finally $\tilde{H}_{-+} (L)$ is the (selfadjoint!) closure of

$$H_0 + H_I(L) + \delta H_{-+}(L).$$

As a consequence of Nelson's Feynman-Kac formula [N1] one has

$$Z_{\underline{++}} (LxT) = \langle \Omega_0 , e^{-T\tilde{H}_{\underline{++}}(L)} \, \Omega_0 \rangle$$

$$Z_{-+} (LxT) = \langle \Omega_0 , e^{-T\tilde{H}_{-+}(L)} \, \Omega_0 \rangle \; ; \tag{3.13}$$

furthermore $\mu_{\pm,g,L}$ is the path space measure of a Markov process on $\mathcal{S}'_{real}$ (R) with transition function $\exp\left[-t \, H_{\underline{++}} (L)\right]$.

We may now define an interacting quantum field with space cutoff by the equation

$$\phi^{(L,\pm)} (\underline{x},t) = e^{itH_{\pm\pm}(L)} \, \phi_0 \, (\underline{x}) \, e^{-itH_{\pm\pm}(L)},$$

an equation that has to be interpreted in the sense of operator valued distributions on $\mathcal{F}$.

It follows from the reconstruction theorems of Section 2.1, Part 1, that the moments

$$\langle \prod_{j=1}^{n} \phi(x_j) \rangle_{\pm,g} \, (L)$$

are the Euclidean Green's functions of spatially cutoff Wightman distributions

$$\langle \Omega_\pm \, (L), \prod_{j=1}^{n} \phi^{(L,\pm)} \, (\underline{x}_j, \, t_j) \, \Omega_\pm \, (L) \rangle \quad ,$$

i.e., for $t_1 < t_2 < \ldots < t_n$,

$$< \prod_{j=1}^{n} \phi(\underline{x}_j, t_j)>_{\pm,g} (L) = <\Omega_\pm(L), \prod_{j=1}^{n-1} \phi_o (\underline{x}_j)\, e^{-(t_{j+1} - t_j)\, H_{\pm\pm}(L)} \phi_o(\underline{x}_n)\Omega_\pm(L)> , \tag{3.14}$$

see also eqn. (1.9), Section 1, Part 1.

This is an extended version of the Feynman-Kac formula; see [S5].

It has been shown [G9 , G8] that

$$\pm\phi^{(L,\pm)} (f) \leqq |f|_{\mathcal{S}} (H_{\pm\pm}(L) + 1), \tag{3.15}$$

for all $f \in \mathcal{S}$, in the sense of quadratic forms on $\mathcal{F}$. Here $|\cdot|_{\mathcal{S}}$ is a norm continuous on $\mathcal{S}(R^2)$ which is <u>independent</u> of L!

From (3.14), (3.15), Theorem 3.1 and a well known theorem concerning convergence of boundary values in $\mathcal{S}'(R^{2n})$ of a convergent sequence of holomorphic functions of several complex variables it then follows that

$$W_{n,\pm} (x_1 ,\dots, x_n) = \lim_{L\to\infty} <\Omega_\pm (L), \prod_{j=1}^{n} \phi^{(L,\pm)} (x_j)\, \Omega_\pm(L)> \tag{3.16}$$

exists in $\mathcal{S}'(R^{2n})$.

The limiting distributions $\{W_{n,\pm}\}_{n=o}^{\infty}$ satisfy the Wightman axioms (WO) - (W4), (including uniqueness of the physical vacuum).

From Section 2.1, Part 1, it can be inferred that

$$\{< \prod_{j=1}^{n} \phi(x_j)>_{\pm,g}\}_{n=o}^{\infty}$$

are the Euclidean Green's functions of the Wightman distributions $\{W_{n,\pm}\}_{n=o}^{\infty}$ (in the sense of equations (1.8), (1.9), Section 1, Part 1).

Let $\phi^{(\pm)}(x)$, resp. $H_\pm$ be the relativistic quantum field, resp. the Hamiltonian (energy operator) reconstructed from

$$\{< \prod_{j=1}^{n} \phi(x_j)>_{\pm,g}\}_{n=o}^{\infty}, \text{ resp. } \{W_{n,\pm}\}_{n=o}^{\infty}$$

(as in Section 2.1, Part 1).

Inequality (3.15) can be transferred to the limit $L = \infty$ and gives

$$\phi^{(\pm)} (f) \leqq |f|_{\mathcal{S}} (H_\pm + 1), \tag{3.17}$$

in the sense of quadratic forms on the physical Hilbert space $\mathcal{H}_W \equiv \mathcal{H}_\pm$.

Hence the hypotheses of Theorem 2.4, Section 2.2, Part 1 (<u>reconstruction of local algebras</u>) are satisfied by the $g\phi_2^4$ - theory in the infinite volume limit $L = \infty$. Therefore we may define local von Neuman algebras $\mathfrak{A}_\pm(\mathcal{O})$, an algebra $\mathfrak{A}_\pm$ of local observables, etc., as explained in Section 2.2, Part 1.

Let $\mathring{\mathfrak{A}}_{\mathfrak{F}}(\mathcal{O})$, $\mathring{\mathfrak{A}}_{\mathfrak{F}}$ denote the corresponding algebras generated by the free field (reviewed in Section 2.2, (a)) on Fock space $\mathfrak{F}$.

The following is a basic theorem, due to Glimm and Jaffe, [G14].

Theorem 3.2

For all bounded double cones $\mathcal{O}$ in M^2 the algebras $\mathring{\mathfrak{A}}_{\pm}(\mathcal{O})$ and $\mathring{\mathfrak{A}}_{\mathfrak{F}}(\mathcal{O})$ are isomorphic and unitarily equivalent.

This result (one of the very difficult and deep results of c.q.f.t.) permits us to identify $\mathring{\mathfrak{A}}_{\pm}(\mathcal{O})$ with $\mathring{\mathfrak{A}}_{\mathfrak{F}}(\mathcal{O})$, henceforth denoted $\mathfrak{A}(\mathcal{O})$, and $\mathring{\mathfrak{A}}_{\pm}$ with $\mathring{\mathfrak{A}}_{\mathfrak{F}}$, henceforth denoted $\mathfrak{A}$.

From (3.15) - (3.17) and Theorem 3.2 we obtain

$$\omega_{\pm}(A) \equiv \langle\Omega_{\pm}, A\Omega_{\pm}\rangle = \lim_{L\to\infty} \langle\Omega_{\pm}(L), A\Omega_{\pm}(L)\rangle$$

exists, for all $A \in \mathfrak{A}$. The states $\omega_{\pm}$ are pure Wightman states on $\mathfrak{A}$, in the sense of Section 2.3, Part 1.

It is believed that, for $0 < g \ll 1$, ω_+ and ω_- are the only pure states in $\mathcal{W}$, the class of all Wightman states on $\mathfrak{A}$.

Next we study the internal symmetries of the $g\phi_2^4$ - theory. Consider the representation of the group Z_2 as a * automorphism group $\{\tau_1, \tau_{-1}\}$ on $\mathfrak{A}$, defined by

$$\tau_{\pm 1}(e^{i\phi(f)}) = e^{\pm i\phi(f)}, \quad f \in \mathcal{S} \tag{3.18}$$

These automorphisms commute with the representation $\{\tau_\xi : \xi \in \mathcal{P}_+^\uparrow\}$ of the Poincaré group on $\mathfrak{A}$, i.e. Z_2 is an internal symmetry group of the $g\phi_2^4$ - theory. However

$$\omega_+ \circ \tau_{-1} = \omega_- \quad , \quad (\text{see } (3.3)), \tag{3.19}$$

so that this internal symmetry group is spontaneously broken by the pure Wightman states ω_+ and ω_-.

As mentioned at the end of Section 2.3 and in Section 2.4 of Part 1, this suggests that there exist charged super selection (soliton) sectors "orthogonal" to the vacuum sectors.

2.4 The charged super selection (soliton) sectors of $g\phi_2^4$

In this section we propose to construct two charged states $\omega_+ o\sigma$ and $\omega_- o\sigma$ (a "soliton-" and an "anti-soliton" state) for the $g\phi_2^4$ - quantum field model with the properties

$$\lim_{\underline{a}\to-\infty} \omega_\pm o\sigma (\tau_{\underline{a}} (A)) = \omega_\mp (A)$$
$$\lim_{\underline{a}\to+\infty} \omega_\pm o\sigma (\tau_{\underline{a}} (A)) = \omega_\pm (A) \quad , \tag{4.1}$$

for all $A \in \mathfrak{A}$; (of course $\tau_{\underline{a}}$ (A) is the translate of A by the vector $(\underline{a},o)$, and σ will turn out to be a charged * automorphism of $\mathfrak{A}$).

Let $\tilde{\phi}_\pm \equiv \omega_\pm (\phi^\pm (x)) = W_{1,\pm} (x) \simeq \phi_\pm$.

Clearly (4.1) yields the formal equations:

$$\lim_{\underline{x}\to\pm\infty} \omega_+ o\sigma (\phi^{(+)} (\underline{x},t)) = \pm \tilde{\phi}_+$$
$$\lim_{\underline{x}\to\pm\infty} \omega_- o\sigma (\phi^{(-)} (\underline{x},t)) = \pm \tilde{\phi}_- = \mp \tilde{\phi}_+ \tag{4.2}$$

Thus the functions $g^{\frac{1}{2}} \omega_\pm o\sigma (\phi^{(\pm)} (\underline{x},t))$ resemble the soliton solutions $\phi_s (x)$, $\phi_{\bar{s}} (x)$ of the classical $g\phi_2^4$ - field theory given in equation (1.8) of Section 1, Part 2.

The classical theory suggests that $\omega_+ o\sigma$ and $\omega_- o\sigma$ ought to be the only charged states of the $g\phi_2^4$ quantum field model.

Equations (4.1) say that, as the observable A is moved from space-like left $\underline{a} = -\infty$, to space-like right, $\underline{a} = +\infty$, the states $\omega_\pm o\sigma$ interpolate between the two pure Wightman states ω_+ and ω_-. In other words, σ interpolates between the automorphism τ_{-1} of $\mathfrak{A}$ and the identity automorphism τ_1 of $\mathfrak{A}$ introduced in (3.18).

In order to construct a charged automorphism σ of $\mathfrak{A}$ with these properties we follow the philosophy described in Section 2.4 of Part 1, (see Theorem 2.16), i.e. we propose to construct a non-trivial, local Poincaré cocycle Γ from which σ can be constructed, (see Theorem 2.9, Section 2.4, Part 1). Because of (4.1) and the above remarks we expect that Γ is indexed by the internal symmetry group Z_2 of the $g\phi_2^4$ - theory, (see equations (2.60) - (2.63), Section 2.4, Part 1).

Of course we try to first construct Γ in a representation of $\mathfrak{A}$ in which it is trivial, i.e. given by unitary operators. The obvious guess for a convenient representation is the Fock representation; and this is what we shall do next.

The following construction of Γ is patterned after the one presented in the author's papers [F2 , F3] where similar, two dimensional quantum field models were studied. The reader ought to consult these references for motivation and details. For the $g\phi_2^4$ - theory the construction of a local Poincaré cocycle Γ outlined below is new and turns out to be somewhat more complicated than what we have done in [F2 , F3].

The short digression which now follows is oriented towards the expert and represents, in some sense, the only new result in these notes. The end of this slightly difficult digression is marked by (*) to which point we refer the non-specialist.

Let $\mathcal{O}$ be a bounded double cone in M^2, $\mathcal{O}_L$ its causal complement to the left and $\mathcal{O}_R$ the one to the right of $\mathcal{O}$. Let $\mathfrak{A}(\mathcal{O}_L)$ and $\mathfrak{A}(\mathcal{O}_R)$ be the local von Neumann algebras generated by the free field on $\mathfrak{F}$, associated with $\mathcal{O}_L$, resp. $\mathcal{O}_R$.

Let $\mathfrak{B}$ be the von Neumann algebra generated by $\mathfrak{A}(\mathcal{O}_L)$ and $\mathfrak{A}(\mathcal{O}_R)$ which is also denoted by $\mathfrak{A}(\mathcal{O}_L) \vee \mathfrak{A}(\mathcal{O}_R)$. It is well known that the bare vacuum Ω_o(4.3) is a cyclic and separating vector for $\mathfrak{B}$, (Reeh-Schlieder thm., [S2]). (4.3).

It follows from ref. [F2] (Lemma 2 and § 5) that there exists a faithful,normal state ρ on $\mathfrak{B}$ such that

$$\rho(A \cdot B) = \rho(A)\,\rho(B), \tag{4.4}$$

for all $A \in \mathfrak{A}(\mathcal{O}_L)$ and $B \in \mathfrak{A}(\mathcal{O}_R)$. (An explicit construction of a ρ satisfying (4.4) can be achieved by considering the two fold tensor product of the free field on $\mathfrak{F} \otimes \mathfrak{F}$ and using the Bogoliubov transformation constructed in Lemma 2 and applied in § 5 of [F2]). It is then easy to see that ρ can be chosen such that

$$\rho(A) = (\Omega_o, A\Omega_o),$$

for all $A \in \mathfrak{A}(\mathcal{O}_L)$ and all $A \in \mathfrak{A}(\mathcal{O}_R)$; see § 5 of [F2]. An alternate construction of a related state can be given by adapting the methods of Buchholz [B6], whose results are important in what follows. By property (4.3) and Lemma 2.1 of [B6] there exists a unit vector $\xi \in \mathfrak{F}$ such that

$$\rho(C) = (\xi, C\,\xi), \text{ for } C \in \mathfrak{B} \quad . \tag{4.5}$$

As noted in [B6], (4.4) and (4.5) imply that there exists a factor $M_{\mathcal{O}}$ of type I_∞ such that

$$\mathfrak{A}(\mathcal{O}_L) \subset M_{\mathcal{O}} \subset \mathfrak{A}(\mathcal{O}_R)' \quad . \tag{4.6}$$

Next consider the isomorphism ι of $\mathfrak{A}(\mathcal{O}_L)$ defined by

$$\iota(A) = \tau_{-1}(A), \text{ for all } A \in \mathfrak{A}(\mathcal{O}_L);$$

(i.e. $\iota(e^{i\phi(f)}) = e^{-i\phi(f)}$, for all $f \in \mathcal{S}$ with supp $f \subset \mathcal{O}_L$).

Since

$$(\Omega_o, \tau_{-1}(A)\,\Omega_o) = (\Omega_o, A\Omega_o),$$

for all $A \in \mathfrak{A}$, τ_{-1} is implemented on $\mathfrak{F}$ by a unitary operator, and hence so is ι.

As noted in [B6] it then follows that there exists a unitary operator $U_{\mathcal{O}} \in M_{\mathcal{O}}$ such that

$$U_{\mathcal{O}}^* \, A \, U_{\mathcal{O}} = \iota(A), \tag{4.7}$$

for all $A \in \mathfrak{A}(\mathcal{O}_L)$. Since $M_{\mathcal{O}} \subset \mathfrak{A}(\mathcal{O}_R)'$,

$$U_{\mathcal{O}}^* \, A \, U_{\mathcal{O}} = A \quad , \tag{4.8}$$

for all $A \in \mathfrak{A} \ (\mathcal{O}_R)$.

It is not hard to show that

$$\sigma_{\mathcal{O}} \ (A) \equiv U_{\mathcal{O}}^* \, A \cdot U_{\mathcal{O}} \tag{4.9}$$

defines a * automorphism of $\mathfrak{A}$ with the property that, given a bounded double cone $\mathcal{O}_1$ there exists a bounded double cone $\mathcal{O}_2 \supseteq \mathcal{O}_1$ (and $\mathcal{O}_2 = \mathcal{O}_1$ if $\mathcal{O}_1 \cap \mathcal{O} = \emptyset$) such that

$$\sigma_{\mathcal{O}} \ (A) \in \mathfrak{A}(\mathcal{O}_2), \text{ for all } A \in \mathfrak{A} \ (\mathcal{O}_1). \tag{4.10}$$

This is an immediate consequence of the facts that $\sigma_{\mathcal{O}} \ (\mathfrak{A}(\mathcal{O}_L)) = \mathfrak{A}(\mathcal{O}_L)$, $\sigma_{\mathcal{O}} \ (\mathfrak{A}(\mathcal{O}_R)) = \mathfrak{A} \ (\mathcal{O}_R)$ and <u>duality</u> (i.e. $\mathfrak{A} \ (\bar{\mathcal{O}})' = \mathfrak{A} \ (\sim \bar{\mathcal{O}})$, for arbitrary, open $\bar{\mathcal{O}}$) (*).

We summarize this discussion in

<u>Theorem 4.1</u>:

Given a bounded double cone $\mathcal{O}$ there exists a * automorphism $\sigma_{\mathcal{O}}$ of the algebra $\mathfrak{A}$ of all local observables generated by the <u>free field</u> with the properties

(1) $\sigma_{\mathcal{O}} \ (A) = \tau_{-1} \ (A)$, for all $A \in \mathfrak{A}(\mathcal{O}_L)$
$\sigma_{\mathcal{O}} \ (A) = A$, for all $A \in \mathfrak{A}(\mathcal{O}_R)$

(2) Given a bounded double cone $\mathcal{O}_1$, there exists a bounded $\mathcal{O}_2 \supseteq \mathcal{O}_1$ such that

$$\sigma_{\mathcal{O}} \ (A) \in \mathfrak{A}(\mathcal{O}_2), \text{ for all } A \in \mathfrak{A}(\mathcal{O}_1).$$

(3) On Fock space $\mathfrak{F}$, $\sigma_{\mathcal{O}}$ is implemented by a <u>unitary operator</u> $U_{\mathcal{O}} \in \mathfrak{A} \ (\mathcal{O}_R)'$ $(= \mathfrak{A} \ (\sim \mathcal{O}_R))$.

Theorem 4.1 is the key ingredient in the following construction of <u>$\mathcal{O}$-local Poincaré cocycles</u> for the $g\phi_2^4$ - theory.

As an example we briefly discuss the space-time translation cocycles; (but see [F2 , F3] for details):

Let $\{\tau_{\underline{a}} : \underline{a} \in R\}$ denote the space translation automorphisms of the free field on $\mathfrak{F}$.

We set

$$\Gamma(\underline{a}) = U_{\Theta}^{*} \tau_{\underline{a}} (U_{\Theta}). \qquad (4.11)$$

By duality, $\Gamma(\underline{a}) \varepsilon \mathfrak{A} (\mathcal{O}_{\underline{a}})$. (4.12)

Moreover, $\Gamma(\underline{a})$ is strongly continuous in $\underline{a}$, (since U_{Θ} is unitary, and $\tau_{\underline{a}}$ is implemented on $\mathcal{F}$ by a continuous, unitary group).

Given some positive number $T < \infty$, let $L(T)$ be so large that the causal shadow of $\Theta((\underline{0},T))$ at time $t = 0$ is contained in the interval $\left[-\frac{L(T)}{2}, \frac{L(T)}{2}\right]$. We recall that

$\Theta((\underline{0},T)) = \{x: x - (\underline{0},T) \ \varepsilon\ \Theta\}$, and the causal shadow of $\Theta((\underline{0},T))$ at $t = 0$ is given by

$$\{x = (\underline{x},0) : (x - y)^2 \geqq 0, \text{ for all } y \ \varepsilon\ \Theta((\underline{0},T))\}).$$

For $|t| < T$, we define

$$\Gamma(t) = U_{\Theta}^{*} e^{itH_{-+}(L)} U_{\Theta} e^{-itH_{++}(L)} , \qquad (4.13)$$

where $L \geqq L(T)$. Since U_{Θ} is unitary, and $\exp(itH_{-+}(L))$, $\exp(-itH_{++}(L))$ are strongly continuous in t, Γ is strongly continuous in t. Using the Trotter product formula for $\exp(itH_{-+}(L))$ and $\exp(-itH_{++}(L))$ and the fact that σ_{Θ} converts - boundary conditions into + boundary conditions one shows that

(i) $\Gamma(t)$ is independent of L, for $|t| \leqq T$ and $L \geqq L(T)$.
(ii) $\Gamma(t) \varepsilon \mathfrak{A}(\Theta_{(\underline{0},t)})$.

This is a basic result of [F2], stated in this form in [B5].

Finally we define

$$\Gamma(\underline{a},t) = \Gamma(\underline{a}) \tau_{\underline{a}} (\Gamma(t)). \qquad (4.14)$$

Theorem 4.2, [F2]:

The operators $\Gamma(\underline{a},t)$ are local observables in $\mathfrak{A}(\Theta_{(\underline{a},t)})$, strongly continuous in $(\underline{a},t)$. They are Θ-local space-time translation cocycles for the $g\phi_2^4$ - theory, in the sense of Section 2.4, Part 1 (in particular Definition 9).

Remarks:

1. This result is a rather straightforward consequence of Theorem 4.1 and (4.11) - (4.14); see [F2].

2. It can be inferred from the results of [F3] (Sections 9, 10) and Theorem 4.1 that there exists an $\mathcal{B}$-local Poincaré cocycle Γ for the $g\phi_2^4$ - theory which for $\xi = (1, (\underline{a},t))$ coincides with $\Gamma(\underline{a},t)$; (see also [F15]).

3. It is easy to see that

$$\lim_{\underline{a}\to-\infty} \Gamma(\underline{a},t)\, A\, \Gamma(\underline{a},t)^* = \sigma_{\mathcal{B}}(A) \equiv \sigma_-(A) \quad , \tag{4.15}$$

for all $A \in \mathfrak{A}$.

We set

$$\sigma_+(A) = \lim_{\underline{a}\to+\infty} \Gamma(\underline{a},t)\, A\, \Gamma(\underline{a},t)^* \quad , \tag{4.16}$$

$A \in \mathfrak{A}$.

It is not hard to see that σ_+ is a * automorphism of $\mathfrak{A}$ with the properties

$$\begin{aligned} &\sigma_+(A) = A, \text{ for } A \in \mathfrak{A}(\mathcal{B}_L) \\ &\sigma_+(A) = \tau_{-1}(A), \text{ for } A \in \mathfrak{A}(\mathcal{B}_R); \end{aligned} \tag{4.17}$$

compare this with Theorem 2.10 and (2.62) - (2.63), Section 2.4, Part 1.

We now define soliton - and anti-soliton states ω_s and $\omega_{\bar{s}}$ by

$$\begin{aligned} \omega_s(A) &= \omega_+(\sigma_-(A)) \\ \omega_{\bar{s}}(A) &= \omega_-(\sigma_-(A)) \end{aligned} \tag{4.18}$$

By (4.17) ω_s and $\omega_- \circ \sigma_+$, and $\omega_{\bar{s}}$ and $\omega_+ \circ \sigma_+$ determine equivalent representations of $\mathfrak{A}$.

Since $\lim_{\underline{a}\to\pm\infty} \omega_s(\tau_{\underline{a}}(A)) = \omega_\pm(A)$,

and

$$\lim_{\underline{a}\to\pm\infty} \omega_{\bar{s}}(\tau_{\underline{a}}(A)) = \omega_{\mp}(A), \tag{4.19}$$

for all $A \in \mathfrak{A}$, and since $\omega_+ \neq \omega_-$ (!), the representations π_s and $\pi_{\bar{s}}$ of the algebra $\mathfrak{A}$ determined by ω_s, resp. $\omega_{\bar{s}}$ are disjoint, and they are disjoint from the vacuum representations $\pi_\pm$ determined by the Wightman states $\omega_\pm$; (see also Theorem 4 of [F2]). Hence the cocycle Γ constructed above is non-trivial in the representations $\pi_\pm$ of $\mathfrak{A}$, (although it is trivial in the Fock representation).

By Theorems 2.12, 2.13 and Corollary 2.15, Section 2.4, Part 1, the states ω_s and $\omega_{\bar{s}}$ are charged states of the $g\phi_2^4$ - theory, i.e. they are Poincaré - covariant, satisfy the relativistic spectrum condition and are not vacuum states. Let $\mathcal{H}_\pm$, $\mathcal{H}_s$, $\mathcal{H}_{\bar{s}}$ denote the super selection sectors reconstructed from $(\mathfrak{A}, \omega_\pm)$, $(\mathfrak{A}, \omega_s)$, $(\mathfrak{A}, \omega_{\bar{s}})$ respectively. The total Hilbert space of the $g\phi_2^4$ - theory is given by

$$\mathcal{H}_{tot.} = \mathcal{H}_+ \oplus \mathcal{H}_- \oplus \mathcal{H}_s \oplus \mathcal{H}_{\bar{s}} \tag{4.20}$$

There exists a selfadjoint operator Q on $\mathcal{H}_{tot.}$, formally given by

$$Q = \int_{-\infty}^{+\infty} d\underline{x}\, (\partial_{\underline{x}}\, \phi)\, (x) \tag{4.21}$$

with eigenvalues 0, $q_s \equiv \omega_+ (\phi) - \omega_- (\phi)$ and $q_{\bar{s}} \equiv \omega_- (\phi) - \omega_+ (\phi) = - q_s$, such that $\mathcal{H}_+ \oplus \mathcal{H}_-$ is the eigenspace for the eigenvalue 0, $\mathcal{H}_s$ the one for q_s and $\mathcal{H}_{\bar{s}}$ the one for $q_{\bar{s}}$.

The Poincaré automorphisms on $\mathfrak{A}$ are implemented on $\mathcal{H}_{tot.}$ by a continuous, unitary group. The infinitesimal generators of the translation subgroup, the energy-momentum operator, are denoted $(H, \underline{P})$. The spectrum of $(H, \underline{P}) \upharpoonright (\mathcal{H}_s \oplus \mathcal{H}_{\bar{s}})$ is contained in the forward light cone V_+ and is purely continuous (no vacuua in $\mathcal{H}_s$, $\mathcal{H}_{\bar{s}}$!)

The representations $\pi_{\mathcal{H}_+ \oplus \mathcal{H}_s}$ and $\pi_{\mathcal{H}_- \oplus \mathcal{H}_{\bar{s}}} \circ \tau_{-1}$ of $\mathfrak{A}$ are obviously unitarily equivalent.

Let $\mathfrak{A}_e = \mathfrak{A}^{Z_2}$ be the subalgebra of $\mathfrak{A}$ invariant under τ_{-1}, i.e. $\mathfrak{A}_e$ is the algebra of even functions of ϕ. Then the representations of $\mathfrak{A}_e$ on $\mathcal{H}_+ \oplus \mathcal{H}_s$ and on $\mathcal{H}_- \oplus \mathcal{H}_{\bar{s}}$ are identical.

Let $T_s \equiv T_{\sigma_-}$ be the unitary intertwiner mapping $\mathcal{H}_{+/-}$ to $\mathcal{H}_{s/\bar{s}}$; see the end of Section 2.4, Part 1.

Let

$$T_s\, (\underline{x},t) = e^{i(tH - \underline{x}\cdot\underline{P})}\, T_s\, e^{-i(t H - \underline{x}\cdot\underline{P})} \tag{4.22}$$

This object plays the role of a charged field; (its commutator with Q is ≠ 0). It is easy to see (Lemma 8 of [F2]) that

$$\prod_{j=1}^{n} T_s(x_j) : \begin{cases} \mathcal{H}_{+/-} \rightarrow \mathcal{H}_{+/-} \\ \mathcal{H}_{s/\bar{s}} \rightarrow \mathcal{H}_{s/\bar{s}} \end{cases} \text{for n even} \quad \begin{cases} \mathcal{H}_{+/-} \rightarrow \mathcal{H}_{s/\bar{s}} \\ \mathcal{H}_{s/\bar{s}} \rightarrow \mathcal{H}_{+/-} \end{cases} \text{for n odd.} \tag{4.23}$$

Hence $\mathcal{H}_+ \oplus \mathcal{H}_s$ and $\mathcal{H}_- \oplus \mathcal{H}_{\bar{s}}$ are <u>invariant subspaces</u> for the "<u>field algebra</u>" $\mathfrak{F}$ generated by $\mathfrak{A}_e$ and $\{T_s(x) : x \in M^2\}$. It is expected (but to date unproven) that

$$\left.\begin{array}{l}\Omega_+ \text{ is \underline{cyclic} for } \mathfrak{F} \text{ in } \mathcal{H}_+ \oplus \mathcal{H}_s, \text{ and} \\ \Omega_- \text{ is cyclic for } \mathfrak{F} \text{ in } \mathcal{H}_- \oplus \mathcal{H}_{\bar{s}}.\end{array}\right\} \tag{4.24}$$

Furthermore, note that $T_s(x)$ and $:\phi^2:(x)$ (the normal ordered square of the relativistic quantum field $\phi \equiv \phi^{(\pm)}$) are almost relatively local [S2], almost local fields:

$$\begin{aligned}\left[T_s(x), T_s(y)\right] &= 0, \text{ for } \mathcal{O}(x) \subseteq \sim\mathcal{O}(y) \\ \left[T_s(x), :\phi^2:(y)\right] &= 0, \text{ for } y \in \sim\mathcal{O}(x),\end{aligned} \tag{4.25}$$

but the commutation relations for $T_s(x)$ and $\phi(x)$ are "<u>strange</u>"; see [F2].

Next, we define the <u>mass operator</u>

$$M = \sqrt{H^2 - P^2} \tag{4.26}$$

The following facts about the spectrum of M (the "mass spectrum") are established.

<u>Theorem 4.3:</u>

For $0 < g << 1$ ($\sigma = -\frac{1}{2}$)

(1) [G7] $\text{spec}(M \restriction \mathcal{H}_+) = \text{spec}(M \restriction \mathcal{H}_-) \subseteq \{0\} \cup [m,\infty)$, for some $m = m(g) > 0$, with $m(g) \to 1$, as $g \searrow 0$.

(2) [B5] $\text{spec}(M \restriction \mathcal{H}_s) = \text{spec}(M \restriction \mathcal{H}_{\bar{s}}) \subseteq [m_s,\infty)$, for some $m_s = m_s(g) > 0$, with

$$m_s(g) = O(g^{-1}), \text{ as } g \searrow 0.$$

<u>Remark</u>:

The key estimate in the proof of (2) is the inequality

$$m_s(g) \geqq \tau(g), \tag{4.27}$$

where $\tau(g) = \overline{\lim_{L\to\infty}} \inf \text{spec}(\tilde{H}_{-+}(L) - E_{++}(L))$, and $\inf \text{spec}(\tilde{H}_{-+}(L) - E_{++}(L)) = \lim_{T\to\infty} -\frac{1}{T}\log\frac{Z_{-+}(L\times T)}{Z_{++}(L\times T)}$, see (3.13) and [B5] (Theorem A'). The quotient $Z_{-+}(L\times T)/Z_{++}(L\times T)$ can be estimated by probabilistic methods from Euclidean field theory (a "Peierls argument") yielding the lower bound $\tau(g) = O(g^{-1})$, as $g \searrow 0$. The mass spectrum spec $(M \upharpoonright \mathcal{H}_s)$ can be analyzed by Euclidean (imaginary time) methods similar to the ones applied in the construction of the Euclidean Green's functions of this model, even though the constructions of $\mathcal{H}_s$, Γ, etc., involve rather abstract C^* algebra techniques.

Finally, we assume now that the points m and m_s in the spectrum of M are isolated eigenvalues. (This has not be proven, yet). Then m and m_s are the masses of stable particles, the "meson" and the "quantum soliton".

A multi-meson - multi-soliton scattering theory can then be developed by the standard Haag-Ruelle procedure [J2]: One can prove that the field $:\phi^2:$ has non-vanishing matrix elements between Ω_+ and the one-meson states, and a strictly local element of $\mathfrak{F}$, linear in T_s, has non-vanishing matrix elements between Ω_+ and the one-soliton states, [F14, F15]. (Thus, by (4.25), the Haag-Ruelle theory applies).

The particle spectrum and the scattering operator on $\mathcal{H}_+ \oplus \mathcal{H}_s$ and on $\mathcal{H}_- \oplus \mathcal{H}_{\bar{s}}$ are identical. Unitarity of the scattering operator is a completely open problem; (to settle this question one must e.g. prove or disprove conjecture (4.24)).

We have now achieved a very detailed illustration of the abstract formalism of r.q.f.t. and c.q.f.t. reviewed in Part 1 within the context of the $g\phi_2^4$ - quantum field model. The results reported in Part 2 for this model, however incomprehensible they may appear to the reader, prove that the basic axioms of r.q.f.t. are compatible with non-trivial scattering (at least in space-time dimensions 2 and 3) and with non-trivial, charged super-selection rules (at least in dimension 2).

The qualitative predictions of the classical ϕ_2^4 - field theory (Introduction and Section 1, Part 2) have turned out to be reliable guides in the analysis of the $g\phi_2^4$ quantum field model, at least when $g \equiv \hbar$ is small. As an important, partly open problem we propose to set up asymptotic expansions in $g^{1/2}$ about the minima $\phi_\pm$, ϕ_s and $\phi_{\bar{s}}$ of the classical Hamilton functional; see also Gervais' lectures.

For the Euclidean Green's functions and the mesons → mesons scattering amplitudes this problem is solved in [G7, E1]. For the mass $m_s(g)$ of the quantum soliton a solution appears to be possible by using known techniques, but has not been given, yet. The expansion for $m_s(g)$ would have the form

$$m_s(g) = \alpha_{-1}\, g^{-1} + \alpha_0 + \sum_{n=1}^{\infty} a_n\, g^{\frac{n}{2}}$$

For the (mesons, solitons) → (mesons, solitons) scattering amplitudes a solution to the problem of finding asymptotic expansions in $g^{1/2}$ (?) about $g = 0$ seems to require new ideas. A combination of the rigorous techniques developed in this section with the results of Hepp [H7] concerning the classical limit of the $g\phi_2^4$ - theory and with [G7, S14] may yield useful hints. But, to date, there are no entirely convincing proposals. The reader is referred to Gervais' lectures in these proceedings for further discussion and references. May we also recommend the references collected under "Further topics in c.q.f.t. and reviews" to the reader's attention.

Acknowledgements:

Useful discussions with E. Seiler and A.S. Wightman and encouragement are gratefully acknowledged.

References

A1 S. Albeverio and R. Høegh-Krohn, J. Funct. Anal. 16, 39, (1974).
A2 H. Araki, Progr. Theor. Phys. 32, 844, (1964).

B1 A. Belavin, A.M. Polyakov, A.S. Schwarz, Y. Tyupkin, Physics Letters B59, 85, (1975).
B2 J. Bisognano and E. Wichmann, J. Math. Phys. 16, 985, (1975).
B3 H. -J. Borchers, Nuovo Cimento 24, 214 (1962).
B4 O. Bratteli, Comm. Pure Appl. Math. 25, 759, (1972).
B5 J. Béllissard, J. Fröhlich and V. Gidas, Preprint 1977, submitted to Commun. Math. Phys.
B6 D. Buchholz, Commun. Math. Phys. 36, 287, (1974).

C1 C. Callan, R. Dashen and D. Gross, Princeton Preprint COO-2220-115, 1977.
C2 C. Callan, R. Dashen and D. Gross, Princeton Preprint, 1977, to appear in Phys. Rev. D (1977).
C3 C. Callan, R. Dashen and D. Gross, Physics Letters 63B, 334 (1976).
C4 S. Coleman, Erice lectures 1975, to appear in the proceedings of the 1975 Int. School on Subnuclear Physics "Ettore Majorana", . Zichichi (ed.).

D1 W. Driessler and J. Fröhlich, Ann. Inst. H. Poincaré, Section A, 27, 221 (1977).
D2 J. Dimock and J. Glimm, Adv. Math. 12, 58, (1974).
D3 S. Doplicher, R. Haag and J. Roberts, Commun. Math. Phys. 23, 199, (1971) and 35, 49, (1974); and refs. to their earlier work given there.
D4 F. Dyson, E.H. Lieb and B. Simon, Princeton Preprint 1977, to appear in J. Stat. Phys.
D5 J. Dimock, Commun. Math. Phys. 35, 347, (1974).

E1 J. -P Eckmann, H. Epstein and J. Fröhlich, Ann. Inst. H. Poincaré, Section A, 25, 1, (1976).
E2 J. -P Eckmann, J. Magnen and R. Sénéor, Commun. Math. Phys. 39, 251 (1975).
E3 H. Ezawa and A. Swieca, Commun. Math. Phys. 5, 330 (1967).
E4 J. -P. Eckman, Preprint, Université de Genève, 1976, to be published.

F1 J. Fröhlich, to appear in the Proceedings of the Int. Conf. on Math. Problems in Theor. Phys., Rome 1977.
J. Fröhlich, and E. Seiler, in preparation.
F2 J. Fröhlich, Commun. Math. Phys. 47, 269, (1976).
F3 J. Fröhlich, Acta Physica Austrica, Suppl 15, 133, (1976).
F4 J. Fröhlich, Ann. Phys. 97, 1, (1976).
F5 J. Fröhlich and B. Simon, Ann. Math. 105, 493, (1977).
F6 J. Fröhlich, Bielefeld lectures 1976, to appear in the Proceedings of the Bielefeld symposium, fall 1976, L. Streit (ed.).
F7 J. Fröhlich, Adv. Math. 23, 119, (1977).
F8 J. Fröhlich, in "Renormalization Theory", G. Velo and A.S. Wightman (eds.), Nato Adv. Stud. Inst. Series C, Reidel, Dordrecht-Boston, 1976 .
F9 J. Fröhlich, and E. Seiler, Helv. Phys. Acta, 49, 889, (1976).
F10 J. Fröhlich and Y.M. Park, Helv. Phys. Acta 50, 315,(1977).
F11 J. Feldman and K. Osterwalder, Ann. Phys. 97, 80, (1976).
F12 J. Fröhlich, Commun. Math. Phys. 54, 135, (1977).
F13 J. Fröhlich, B. Simon, and T. Spencer, Commun. Math. Phys. 50, 79, (1976).
F14 J. Fröhlich and T. Spencer, Cargèse lectures 1976, to appear in the Proceedings of the 1976 Cargèse Summer School in Theor. Phys.
F15 J. Fröhlich, unpubl. notes.

G1 J. Glimm and A. Jaffe, in "Statistical Mechanics and Quantum Field Theory", Les Houches 1970, C. DeWitt and R. Stora (eds.), Gordon and Breach, New York - London - Paris, 1971.
J. Glimm and A. Jaffe, in "Mathematics of Contemporary Physics", R. Streater (ed.), Academic Press, London - New York, 1972.

G2 V. Glaser, Commun. Math. Phys. 37, 257, (1974).
G3 J. Glimm, A. Jaffe and T. Spencer, Ann. Math. 100, 585 (1974).
J. Glimm, A. Jaffe and T. Spencer, Erice Lectures, 1973, in ref. [V1] .
G4 F. Guerra, L. Rosen and B. Simon, Ann. Math. 101, 111, (1975).
G5 J. Glimm and A. Jaffe, Fortschritte der Physik, 21, 327 (1973).
G6 J. Glimm, A. Jaffe and T. Spencer, Commun. Math. Phys. 45, 203, (1975).
G7 J. Glimm, A. Jaffe and T. Spencer, Ann. Phys. 101, 610 (1976), and Ann. Phys. 101, 631, (1976).
G8 F. Guerra, L. Rosen and B. Simon, Commun. Math. Phys. 27, 10, (1972) and 29, 233, (1973); and paper to appear in Ann. Inst. H. Poincaré, Section A, (1977).
G9 J. Glimm and A. Jaffe, J. Math. Phys. 13, 1568, (1973).
G10 K. Gawedzki , Preprint, Warsaw University, 1977.
G11 J. Glimm and A. Jaffe, Phys. Rev. D11, 2816, (1975).
G12 J. Glimm, Commun. Math. Phys. 8, 12, (1968).
G13 G. Gallavotti et al., Preprints, Università di Roma and I.H.E.S., 1977, and paper to appear in Commun. Math. Phys.
G14 J. Glimm and A. Jaffe, Acta Math. 125, 203, (1970); (and refs. to their earlier work given there).

H1 G. 't Hooft, Erice lectures 1977, to appear in the Proceedings of the Int. School on Subnuclear Physics "Ettore Majorana", A. Zichichi (ed.).
H2 G. 't Hooft, Phys. Rev. Letters, 37, 8, (1976), Phys. Rev. D14, 3432, (1976).
H3 R. Haag and D. Kastler, J. Math. Phys. 5, 848, (1964).
H4 K. Hepp, "Théorie de la Renormalisation", Lecture Notes in Physics 2, Springer-Verlag, Berlin-Heidelberg-New York, 1969.
H5 P. Halmos, "Measure Theory", Graduate Texts in Mathematics 18, Springer-Verlag, Berlin-Heidelberg-New York, 1974.
H6 G. Hegerfeldt, Commun. Math. Phys. 45, 133, (1975).
H7 K. Hepp, Commun. Math. Phys. 35, 265, (1974).

J1 R. Jackiw and C. Rebbi, Phys. Rev. Letters 37, 172, (1976).
J2 R. Jost, "The General Theory of Quantized Fields," Amer. Math. Soc. Publ., Providence, R.I. 1965.
J3 A. Jaffe, Commun. Math. Phys. 1, 127, (1965).

K1 D. Kastler, D.W. Robinson and A. Swieca, Commun. Math. Phys. 2, 108, (1966).

L1 O.E. Lanford III, in "Statistical Mechanics and Quantum Field Theory", Les Houches 1970, C. DeWitt and R. Stora (eds.), Gordon and Breach, New York-London-Paris, 1971.
L2 J.L. Lebowitz, to appear in the Proceedings of the Int. Conf. on Math. Problems in Theor. Phys., Rome 1977; see also: J.L. Lebowitz and A. Martin-Löf, Commun. Math. Phys. 25, 276, (1972); J.L. Lebowitz, J. Stat. Phys. (1977).

M1 R.A. Minlos, Trans. Moscow Math. Obs. 8, 471, (1959); see also ref. (G1).
M2 J. Magnen and R. Sénéor, Harvard University, Preprint 1977, to appear in Commun. Math. Phys.; see also J. Magnen, Thesis, Orsay University, 1976.
M3 J. Magnen and R. Sénéor, Ann. Inst. H. Poincaré, Section A, 24, 95, (1976).
M4 O. McBryan, in "Les Méthodes Mathématiques de la Théorie Quantique des Champs". Éditions du C.N.R.S., Paris, 1976.

N1 E. Nelson, in "Partial Differential Equations", D. Spencer (ed.), Symposium in Pure Math., 23, Amer. Math. Soc. Publ., Providence, R.I. 1973;
E. Nelson, J. Funct. Anal. 12, 97, (1973).
N2 E. Nelson, J. Math. Phys. 5, 332, (1964).
N3 E. Nelson, J. Funct. Anal. 12, 211, (1973).
N4 E. Nelson, Erice Lectures, 1973, in ref. [V1] .
N5 E. Nelson, J. Funct. Anal. 11, 211, (1972).

N6 E. Nelson, in "Mathematical Theory of Elementary Particles", R. Goodman and I. Segal (eds.), M.I.T. Press, Cambridge, Mass. 1966.

O1 K. Osterwalder and R. Sénéor, Helv. Phys. Acta 49, 525, (1976).
O2 K. Osterwalder and R. Schrader, Commun. Math. Phys. 31, 83, (1973) and 42, 281, (1975).

P1 A.M. Polyakov, Physics Letters 59B, 82, (1975); see also ref. [B1].
P2 A.M. Polyakov, Nuclear Physics B120, 429, (1977).
P3 K. Pohlmeyer, Commun. Math. Phys. 25, 73, (1972).

R1 M. Reed and B. Simon, "Methods of Modern Mathematical Physics", Vol. IV, Academic Press, New York, to appear.
R2 J.E. Roberts, Commun. Math. Phys. 51, 107, (1976).

S1 A. Sokal, Princeton University, Preprint 1977, submitted to Physics Letters.
S2 R.F. Streater and A.S. Wightman, "PCT, Spin and Statistics and All That", Benjamin, New York, 1964; (new edition to appear).
S3 T. Spencer and F. Zirilli, Commun. Math. Phys. 49, 1, (1976).
S4 K. Symanzik, "A Modified Model of Euclidean Quantum Field Theory", New York University, Report, 1964.
K. Symanzik, J. Math. Phys. 7, 510 (1966).
K. Symanzik, in "Local Quantum Theory", R. Jost (ed.), Academic Press, New York, 1969.
S5 B. Simon, "The $P(\phi)_2$ Euclidean (Quantum) Field Theory", Princeton Series in Physics, Princeton University Press, Princeton, N.J. 1974.
S6 T. Spencer, Commun. Math. Phys. 39, 63, (1974).
S7 E. Seiler and B. Simon, Ann. Phys. 97, 470, (1976).
S8 J. Slawny, Commun. Math. Phys. 35, 297, (1974).
S9 B. Simon and R. Griffiths, Commun. Math. Phys. 33, 145, (1973).
S10 K. Symanzik, Commun. Math. Phys. 6, 288, (1967).
S11 F. Strocchi, Lectures delivered at Princeton University, 1974 (unpubl.).
S12 R.F. Streater and I.F. Wilde, Nuclear Physics B24, 561 (1970).
S13 T. Spencer, Commun. Math. Phys. 39, 77, (1974).
S14 T. Spencer, Commun. Math. Phys. 44, 143, (1975).

T1 B. Berg, M. Karowski and H.J. Thun, Physics Letters 62B, 63 (1976) and 62B, 187, (1976), Erice Lectures 1977; (see [H1]).

V1 G. Velo and A.S. Wightman, (eds.), "Constructive Quantum Field Theory", (1973 "Ettore Majorana" Int. School of Math. Physics, Erice, Sicily), Lecture Notes in Physics 25, Springer-Verlag, Berlin-Heidelberg-New York, 1973.

Z1 A.B. Zamolodchikov, Commun. Math. Phys. 55, 183, (1977).

Further topics in c.q.f.t. and reviews:

A. Models with Fermions (Y_2):
E. Seiler, Commun. Math. Phys. 42, 163 (1975); ref. [S7] and refs. given there; ref. [M4].
J. Magnen and R. Sénéor, Commun. Math. Phys.(1977).
A. Cooper and L. Rosen, to appear in Trans. Amer. Math. Soc.
See also [F8] and [F9] (Q E D_2).

B. Lattice gauge theories:
K. Wilson, Phys. Rev. D10, 2445, (1974).
R. Balian, J.M. Drouffe and C. Itzykson, Phys. Rev. D10, 3376, (1974), D11, 2098, (1974), D11, 2104, (1975).

K. Osterwalder and E. Seiler, Harvard University, Preprint 1977.

J. Glimm and A. Jaffe, Physics Letters 66B, 67 (1977); Harvard University Preprint 1977.

G.F. De Angelis, D. deFalco and F. Guerra, Physics Letters 68B, 255, (1977); Lettere al Nuovo Cimento, to appear.

Ref. [F1]

C. Reviews of C.Q.F.T.

Refs. [G1], [V1], [S5], [F14] and

J. Glimm and A. Jaffe, in "Les Méthodes Mathématiques . . . ", (see [M4]).

J. Glimm and A. Jaffe, in "Quantum Dynamics: Models and Mathematics," L. Streit (ed.), Springer-Verlag, to appear.

J. Glimm and A. Jaffe, Cargèse Lectures 1976, (see [F14]).

J.-P. Eckmann, "Relativistic Boson Quantum Field Theories in Two Space-Time Dimensions", Quaderni del Cons. Naz. delle Ricerche, Bologna, 1977.

THE S-OPERATOR FOR SPIN-0 AND SPIN-1/2 PARTICLES IN TIME-DEPENDENT EXTERNAL FIELDS

S.N.M. Ruijsenaars
Department of Physics
Princeton University
Princeton, N.J. 08540

The following is a brief account of two results on the external field problem. Detailed proofs can be found in a forthcoming paper. Consider the Klein-Gordon equation

$$([\partial_\mu - iA_\mu(x)][\partial^\mu - iA^\mu(x)] + m^2 - A_4(x))\phi(x) = 0 \tag{1}$$

and the Dirac equation

$$(-i\partial_t - i\vec{\alpha}\cdot\vec{\nabla} + \beta m - V(x))\psi(x) = 0 , \tag{2}$$

where $A_0,\ldots,A_4$ are real-valued functions in $S(R^4)$ resp. where $V(x)$ is a function from R^4 to the Hermitean 4×4 matrices, whose entries are in $S(R^4)$. The first result is that the classical S-operator S associated with (1) resp. (2) is gauge invariant, in the sense that it remains the same if the vector field A_μ is replaced by $A_\mu + \partial_\mu\Lambda$, where Λ is a complex-valued function in $S(R^4)$ (in (2) this amounts to replacing V by $V + \gamma^0\partial\!\!\!/\Lambda$) . This implies that the Fock space S-operator $\mathcal{S}$ is gauge invariant as well, if it exists. Its existence is guaranteed if S_{+-} and S_{-+} are Hilbert-Schmidt (cf. Seiler's contribution). That this is the case constitutes the second result.

In the spin-1/2 case existence of $\mathcal{S}$ has been recently proved by Palmer [1] for a somewhat more general situation. His proof is very involved, but it is based on a simple idea, viz. estimating the multiple commutator of the free Hamiltonian with S to show the Hilbert-Schmidt (H.S.) property. This idea can be used to give a proof that is considerably shorter and simpler and can be easily applied to the spin-0 case as well. In the latter case the existence result extends results of [2] and [3]. In [2] Bellissard proves the H.S. property for $\lambda A_\mu,\lambda A_4$ with λ in a neighborhood of the origin, while in [3] it is shown that $\mathcal{S}$ exists if $\vec{A} = 0$ in (1). The present result lifts the restrictions on the coupling constant and the type of field.

1. Gauge Invariance

The gauge invariance of S is a consequence of the following relation between the (interaction picture) time evolution operators in the presence resp. absence of the gauge term:

$$\check{U}(t,-\infty;V,\Lambda) = \exp(i\check{H}_0t)\exp(i\Lambda(t,\cdot))\exp(-i\check{H}_0t)\check{U}(t,-\infty;V) \quad \forall t \in R . \tag{3}$$

Indeed, by taking the limit $t \to \infty$ one obtains from this

$$S(V,\Lambda) = S(V) . \tag{4}$$

In Equation (3) we have used the notation of [3]. It is proved by showing that the r.h.s. satisfies the integral equation of which $\check{U}(t,-\infty;V,\Lambda)$ is the unique solution that is bounded in norm and norm continuous on $R \cup \{-\infty,\infty\}$.

2. Existence of $\mathcal{S}$

By employing a spectral representation of $\check{H}_0$ the theory can be formulated in a Hilbert space $\mathcal{H} = \mathcal{H}_+ \oplus \mathcal{H}_-$, where $\mathcal{H}_+$ and $\mathcal{H}_-$ equal $L^2(R^3,d\vec{p})$ in the spin-0 case, $L^2(R^3,d\vec{p})^2$ in the spin-1/2 case. On $\mathcal{H}$ the S-operator is given by

$$(Sf)_\varepsilon(\vec{p}) = f_\varepsilon(\vec{p}) + \sum_{n=1}^{\infty} \sum_{\varepsilon'=+,-} \int d\vec{q}\, R^{(n)}_{\varepsilon\varepsilon'}(\vec{p},\vec{q}) f_{\varepsilon'}(\vec{q}) \qquad \varepsilon = +,- \tag{5}$$

resp.

$$(Sf)^i_\varepsilon(\vec{p}) = f^i_\varepsilon(\vec{p}) + \sum_{n=1}^{\infty} \sum_{\substack{\varepsilon'=+,-\\ i'=1,2}} \int d\vec{q}\, R^{(n)ii'}_{\varepsilon\varepsilon'}(\vec{p},\vec{q}) f^{i'}_{\varepsilon'}(\vec{q}) \qquad \varepsilon=+,- \quad i=1,2 . \tag{6}$$

Here, $R^{(n)(ii')}_{\varepsilon\varepsilon'}(\vec{p},\vec{q})$ are functions that can be explicitly expressed in A_μ, A_4 resp. V . For $S_{\substack{+-\\-+}}$ to be H.S. it is sufficient that the $R^{(n)(ii')}_{\substack{+-\\-+}}$ are in $L^2(R^6)$ and that the functions $\sum_{n=1}^{N} R^{(n)(ii')}_{\substack{+-\\-+}}$ converge in $L^2(R^6)$. For brevity we only sketch the proof of the latter statement for integral operators $T^{(n)}$ on $L^2(R^3)$ with kernels

$$\begin{aligned} T^{(n)}(\vec{p},\vec{q}) = \int \prod_{j=1}^{n-1} d\vec{k}_j \int_{-\infty}^{\infty} dt_1 \exp(iE_p t_1)\hat{F}(t_1,\vec{p}-\vec{k}_1) \int_{-\infty}^{t_1} dt_2 \exp(iE_{k_1}(t_1-t_2)) \\ \hat{F}(t_2,\vec{k}_1-\vec{k}_2) \ldots \int_{-\infty}^{t_{n-1}} dt_n \exp(iE_{k_{n-1}}(t_{n-1}-t_n))\hat{F}(t_n,\vec{k}_{n-1}+\vec{q})\exp(iE_q t_n) . \end{aligned} \tag{7}$$

Here, $E_k \equiv (k^2+m^2)^{\frac{1}{2}}$ and $\hat{F}$ is the partial Fourier transform of a function $F(t,\vec{x}) \in S(R^4)$. ($R^{(n)(ii')}_{+-}(\vec{p},\vec{q})$ is given by a sum of terms having this structure, the only differences being that the E_{k_i} can have a minus sign, that instead of F one has A_μ, A_4 resp. $V_{\alpha\beta}$, and that additional time-independent functions occur in the integrand.) The crux of the proof is the formula

$$\begin{aligned} (iE_p+iE_q)^m T^{(n)}(\vec{p},\vec{q}) = (-)^m \sum_{|\alpha|=m} \binom{m}{\alpha} \int \prod_{j=1}^{n-1} d\vec{k}_j \int_{-\infty}^{\infty} dt_1 \exp(iE_p t_1)\hat{F}^{(\alpha_1)}(t_1,\vec{p}-\vec{k}_1) \\ \int_{-\infty}^{t_1} dt_2 \exp(iE_{k_1}(t_1-t_2))\hat{F}^{(\alpha_2)}(t_2,\vec{k}_1-\vec{k}_2) \ldots \int_{-\infty}^{t_{n-1}} dt_n \exp(iE_{k_{n-1}}(t_{n-1}-t_n)) \\ \hat{F}^{(\alpha_n)}(t_n,\vec{k}_{n-1}+\vec{q})\exp(iE_q t_n) , \end{aligned} \tag{8}$$

where $F^{(i)} \equiv \partial_t^i F$. This relation is proved by setting

$$iE_p T^{(n)}(\vec{p},\vec{q}) = \int \ldots \int_{-\infty}^{\infty} dt_1 (\partial_{t_1} \exp(iE_p t_1))\ldots \tag{9}$$

and partially integrating the r.h.s. n times, after which one obtains (8) with m=1 . Repeating this m-1 times one gets (8). By estimating the r.h.s. of (8), it can now be seen that for any $\ell \in N$ there is a positive C not depending on n such that

$$|(E_p+E_q)^m T^{(n)}(\vec{p},\vec{q})| \leq C^n (n!)^{-1} (1+|\vec{p}+\vec{q}|^2)^{-\ell} . \tag{10}$$

From this our statement easily follows.

References

[1] Palmer, J.: Scattering automorphisms of the Dirac field, to be published in Jour. Math. Anal. and Appl.

[2] Bellissard, J.: Commun. Math. Phys. 46, 53-74 (1976)

[3] Ruijsenaars, S.N.M.: J. Math. Phys. 18, 720-737 (1977)

Selected Issues from

Lecture Notes in Mathematics

Vol. 431: Séminaire Bourbaki – vol. 1973/74. Exposés 436–452. IV, 347 pages. 1975.

Vol. 433: W. G. Faris, Self-Adjoint Operators. VII, 115 pages. 1975.

Vol. 434: P. Brenner, V. Thomée, and L. B. Wahlbin, Besov Spaces and Applications to Difference Methods for Initial Value Problems. II, 154 pages. 1975.

Vol. 440: R. K. Getoor, Markov Processes: Ray Processes and Right Processes. V, 118 pages. 1975.

Vol. 442: C. H. Wilcox, Scattering Theory for the d'Alembert Equation in Exterior Domains. III, 184 pages. 1975.

Vol. 446: Partial Differential Equations and Related Topics. Proceedings 1974. Edited by J. A. Goldstein. IV, 389 pages. 1975.

Vol. 448: Spectral Theory and Differential Equations. Proceedings 1974. Edited by W. N. Everitt. XII, 321 pages. 1975.

Vol. 449: Hyperfunctions and Theoretical Physics. Proceedings 1973. Edited by F. Pham. IV, 218 pages. 1975.

Vol. 458: P. Walters, Ergodic Theory – Introductory Lectures. VI, 198 pages. 1975.

Vol. 459: Fourier Integral Operators and Partial Differential Equations. Proceedings 1974. Edited by J. Chazarain. VI, 372 pages. 1975.

Vol. 461: Computational Mechanics. Proceedings 1974. Edited by J. T. Oden. VII, 328 pages. 1975.

Vol. 463: H.-H. Kuo, Gaussian Measures in Banach Spaces. VI, 224 pages. 1975.

Vol. 464: C. Rockland, Hypoellipticity and Eigenvalue Asymptotics. III, 171 pages. 1975.

Vol. 468: Dynamical Systems – Warwick 1974. Proceedings 1973/74. Edited by A. Manning. X, 405 pages. 1975.

Vol. 470: R. Bowen, Equilibrium States and the Ergodic Theory of Anosov Diffeomorphisms. III, 108 pages. 1975.

Vol. 474: Séminaire Pierre Lelong (Analyse) Année 1973/74. Edité par P. Lelong. VI, 182 pages. 1975.

Vol. 484: Differential Topology and Geometry. Proceedings 1974. Edited by G. P. Joubert, R. P. Moussu, and R. H. Roussarie. IX, 287 pages. 1975.

Vol. 487: H. M. Reimann und T. Rychener, Funktionen beschränkter mittlerer Oszillation. VI, 141 Seiten. 1975.

Vol. 489: J. Bair and R. Fourneau, Etude Géométrique des Espaces Vectoriels. Une Introduction. VII, 185 pages. 1975.

Vol. 490: The Geometry of Metric and Linear Spaces. Proceedings 1974. Edited by L. M. Kelly. X, 244 pages. 1975.

Vol. 503: Applications of Methods of Functional Analysis to Problems in Mechanics. Proceedings 1975. Edited by P. Germain and B. Nayroles. XIX, 531 pages. 1976.

Vol. 507: M. C. Reed, Abstract Non-Linear Wave Equations. VI, 128 pages. 1976.

Vol. 509: D. E. Blair, Contact Manifolds in Riemannian Geometry. VI, 146 pages. 1976.

Vol. 515: Bäcklund Transformations. Nashville, Tennessee 1974. Proceedings. Edited by R. M. Miura. VIII, 295 pages. 1976.

Vol. 516: M. L. Silverstein, Boundary Theory for Symmetric Markov Processes. XVI, 314 pages. 1976.

Vol. 518: Séminaire de Théorie du Potentiel, Proceedings Paris 1972–1974. Edité par F. Hirsch et G. Mokobodzki. VI, 275 pages. 1976.

Vol. 522: C. O. Bloom and N. D. Kazarinoff, Short Wave Radiation Problems in Inhomogeneous Media: Asymptotic Solutions. V. 104 pages. 1976.

Vol. 523: S. A. Albeverio and R. J. Høegh-Krohn, Mathematical Theory of Feynman Path Integrals. IV, 139 pages. 1976.

Vol. 524: Séminaire Pierre Lelong (Analyse) Année 1974/75. Edité par P. Lelong. V, 222 pages. 1976.

Vol. 525: Structural Stability, the Theory of Catastrophes, and Applications in the Sciences. Proceedings 1975. Edited by P. Hilton. VI, 408 pages. 1976.

Vol. 526: Probability in Banach Spaces. Proceedings 1975. Edited by A. Beck. VI, 290 pages. 1976.

Vol. 527: M. Denker, Ch. Grillenberger, and K. Sigmund, Ergodic Theory on Compact Spaces. IV, 360 pages. 1976.

Vol. 532: Théorie Ergodique. Proceedings 1973/1974. Edité par J.-P. Conze and M. S. Keane. VIII, 227 pages. 1976.

Vol. 538: G. Fischer, Complex Analytic Geometry. VII, 201 pages. 1976.

Vol. 543: Nonlinear Operators and the Calculus of Variations, Bruxelles 1975. Edited by J. P. Gossez, E. J. Lami Dozo, J. Mawhin, and L. Waelbroeck, VII, 237 pages. 1976.

Vol. 552: C. G. Gibson, K. Wirthmüller, A. A. du Plessis and E. J. N. Looijenga. Topological Stability of Smooth Mappings. V, 155 pages. 1976.

Vol. 556: Approximation Theory. Bonn 1976. Proceedings. Edited by R. Schaback and K. Scherer. VII, 466 pages. 1976.

Vol. 559: J.-P. Caubet, Le Mouvement Brownien Relativiste. IX, 212 pages. 1976.

Vol. 561: Function Theoretic Methods for Partial Differential Equations. Darmstadt 1976. Proceedings. Edited by V. E. Meister, N. Weck and W. L. Wendland. XVIII, 520 pages. 1976.

Vol. 564: Ordinary and Partial Differential Equations, Dundee 1976. Proceedings. Edited by W. N. Everitt and B. D. Sleeman. XVIII, 551 pages. 1976.

Vol. 565: Turbulence and Navier Stokes Equations. Proceedings 1975. Edited by R. Temam. IX, 194 pages. 1976.

Vol. 566: Empirical Distributions and Processes. Oberwolfach 1976. Proceedings. Edited by P. Gaenssler and P. Révész. VII, 146 pages. 1976.

Vol. 570: Differential Geometrical Methods in Mathematical Physics, Bonn 1975. Proceedings. Edited by K. Bleuler and A. Reetz. VIII, 576 pages. 1977.

Vol. 572: Sparse Matrix Techniques, Copenhagen 1976. Edited by V. A. Barker. V, 184 pages. 1977.

Vol. 44: R. A. Breuer, Gravitational Perturbation Theory and Synchrotron Radiation. VI, 196 pages. 1975.

Vol. 45: Dynamical Concepts on Scaling Violation and the New Resonances in e^+e^- Annihilation. Edited by B. Humpert. VII, 248 pages. 1976.

Vol. 46: E. J. Flaherty, Hermitian and Kählerian Geometry in Relativity. VIII, 365 pages. 1976.

Vol. 47: Padé Approximants Method and Its Applications to Mechanics. Edited by H. Cabannes. XV, 267 pages. 1976.

Vol. 48: Interplanetary Dust and Zodiacal Light. Proceedings 1975. Edited by H. Elsässer and H. Fechtig. XII, 496 pages. 1976.

Vol. 49: W. G. Harter and C. W. Patterson, A Unitary Calculus for Electronic Orbitals. XII, 144 pages. 1976.

Vol. 50: Group Theoretical Methods in Physics. 4th International Colloquium. Nijmegen 1975. Edited by A. Janner, T. Janssen, and M. Boon. XIII, 629 pages. 1976.

Vol. 51: W. Nörenberg und H. A. Weidenmüller. Introduction to the Theory of Heavy-Ion Collisions. IX, 273 pages. 1976.

Vol. 52: M. Mladjenović, Development of Magnetic β-Ray Spectroscopy. X, 282 pages. 1976.

Vol. 53: D. J. Simms and N. M. J. Woodhouse, Lectures on Geometric Quantization. V, 166 pages. 1976.

Vol. 54: Critical Phenomena. Sitges International School on Statistical Mechanics, June 1976. Edited by J. Brey and R. B. Jones. XI, 383 pages. 1976.

Vol. 55: Nuclear Optical Model Potential. Proceedings 1976. Edited by S. Boffi and G. Passatore. VI, 221 pages. 1976.

Vol. 56: Current Induced Reactions. International Summer Institute, Hamburg 1975. Edited by J. G. Körner, G. Kramer, and D. Schildknecht. V, 553 pages. 1976.

Vol. 57: Physics of Highly Excited States in Solids. Proceedings 1975. Edited by M. Ueta and Y. Nishina. IX, 391 pages. 1976.

Vol. 58: Computing Methods in Applied Sciences. Proceedings 1975. Edited by R. Glowinski and J. L. Lions. VIII, 593 pages. 1976.

Vol. 59: Proceedings of the Fifth International Conference on Numerical Methods in Fluid Dynamics. 1976. Edited by A. I. van de Vooren and P. J. Zandbergen. VII, 459 pages. 1976.

Vol. 60: C. Gruber, A. Hintermann, and D. Merlini, Group Analysis of Classical Lattice Systems. XIV, 326 pages. 1977.

Vol. 61: International School on Electro and Photonuclear Reactions I. Edited by C. Schaerf. VIII, 650 pages. 1977.

Vol. 62: International School on Electro and Photonuclear Reactions II. Edited by C. Schaerf. VIII, 301 pages. 1977.

Vol. 63: V. K. Dobrev et al., Harmonic Analysis on the n-Dimensional Lorentz Group and Its Application to Conformal Quantum Field Theory. X, 280 pages. 1977.

Vol. 64: Waves on Water of Variable Depth. Edited by D. G. Provis and R. Radok. 231 pages. 1977.

Vol. 65: Organic Conductors and Semiconductors. Proceedings 1976. Edited by L. Pál, G. Grüner, A. Jánossy and J. Sólyom. 654 pages. 1977.

Vol. 66: A. H. Völkel, Fields, Particles and Currents. VI, 354 pages. 1977.

Vol. 67: W. Drechsler and M. E. Mayer, Fiber Bundle Techniques in Gauge Theories. X, 248 pages. 1977.

Vol. 68: Y. V. Venkatesh, Energy Methods in Time-Varying System Stability and Instability Analyses. XII, 256 pages. 1977.

Vol. 69: K. Rohlfs, Lectures on Density Wave Theory. VI, 184 pages. 1977.

Vol. 70: Wave Propagation and Underwater Acoustics. Edited by J. Keller and J. Papadakis. VIII. 287 pages. 1977.

Vol. 71: Problems of Stellar Convection. Proceedings 1976. Edited by E. A. Spiegel and J. P. Zahn. VIII, 363 pages. 1977.

Vol. 72: Les instabilités hydrodynamiques en convection libre forcée et mixte. Edité par J. C. Legros et J. K. Platten. X, 202 pages. 1978.

Vol. 73: Invariant Wave Equations. Proceedings 1977. Edited by G. Velo and A. S. Wightman. VI, 416 pages. 1978.